Sustainable Energy Technologies

Sustainable Energy Technologies

Edited by
Eduardo Rincón-Mejía and
Alejandro de las Heras

CRC Press
Taylor & Francis Group
Boca Raton London New York

CRC Press is an imprint of the
Taylor & Francis Group, an **informa** business

CRC Press
Taylor & Francis Group
6000 Broken Sound Parkway NW, Suite 300
Boca Raton, FL 33487-2742

First issued in paperback 2020

ISBN-13: 978-0-367-57267-9 (pbk)
ISBN-13: 978-1-138-03438-9 (hbk)

Visit the Taylor & Francis Web site at
http://www.taylorandfrancis.com

and the CRC Press website at
http://www.crcpress.com

To my wife, Gabriela Carolina, and our daughters, Gabi, Almita, and Vero.

Contents

Foreword

In December 2015, nearly all nations around the world came together in Paris to sign a historic agreement to tackle human-caused climate change. These nations pledged to take dramatic steps to reduce anthropogenic greenhouse gas emissions. The Paris Agreement sets a target of limiting global average temperature warming to no more than 2°C and ideally to no more than 1.5°C above preindustrial levels. Already, by 2016, global warming has approached nearly 1°C since reliable temperature measurements began in the mid-1800s.

The timing of this agreement was critical in many ways. In the previous year, the Intergovernmental Panel on Climate Change (IPCC) had released its fifth Assessment Report,[*] which showed that unless drastic greenhouse gas emission reductions started within the next decade or so, potentially catastrophic impacts of human-caused climate change on the earth's weather and ecosystems will occur, including unprecedented sea-level rise and coastal flooding, changes in weather patterns that would result in increase in severe weather events and flood and drought cycles, significant shifts in agricultural production and vector-borne diseases, and untold other societal consequences. The IPCC report shows that global warming must be limited to 2°C to avoid these consequences. However, the IPCC report also shows that in order to achieve such a goal, society must implement an ambitious emissions reduction scenario, which allows for only another decade or so of increasing anthropogenic emissions before requiring a rapid decline to levels near 0 within 50 years.

As part of their signing agreement in Paris, countries were required to submit "Nationally Determined Contributions" (NDCs), with specific greenhouse gas emission reduction commitments over the coming decades. Although not legally binding, the current NDCs represent a strong global commitment to tackle climate change through a number of approaches, including clean energy technology deployments. Implementing the NDCs would result in year-over-year growth in renewable energy exceeding even the current numbers summarized in this book, *Sustainable Energy Technologies*. However, the NDCs in and of themselves would likely fall short in achieving the goals of the Paris Agreement. If no additional action were taken, global warming could exceed 3.6°C by the end of this century, well above the level recommended by the IPCC.

Achieving a maximum of 2°C, and especially 1.5°C, from the energy perspective alone, would require an unprecedented transformation in the way we produce and consume energy. Although the Paris Agreement comes at a time when clean, renewable energy is experiencing significant growth throughout the world, the reality is that nearly 80% of our end-use energy consumption is still based on the production and consumption of carbon-based fuels.[†] Over the past decade, global R&D investments in renewable energy technologies have grown substantially to a current level of nearly a quarter of a trillion U.S. dollars per year. These investments, along with significant R&D in technology performance and reliability and favorable government policies, have resulted in substantial decrease in the costs of these technologies, especially hydropower, solar PV, and onshore wind, so that by 2016 the median-levelized cost of energy from onshore wind farms is competitive with electricity generated by fossil fuels, despite recent drops in fossil fuel prices. Solar PV costs are dropping so rapidly that these technologies will also be competitive with traditional power generation in

---

[*] c.f., Intergovernmental Panel on Climate Change, 2014: *Climate Change 2014, 5th Assessment Report, Synthesis Report.* http://ipcc.ch/report/ar5/syr/.

[†] IRENA, 2017: *Rethinking Energy 2017.* International Renewable Energy Agency Headquarters, Abu Dhabi, UAE. www.irena.org. Also REN21 (Renewable Energy Network for the 21st Century) 2016: *Renewables 2016 Global Status Report.* http://www.ren21.net/gsr.

the near future. These cost drops have largely contributed to renewables exceeding nonrenewable power expansion since 2012. Globally, Photovoltaic (PV) capacity has grown by 20%–30% for the past 10 years and added approximately 75 GW of capacity in 2016, so that current global capacity is surpassing 300 GW. In a recent IRENA study,* jobs in the renewable energy sector surpassed 8 million worldwide in 2015, with 2.8 million working in the solar PV sector; overall, annual job growth in the renewable sector is 5% (over 11% in solar PV).

With the adoption of the Paris Agreement and the rapid emergence of renewable energy technologies as a viable and significant global business enterprise, *Sustainable Energy Technologies* is a timely and important volume. Written primarily by engineers, the book addresses the overall status and trends of renewable energy technologies (including all of the broad categories of solar energy, geothermal, wind, tidal, biofuels, and related technologies) and how these technologies can and must all work together to achieve a clean and resilient energy supply by the end of this century. Besides adding resiliency to the rapid expansion of electricity supply, renewable technologies can contribute to all end-use energy sector demands, including heat and transport. This concept of renewables working together to achieve a 100% renewable energy supply for all end-use energy needs is a major theme of the International Solar Energy Society.

Sustainable Energy Technologies addresses significant questions and concerns regarding certain low-carbon emission technologies such as nuclear and fusion energy, which, although on principle can offer a substantial contribution to clean energy supply, are also bereft with safety, international security, and cost concerns that may and perhaps should exclude them from a future energy mix. Further, this volume takes a close look at the real sustainability of energy sources that are typically viewed as renewable, such as large hydropower and biomass combustion.

Sustainable Energy Technologies supports the energy transformation that is being stimulated by the rapid growth in renewable energy business opportunities and by the challenges set forth in the Paris Agreement. The United Nations' Sustainable Development Goal #7, "Affordable and Clean Energy,"† calls for doubling the world's renewable energy supply (to 36% of total consumption) and making energy accessible to all, including the 1.2 billion people currently lacking access to reliable energy sources, by the year 2030. To achieve such a goal, global investments in renewable energy must quadruple from their current level to nearly 1 trillion USD/year. What makes *Sustainable Energy Technologies* especially important is that renewable energy technologies can address *all* end-use energy sectors: power, heat, and transport. Of these sectors, the power sector is experiencing the most rapid growth, as electricity becomes the energy carrier of choice to meet an ever-expanding array of end-use energy demands. However, even today, the power sector still represents only about 20% of total end use energy demand. The heating sector represents 47% of our end use energy consumption, and the transport sector represents 37%.

Thus, in *Sustainable Energy Technologies*, you will see thoughtful reports on the broad range of renewable technologies that address all three sectors. It is essential that these technologies all work together to achieve a total carbon-free energy supply over the decades to come. The hybridizing of these technologies is addressed in Part I of this book, and further bioengineering solutions are addressed in Part II. The book then addresses, in Part III, the technological issues related to security, safety, and geopolitical context of energy systems, which of course have national implications, as well as implications on the energy mixes that will ultimately be chosen by individual countries.

---

* IRENA, 2016: *Renewable Energy and Jobs Annual Review 2016*. International Renewable Energy Agency Headquarters, Abu Dhabi, UAE. http://www.irena.org.
† http://www.un.org/sustainabledevelopment/sustainable-development-goals/.

All of this material is presented in a way that bridges the dialogue between sustainable energy practitioners and the related relevant sciences, security professionals, and experts working in the fields of socioeconomics and politics. *Sustainable Energy Technologies* is truly a timely and critically important volume to support the work of the vast array of professionals and decision-makers involved in the renewable energy transformation that is sweeping through society.

Dr. David Renné
President, International Solar Energy Society, and Owner, Dave Renné Renewables LLC

Preface

WHAT 'SUSTAINABILITY' MIGHT MEAN

American President Theodore Roosevelt said "We have become great because of the lavish use of our resources. But the time has come to inquire seriously what will happen when our forests are gone, when the coal, the iron, the oil, and the gas are exhausted, when the soils have still further impoverished and washed into the streams, polluting the rivers, denuding the fields and obstructing navigation. These issues are not just about the next century or generation." "The nation behaves well if it treats the natural resources as assets which it must turn over to the next generation increased, and not impaired, in value" (1910).

This far-sighted comment warrants reevaluation of our culture. Sustainable development has now become a household word, but the term "development" hides phenomena such as the concentration and accumulation of capital, the ruthless destruction of nature, and the alienation of individuals (Latouche, 2009). Economics must rethink the idea that consumption of unlimited resources is the ultimate goal in life, as Georgescu-Roegen did, linking economy and thermodynamics. In the real world, all processes obey the laws of physics and biology (Bejan & Llorente, 2010).

In practice, a sustainable energy system should have the following features:

1. Energy reserves should last for as long as man exists on the planet.
2. The Hartwick rule (1977) should apply to exhaustible reserves. For example, fossil natural gas can be sustainable only if it is entirely dedicated to producing devices that tap much more renewable energy than is consumed in the production of that device and if pollution is offset.
3. Waste materials should all be treated as resources in recycling processes.
4. Working efficiencies should approximate thermodynamic limits.
5. The system should be resilient and dominated by maximally diverse modular and decentralized applications.
6. Energy should be used sparingly by all.
7. All subsystems should have very high energy return on energy invested.
8. Energy and matter transportation and conduction should be minimal thanks to local use of resources.
9. Profitability must consider both monetary and environmental numeraires on equal terms.
10. Sustainability can only be attained if ecosystems are restored.

These principles are embodied in the applications shown in this book.

BOOK OVERVIEW

This book begins with a reminder in Renné's Foreword of the 2015 Paris Agreements to further reduce carbon dioxide in the atmosphere. A few months after the Paris Agreements, news came that the price of electricity from renewable energy sources was, for the first time, lower than the nuclear MWh price.

In Chapter 1, Rincón-Mejía and colleagues link current carbon dioxide levels to the global mix of energy technologies. They caution against dead-end engineering ideas that could further imbalance the atmosphere. They also make the essential distinction between renewable and sustainable energy technologies, thereby questioning nonwaste agro-fuels. Finally, they show the momentous surge of sustainable energy sources in the worldwide market.

Chapter 2, by de las Heras, and colleagues, explains that fusion in the Sun and radioactive decay in Earth's crust and mantle are safe and easy to use while artificial fission poses intractable technical issues. Further, Sani's firsthand experience in nuclear power plants points to the intricate relationships of the oil and nuclear industries with regional and global military interests.

The lack of transparency of nuclear power resonates with citizen concern and involvement, in Chapter 3 by Leon-Grossmann; firsthand experience in California shows to the world that in one of the most democratic areas of the world, corporations and highly regarded politicians vie for power at the expense of the environment and democracy, two pillars of sustainability.

Having stated the current bias in the supply side of energy and the human flaws in energy systems, engineering takes over with a theoretical viewpoint on technology evolution: dos Santos and colleagues very didactically expound the intricacies and extremely wide field of application of the Constructal physical law in Chapter 4. This theory pinpoints many practical applications in sustainable energy systems.

In Chapter 5, Walker addresses one of the main challenges in the energy transition toward sustainability, namely the integration of more intermittent renewable energy to grids feeding cities and buildings. Although energy storage is the main solution, adaptations of sustainable energy sources are also needed, such as photovoltaic solar distribution over larger areas or different orientations than south-facing arrays, to even out the effects of partly cloudy days and morning and evening peak demand times. Hardware standards and load forecasting methods are also enhancing operation in grids coupling conventional and sustainable energy sources during the energy transition.

Changes in the global energy subsystems in the last four decades are described by Ruiz-Hernández in Chapter 6. The most salient feature of the global systems is the rise of renewable sources, and the resistance of the International Energy Agency to acknowledge this fact. Solar energy in particular has a potential for *direct* thermal and photovoltaic applications that bolster their efficiency and, consequently, their economic competitiveness. Ruiz-Hernández draws on the experience of Spanish top-of-the-line facilities to explore upcoming developments in solar concentrating thermal and electric applications.

Fundamental physics, mathematical tools, and the economic aspects of energy balance are used by González in Chapter 7 to show that technology has solved one of the key issues in tropical energetics: sustainable cooling in the face of high humidity. These results also point to the possibility of sustainable energy independence in islands around the world, and coastal areas, the most endangered areas in relation to global warming and sea level rise.

Taking advantage of a warming atmosphere, absorption thermodynamic cycles now allow for heat, cold or electricity applications, depending on the needs of the end user, as demonstrated by Lecuona-Neumann and colleagues in Chapter 8. These applications totally supersede systems that use ozone-depleting substances.

Weber takes over cooling and heating applications in Chapter 9. He shows the efficiency of hybridized solar and heat-pump systems. In these systems, the sun is the energy source, and heat is stored underground. These applications have large potentials in temperate and cold climes, where seasonal soil-air temperature gradients are elevated.

The field of solar energy storage applications is further explored by Solé and colleagues in Chapter 10, with a focus on recent developments in thermochemical materials. The full spectrum of solar energy storage is covered in that chapter.

Solar energy storage and the chemistry thereof are approached from another angle by Cabrera-Lara in Chapter 11. There, solar photocatalysis is used for hydrogen production, the energy carrier used for storage. The role of semiconductor catalysts and photoelectrochemical cells is elucidated, and key parameters are highlighted.

Dispatchability (i.e., use on demand) is fundamental in the competition of renewable energy systems against fossil fuels. Sattler and colleagues in Chapter 12 also deal with the conversion of solar energy into solar fuels at higher, more efficient, temperatures, using solid oxide and molten carbonate electrolyses. The focus is on concentrating solar power thermochemical H_2O and CO_2 transformation.

Sustainable electrochemical energy storage can also take place in supercapacitors, whose use in addition to current storage batteries is developing fast. Fierro and colleagues in Chapter 13 explain the fundamentals and applications of tannins, some cheap, inexpensive, non-toxic and renewable compounds as precursors of supercapacitor carbon electrodes.

Transportability is another key parameter in sustainable energy carriers. Hydrogen in particular lends itself to fuel cell applications, as shown by Reyes-Rodríguez and colleagues in Chapter 14. Fuel cell thermodynamics, components, and perspectives are dealt with in that chapter.

Transportability is of the essence in aeronautical applications. Iturbe and colleagues in Chapter 15 demonstrate the potential of sustainable biofuels in jet propulsion turbines and the solutions to the higher viscosities of biofuels. Sprays and droplet-size distribution in ultrasonic actuated fuel injection are treated in detail.

Global society is highly dependent on air travel, but for billions of people, gathering cooking fuel is still a highly energy-demanding activity, especially since wood fuel is becoming sparse due to deforestation. As shown by Lecuona-Neumann and colleagues in Chapter 16, solar cooking is a powerful alternative to combustion in most kitchens of the world and a climate-change mitigator. Concentrating solar cookers are now being hybridized with thermal or electrical solar energy storage, for enhanced nighttime and cloudy-day dispatchability.

Both large-scale and small autonomous systems are likely to require windpower in their energy mix: windpower is one of the lower-cost energy sources and has been harnessed for centuries, a sure indicator of the robustness of even less-efficient applications. Rincón-Mejía in Chapter 17 gives account of the principles of windpower and the most viable, up-to-date, applications.

Offshore windpower is but one of the sustainable energy sources that can be drawn from marine environments. Magar in Chapter 18 shows how tides and their interaction with the sea floor generate strong tidal currents, close by the sea shores, owing to Venturi effects. Numerical tidal resource assessment, based on computational fluid dynamics and technology developments in the last 20 years, is explicated.

Any discussion of current energy supply mixes should probably include Brazil, perhaps the foremost user of renewable energy sources: by 2015, an approximate 44% of the energy mix was obtained from renewable sources. Several of these however are not sustainable: wood fuel, vegetal coal, sugar cane ethanol, and hydroelectricity have large environmental and social footprints. Korys and Latawiec in Chapter 19 cover the environmental and social implications of large-scale hydropower dams, with implications of any such future projects in the world. They also discuss small-scale and run-of-river alternatives to making hydropower more sustainable.

Sustainable energy systems will not be complete if they do not tap the vast stores of energy in the billions of extant human bodies and the huge amounts of biowaste that they generate, as explained by Islas-Espinoza and de las Heras in Chapter 20. Starting with fundamentals of bioenergy, they go on to highlight the main application branches, human power, and waste-based biomethane. Hybrid applications are also illustrated.

The importance of biomethane can hardly be overestimated. Challenges remain, but Aydin and colleagues in Chapter 21 highlight solutions. They cover the complex biological essentials of the topic, key control parameters, and pathways to enhanced biomethane generation, including control of carbon dioxide and hydrogen sulfide.

As to fundamentals of human bodily energy, they are provided by Aguilar Becerril and colleagues in Chapter 22. These include (an)aerobic biochemical pathways of energy generation, which interestingly show the hybrid character of human energetics. Food energy substrates are covered, as is the essential concept of energy recovery via sleep and rehydration. Ergometric, spirometric, and anaerobic aspects of performance are covered, using as a model athletic performance.

As increasing numbers of humans settle in cities, and emerging economies grow, the already huge energy demand from homes is bound to soar. Energy generation is not a solution. The Net Zero Energy concept in building design has emerged as an alternative. Morillón Gálvez and Ceballos

Ochoa in Chapter 23 develop an energy balance account of a bioclimatic and photovoltaic Net Zero Energy architectural design in three of the world's bioclimates.

REFERENCES

Bejan, A. and Llorente, S. (2010). The constructual law of design and evolution in nature. *Philosophical Transactions of the Royal Society B*, *365*, 1335–1347.

Hartwick, J. M. (1977). Intergenerational Equity and the Investing of Rents from Exhaustible Resources. *Review of Economic Studies*, *67*, 973–974.

Latouche, S. (2009). *Farewell to Growth*. Cambridge, UK: Polity Press.

Roosevelt, T., 26th President of the US (1910). Speech before the Colorado Livestock Association, Denver, Colorado, August 19, 1910. Retrieved from quotationspage.com/quote/41325.html.

Acknowledgments

The editors are grateful for the effort dedicated by all the authors to this book. We also wish to thank Irma Britton, our Purchasing Editor at CRC Press.

Editors

Eduardo A. Rincón-Mejía, Eng, PhD, Engineering Professor.

American Society for Mechanical Engineers, Chair of the Solar Energy Division (2013–2014), Chair of the Solar Heating and Cooling of the SED Technical Committee (2009–2013). International Solar Energy Society, Board of Directors Member (2005–2013). Secretary (2008–2009). Mexican Solar Energy Association, President (2002–2004). International Energy Foundation, Board of Directors Member (2002–2009). Member of the Joint Public Advisory Committee of the Commission for Environmental Cooperation of North America (2004–2008). Mexican Renewable Energy Award (Ministry of Energy, Mexico) (2004).

Professor Rincón-Mejía works primarily in the area of Renewable Energy Technologies, with an emphasis on affordable solar technologies. His current research focus is the development of solar concentrators using nonimaging optics for applications such as solar cooking, water distillation, space heating, steam generation for industrial and residential uses, and high-flux research. Some of his latest developments include affordable small-scale wind generators, high-efficiency solar ovens, and solar hotplates.

He has authored over 40 articles in scientific and engineering journals. He has supervised over 60 graduate theses and taught thermal science as well as specific courses on solar energy applications, wind generators, and renewable energy systems.

Alejandro de las Heras, PhD, has worked with grassroots movements and nongovernmental organizations in Mali, France, and Mexico. He currently works in R&D related to hybridizing permaculture and water and energy appropriate technologies. With CRC Press, he has edited *Sustainable Science and Technology: An Introduction*, Boca Raton, FL, 2014.

Contributors

José Antonio Aguilar Becerril
Medicine Department
Mexico State University
Toluca, Mexico

Jaime Manuel Aguilar Becerril
Mexican Social Security Institute
Toluca, Mexico

Sevcan Aydin
Environmental Biotechnology Department
Istanbul Technical University
Istanbul, Turkey

Luisa F. Cabeza
GREA Innovació concurrent, INSPIRES
    Research Centre
University of Lleida
Lleida, Spain

Lourdes Isabel Cabrera-Lara
Independent Researcher
Mexico City, Mexico

Francisco Javier Ceballos Ochoa
Institute of Engineering
National Autonomous University of Mexico
Mexico City, Mexico

Alain Celzard
Jean Lamour Institute
University of Lorraine
Epinal, France

Heriberto Cruz-Martínez
Department of Chemistry
National Polytechnic Institute
Mexico City, Mexico

Alejandro de las Heras
Independent Researcher
Burgos, Spain

Elizaldo Domingues dos Santos
Ocean Engineering Graduate Program
and
Computational Modeling Graduate Program
Federal University of Rio Grande
Rio Grande, Rio Grande do Sul, Brazil

Antonio Famiglietti
Thermal and Fluids Engineering Department,
    ITEA Research Group
Carlos III University
Madrid, Spain

Vanessa Fierro
Jean Lamour Institute
National Scientific Research Centre
Paris, France

Mateus das N. Gomes
Computational Modeling Graduate Program
Federal University of Rio Grande
Rio Grande, Rio Grande do Sul, Brazil
and
Science, Technology and Society Graduate
    Program
Federal Institute of Paraná
Paranaguá, Paraná, Brazil

Jorge E. González
Department of Mechanical Engineering
City College of New York
New York, New York

Alvaro de Gracia
Mechanical Engineering Department
Rovira i Virgili University
Tarragona, Spain

Bahar Yavuzturk Gul
Environmental Biotechnology Department
Istanbul Technical University
Istanbul, Turkey

J. E. V. Guzmán
Institute of Engineering
National Autonomous University of Mexico
Mexico City, Mexico

Anis Houaijia
Institute of Solar Research
German Aerospace Center (DLR)
Cologne, Germany

Marina Islas-Espinoza
Engineering Department
Mexico State University
Toluca, Mexico

Liércio A. Isoldi
Ocean Engineering Graduate Program
and
Computational Modeling Graduate
  Program
Federal University of Rio Grande
Rio Grande, Rio Grande do Sul, Brazil

Aris Iturbe
Aeronautic University of Queretaro
Colón, Mexico

Paola Yazmín Jiménez Colín
Medicine Department
Mexico State University
Toluca, Mexico

Katarzyna Anna Korys
International Institute for Sustainability
Rio de Janeiro, Brazil

Agnieszka Ewa Latawiec
International Institute for Sustainability
Rio de Janeiro, Brazil
and
Pontifical Catholic University of Rio de Janeiro
Rio de Janeiro, Brazil
and
University of Agriculture in Krakow
Krakow, Poland

Antonio Lecuona-Neumann
Thermal and Fluids Engineering Department,
  ITEA Research Group
Carlos III University
Madrid, Spain

Mathieu Legrand
Thermal and Fluids Engineering Department,
  ITEA Research Group
Carlos III University
Madrid, Spain

Andrea Leon-Grossmann
Food and Water Watch
Los Angeles, California

Vanesa Magar
Physical Oceanography Department
Centre of Scientific Research and Higher
  Education of Ensenada (CICESE)
Ensenada, Mexico

David Morillón Gálvez
Institute of Engineering
National Autonomous University of Mexico
Mexico City, Mexico

José I. Nogueira
Thermal and Fluids Engineering Department,
  ITEA Research Group
Carlos III University
Madrid, Spain

Diana Gabriela Pinedo Catalán
Medicine Department
Mexico State University
Toluca, Mexico

José Luis Reyes-Rodríguez
Department of Chemistry
National Polytechnic Institute
Mexico City, Mexico

Eduardo Rincón-Mejía
Energy Program
Autonomous University of Mexico City
México City, Mexico

Ana Rincón-Rubio
School of Engineering
Mexico State University
Toluca, Mexico

Luiz A. O. Rocha
Mechanical Engineering Graduate Program
University of Vale do Rio dos Sinos
São Leopoldo, Rio Grande do Sul, Brazil

Pedro A. Rodríguez-Aumente
Thermal and Fluids Engineering Department,
    ITEA Research Group
Carlos III University
Madrid, Spain

Martin Roeb
Institute of Solar Research
German Aerospace Center (DLR)
Cologne, Germany

Valeriano Ruiz-Hernández
Advanced Technology Centre for Renewable
    Energies (CTAER)
Tabernas-Almería, Spain

Angela Sánchez-Sánchez
Jean Lamour Institute
National Scientific Research Centre
Epinal, France

Behrooz Sani
Independent Researcher
Toluca, Mexico

Christian Sattler
Institute of Solar Research
German Aerospace Center (DLR)
Cologne, Germany

Aiyoub Shahi
Animal Biology Department
University of Tabriz
Tabriz, Iran

Aran Solé
Department of Mechanical Engineering and
    Construction
Jaume I University
Castellón de la Plana, Spain

Omar Solorza-Feria
Department of Chemistry
National Polytechnic Institute
Mexico City, Mexico

Miriam Marisol Tellez-Cruz
Department of Chemistry
National Polytechnic Institute
Mexico City, Mexico

Adrián Velázquez-Osorio
Department of Chemistry
National Polytechnic Institute
Mexico City, Mexico

W. Vicente
Institute of Engineering
National Autonomous University of Mexico
Mexico City, Mexico

Andy Walker
Integrated Applications Office
National Renewable Energy Laboratory
Golden, Colorado

Bernd Weber
Engineering Department
Mexico State University
Toluca, Mexico

1 Introduction

Eduardo Rincón-Mejía
Autonomous University of Mexico City

Alejandro de las Heras
Independent Researcher

Marina Islas-Espinoza
Mexico State University

CONTENTS

1.1 STATE OF THE PLANET

Air records trapped in millenary ices show that CO_2 levels before the Industrial Revolution remained stable around 280 ppm for tens of millennia. However, on April 18, 2017, 410 ppm was exceeded for the first time in hundreds of thousands of years (Figure 1.1). A rapid estimate of the mass quantity of this gas gives us about 7.749 billion metric tons per ppm (Rincón-Mejía 2011), so that sequestering a little more than one Peta kilogram (1 Pkg = 10^{15} kg) of this gas is required to recover the 280 ppm level of preindustrial times, or 465 Gt CO_2 (Giga tons CO_2) to return to the 350 ppm required to keep the Earth's atmosphere thermally stable (Hansen et al. 2008).

Earth's vegetation has a wondrous capacity to sequestrate CO_2: the magnitude of the seasonal fluctuation (7 ppm per annual cycle, Figure 1.1) is about 54 Gt CO_2, more than the annual emission of this gas for fossil fuel consumption and cement production (Olivier et al. 2016) and more than one-ninth of the amount of CO_2 that has to be removed from the atmosphere in order to return to 350 ppm CO_2 concentration. Another point to emphasize is that the 21% of oxygen content in the atmosphere's composition is due to photosynthetic processes working during several billion years. Fossil fuels consume this vital oxygen when they are burned, so besides emitting GHG, their combustion decreases the oxygen concentration.

The only viable way to sequestrate CO_2 from the atmosphere and reduce its concentration is via large-scale reforestation and afforestation to restore original forests, mangroves, algae, and many other photosynthetic communities.

1.2 GEOENGINEERING: CAUTIONARY WORD

The alleged carbon sequestration by technological means has not yet demonstrated its viability. Geoengineering is meant to achieve large-scale modifications of Earth's energy balance through

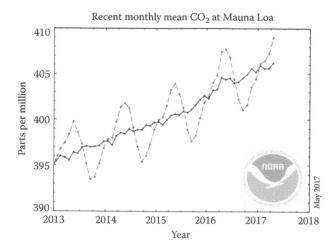

FIGURE 1.1 Keeling curve of the global atmospheric co_2 concentration, measured at Mauna Loa, Hawaii. May 5, 2017 update (Tans and Keeling n.d.). Seasonal variations are due to the photosynthesis cycle in the Northern Hemisphere, but the salient feature is the unabated increase in the average yearly value. CO_2 global levels are growing 3% per year.

human interventions. Very few studies have been published on the environmental effects, economical costs, sociopolitical impacts, and legal implications of geoengineering. Deployment of these technologies has the potential to cause significant negative effects (Williamson and Bodle 2016).

Generally speaking, geoengineering technologies are categorized as either carbon dioxide removal (CDR) methods or albedo-modification or solar radiation management (SRM) methods. CDR methods include potentially perilous ocean "fertilization" and "carbon capture and sequestration" in big caverns. The only reasonable option among the "bioenergy with carbon capture and storage" proposals is, again, those consisting of afforestation and massive reforestation. In turn, SRM methods try to address climate change by increasing the reflectivity of the Earth's atmosphere or surface, the way ice covers do. Aerosol injection and space-based reflectors are unfounded examples of SRM methods. SRM methods do not remove greenhouse gases from the atmosphere, but they could be deployed more quickly with relatively immediate global cooling results compared to CDR methods. SRM methods run counter to improving our capacity to harvest solar energy, our most abundant, available, and clean source of energy.

1.3 ENERGY CONSUMPTION INCREASES WITH WASTE

Well into the first quarter of the twenty-first century, the world economy is still based on the consumption of fossil and nuclear fuels (80% of the energy consumption in realistic assessments, or 90% according to Table 1.1). Discussion on the many shortcomings of fossil fuels is moot; the reader will find accumulating consensus information by looking up the physics of molecules vibrating to infrared waves and the Intergovernmental Panel on Climate Change reports. Both show the central role of anthropogenic CO_2 emissions in global climate change. Any day's landslide of environmental information suffices to illustrate the detriments of oil, the much-touted natural gas, and transportation.

Much less known are the detrimental effects of using electricity: as a secondary energy source, it is generated with a high increase in entropy (loss of energy mostly as unrecovered heat). Actually, electricity, closely followed by oil used in transportation, totals 71% of all U.S. rejected energy (Figure 1.2).

TABLE 1.1
Primary Energy Consumption by Source, 2015

Energy Source	Consumption (EJ)	Contribution to Total Consumption (%)
Oil	181.34	32.94
Coal	160.77	29.21
Natural Gas	131.26	23.85
Hydropower	23.91	4.34
Nuclear	19.59	3.56
Other renewables	33.58	6.10
Total	550.45	100.00

Sources: BP, *Statistical Review of World Energy,* 65th ed., BP p.l.c., London, 2016; IRENA, *Renewable Capacity Statistics 2017*, International Renewable Energy Agency, United Arab Emirates, Abu Dhabi, 2017.

The current energy system based on fossil and nuclear fuels cannot be sustained even in the medium term due to three fundamental factors: (1) their very limited availability, given that they are scarce and that it is expected that energy consumption will continue to grow in the foreseeable future; (2) the environmental problem caused by the emission of greenhouse and toxic gases and long-term persistent radioactive waste; and (3) their increasing economic, environmental, and social costs. The reason these nonrenewable energies continue to dominate global energy supply is that these are actually heavily subsidized by governments that consciously or unconsciously serve powerful interests in the oil, gas, and military industries. These companies maintain expensive advertising campaigns to deny the unacceptable risk their activities pose.

The top 10 countries in energy consumption are industrial or technological powerhouses (Table 1.2). What can they do to curb consumption, losses, and global warming and its consequences? First, there should be subsidy reduction for all energy sources and waste reduction in animal and plant husbandry since agricultural production in developed countries is energy intensive. These countries should uncouple economic accounts from consumption statistics and fossil fuel consumption in particular. At the same time, regulations and taxes should not impede local citizen energy organization or neighborhood grids.

1.4 BIOMASS: DRAWBACKS APLENTY

Official energy consumption figures do not include the metabolic energy consumed by 7.5 billion human beings, equivalent to 23.65 EJ, assuming a mean rate of human metabolic consumption of about 100 W. It is therefore ill-advised to dedicate crops to obtain agro fuels when they compete with food production. This is especially true as it is already much more efficient to obtain power by other means, like direct sunlight-to-electricity conversion by solar photovoltaic (PV) cells: PV cell conversion efficiency is already one order of magnitude greater than most land crops.

Agro fuels may be far from sustainable because of the following drawbacks and deleterious effects:

- Agro fuels produce acroleins and acrylamides, volatile organic compounds, and polycyclic aromatic hydrocarbons. All of them are directly related to cancer according to the International Agency for Research on Cancer (IARC 2015). Like so many other organic compounds, agro fuels produce dioxins upon incomplete combustion, which are often related to the evolution of cancer.
- The worst use for wood is combustion: so far, wood is the only renewable material produced on a large scale. The jungles now compete for land and water with agro fuels, crop-based bioplastics, and livestock production. Phosphate fertilizers for agro fuel crops have radionuclide contents, so biofuels do not free us from increased exposure to radiation.

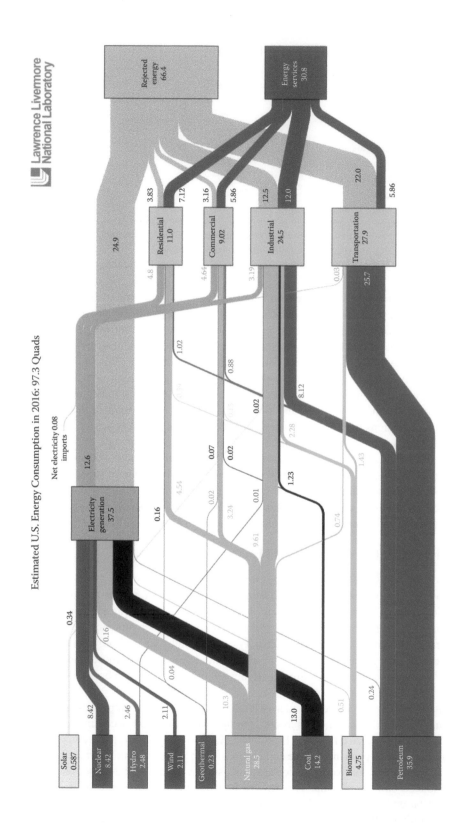

FIGURE 1.2 The 2016 energy flowchart for the U.S. 1 Quad ≈ 1.055 EJ. (From LLNL, Estimated USA Energy Consumption, Lawrence Livermore National Laboratory, Livermore, CA, http://flowchart.lllnl.gov, 2016.)

TABLE 1.2

Top 10 Countries with Highest Primary Energy Consumption during 2014

Country	Primary Energy Consumption (EJ)	Percentage of Global Consumption (%)
China	104.75	20.30
U.S.	99.31	19.25
Russian Fed.	30.14	5.84
India	24.95	4.84
Japan	18.51	3.59
Canada	12.64	2.45
Germany	11.97	2.32
Brazil	11.10	2.15
Rep. of Korea	11.01	2.13
Iran	10.68	2.07

Source: EIA, *International Energy Statistics 2017,* US Energy Information Administration, Washington DC, 2017 b.

- In Brazil, the pioneer in agro fuels, giant clouds of smoke from sugarcane preharvest fires hover over the most populated region around Sao Paulo. Their equivalent, emitted from China and India, are the atmospheric brown clouds (ABC) traveling to North America each year. Brownish ABCs are the product of biomass combustion.
- Residual vinasse from ethanol production contaminates water bodies with high concentrations of organic matter, metal additives (magnesium, aluminum, iron, manganese), and chlorides. Treating such waste takes a lot of energy and water. Emissions increase the air concentrations of aldehyde and peroxyacyl nitrate (PAN); they are toxic and possibly carcinogens in animals. PAN is highly oxidizing and appears in the photochemical smog, along with tropospheric ozone. Tropospheric ozone increases with biofuel use due to higher levels of volatile organic compounds and nitrogen oxides (De Oliveira et al. 2005).
- Methanol emits irritants and unburned methanol (which is poisonous and may produce blindness). Methanol produces two times more aldehydes than gasoline (Bromberg and Cheng 2010). Even waste cooking oil derived biodiesel (WCOB) seems to lead to more particle emissions than diesel. Of particular health concern are ultrafine metallic particles from WCOB exhaust (Betha and Balasubramanian 2014).
- Finally, pyrolysis, claimed to be a geoengineering solution, produces acid gases, ashes, dioxins, and nitrogen and sulfur oxides.

1.5 WIND AND SUN TAKE MARKETS BY STORM

1.5.1 COST REDUCTIONS

Solar and wind in 2016 were already cheaper than fossil fossils in 30 countries; even without subsidies they were on par with fossil fuels in a majority of countries (Bleich et al. 2016). This will further improve, as subsidies for renewables are on par with USD 5.6 trillion yearly global subsidies for fossil and nuclear energy estimated by the International Monetary Fund (Sawin et al. 2015).

Globally, solar PV utilities and wind energy are already cheaper than nuclear energy, and median costs of geothermal and biomass energy are lower than nuclear costs. The explanation is dual: hardware costs have steadily decreased while technology improvements have ramped up efficiency. As to capital costs, all solar PV options have much lower costs, and virtually all median renewable costs are lower than nuclear and coal plants (Lazard 2016).

Similarly, 2016 wind, geothermal, and photovoltaic costs in the U.S. were up to twice as attractive as nuclear energy, so that projected 2018–2022 capacity additions for wind and PV are more than tenfold and thrice those of nuclear energy, respectively (EIA 2017a). In the U.S. in 2016, 125 PV panels were added every minute (Lyons 2017), and wind had outpaced natural gas by 2015 in electric generation capacity additions (U.S. Energy Information Administration 2016).

In the following years, U.S. electricity generation will rise for only two primary sources of energy: natural gas closely followed by renewable sources. However, gas-generated electricity may decrease unless new reserves and technologies are discovered. Further, natural gas prices are likely to increase. Meanwhile, nuclear capacity retirements will continue through 2040, with no new plants (EIA 2017a). Other issues in the global nuclear industry have been Westinghouse filing for bankruptcy (Hals et al. 2017) and the U.S. Watts Bar 2 reactor finally coming online in October 2016, after 43 years under construction, only to go offline five months later and remaining that way. Watts Bar 2 was the first new reactor in the U.S. in 20 years (Hiltzik 2017). Similarly, dwindling onshore wind and utility solar costs in the UK compared to rising nuclear costs have made the latter a more expensive option (The Comptroller and Auditor General 2016).

In the words of International Renewable Energy Agency Director-General (IRENA 2016), "The age of renewable power has arrived. In every year since 2011, renewable power generation technologies have accounted for half or more of total new power generation capacity added globally."

1.5.2 The Generation of Permanent and Well-Paid Jobs

Employment in renewable energies already totaled 9.8 million jobs in 2016 (including 3.1 in the solar PV and 1.5 in large hydropower industries) (Ferroukhi et al. 2017). A flurry of reports (Bloomberg New Energy Finance 2016; Garrett-Peltier 2016; U.S. Department of Energy 2017) indicates that renewable technologies generate much more permanent, safe, and well-paid jobs than all fossil and nuclear industries. In the U.S. alone, renewable energy is creating jobs 12 times faster than the rest of the economy. Providing healthy, permanent, and well-paid jobs for hundreds of thousands of engineers, technicians, maintenance and installation workers, plant and equipment operators, planners, accountants, ecology economists, and a large number of professionals is undoubtedly one of the pillars of sustainability.

REFERENCES

Betha, R. and Balasubramanian, R. (2014). Emissions of particulate-bound elements from biodiesel and ultra low sulfur diesel: Size distribution and risk assessment. *Chemosphere*, *90*, 1005–1015. doi:10.1016/j.chemosphere.2012.07.052.
Bleich, K., Dantas-Guimaraes, R. (2016). *Renewable Infrastructure Investment Handbook: A Guide for Institutional Investors* (p. 18). Geneva, Switzerland: World Economic Forum.
Bloomberg New Energy Finance. (2016). *New energy outlook: Long-term projections of the global energy sector.* Retrieved from http://first.bloomberglp.com/documents/694813008_BNEF_NEO2016_ExecutiveSummary.pdf?elqTrackId=431b316cc3734996abdb55ddbbca0249&elq=07da292764ec4fe69c7a3adaf7981991&elqaid=3873&elqat=1&elqCampaignId.
BP. (2016). *Statistical Review of World* Energy (65th ed.). London: BP plc.
Bromberg, L. and Cheng, W. (2010). Methanol as an alternative transportation fuel in the US: Options for sustainable and/or energy-secure transportation. Retrieved from https://www.afdc.energy.gov/pdfs/mit_methanol_white_paper.pdf.
De Oliveira, M., Vaughan, B., and Rykiel, J. E. (2005). Ethanol as fuel: Energy, carbon dioxide balances, and ecological footprint. *Bioscience*, *55*, 593–603.
EIA. (2017a). Annual Energy Outlook 2017 with projections to 2050 (p. 64). Retrieved from eia.gov/aeo
EIA. (2017b). *International Energy Statistics 2017.* Washington DC: US Energy Information Administration.

Ferroukhi, R., Khalid, A., Garcia-Baños, C., and Renner, M. (2017). *Renewable Energy and Jobs. Annual Review 2017* (p. 24). Masdar City, Abu Dhabi, United Arab Emirates: International Renewable Energy Agency. Retrieved from irena.org.

Garrett-Peltier, H. (2016). Green versus brown: Comparing the employment impacts of energy efficiency, renewable energy, and fossil fuels using an input-output model. Retrieved from http://www.sciencedirect.com/science/article/pii/S026499931630709X.

Hals, T., Yamazaki, M., and Kelly, T. (2017). Westinghouse bankrupt. *Reuters*. Retrieved from http://www.reuters.com/article/us-toshiba-accounting-board-idUSKBN17006K.

Hansen, J., Sato, M., Kharecha, P., Beerling, D., Berner, R., Masson-Delmotte, V., Pagani, M., Raymo, M., Royer, D. L., and Zachos, J. C. (2008). Target atmospheric CO_2: Where should humanity aim? *Open Atmospheric Science Journal*, 2, 217–231.

Hiltzik, M. (2017). America's first "21st century nuclear plant" already has been shut down for repairs. *Los Angeles Times*. Retrieved from http://www.latimes.com/business/hiltzik/la-fi-hiltzik-nuclear-shutdown-20170508-story.html.

IARC. (2015). Press Release No. 231. January 13, 2015. Most types of cancer not due to "bad luck." IARC responds to scientific article claiming that environmental and lifestyle factors account for less than one third of cancers.

IRENA. (2016). *The Power to Change: Solar and Wind Cost Reduction Potential to 2025*. Bonn, Germany: IRENA Innovation and Technology Centre.

IRENA. (2017). *Renewable Capacity Statistics 2017* (p. 60). Abu Dhabi, United Arab Emirates: International Renewable Energy Agency.

Lazard. (2016). Lazard's levelized cost of energy analysis—Version 10.0. New York: Lazard Ltd.

LLNL. (2016). Estimated USA Energy Consumption. Lawrence Livermore National Laboratory, Livermore, CA. Retrieved from http://flowchart.lllnl.gov

Lyons, M. (2017). Solar Industry Sees Largest Quarter Ever. *Solar Energy Industries Association*. Retrieved May 4, 2017, from http://www.seia.org/blog/solar-industry-sees-largest-quarter-ever.

NBC News. (2010). Hawking: Off Earth by 2110? Retrieved May 18, 2017, from http://cosmiclog.nbcnews.com/_news/2010/08/09/4850998-stephen-hawking-off-earth-by-2110.

Olivier, J. G. J., Janssens-Maenhout, G., Muntean, M., and Peters, J. A. H. W. (2016). Trends in global CO_2 emissions, 2016 Report. PBL Netherlands Environmental Assessment Agency, The Hague, the Netherlands; European Commission, Joint Research Centre, Ispra, Italy.

Rincón-Mejía, E. (2011). Tecnologías solares de cero emisiones de carbono. In L. García-Colín and J. Varela (Eds.), *Contaminación atmosférica y tecnologías de cero emisiones de carbono*. Mexico City, Mexico: Universidad Autónoma Metropolitana.

Sawin, J. L., Sverrisson, F., Rickerson, W., Lins, C., Musolino, E., Petrichenko, K., Seyboth, K., Skeen, J., Sovacool, B., and Williamson, L. E. (2015). *Renewables 2015 Global Status Report* (p. 251). Paris, France: REN21 Secretariat.

Tans, P. and Keeling, R. (n.d.). Recent monthly mean CO_2 at Mauna Loa. *NOAA/ESRL, Scripps Institution of Oceanography*. Retrieved May 2, 2017, from www.esrl.noaa.gov/gmd/ccgg/trends; scrippsco2.ucsd.edu.

The Comptroller and Auditor General. (2016). *Nuclear Power in the UK* (p. 48). London. Retrieved from nao.org.uk.

U.S. Department of Energy. (2017). *U.S. Energy and Employment Report*. Retrieved from https://energy.gov/sites/prod/files/2017/01/f34/2017 US Energy and Jobs Report_0.pdf.

U.S. Energy Information Administration. (2016). Wind adds the most electric generation capacity in 2015, followed by natural gas and solar. *Preliminary Monthly Electric Generator Inventory*. Retrieved May 4, 2017, from https://www.eia.gov/todayinenergy/detail.php?id=25492.

Williamson, P. and Bodle, R. (2016). *Update on Climate Geoengineering in Relation to the Convention on Biological Diversity: Potential Impacts and Regulatory Framework* (Technical Series No. 84). Montreal, QC: Secretariat of the Convention on Biological Diversity.

2 Solar and Geothermal Energies Are Sustainable; Nuclear Power Is Not

Alejandro de las Heras
Independent Researcher

Eduardo Rincón-Mejía
Autonomous University of Mexico City

Behrooz Sani
Independent Researcher

Marina Islas-Espinoza
Mexico State University

CONTENTS

2.1   INTRODUCTION

Nuclear *fusion* taking place in the sun is safe to use directly as photovoltaic (PV) and concentrated thermal energy or indirectly as wind, wave, or tide energy. Nuclear *decay* in the Earth's interior is also harnessed, with Iceland setting the pace for reliable whole-country supply of geothermal energy. Conversely, nuclear *fission* is an ill-understood process making nuclear technologies' failures surprisingly regular.

This chapter addresses the sustainability of the foregoing types of radiation energy sources. It first highlights the supply of incident solar energy that can power the geosphere and biosphere and cater to the needs of a thriving future human civilization. Second, the technological conditions for sustainable geothermal energy (SGE) are examined. Third, while discounting weak antinuclear arguments (public perceptions), it shows the absence of technical conditions for sustainable fission nuclear power.

2.2   ENOUGH SOLAR ENERGY TO FUEL 10,000 EARTHS

In its orbit around the sun, during each nonleap year Earth intercepts an amount of energy equal to the solar constant $G_S = 1360.8 \pm 0.5 \, \dfrac{W}{m^2}$ (Kopp and Lean, 2011) multiplied by the area of a circle with Earth's mean radius $R_T = 6.371 \times 10^6$ m, times the duration of a year in seconds

$$E_S = G_S \left[\pi R_T^{\,2} \right][year] = \left(1360.8 \, \frac{W}{m^2} \right) \left[\pi \left(6.371 \times 10^6 \text{ m} \right)^2 \right] \left(3.1536 \times 10^7 \text{ s} \right)$$

$$= 5.472 \; 251 \times 10^6 \text{ EJ} \tag{2.1}$$

This is almost 10,000 times more than the energy consumed globally ($\sim 5.504 \times 10^2$ EJ in 2015) (BP, 2016), enough to meet present and future needs with very few or no emissions to the geosphere and the biosphere. As a reminder, 1 EJ is 10^{18} J.

But ~30% of this solar energy is reflected by clouds, ice, and Earth's surface (a factor known as albedo), so the solar energy caught by Earth is

$$E_{S \; caught} \cong \left(5.472251 \times 10^6 \text{ EJ} \right) \left(1 - 0.30 \right) = 3830576 \text{ EJ} \tag{2.2}$$

And since ~19% of the solar radiation is absorbed by the atmosphere, the solar radiation that finally reaches the Earth's surface, the gross solar potential (GSP) is

$$GSP \cong \left(5472251 \text{ EJ} \right) \left(1 - 0.30 - 0.19 \right) \approx 2,790,848 \text{ EJ} \tag{2.3}$$

Most of this energy, in thermal form, is required to move Earth's oceans and atmosphere. The photosynthetically active radiation (i.e., the visible range) is used by algae, plants, and bacteria.

If we assume conservatively that only one-hundredth of this energy could be sustainably utilized with future efficiencies close to the thermodynamic limit, the sustainable solar potential would be ~27,900 EJ, more than 50 times the current world energy consumption.

The technical potential is an even lower figure given by the second law of thermodynamics (no thermal machine is 100% efficient); by current mature and emerging technologies, not speculative ones; and by other practical limitations, including the environmental ones. Considering that 29% of the terrestrial surface is mainland, and assuming that 0.1% of this surface can be populated with PV modules with 16% mean sunlight-to-electricity conversion efficiency, and another 0.1% can be populated with solar thermal collectors with 60% mean thermal efficiency, the current technical solar potentials (TSPs) $TSP_{electric}$ and $TSP_{thermal}$ would be in the order of

$$TSP_{electric} \cong \left(2,790,848 \text{ EJ} \right) \left(0.29 \right) \left(0.001 \right) \left(0.16 \right) \cong 129.50 \text{ EJ} \tag{2.4}$$

and

$$TSP_{thermal} \cong (2{,}790{,}848 \text{ EJ})(0.29)(0.001)(0.60) \cong 485.60 \text{ EJ} \qquad (2.5)$$

These two conservative technical potentials combined exceed the current world energy consumption, so in principle, all fossil and nuclear plants could be abandoned much before 2050 by using only conventional solar technologies. Assuming a capacity factor of 0.20, an acceptable mean value encompassing all latitudes for current PV technologies, the $TSP_{electric} = 129.50$ EJ shown in equation (2.4) would correspond to a ~20.5 TW installed capacity. Considering all technical, infrastructure, economic, and policy barriers to be overcome, it is feasible to reach 10 TW by 2030 (Haegel et al., 2017).

Green plants convert solar energy with sunlight-to-bioenergy efficiencies greater than 2% (Bonner, 1962), but much larger efficiencies have been recorded in algae and phytoplankton (Ryther, 1959). The thermodynamic knowledge of photochemical solar energy conversion efficiencies is now at an advanced stage (De Vos, 1995). A very conservative estimate of a mean photochemical solar efficiency would be 1% for all photosynthetic organisms. Considering a photosynthetic cover on 50% of the oceans and land, the gross bioenergy potential (GBP) can be in the order of

$$GBP \approx (GSP)(0.01)(0.21)(0.5)(2) = 5860.78 \text{ EJ} \qquad (2.6)$$

The current rate of energy capture by photosynthetic processes seems to be ~130 TW (Steger et al., 2005), which corresponds to 4099.68 EJ, 30% less than the estimate in Equation 2.6. This potential must factor in reforestation and sustainable agroforestry using perennial plants. It has just been discovered (Bastin et al., 2017) that drylands, which occupy more than 40% of Earth mainland, have much more extensive forest than previously reported and cover a total area similar to that of tropical rainforests or boreal forests. This increases estimates of global forest cover by at least 9%, approaching the GBP to that given in Equation 2.6. This does not even account for the very extensive soil biological crusts and their photosynthetic role. The current technical bioenergy potential could conservatively be as large as 1% of GBP, which would amount to 59 EJ.

To understand the amount of photosynthetic activity on Earth, one must realize that it converts ~110 billion metric tons of carbon into biomass each year (Field et al., 1998). Fossil fuels were all formed during hundreds of million years with biomasses nourished by solar energy via photosynthesis with relatively low solar–biomass energy efficiencies.

Most of the bioenergy potential is used in endoenergetic (metabolic) needs, and only a residual portion could be used for human exoenergetic applications. The annual human endoenergetic consumption of 10 billion people much before 2100, assuming 100 W per capita metabolic consumption would be

$$HEEC \cong \left(100 \ \frac{W}{inhabitant}\right)\left(10^{10} \ inhabitant\right)\left(3.1536 \times 10^7 \text{ s}\right) = 31.54 \text{ EJ} \qquad (2.7)$$

This is ~0.5% of the GBP, but the energy needs of thousands of other species limit the amounts of bioenergy that could be dedicated to human exoenergetic consumption. Finally, deforestation and land use emissions since 1850 have totaled ~660 $Gt \ CO_2$; this figure provides an upper limit to the physical potential for reforestation to capture CO_2 (National Research Council, 2015). Estimating that complete afforestation is rather unrealistic, and that deforestation emissions have been a tenth of those from fossil fuel consumption, we must not rely solely on reforestation to return to 350 ppm atmospheric CO_2 in the medium term.

Now, as mentioned in Chapter 17, ~1% of the solar energy caught by Earth is converted into wind energy (Peixoto and Oort, 1992). Then the gross wind potential (GWP) can be estimated as

$$GWP \cong 0.01 \ E_{S \ caught} = 38{,}306 \text{ EJ} \qquad (2.8)$$

Using wind turbines with power coefficients Cp=0.50, harvesting 0.5% of this GWP gives the approximate current technical wind potential (TWP):

$$TWP \cong (0.50)(0.005)GWP = 95.77 \text{ EJ} \qquad (2.9)$$

These and other gross and technical potentials for most renewable sources of energy as well as their current utilization are summarized in Table 2.1.

In Table 2.2, the proved reserves of oil, natural gas, and coal are shown. By total proved reserves, we mean those quantities that geological and engineering information indicate, with reasonable certainty, can be recovered under existing economic and operating conditions. The reserves/extraction (R/E) ratio is the expected time that a proved reserve would last if extraction were to continue at the rate of the last year of extraction.

If those proved reserves are expressed in their energetic content in EJ, the magnitude of the oil reserves is 8,877.10 EJ, the gas reserves contain 6,933.15 EJ, and the carbon ones have 18,327.78 EJ. Altogether, they have an energetic potential of 34,138 EJ. The GSP of 1 year is more than 80 times larger. In other words, in just 5 days, the terrestrial surface receives more solar energy than is contained in all fossil proved reserves.

Knowledge of thousands of main sequence stars like our sun, also known as dwarves, indicates that it will not start growing for at least another 3 billion years, following which life on Earth will be wiped out. And so, the potential of solar energy is, for all practical purposes, infinitely greater than that of all fossil energy sources. Nuclear materials, like uranium and thorium, have even smaller potentials than those of fossils: all nonrenewable sources of energy are insignificant compared to solar and other renewable sources. According to Table 2.2, fossil (and nuclear) sources cannot supply energy for much more than half a century. Therefore, their consumption at current rates, even ignoring the deadly pollution that their use entails, is unsustainable.

TABLE 2.1

Estimated Gross and Technical Potentials of Main Renewable Energy Sources

Energy Source	Gross Potential (EJ/year)	Current Technical Potential (EJ/year)	Supply in 2016 (EJ)
Solar	2,790,848	$TSP_{electric} \approx 129.5$	1.5
		$TSP_{thermal} \approx 485.6$	1.38 [12]
Wind	38,306	$TWP \cong 95.77$	3.15
Geothermal	1,400 [13]	60	0.34
Bioenergy	5,860.78	59	9.41 [12]
Ocean tidal	75.69 [14]	4.67 [15]	0.004
Ocean surface currents	255.44 [16]	0.473 [17]	–
Hydropower	365.82 [18]	51.40 [19]	37.38 [1]
Total	2,837,111.73	886.41	53.16

TABLE 2.2

Proved Reserves of Fossil Fuels and Relation Reserve/Extraction

Energy Source	Total Proved Reserves	Extraction in 2015	R/E (Years)
Oil	1697.6×10^9 barrels	34.6779×10^9 barrels	48.95
Natural gas	$186.9 \times 10^{12} \text{ m}^3$	$3538.6 \times 10^9 \text{ m}^3$	52.82
Coal	437,749 mtoe	3839.9 mtoe	114

Source: BP, *Statistical Review of World Energy,* 65th ed., BP plc, London, 2016.

2.3   GEOTHERMAL ENERGY

Geothermal energy is Earth's heat that can be exploited by technology. This heat source is mostly due to residual planetary accretion heat and radioactive decay of uranium isotopes (^{238}U, ^{235}U), thorium (^{232}Th), and potassium (^{40}K) (Lubimova, 1968). At present, the planet undergoes slow net thermal loss to space (minus 300°C–350°C every 3 billion years, with a current temperature of 4000°C at the bottom of Earth's mantle). Outgoing heat flows are 44.5 TW from Earth, out of which 19% escape from the radioisotope-richer crust, 77% from the mantle, and the remainder from the nonradioactive core (Stacey and Loper, 1988). Half of the mantle's heat is radiogenic. The total heat content of Earth is 12.6×10^{24} MJ (Armstead, 1983), equivalent to 3.3 years of solar energy captured by Earth.

Current technologies tap convective hydrothermal systems (liquids or steam) and conductive geothermal systems (hot rock). Larderello, Italy, in 1904 was the first spot to be utilized, both as a source of direct thermal energy and in electricity generation. By 1928, Iceland started to exploit hydrothermal sources.

2.3.1   Sustainable Geothermal Energy

Several conditions that appear to be linked to SGE are as follows:

- Although renewable, crust heat is often withdrawn 10 times faster than its replacement rate. Pressure can decline and cold water can permeate into the geothermal reservoir (World Energy Council, 2016).
- SGE microgeneration by end users (mostly ground source heat pumps) reduces transport costs and attendant heat losses but often requires artificial wells. Drilling economic and environmental costs need to be gauged against longer-term sustainability. These costs can be modest since drilling is limited to shallow depths where the sun drives the temperature difference that allows for ground source geoexchange technologies.
- Natural sources are frequently large and become the hub of centralized networks with extensive piping subject to corrosion.
- Pollutants threaten aquifers. SGE must differ from practices in unsustainable technologies: fracking (shale gas hydraulic fracturing) uses many pollutants, such as biocides that limit microbiologically enhanced corrosion of piping and equipment (Sovacool, 2014). Such biocides only work until microbes develop resistance and contribute to the global health problem of antibiotic resistance.
- Natural wells and deep artificial ones often bring to the surface brine containing potentially toxic metals (Baba et al., 2008), including naturally occurring radioactive material (IAEA, 2013). The brine must be recycled in situ, after recovery of pollutants (Premuzic et al., 1995), all of which are useful. Brine recycling has the advantage of reducing its mineralization thereby augmenting the amount of available heat (World Energy Council, 2016).
- Stock must be taken of seismicity lessons from other energy technologies (nuclear power, large hydropower, fracking): local seismicity hazards endanger both the environment and utilities; these in turn affect natural seismicity.
- Invasive methods aiming at reservoir stimulation (chemically enhancing permeability, thermal cracking, or deflagration) (World Energy Council, 2016) put reservoirs and aquifers at chemical and mechanical risk.

Solutions exist for all the above. In particular, Enel S.p.A.'s stillwater solar PV-concentrating solar–geothermal hybrid plant in Nevada has extended the life of the geothermal reservoir, reduced injections of water in a water-scarce area, and recycled and reheated spent brine into the reservoir. Also, coiled tube and wirelines can successfully reduce costs of mineral scale removal (World Energy Council, 2016) without using biocides.

2.4 NUCLEAR POWER: DREAM FOILED

The Anthropocene Epoch started around 1950 when a layer of radioactive plutonium and ^{14}C fallouts covered, indelibly, the entire globe, following nuclear arms experiments (Clarke et al., 2012; University of Leicester Press Office, 2016). As to civil nuclear power, it would only be acceptable if it met proper conditions. But does it?

According to the nuclear industry, the most lethal nuclear meltdown so far, Chernobyl in 1986, led to 58 deaths (World Nuclear Association, 2016). However, independent researchers report that in Belarus, Russia, and Ukraine alone, an estimated 200,000 additional deaths resulted from the catastrophe during 1990–2004 (Greenpeace, 2006). This order of magnitude difference between the industry and independent assessments forces a step back and a thorough examination of sustainability of nuclear power.

There is overwhelming evidence that current and proposed nuclear power technologies should end as sources of energy (Rincón, 2011). For Arnie Gundersen, a former nuclear industry senior vice president and licensed reactor operator "the evidence proves that new nuclear power plants will make global climate change worse due to huge costs and delayed implementation periods. Lift the CO_2 smoke screen and implement the alternative solutions that are available now, faster to implement and much less expensive" (Gundersen, 2016).

The nuclear power industry in fact is facing economic challenges, with renewables reaching lower electricity costs than nuclear power (Chapter 1) and resistance to protracted nuclear projects that have much hindered the expansion of the industry since around 2000 in Europe and Eurasia, the U.S., and the Asia–Pacific regions (Figure 2.1). Biological unsustainability of nuclear power is addressed in Chapter 20. In the remainder of this section, the main scientific and technological issues of nuclear power are reviewed.

2.4.1 QUANTUM PHYSICS UNCERTAINTIES

The current explanation of nuclear fission is a quantum tunneling process. As such, it has a Schrödinger equation of the probability (>0) of anything happening, in this case neutrons escaping

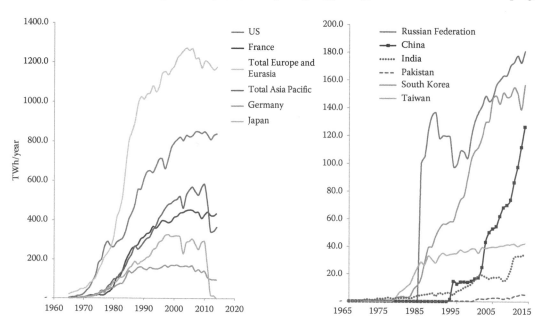

FIGURE 2.1 Nuclear power plants: installed capacity and stalled growth trend among main nuclear power users. (From BP, *Statistical Review of World Energy*, 65th ed., BP plc, London, 2016.)

the strong nuclear force binding them to protons. Knowledge gaps are still disturbing: there is a dearth of "complete fission theory, many puzzles are yet unsolved" (Al-Adili et al., 2015). "Quantitative understanding [of the nuclear fission process] remains elusive" even though "many nuclear applications as well as fundamental research rely on a precise description of fission data" (Talou et al., 2015). "Exact models are seldom available in nuclear physics. Good agreement between different models (precision) does not guarantee that they are near the truth; all models can be systematically off due to missing knowledge," and "systematic uncertainties, on the other hand, are very tough to assess" (Ireland and Nazarewicz, 2015).

2.4.2 On Uncertainties and Catastrophes

The Schrödinger probability argument also applies to accidents: The clearest example of a nonnull (Schrödinger) probability for any event, however complex and unlikely, was the catastrophic loss between March 12 and March 15, 2011, of Fukushima Daiichi reactors 1–4 (in that order): following the earthquake, steel towers collapsed, external power was interrupted; then following the tsunami, the emergency diesel generators stopped and the diesel tanks were washed away. Battery power was then exhausted and the "normal" heat release to the sea was cut off; the cooling function in spent fuel pools continued to be lost until the fuel rods became exposed to air; they heated rapidly and their cladding reacted with water producing vast amounts of hydrogen, raising the pressure; not even venting impeded a hydrogen explosion that blew the reactor 1 building apart. At reactor 2, unstable cooling via the steam-driven Reactor Core Isolation Cooling system stopped and the fuel became exposed to air and totally released. At reactor 3, coolant injection failed, not permitting the Reactor Core Isolation Cooling to operate and raising the pressure; later, when the spent fuel pool water evaporated, the rods became exposed to air, producing a hydrogen explosion that destroyed the building. At reactor 4, a similar spent fuel pool cooling failure led to a hydrogen explosion that also tore apart the building (Great East Japan Earthquake Taskforce & Science Council of Japan, 2011). Clearly, the builders of Fukushima's redundant systems did not manage to make them independent of each other or able to withstand a common (seismic) shock, despite decades of modeling the effects of common cause failures on redundant systems.

2.4.3 Uncertainties and Widespread Failures

2.4.3.1 Criticality and Load Factor

Theories also have failed to predict normal reactor behavior: unpredictable variations in parameter K (the reproduction rate of the fission reaction) seem linked to still ill-understood load factors of nuclear reactors below 65% on average (the load factor is the ratio of the power produced in a year to the nominal capacity without downtimes) (Cowan, 1990). In other words, technology is not able to control fission at the moment; at best, criticality ($K \geq 1$) can be controlled by stopping the reaction. There are more technological experimentation and unsolved issues in the nuclear industry than sound science to pinpoint them. Chernobyl was a case in point of a failed experiment.

2.4.3.2 Leaks

The drawbacks of technological experimentation are illustrated by widespread leaks demonstrated by using iodine-129 as a marker. ^{129}I is formed by uranium and plutonium fission in nuclear power plants and in reprocessing facilities. The nuclear reprocessing plants in La Hague (France) and Sellafield (UK) are globally the main sources of ^{129}I. But ^{129}I reemission from the sea surface in the English Channel, and the Irish, North, and Norwegian seas, is the main ^{129}I source washed out in precipitation farther away, in Denmark (Hou et al., 2009a). Atmospheric transport from the Marcoule (France) nuclear reprocessing plant (closed in 1998 after 40 years of emissions) is evidenced by enriched levels of ^{129}I, ^{238}Pu, and $^{239+240}Pu$ in collected terrestrial vegetation samples in Marcoule's vicinity. Farther away, activity and depositions follow a North–South direction related to atmospheric discharges; these eventually meet the Rhone River (Duffa and Fréchou, 2003) and so enter another environmental compartment in a highly populated area. The same pattern of surface water ^{129}I pollution occurs in Western Europe, North America, and Central Asia, especially in terrestrial surface water of the Northern hemisphere (50°N). Topmost ^{129}I concentrations are found in England, followed by Israel, Europe, and North America. Direct gaseous emissions from the nuclear facilities and marine atmospheric reemissions are the most probable sources of ^{129}I polluting terrestrial surface waters (Chen et al., 2015).

Organic forms of iodine are highly bioavailable to humans through ingestion (via water and plants). After transport to the thyroid, this organ is exposed to long-term, low-dose, beta radiation from decaying ^{129}I (half-life 15.7 million years). More nuclear power plants and more spent fuel reprocessing would largely increase ^{129}I levels (Hou et al., 2009b). As for uranium, cleanup of mill sites has cost U.S. taxpayers in excess of USD 2 billion; in its U(IV) state, it accumulates in cool anoxic sediments where it adsorbs to natural organic matter (Bone et al., 2016).

2.4.3.3 Waste

Perhaps the most difficult, and as yet unsolved, issue for the global nuclear power industry is final confinement of spent nuclear fuel. This is a limiting factor to the expansion of that industry. No solution has yet been found, and no alternative to the leaks from La Hague, Sellafield, and similar facilities. Another kind of uncertainties, geological this time, are met by permanent disposal of waste, according to the U.S. National Academy of Sciences: not all uncertainties can be allayed by research and development as, over geological time, volcanism, behavior of the (un)saturated zones, geochemistry, and climate will evolve. Fundamental properties of the rock are further disrupted by tunnels and the fractured zone they create. Moreover, a radioactive heat pulse creates pore water movements (Ramana, 2017).

These factors led to the closure in 1998 of the Morsleben (Germany) nuclear waste storage since salt domes are on the verge of collapse despite temporary stabilization. In Asse II, the overlying rock shifts 15 cm/year, weakening the mine; the increasing rate of new water breaches predicts uncontrollable water inflow with ensuing collapse hazard. Brine has to be captured before it reaches the storage canisters and accelerates radioactivity-induced corrosion.

Geology is not the only set of risks in final confinement. In the WIPP facility in New Mexico, an alleged error in packaging material (Figure 2.2) caused an explosive release of plutonium and americium, which reached the surface. The accident, one of the costliest in the U.S. (Ramana, 2017), provoked the closure of the facility in 2014.

2.4.4 Accidents and the Human Factor

Regarding the WIPP accident, one assertion by the U.S. Department of Energy was that some of the organizations managing WIPP had let the safety culture deteriorate (Ramana, 2017). The case of the Indian nuclear industry is also well documented: incidents occur in reprocessing plants, involving the security systems and widespread lack of learning culture, safety culture, and justice (blame shifting from managers to operators) (Ramana and Kumar, 2013). Recurrent failures occur in the same components of the reactors (Figure 2.3).

2.4.5 Nondemocratic Practices

The nuclear power industry is therefore exposed to design, human, and institutional failures. Two resounding cases of institutional failures are the San Onofre nuclear power plant scheduled for decommissioning by 2030–31, and the Hinkley C reactor currently in the early construction phases. San Onofre was closed in 2012 following a USD 680 million investment aimed at adding 40 years to its useful life and instead precipitating its demise. As per an open records request to the U.S. Nuclear Regulatory Commission, San Onofre's operators seem to have recklessly pushed the reactor for the

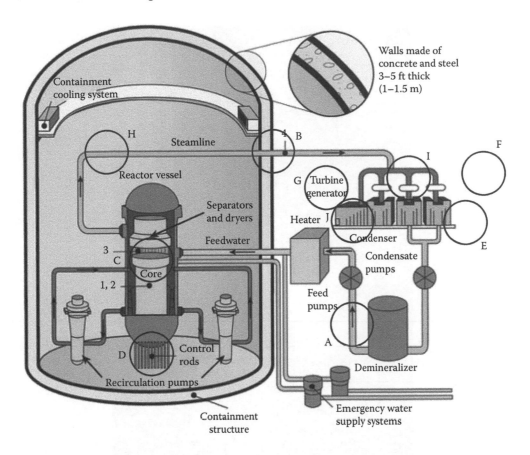

FIGURE 2.2 Typical failures and incidents in Indian nuclear reactors. (a) Madras 2 cool water plug slipped away (1999) and a large amount (4–14 tons) of heavy water leaked out. (b) Large tritium release to the atmosphere in Rajasthan (2004): high levels of tritium in the liquid discharges in Narora and Kakrapar (2003). (c) Fortuitous hole in the primary heat transfer system as a result of an eroded wall in Rajasthan 2 (2007). Three heavy-water leaks elsewhere (2009). (d) Failure of the moderator inlet receiving heavy water at high speed, in Madras (1989). (e) Excessive vibration in the turbine bearings and oil leaks in Narora (1993). This led to sparking (1981). (f) Vibrations in turbine bearings and failure in turbine blades (1982). Vibrations in the turbine generator bearings and blades sheared off at the base (1983). (g) High bearing vibrations in the turbine generator in Madras 1 (1986), similar to Rajasthan 1 (1985, 1989, and 1990). (h) Oil leak in the generator transformer in Madras 1 (1988), with heavy sparking in Madras 1 (1989) and twice in the turbine generator in Narora 1 (1992). (i) Two fires in the primary heat transport system, oil leak in the turbine valve. (j) Hydrogen gas leak in the generator cooling system in Madras 2 (1991). (From Nuclear Regulatory Commission, 2015. Boiling Water Reactors. Retrieved May 2, 2017, from nrc.gov/reactors/bwrs.html, 2015; Ramana, M. V., and Kumar, A., *J. Int. Stud.*, *1*, 49–72, 2013.)

sake of profit, leading to definitive damage. The operators later sued the Japanese provider as allegedly responsible for the damage. Then, the USD 4.7 billion settlement between the operators and the California Public Utilities Commission (CPUC) included a clause, which left USD 3.3 billion to be paid by the ratepayers. California's attorney general has opened a criminal investigation against the president of CPUC on grounds of possible collusion between the state regulator and San Onofre operators. This is also the cause for a USD 16.7 million fine to one of the operators (Herchenroeder, 2016; McDonald, 2017). The California Coastal Commission then unanimously approved underground confinement of San Onofre's spent fuel after 15 months of negotiations behind closed doors with the operators. While legal, the closed-door sessions gave a lengthy lead time to the operators

FIGURE 2.3 Damaged drum with radioactive waste inside WIPP Panel 7, Room 7, New Mexico, USA, 2014–05–15 during investigations on the cause of radioactive contamination. (DoE Photographer, damaged drum with radioactive waste inside WIPP Panel 7, Room 7, New Mexico, USA, 2014-05-15 during investigations on the cause of radioactive contamination. Retrieved June 4, 2017, from https://en.wikipedia.org/wiki/File:WIPP_DoE_2014-05-15_5_15_Image_lrg.jpg, 2014.)

but only a week's notice to the public, before the hearings and decision. As a consequence, a lawsuit against the Coastal Commission was allowed to proceed (Nikolewski, 2017; Sharma, 2017). These issues have prompted Californian municipalities to opt for community choice aggregation (CCA), a scheme giving more say to citizens in energy matters and the possibility to transition to renewable energy; in turn this has led to the emergence of a lobby group against CCA, funded by one San Onofre operator (Smith, 2017).

In Britain's Hinkley C project, the government seems to have ended wind power subsidies and cut back solar subsidies while giving unwavering support to nuclear power. The decision of the government was made despite higher costs of nuclear electricity, faulty components in the same-design plants having led the French authorities to stop several plants, and an indictment by U.S. authorities against the main Chinese financial backer of Hinkley C. Chinese financial support was agreed to at the top executive level in exchange of Britain purchasing a small modular reactor (SMR), and the indictment was caused by unauthorized technology exports from the U.S. to China in relation to said modular reactor. In secret documents that the British government refused to disclose under a Freedom of Information request, the government also pledged to bear overrun costs, in particular waste disposal costs. While the government stands by the French and Chinese corporations involved in Hinkley C, the UN Economic Commission for Europe asked the UK to suspend work on the site, pending notification of potential impacts of Hinkley C to Germany, Norway, and Holland in line with transboundary pollution treaties (Dombey, 2016; Doward, 2016; Vaughan, 2016, 2017; Vaughan and Willsher, 2016).

2.4.6 SMALL MODULAR REACTORS: KILLED IN THE U.S., A RETURN WOULD SPREAD THE ISSUES

SMRs like all nuclear power are unprofitable since the raw material, construction, production, and maintenance are globally dangerous and are extremely costly compared to other industries. Even the leading SMR developers Babcock & Wilcox and Westinghouse, who had obtained increasing federal SMR subsidies, decided to stop the development of SMR technologies for lack of customers and investors. Westinghouse decided to focus on decommissioning existing reactors (Cooper, 2014). Despite the lack of market in the U.S., other countries pursue SMR development probably due more to government subsidies than market interest. SMR designs claim that there are only four problems identified with nuclear power today: costs, safety, waste, and proliferation. Countries like the U.S., Russia, China, France, Japan, South Korea, India, and Argentina intend to develop SMRs based on subsidies; however, the characteristics of the different SMR designs under development suggest that none of the designs meet all four of these challenges simultaneously (Ramana and Mian, 2014).

2.4.7 MEDICAL AND SCIENTIFIC USES REQUIRE FEW REACTORS

The global demand for medical isotopes requires only a couple of experimental reactors, since at one time the small Chalk River reactor in Canada supplied half the world's requirements (Harris, 2014). As for fundamental physics requirements, they would be better served by a few well-characterized hence comparable reactors. At the turn of the century, the U.S. Department of Energy recommended only the use of existing facilities to further science and supply civil nuclear uses (DoE, 2000).

2.5 MILITARY DRIVERS OF NUCLEAR POWER

The top global strategic topics are, in decreasing importance, energy and weapons. Indeed, one justification for civil nuclear power has been to obtain weapons-grade uranium or plutonium, through at least four pathways: uranium enrichment, heavy water reactors that do not require enrichment, plutonium production reactors, or plutonium waste from a civil reactor followed by reprocessing. The UN Security Council 5+1 countries have tried to monitor these pathways to curtail nuclear weapons proliferation, with either bias or lack of success: the production of atomic weapons by India, Pakistan, and Israel has complicated the political and military arena in Western and Southern Asia. Tension escalation in Eastern Asia following tests in North Korea and a standoff in the South China Sea might lead Japan and South Korea, caught in the regional game for supremacy between the U.S. and China, to a rather simple transfer of civil to military nuclear wherewithal. Germany might also try to compensate for lessened protection from post-Brexit UK and U.S. withdrawal from NATO, although it may become patent that Helmut Kohl's Germany had purchased Soviet missiles and they were left behind when the Red Army left Eastern Germany.

However, the usefulness of nuclear deterrents after the Cold War is now denied by a bipartisan group of former U.S. State and Defense Secretaries (Shultz et al., 2007–2011). They call for urgent and complete nuclear disarmament, prevention of accidental detonations, phaseout of commercial and scientific uses of highly enriched and weapons-grade materials, and prevention of regional conflicts. Other former defense and state secretaries, national security advisors, and Western and Asian former heads of state endorse their views. Additional arguments against nuclear weapons are, on the one hand, technical advances in conventional arms that make several nuclear goals and missions obsolete (Oelrich, 2005). On the other hand, new warfare fields, namely terrorism and cyberattacks, are exposing vulnerabilities of nuclear defenses.

At the time of writing, higher risks of nuclear proliferation exist under the new Trump administration. The remainder of this chapter will nevertheless address the hitherto little-studied field of nuclear power security from a military point of view (drawing on firsthand experience in the Iranian civilian program). This is motivated by two facts: power generation is the first target of any military attack, and a provoked nuclear power plant accident can be an extremely contaminant event, able to impair a country's defenses in a large region.

2.5.1 NUCLEAR POWER IS OIL DEPENDENT AND SECURITY INTENSIVE

Most of the nuclear energy life cycle is fossil fueled, from mining and transportation to initial centrifugation, many backup systems, massive concrete and burrowed buildings to shelter the nuclear reactors, the manufacture of all reactor components, their maintenance, and finally plant decommissioning, dismantlement, and cleanup. In 14 U.S. states, there are around 15,000 abandoned uranium mines still not cleaned up.

The security of the nuclear industry is also very energy intensive. Nuclear energy is cumbersome: aerospace experiments are few, and along with nuclear submarines and carriers are the preserve of the major nuclear powers. Air strikes and troop deployment critically depend on oil. Additional oil consumption stems from military vehicles and systems protecting nuclear power plants around the world. Security and safety (2S) costs also include multilateral agencies (UN Security Council and IAEA), nuclear industry public representation, and diplomacy. Other costs are the physical enforcement of technology secrecy and safekeeping of the always weaponizable waste. The recovery of lost/stolen devices and material, which seems to occur or be reported more often in the U.S., is an additional military and police task (CNS, 2017).

Although nuclear power is probably not viable beyond the exhaustion of oil reserves, this dependency of nuclear power on oil exacerbates the armed conflicts around the dwindling yet ever less profitable, oil reserves.

2.5.2 CYBERWARFARE, INTELLIGENCE, AND COMPLEXITY

In 2010, a computer program later nicknamed Stuxnet was introduced by a mole using a USB drive in the computer system of the Natanz nuclear enrichment plant in Iran. The program remained dormant for a month and then took control of the Siemens system driving the centrifugation enrichment process; all displays appeared to be normal until the backup system kicked in. The attack seemed to pertain to the U.S. Olympic Games covert operation carried out by U.S. and Israeli agencies. Retaliation was not direct and instead targeted the Saudi oil company Aramco network of computers. Internationally, these are construed as attacks. In nuclear power plants, exploiting code vulnerabilities is a chief factor complicating security, as many systems are computerized in the man–computer–robot environment of a plant (Figure 2.4).

Cyberwarfare adds complexity to secrecy, an indispensable element of security, but one that contributes to noncivilian, nondemocratic control of nuclear power. In turn, intelligence has two components that can easily conflict with one another, although both are concerned with monies: on the one hand, military intelligence is the part of the military that checks, controls, and rechecks the planning and programming of an operation so it can meet several objectives, within a stated uncertainty (Figure 2.5). On the other hand, intelligence in the sense of espionage is much less

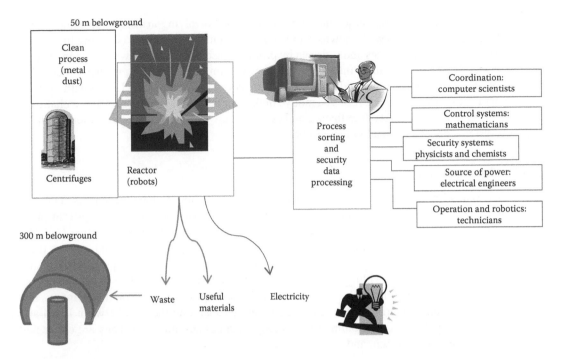

FIGURE 2.4 Vulnerability areas in the human–computing interface of a nuclear power plant (thick lines and outlines). Vulnerabilities are numerous in real life. The main risks are as follows: (1) nuclear fission energy > electricity output; (2) attack to the centrifuges; (3) external cyber- or physical attack during the recovery from heavy seismic event, and/or technical failure and/or (cyber) attack; and (4) simultaneous electromagnetic pulse (via nuclear detonation) to knock down all electronic systems and neutron bomb (to kill all human operators). Issues can occur simultaneously as shown in Fukushima.

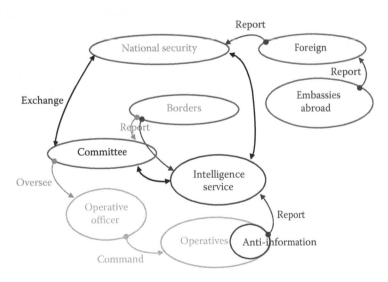

FIGURE 2.5 Intelligence in the Iranian nuclear context. Schematic of the complex information flows between nuclear plant operatives and intelligence services. The information is both administrative (concerning approved monies and reporting deadlines) and operative. The tricky part for operatives is to always keep a flexibility margin to accommodate for nonplanned events. Information from abroad is a very early warning but can in fact be disinformation.

predictable, as its stock-in-trade is mediation in transactions involving arms, politics, and, often, fuels. The case of Irangate is well known. Another well-documented example (showing that espionage overcomes national loyalties and is driven by profit) is the arms trade that was set up in 2010 by Arthur Porter, head of the Security Intelligence Review Committee in Canada; former Israeli intelligence officer Ari Ben-Menashe; and Russian arms dealers to provide Sierra Leone with arms of unknown origins (Harris, 2014). Profit as a motive makes for fickle allegiances and so is a challenge to the security of national assets. Intelligence and cyberwarfare demonstrate that ever more complex security systems beget additional vulnerabilities.

2.5.3   IDEOLOGY AND OIL

Ideologies, *sensu* Immanuel Wallerstein, are strategies in the pursuit of global hegemony. The U.S. now dominates the world in military terms. U.S. national security is based on long-term plans elaborated by the CIA with a say from the American Israel Public Affairs Committee (AIPAC) lobby; in such plans, the probability of remaining the only superpower is gauged against time: the U.S. will retain the hegemony with a 100% probability within 2 years, 50% probability within 25 years, and 25% probability within 50 years. These plans were uncovered when the U.S. embassy in Tehran was taken over in 1979. But the Middle East challenges the U.S. global hegemony. The region has grown into a matrix of interconnected countries spanning the area between Israel and India, Southern Republics of the Community of Independent States, and Iran. They are connected by fossil fuel reserves, terrorism, and nuclear threats.

2.5.3.1   Hegemony and Instability (Divide and Rule)

Terrorism has not yet materialized as a threat, but there is enough material being lost or stolen for the danger of radiological dispersal devices to be considered. More acute is the complexity added by terrorism to the global geopolitical situation. Terrorism is the protracted outcome of the Cold War and the struggle of the U.S. toward global hegemony: it was the brainchild of Zbigniew Brzezinski, National Security advisor to U.S. President Jimmy Carter, aimed at making the Soviet invasion of Afghanistan difficult; it was called the Green Belt. The USSR wanted to follow up on the tsars' geopolitical view: we must reach the Persian Gulf. Ethnic groups from the USSR retreated to Pakistan. Ronald Reagan followed up on Brzezinski and used Mujahidin, revolution in Iran, radicalism in Turkey, Saddam Hussein in Iraq, and Shiites vs. Sunnites in Lebanon. The reaction from a segment of the population in those countries was to emigrate to Europe. The problem was the same for them as for North Africans: second-generation "migrants" (i.e., European citizens) were discriminated against and disenfranchised. Years later, a marginal fraction of these people became the network of Daesh terrorists sent from Syria to Europe.

Brzezinski's strategies also damaged the SENTO confederation that included Iran, Jordan, Oman, Egypt, Turkey, and later Iraq, Syria, and South Yemen. It was a counterweight to NATO and the Warsaw Pact. Its demise coincided with U.S. support to the Mujahidin and the accession to power of Khomeini and Bhutto in Iran and Pakistan, respectively.

In addition, the U.S. has fueled the religious conflict between Iran and Saudi Arabia (SA), to divide OPEC and weaken Iran's position, formerly OPEC's top oil producer. SA has used its oil production to decrease the oil prices in order to slow Iran's development. In the Iran–Iraq war, Iraq was supported by SA, vying for regional resource control. At one time, SA offered to pay Iran to leave Iraq and let go of 250–300 km of land gained during the conflict. More recently, the armed conflicts by proxy between SA and Iran are taking place in Yemen and Syria.

2.5.3.2   The Next Oil Wars

George W. Bush changed strategy: the U.S. took direct military action in Iraq to gain control over energy reserves. Further control was sought by the presence in the Persian Gulf of the U.S. Navy, to curtail the control by Iran of the Strait of Hormuz. Through the Strait, a vast amount of all the oil in the world transits; 90%, 60%, and 40% of the Chinese, Japanese, and Korean fossil fuel supplies come from Iran.

When the Democrats took over from G.W. Bush, they returned to action by proxy, such as in the 2011 Arab Spring, but the end result there is that Libya is now divided between Daesh and the rebels; Egypt is led by Eghnan El Muslemi after two successive coups; Syria is struggling against Daesh (which has received help from SA, Turkey, and Qatar, a fact well known by Department of State secretary Hillary Clinton). One must note that after the destruction of Saddam Hussein's army by the U.S. and allies, removed Iraqi officers went on to constitute half of Daesh's forces; 25,000 warriors were also brought in from Europe. At stake in the conflict in Syria is the control of oil reserves, where Daesh receives support from the Erdogan regime in Turkey, acting as mediator of BP's oil interests.

Future oil wars will depend on the outcome of the conflict surrounding Syria, but likely next oil conflicts will include Sudan (oil) and then Somalia (gas). Already, gas reserves motivated the annexation of Crimea by Russia in 2014. The region holds the most gas reserves in the world (Iran > Russia > Qatar).

2.5.4 REALPOLITIK AMONG THE U.S., ISRAEL, AND IRAN

One can only understand the U.S.–Israel relationship under the prism of pragmatism on both sides. Israel's foreign policy relies on support from abroad and especially lobbying in the U.S. through AIPAC. Often, but not always, the U.S. and Israel are strategic partners: Israel also buys military technology from Russia and Europe as a way of gaining political and military support. Moreover, Israel has sold U.S. weapons to China.

Conversely, the U.S. does not depend on Israel alone in the Middle East: they also partner with NATO member Turkey, Bahrain (to station the fifth U.S. fleet), Kuwait, SA, and occasionally with Egypt. The link between SA and the U.S. hinges on SA's holding USD 800 billion in the U.S. stock exchange and USD 900 billion in the treasury in Ryad. As to Pakistan, which harbors terrorist groups, it receives around USD 15 billion each year, because the U.S. pursues the logic of "divide and rule" in Afghanistan.

Similarly, the relationship between Israel and Iran has been marked by the will of Menachem Begin to use nuclear weapons against Iran, stopped by Israel's Defense Minister in 1967. Later, 15,000 Israelis were allowed to volunteer in Iran against Iraq, but Iran refused the offer. During the Iran–Iraq war, Iran and Israel successively bombed an Iraqi nuclear power plant. As for Irangate, Iran helped the U.S. control Saddam Hussein, since the U.S. could not fly above the Oman Sea; sanctions motivated by the Iranian nuclear program were lifted; and the U.S. intended to buy nuclear fuel and waste from Iran; this all demonstrates the variability of the mutual attitude of the U.S. and Iran.

2.5.5 THE QUEST FOR ENERGY AUTONOMY AND NONALIGNMENT

History has shown that treatises and alliances have many times been flouted. History also shows the very high costs of keeping nonaligned. In particular, 2S costs of nuclear power are very high.

2.5.5.1 Security Lessons from Chernobyl

Assessments of the Chernobyl catastrophe have failed to ascertain the strategic facet. The collapse of the reactors partly resulted from Soviet security decisions to suspend support to Ukraine. To understand the rationale of Soviet nuclear security, three principles are worth recalling: first, that to withstand attacks, the technology had to be deployed in several Republics; second, that Russia was central, Ukraine peripheral, and Kazakhstan marginal; and third, that technology needed to be only partially transferred to be controlled. Accordingly, uranium from the reactors in Kazakhstan and

Ukraine left with 70% purity for Leningrad to be upgraded to 90% and then used in warheads; in the opposite direction, nuclear waste flowed from Russia to Kazakhstan. The Soviet technological withdrawal was completed in 1988, when the USSR returned the nuclear scientists to Russia. This left Chernobyl operating until international pressure forced its closure in 2000.

The lessons learned by Iran from Chernobyl include the following:

- Security is often sought at the expense of safety (i.e., public health and the environment).
- To operate a technology is not to learn and master it.
- Energy independence depends on diversifying energy sources and technologies. This is why Iran has developed technologies for the whole nuclear and oil production chains and invested in wind power (e.g., in Manjil near the Caspian Sea).
- A host of 2S criteria needs thorough and continually updated seismology, oceanology, climate, weather, geology, soils, transportation, and military risk maps.
- Corruption is incompatible with 2S and is avoided via supervision of a nuclear plant by a consortium where universities sit alongside the military and the Institute of Atomic Energy (in contact with the IAEA in relation to basic standards). In turn, the military is controlled, in Iran, by three intelligence agencies belonging to the military, the Presidency, and the High Committee for National Security. Foreigners are only consultants with short contracts.
- The politicians do not supervise the nuclear power plants; these respond to technical and 2S principles, although building a plant is decided by politicians.

2.5.5.2 Security Lessons from Furdu

- Trade-offs are difficult to achieve: the Furdu nuclear plant in Iran was located in the mountains, 300 m below ground, but far from where electricity output could be consumed.
- Furdu was deeper inside Iran and better than the Busher nuclear plant that, if bombed, would irradiate all of the Persian Gulf.
- Furdu is located within a very narrow viewshed, to delay immediate radiation propagation, and reduce the time frame available to supersonic bombers. Furdu was also equipped with the latest Russian S300 antiaircraft defense.
- The dedication of scientists has to border on total seclusion.
- 2S decisions can be misinterpreted and cause security tensions with other countries: Furdu's seclusion was one element presented by the U.S. suggesting that Iran might want to avoid scrutiny and pursue military use of Furdu. This opened a possibility for an Israeli attack.
- In 2008 the U.S. promoted a vote in the UN Security Council to attack Iran, which was vetoed by Russia and China; this cost Iran a wide market opening to Chinese products and the Iranian industries, and jobs suffered substantially as a consequence.

2.5.6 Nuclear Security and Sustainability

Peace is essential to sustainability. Proliferation control will not make nuclear energy a solution for all nations. Nuclear power itself breeds instability: waste can lead to both proliferation and terrorism. At present, proliferation is not checked but used in the triple confrontation among U.S., Russia, and China, each of which supports different nations in the Middle East and the Korean peninsula.

An enemy abiding by Geneva rules will not attack civilian populations but rather strike the energy system (nuclear, oil, and electric supplies); then, if electricity is down, communications are down; if petroleum is down, transportation is down. Paradoxically, within Geneva rules, preemptive strikes and open war make nuclear power plants very large liabilities in terms of nuclear fallouts affecting human and natural populations.

Natural phenomena occur in energy ranges that challenge any human structure, and so there is hubris in claims that plants can withstand bombing, technical failure, or extreme environmental events in close succession. This is now complicated by cyberwarfare. Attacks would likely profit from concurring challenges to a nuclear plant.

There is often a security–safety balance to strike. For example, an aboveground oil duct is good for safety, since leaks are readily visible but very vulnerable in security terms. Often the environment pays the higher costs. The intuition of isolation as source of security and safety hardly withstands an examination of material flows (oil, uranium, and other wares) that cross borders. Isolation from natural phenomena is even more questionable.

Energy is a national security matter, and civil nuclear power without military security management is unfeasible. This also moves the nuclear industry beyond the pure market economy. Also, nuclear technology and operations cannot be transparent. Hence, decisions thereupon cannot be democratic, for security reasons. Complex systems are ideal for nondisclosure and for heavy subsidies. Nondisclosure is good for security but very bad for safety and democracy. Population control by intelligence agencies on grounds of terrorism prevention and covert operations motivated by profit are causes for concern. Where information is lacking, extremism can flourish. 2S is all about protocols, best kept secret. But many crucial reactions to alerts draw on experience rather than protocol.

Control systems, including 2S, are themselves subject to failure. The inherently high complexity of nuclear fission technology is thus amplified by complex but fallible security systems and a dynamic geopolitical environment. In the future, heightened complexity can make more issues intractable.

REFERENCES

Al-Adili, A., Alhassan, E., Gustavsson, C., Helgesson, P., Jansson, K., Koning, A., Lantz, M., Mattera, A., Prokofiev, A. V., Rakopoulos, V., and Sjöstrand, H. (2015). Fission activities of the nuclear reactions group in Uppsala. *Physics Procedia, 64*, 145–149. doi:10.1016/j.phpro.2015.04.019.

Armstead, H. (1983). *Geothermal Energy* (p. 404). London: E. & F. N. Spon.

Baba, A., Deniz, O., Ozcan, H., Erees, S. F., and Cetiner, S. Z. (2008). Geochemical and radionuclide profile of Tuzla geothermal field, Turkey. *Environmental Monitoring and Assessment, 145*, 361–374. doi:10.1007/s10661-007-0045-0.

Bastin, J.-F., Berrahmauni, N., Grainger, A., Maniatis, D., Mollicone, D., Moore, R., Patriarca, C., Picard, N., Sparrow, B., Abraham, E. M., and Aloui, K. (2017). The extent of forest in dryland biomes. *Science, 356*, 635–638.

Bone, S. E., Dynes, J. J., Cliff, J., and Bargar, J. R. (2016). Uranium(IV) adsorption by natural organic matter in anoxic sediments. *PNAS, Early Edit*, 1–6. doi:10.1073/pnas.1611918114.

Bonner, J. (1962). The upper limit of crop yield. *Science, 137*, 11–15.

BP. (2016). *Statistical Review of World Energy* (65th ed.). London: BP plc.

Chen, X., Gong, M., Yi, P., Aldahan, A., Yu, Z., Possnert, G., and Chen, L. (2015). Nuclear Instruments and Methods in Physics Research B I in terrestrial surface water environments. *Nuclear Instruments and Methods in Physics Research B*. doi:10.1016/j.nimb.2015.04.073.

Clarke, L., Robinson, S., Hua, Q., Ayre, D., and Fink, D. (2012). Radiocarbon bomb spike reveals biological effects of Antarctic climate change. *Global Change Biology, 18*, 301–310. doi:10.1111/j.1365-2486.2011.02560.x.

CNS. (2017). *Global Incidents and Trafficking Database. 2016 Annual Report* (pp. 1–24).

Cooper, M. (2014). Small modular reactors and the future of nuclear power in the United States. *Energy Research and Social Science, 3*, 161–177. doi:10.1016/j.erss.2014.07.014.

Cowan, R. (1990). Nuclear power reactors: A study in technological lock-in. *The Journal of Economic History, 50*, 541–567.

De Vos, A. (1995). Thermodynamics of photochemical solar energy conversion. *Solar Energy Materials and Solar Cells, 38*, 11–22.

DoE Photographer. (2014). Damaged drum with radioactive waste inside WIPP Panel 7, Room 7, New Mexico, USA, 2014-05-15 during investigations on the cause of radioactive contamination. Retrieved June 4, 2017, from https://en.wikipedia.org/wiki/File:WIPP_DoE_2014-05-15_5_15_Image_lrg.jpg.

Dombey, N. (2016 August 16). Letter. *The Guardian*. London.

Doward, J. (2016 October 30). Secret government papers show taxpayers will pick up costs of Hinkley nuclear waste storage. *The Guardian*. London.

Duffa, C., and Fréchou, C. (2003). Evidence of long-lived I and Pu isotopes enrichment in vegetation samples around the Marcoule Nuclear Reprocessing Plant (France). *Applied Geochemistry*, *18*, 1867–1873. doi:10.1016/S0883-2927(03)00148-3.

Field, C. B., Behrenfeld, M. J., Randerson, J. T., and Falkowski, P. (1998). Primary production of the biosphere: integrating terrestrial and oceanic components. *Science*, *281*, 237–240.

Great East Japan Earthquake Taskforce & Science Council of Japan. (2011). *Report to the Foreign Academies from Science Council of Japan on the Fukushima Daiichi Nuclear Power Plant Accident* (pp. 1–25). Retrieved from http://www.scj.go.jp/en/index.html.

Greenpeace. (2006). Executive summary, p. 10. In A. Yablokov, I. Labunska, I. Blokov, R. Stringer, D. Santillo, P. Johnston, and T. Sadownichik (Eds) *The Chernobyl Catastrophe. Consequences on Human Health*. Amsterdam: Greenpeace.

Gundersen, A. (2016). Nuclear Power Is Not 'Green Energy': It Is a Fount of Atomic Waste. *Truthout New Analysis*. Retrieved from truth-out.org/news/item/38326-nuclear-power-is-not-green-energy-it-is-a-fount-of-atomic-waste.

Haegel, N. M., Margolis, R., Buonassisi, T., Feldman, D., Froitzheim, A., and Al., E. (2017). Terawatt-scale photovoltaics: Trajectories and Challenges. *Science*, *356*, 141–143.

Harris, M. (2014). *Party of one* (534 pp). Toronto: Viking.

Herchenroeder, K. (2016). SCE Fully Responsible for SONGS Shutdown: Report. *RadWaste Monitor*. Retrieved from exchangemonitor.com/publication/rwm/sce-fully-responsible-for-songs-shutdown-report/.

Hou, X., Aldahan, A., Nielsen, S. P., and Possnert, G. (2009a). Time series of 129I and 127I speciation in precipitation from Denmark. *Environmental Science and Technology*, *43*, 6522–6528.

Hou, X., Hansen, V., Aldahan, A., Possnert, G., Christian, O., and Lujaniene, G. (2009b). A review on speciation of iodine-129 in the environmental and biological samples. *Analytica Chimica Acta*, *632*, 181–196. doi:10.1016/j.aca.2008.11.013.

IAEA. (2013). *Management of NORM Residues. IAEA-TECDOC-1712* (p. 66). Vienna: International Atomic Energy Agency.

Ireland, D. and Nazarewicz, W. (2015). Enhancing the interaction between nuclear experiment and theory through information and statistics. *Journal of Physics G: Nuclear and Particle Physics*, *42*, 1–3. doi:10.1088/0954-3899/42/3/030301.

Kopp, G., and Lean, J. L. (2011). A new, lower value of total solar irradiance: Evidence and climate significance,. *Geophysical Research Letters*, *38*. doi:10.1029/2010GL045777.

Lubimova, E. A. (1968). Thermal history of the Earth. *Geophysical Monograph Series*, *13*, 63—77.

McDonald, J. (2017). 5 dramas still playing out 5 years after San Onofre shutdown. *San Diego Union Tribune*. San Diego.

National Research Council. (2015). *Climate Intervention: Carbon Dioxide Removal and Reliable Sequestration. Committee on Geoengineering Climatic: Technical Evaluation and Discussion of Impacts* (p. 140). Washington, DC: The National Academies Press.

Nikolewski, R. (2017). Judge allows San Onofre waste lawsuit to advance. *San Diego Union Tribune*. San Diego.

Nuclear Regulatory Commission. (2015). Boiling Water Reactors. Retrieved May 2, 2017, from nrc.gov/reactors/bwrs.html.

Oelrich, I. (2005). *Missions for Nuclear Weapons after the Cold War* (p. 75). Washington, DC: Federation of American Scientists.

Peixoto, J. and Oort, A. (1992). *Physics of Climate* (pp. 365–400). New York: American Institute of Physics.

Premuzic, E. T., Lin, M. S., and Lian, H. (1995). Biochemical technology for the detoxification of geothermal brines and the recovery of trace metals. *International Conference on Heavy Metals in the Environment, Hamburg (Germany)*, 18–22 Sep.

Ramana, M. V. (2017). An enduring problem: Radioactive waste from nuclear energy. *Proceedings of the IEEE*, *105*, 415–418.

Ramana, M. V. and Kumar, A. (2013). Nuclear safety in India: Theoretical perspectives and empirical evidence. *Journal of International Studies*, *1*, 49–72.

Ramana, M. V. and Mian, Z. (2014). One size doesn't fit all: Social priorities and technical conflicts for small modular reactors. *Energy Research and Social Science*, *2*, 115–124. doi:10.1016/j.erss.2014.04.015.

Rincón, E. (2011). Por qué las plantas nucleoeléctricas no son una opción conveniente para México. *Revista de Energía*, *11*, 12–13.

Ryther, J. (1959). Potential productivity of the sea. *Science*, *130*, 602–608.

Sharma, A. (2017). Coastal Commission Met Privately with Edison a Year Before Public San Onofre Waste Storage Vote. *KPBS*. Retrieved May 30, 2017, from kpbs.org/news/2017/may/30/public-excluded-key-talks-over-san-onofre-waste-st/.

Shultz, G. P., Perry, W. J., Kissinger, H. A., and Nunn, S. (2007–2011). Toward a World without Nuclear Weapons. The groundbreaking Wall Street Journal op-ed series. Washington, DC: NTI.

Smith, J. E. (2017 May 23). This city could become region's first to offer SDG&E alternative for electricity. *San Diego Union Tribune*. San Diego.

Sovacool, B. (2014). Cornucopia or curse? Reviewing the costs and benefits of shale gas hydraulic fracturing. *Renewable and Sustainable Energy Reviews*, *37*, 249–264. doi:10.1016/j.rser.2014.04.068.

Stacey, F. D. and Loper, D. E. (1988). Thermal history of the Earth: A corollary concerning non-linear mantle rheology. *Physics of the Earth and Planetary Interiors*, *53*, 167–174.

Steger, U., Achterberg, W., Blok, K., Bode, H., Frenz, W., Gather, C., Hanekamp, G., Imboden, G., Jahnke, M., Kost, M., Kurz, R., Nutzinger, H. G., Ziesemer, T. (2005). *Sustainable Development and Innovation in the Energy Sector* (p. 32). Berlin: Springer.

Talou, P., Kawano, T., Chadwick, M. B., Neudecker, D., and Rising, M. E. (2015). Uncertainties in nuclear fission data. *Journal of Physics G: Nuclear and Particle Physics*, *42*, 1–17. doi:10.1088/0954–3899/42/3/034025.

University of Leicester Press Office. (2016). Anthropocene Working Group (AWG). *Media Note*. Retrieved April 10, 2017, from www2.le.ac.uk/offices/press/press-releases/2016/august/media-note-anthropocene-working-group-awg.

Vaughan, A. (2016 August 11). Solar and wind "cheaper than new nuclear" by the time Hinkley is built. *The Guardian*. London.

Vaughan, A. (2017 March 20). UN asks UK to suspend work on Hinkley Point. *The Guardian*. London.

Vaughan, A., and Willsher, K. (2016 December 2). New blow for Hinkley Point contractor EDF after French safety checks. *The Guardian*, pp. 2016–2018. London.

World Energy Council. (2016). *World Energy Resources Geothermal 2016* (p. 54).

World Nuclear Association. (2016). Chernobyl Accident 1986. Retrieved May 15, 2017, from http://world-nuclear.org/information-library/safety-and-security/safety-of-plants/chernobyl-accident.aspx.

3 Renewables—The Politics and Economics Behind Them*

Andrea Leon-Grossmann
Food and Water Watch

CONTENTS

3.1 BREAKING THE VICIOUS CYCLE: POLICY, CONTRIBUTIONS, AND SUBSIDIES FOR FOSSIL FUELS

3.1.1 THE FOSSIL FUEL INDUSTRY AND ITS ONGOING WAR AGAINST ENERGY DEMOCRACY

3.1.1.1 Transportation Was First Electric

In the same manner that most are unaware that Nikola Tesla was mostly to thank for electricity, few know that the first vehicles were electric. The oil companies conspired to turn all cars into internal combustion engines through many political maneuvers and even colluded to ensure that new cities would be designed around such vehicles instead of around public transportation, which was powered by electricity.

Los Angeles used to have a vast network of electric trolleys and streetcars from the 1870s to the 1970s. Companies such as the Pacific Electric Railway with its "Red Cars" and the Los Angeles

---

* I would like to dedicate my chapter to my cousin Rafa Camargo, my friend Gaby Jaen, and all victims of natural disasters. We all deserve clean air, clean energy, and a livable climate. No one deserves to be left out in the dark—whether in the rubble in a building in Mexico City after an earthquake that caused massive gas leaks that hindered rescue operations, in Puerto Rico after Hurricane Irma, or in India and South Asia after the massive floods.

Railway with its "Yellow Cars" were ubiquitous in Southern California. The fossil fuel industry and automobile manufacturers put a halt to that and pushed for individual automobiles. In 1950, the massive construction of freeways began making Los Angeles more and more dependent on automobiles. The electric streetcars could only withstand the pressures of the oil companies in a handful of cities and not in Los Angeles (LA). As the second largest city in the United States, LA has the worst public transportation system, mostly relying on diesel and natural gas-operated buses.[*] The city is now trying to reinstate a better network of electric transportation with expansions of the Metro subway lines at a huge cost.

In the United States, most trains are still diesel-powered, while in Europe they are mostly electric. Until recently, most public transit buses were either diesel or powered by natural gas, but electric buses appear in cities across the country. It will be a battle to switch to an all-electric fleet since the fossil fuel industry will try to keep selling its products.

3.1.2 The Fossil Fuel Industry and Its Ties to Financial Institutions

The financial institutions have been the enablers of the fossil fuel industry, and they continue to be. The divestment movement has been shedding light on the institutions that invest the depositors' money in dirty energy. Such movement has included not only account holders at banks, but also pension funds at labor unions and other public institutions and universities. Initially, the fossil fuel industry was not worried, but as the movement grew, it is pushing back.

The financial institutions finance all aspects of the industry, from extraction equipment to operations and pipelines. https://www.foodandwaterwatch.org/news/who%27s-banking-dakota-access-pipeline. These financial institutions are also able and willing to short the renewables market to ensure their (larger) investments in fossil fuels pay off.

A word of caution should go out to the community at large as the need for renewable energy is greater now more than ever. Some solar companies are themselves becoming financial institutions and engage in creative financing and lending, the same kind that led us to the 2008 financial crisis.[†,‡] Although not as big as the housing sector, the solar industry cannot afford any negative image for renewables or have a leading company go under and leave its customers hanging, without warranty and/or service.

The Solyndra debacle did a lot of damage to renewables, and the fossil fuel industry and its allies capitalized on the company's failure to inflict the most damage and undermine renewables as much as possible. Divesting from fossil fuels by taking money out of the institutions and investments is a step forward to transitioning to renewables. Shifting that capital away from toxic assets (in every sense of the word) to either renewables or efficiency will ensure a faster transition and money well spent.

3.1.3 The Fossil Fuel Industry and Warfare

The fossil fuel industry has managed to get behind or be the cause of most conflicts around the globe to ensure continued access to dirty energy, and governments have engaged accordingly.

Usually, first world countries and their corporations going into third world countries to ensure the flow of oil and gas is ongoing and funneled through certain entities. Other times, it is countries turning against their own citizens, violating human and civil rights to ensure corporate giants get what they desire.

---

[*] https://www.metro.net/about/library/about/home/los-angeles-transit-history/; https://en.wikipedia.org/wiki/General_Motors_streetcar_conspiracy; http://usp100la.weebly.com/history-of-transportation.html.

1. David Brodsly. *L.A. Freeway: An Appreciative Essay.* Berkeley: University of California, 1981. Print.

2. "Los Angeles Transit History." Metro.net.

3. "Streetcar History." Lastreetcar.org.

4. "History of Metrolink." Metrolinktrains.com.

[†] https://www.fool.com/investing/general/2015/09/21/is-solarcity-really-the-next-subprime-disaster.aspx.

[‡] https://seekingalpha.com/article/3764916-solar-lease-like-second-mortgage-solarcity-like-subprime-lender.

Whether oil/gas reserves and/or access to pipelines, the desire to control dirty energy assets often results in armed conflict. Incredibly, fossil fuels still represent a major source of income for many countries and their corporations. With fluctuation of price or production and the fact that fossil fuels are a finite resource, the result is price gouging and speculation that can lead to instability and even the demise of countries. Venezuela is a clear example: with most of its conventional oil gone and oil prices plummeting, the country has no money to subsidize all the programs it once did, due to shortsightedness and reliance on dirty energy to keep afloat.

Once we combined greed with the appetite for energy at any cost, we got a U.S. Congress able and willing to lift the oil export ban. Once in place, the United States started a war with its own people over more extraction and pipelines, like the Dakota Access Pipeline. Since the pipeline operator had a good relationship with the authorities, it has managed to stay in business despite many fines for environmental and workplace violations.* Moreover, despite the argument from Energy Transfer Partners that the crude would be used domestically to "enhance energy independence," some of the crude from the pipeline was exported.[†]

The truth is that fossil fuel companies do not really care about "energy independence" as much as they care about who will be willing to pay the most for their product. After all, they are in the energy business, and they know their product is harmful, even if they publicly deny it.

The appetite for energy is coming from China and India at this moment; the industry knows it, so they are willing to do what it takes to get a piece of the pie. On the other hand, renewables represent true energy democracy and energy independence as they are an infinite and abundant resource, but they need to be sourced locally for local/regional consumption. This way, power is taken away from a few companies, and a few centralized locations, so individuals or communities can produce or buy their own energy independently. It is also harder to shut them down, unlike a pipeline, a train, or a rig; most renewable energy is distributed, and so shutting it all down is virtually impossible.

Wars are very profitable, and fossil fuel companies are often involved, such as Halliburton. Renewables take the power and some of the profits out of the equation, so the fossil fuel industry wages war against the very thing that will put it out of business and potentially end most wars.

3.1.4 THE FIGHT TO KEEP THE STATUS QUO, OLIGARCHY, AND POWER

3.1.4.1 The Beneficiaries and the Climate Change Deniers

The corporate executives who reap the benefits of the current system aim to keep it that way. When they realized the facts about global warming were being exposed, their best strategy was to hire a few scientists willing to plant seeds of doubt, to flat out deny global warming, despite overwhelming evidence to the contrary.

3.1.4.2 The Deception and an Industry Whose Mission Is to Obliterate the Planet for Profit

It has been documented that Exxon not only knew early on how fossil fuel extraction and burning would adversely affect the planet,[‡] but also that it acted to ensure future profits at the expense of humanity. Also, Exxon hires top lobbyists and communicators to ensure politicians are bought and their messaging is as deceiving as it needs to be to continue to thrive.[§]

---

[*] http://violationtracker.goodjobsfirst.org/parent/energy-transfer.

[†] https://theintercept.com/2016/09/01/dakota-access-export/.

[‡] https://www.bloomberg.com/news/articles/2016-09-07/will-exxonmobil-have-to-pay-for-misleading-the-public-on-climate-change.

[§] https://www.scientificamerican.com/article/tobacco-and-oil-industries-used-same-researchers-to-sway-public1/; http://www.eenews.net/assets/2016/07/20/document_cw_02.pdf; http://www.eenews.net/assets/2016/07/20/document_cw_01.pdf; http://www.eenews.net/climatewire/stories/1060040266/; https://energyindepth.org/national/exxonknew-activists-to-re-release-old-documents-in-desperate-bid-for-relevance/; http://www.eenews.net/climatewire/2016/07/20/stories/1060040530.

Luckily, we have records and a few good climate hawks in Congress who are willing to take fossil fuel corporations on and make them accountable, including Rep. Ted Lieu.

As expected, the industry is responding to the bad publicity by engaging in many different tactics to keep selling its dirty products. A great example is methane sold by Sempra Energy. Sempra is not content with selling fracked gas; now it wants to sell gas that is factory farmed and collected from landfills branding it "renewable natural gas" and claiming through some creative math that it has near-zero emissions (sometimes zero)[*] when used in transportation or power generation.

What Sempra and other gas companies fail to acknowledge is that they are incentivizing CAFOs at a time when Americans are eating less meat[†] and meat producers are now promoting their products abroad and exporting more.[‡] Exporting meat not only requires a lot of energy that generates a lot of emissions from transportation, but also in most cases, the meat products need to be refrigerated, which increases the energy use and emissions exponentially. Once the energy and water are taken into account to manufacture all the antibiotics and hormones that are fed to the animals, the emissions go up yet again. In addition, there is the ethical issue of fueling transportation and the grid with animal abuse and the fact that the overuse of antibiotics in those animals threatens public health.

Then there are cities like San Francisco aiming to be a zero waste city by 2020, but corporations like Sempra incentivize waste and could even fight programs like the one in San Francisco to generate "renewable natural gas."

3.1.4.3 The Blame Game and How to Tackle It

The fact remains that fossil fuel producers still receive more subsidies than renewables; regardless, they often go under or file for bankruptcy after a spill, explosion, or other accident, just to reemerge under another name. In other words, the profits are private; the losses are public, along with the cleanup.

We need to ensure the public understands the modus operandi of the energy companies so it can demand the defunding of the fossil fuel industry and a rapid transition to renewables.

Moreover, most of the infrastructure, extraction, and refinement for the fossil fuel industry happen where low-income people of color and immigrants live and, along with many individuals who work directly with fossil fuels, end up sick from asthma, skin conditions, and even cancer.

We must all demand access to clean air, clean drinking water, and soil that are not contaminated by extraction or emissions. No one should live next to a sacrifice zone, no one.

3.2 INFRASTRUCTURE

3.2.1 A Big Obstacle and a Huge Opportunity

3.2.1.1 For the Developed Countries

The case can be made that a rapid and just transition to renewables is doable—by shutting down failing and aging fossil fuel infrastructure to ensure reliable, clean, renewable energy.

Many gas-fired power plants can be retrofitted to use hydrogen, and the hydrogen can be generated with solar power using electrolysis and ocean water. Although a substantial investment is needed upfront, it pays off many times over once the hydrogen generator is built, as ocean water is plentiful and free. Pipelines, storage, fracking, and wastewater disposal will no longer be needed to get natural gas, and no emissions will be generated by extracting it or burning it.

---

[*] http://cngvc.org/ttsi-cummins-westport_release-032217/.

[†] https://www.nytimes.com/2017/03/21/dining/beef-consumption-emissions.html?_r=0; https://www.nrdc.org/sites/default/files/less-beef-less-carbon-ip.pdf.

[‡] http://www.thefencepost.com/news/beef-exports-key-to-market-improvement-in-beef-business/.

For many cities and counties, Community Choice Aggregation is undoubtedly the way to go. It not only provides clean energy produced locally, it is also a great job generator, and the energy is distributed throughout the region, not centralized. Microgrids also stabilize the grid and force utilities to invest in transmission and efficiency. For those concerned about national security, renewables are definitely the way to go; not only is it harder to shut the grid down, but they are not volatile either.

3.2.1.2 For the Developing Countries

In the same manner that most underdeveloped countries skipped landlines and started directly with cell phones, it is fairly simple to start with a clean, renewable grid that is distributed rather than centralized—no need for any extraction, refinement, or burning of fossil fuels.

Unlike fossil fuel infrastructure, people can be trained to install panels and inverters as well as some windmills, so villages can be part of building the infrastructure with few engineers involved. Renewables today are either on par with or cheaper than fossil fuels, even without subsidies when hard costs are considered; once human health and climate change are factored in, renewables are way more affordable than fossil fuels.

3.3 THE CASE TO START THE SWITCH RIGHT NOW

The best time to switch was many decades ago, the next best time is right now, as soon as possible.

Climate change is already here, and we need to act right now. We just cannot afford to keep extracting and burning fossil fuels at the rate we have for the last century; the fact is that the world population is growing, and so is our demand for energy.

It is clear that we need 350 or less CO_2 parts per million to live in harmony; we are well over 400 and climbing. At the Paris Climate Agreement in 2015, countries agreed to hold the increase in the global average temperature below 2°C above preindustrial levels and to pursue efforts to limit the temperature to 1.5°C above preindustrial levels to reduce the impacts and risks of climate change. The problem is that we are on track to get to 2°C sooner than many expect.[*]

3.3.1 CLIMATE CHANGE, AWARENESS, AND PUBLIC PRESSURE

3.3.1.1 Renewables as the Solution

Only renewables can provide us with the energy we need and leave out the emissions we do not. Moreover, renewables are part of the solution for the water crisis we are facing. Whether we face drought or the fact that we are polluting the water we do have, it is a fact that renewables use the least amount of water to produce energy by far.

Even if we were not facing calamity, fossil fuels are a finite resource, and it is our responsibility to start a rapid transition to ensure reliability and availability for all. Clean, renewable energy has been here for a while now. Many naysayers will say that we still have to do research or "that is a technology of the future." The fact is that solar and wind energy have not only been around for some time, their efficiency has gone up while their costs have plummeted.

Geothermal done responsibly and wave and tidal energy as well as hydrogen power produced with solar energy are other technologies that can be considered both clean and renewable. Biogas and/or "renewable gas" must not be considered in the mix for a clean renewable scenario. Besides being methane, it needs to be stored and transported by pipelines, it has emissions when it is burned, and it relies heavily on factory farming.

Many claim that renewables are not clean because they use some fossil fuels to build solar panels and windmills. While that may be true for now, as we transition to 100% renewable energy, any new or replacement technology will be built using renewables.

---

[*] https://insideclimatenews.org/news/22032016/scientists-warn-drastic-climate-change-impacts-will-come-sooner-expected.

3.3.2   Awareness for Efficiency; New Technologies as Part of the Solution

Wenonah Hauter, executive director of Food & Water Watch, puts it best when she says that the bridge to 100% renewables must be efficiency. The fossil fuel industry and the politicians they have bought usually call natural gas a bridge fuel, but natural gas is only a bridge to more global warming, more wasted water, more emissions, more manmade earthquakes, and more sick people.

A great example is a report from CleanTechnica where a preliminary analysis reveals that 4% of Los Angeles buildings use 50% of the city's electricity.[*] Later in this chapter, it is explained how the Los Angeles Department of Water and Power is falling way short of a goal of 100% renewable energy by 2030, but this showcases perfectly how efficiency can work by not having to replace a lot of fossil fuel power with anything. The fact that it took this long to get someone to notice is appalling, but the great news is that we can do something about it right away. We must be creative with the solutions we bring to the table to ensure we use the energy we have wisely. Most urban cities and utilities have programs that subsidize smart thermostats, electric vehicle chargers, rooftop solar, battery backup systems, and even free energy audits to ensure residences are as efficient as possible. Unfortunately, most of those programs are not very well publicized, and utilities don't go out of their way to ensure the public knows about them.

Nowadays we seem to have an app for anything and everything, so it is about time city governments and/or utilities create their own apps and promote them as a way to help people conserve and be more efficient. From "flex alerts" and last-minute notice on outages to alerts on when to charge your electric vehicle, they can do more to engage with the ratepayers and treat them as part of the solution.

In addition, as more low-income households qualify for solar energy and/or an electric vehicle, in many instances they will no longer qualify for combined lifeline and time-of-use rates. Since lifeline ends up being cheaper, they keep that rate and there is no penalty or incentive to charge during peak usage. A combination of lifeline and time-of-use rates should be considered to incentivize off-peak use.

When it comes to electric vehicles, renewable power, and charging, there is much more the automakers and the utilities can do. A simple flyer with basic information on how to get any available rebates for chargers, how to apply for a time-of-use rate, and why going with solar makes sense would help educate consumers.

Some utilities like PG&E that also sell natural gas are perceived as more reluctant to change, as electrification of appliances will effectively put an end to the gas portion of their business.

3.3.3   Demanding Accountability from Legislators and Media Outlets

Mainstream media outlets are still reporting on climate change very sporadically and even more sporadically on the root causes. Eventually, just like tobacco advertisements, fossil fuel ads will be banned, especially the ones from the American Petroleum Institute[†] and American Gas Association. Until they are banned, the media is taking the industry's ad dollars and therefore not covering much news on climate change.

Fortunately, the API has not spent as much money in Spanish-language media, so more reports have been present on Univision to report on climate change, urban drilling, refineries, and even how the Los Angeles Archdiocese is currently leasing two of its properties in South Los Angeles to the industry where unconventional drilling takes place and people get sick.

A similar situation takes place with legislatures. Those who take campaign contributions from the fossil fuel industry are less likely to adopt renewables wholeheartedly and many become flat-out climate change deniers.

Some who take substantial money from the fossil fuel industry, like Assemblymember Sebastian Ridley-Thomas, try to make amends by taking contributions from the renewable industry as well, but

---

[*] https://cleantechnica.com/2016/12/20/preliminary-analysis-reveals-4-of-buildings-use-50-of-energy-in-los-angeles/.
[†] http://fuelfix.com/blog/2016/10/10/whos-behind-the-i-am-an-energy-voter-ads/.

most of his votes and actions have supported and favored the fossil fuel industry. He has been so emboldened by the industry as to shut down many constituents who have asked him to support tough environmental legislation. Ridley-Thomas represents one of the most progressive districts, and yet he was able to run unopposed in the 2016 primary largely of because the campaign contributions of special interests.[*]

Worse yet is Governor Jerry Brown. Most mainstream media outlets consider him an environmentalist leader, but nothing could be further from the truth. The fact is that no human being can be a climate leader and support fracking or any other kind of oil and gas extraction. Brown has opened the state to the industry and gone as far as to fire competent regulators to appoint only those approved [†‡§¶**] of by the industry. California is still in a drought, and the fact that Brown allows the use of potable water for oil and gas extraction and refinement is unacceptable. Brown doesn't stop there; he also allows the oil and gas industry to sell its oil wastewater to farmers for irrigation of America's food. Such wastewater used to be disposed of in wastewater injection wells (many of which reached and contaminated aquifers) at the cost of the producer. Now the industry figured out a way to turn a cost into a profit by selling its toxic waste to water districts.[††]

Brown has also failed to order the decommissioning of Aliso Canyon gas storage facility, after the worst environmental disaster since the BP Deepwater Horizon oil spill. To top it all off, Brown is seeking to build two mega tunnels that will be exempt from the Endangered Species Act to take water from the Bay Delta to California's Central Valley to use in commodity crops and oil extraction. No environmental leader would do what Brown has done.

We need legislators who are climate leaders and understand what is at stake. We do not need to reduce our dependence on fossil fuels; we need to stop it altogether. An "all of the above" energy proposal that includes fossil fuels is not a solution. A solution is 100% renewables.

1. Community organizing

A. *Setting goals, being strategic*

The switch to 100% renewables is unlikely to happen by 2030 or sooner unless we get organized and demand it. To do so we need to be clear about what we want, where we can get it, who can help us get it, and more importantly, how we can get it. The structure as it stands today is to ensure fossil fuels prevail and renewables are at a disadvantage with less subsidies, as they are newly formed corporations and advocates that are not as powerful or wealthy as their counterparts. However, due to the plummeting costs of solar and wind, as well as the high costs of extreme extraction of fossil fuels and the impacts on human health and climate change, the facts and the community (once it is informed) back renewables.

The goal has to be ambitious as the circumstances call for it. Nothing short of 100% renewable energy for all should be acceptable. Groups or organizations with good intentions might want to settle for less, or even agree to a carbon tax, just to get an increase in renewables. We cannot afford to build and/or rebuild more fossil fuel infrastructure. We cannot afford to keep burning fossil fuels, and we must not settle with pay-to-pollute schemes that the industry itself supports.

We need community engagement, education, and support to ensure the success of the campaign to take our cities to 100% clean renewable energy.

---

[*] http://www.couragescore.org/hall-of-shame/.

[†] http://articles.latimes.com/2012/jan/29/local/la-me-oil-20120129.

[‡] http://www.capoliticalreview.com/trending/gov-jerry-brown-fires-states-top-oil-regulators/.

[§] http://www.huffingtonpost.com/2015/07/02/western-states-petroleum-association_n_7715584.html.

[¶] https://www.eastbayexpress.com/SevenDays/archives/2015/10/12/governor-brown-appoints-republican-oil-executive-to-be-industry-regulator.

[**] https://www.counterpunch.org/2017/01/02/green-governor-jerry-brown-appoints-oil-industry-loyalist-to-public-utilities-commission/.

[††] http://www.protectcafood.org/.

B. Targeting elected officials—from city council to congress

Most of the time, elected officials, from the local level all the way to the federal level, can make things happen. Although most constituents believe federal law to be the most important in terms of impact, in reality, state and local ordinances may be just as if not more important.

A state like California tends to be more innovative and progressive (despite its governors) and a trendsetter in many ways. This is attributed to the state legislature and the fact that many community-based organizations and the community at large are very involved. California changed radically after Pete Wilson's time in office and especially after the passage of the anti-immigrant Prop 187.

Latino population and engagement has risen significantly, and the Latino segment of the population cares deeply about environmental issues. Usually, legislators (whose districts have not been gerrymandered to have fewer Latinos) vote in favor of renewables, unless most of the Latinos in their district are disengaged or undocumented.

At the city level, change may come from the City Council and a public utility the way it happens with Los Angeles and LADWP, the biggest public utility in the nation. Recently, there was a referendum (Measure RRR) to allow LADWP to have less oversight from the City Council, which could have led to privatization of the utility. The measure failed as voters felt the reform was not what the department needed.

At the same time, Eric Garcetti, the Mayor of Los Angeles, promised early in 2016 that LADWP would study a "fossil-free scenario by 2036" in its Integrated Resource Planning (IRP), its 20-year plan. By the time they presented it in October, the best-case scenario for 2036 was 65% renewables with 35% investment in fossil fuel infrastructure. Garcetti and the LADWP failed and people told them during public comment, but they seem undeterred. The community needs to keep the pressure on and meet with the City Council's energy and environment committee.

Even more local than the City Council are the neighborhood councils, and now in the city of Los Angeles they can submit community impact reports and environmental impact reports for any given project. Los Angeles also has the Neighborhood Council Sustainability Alliance to advance sustainability and resilience throughout the city. Anyone can attend neighborhood council meetings, usually once a month on a weeknight, and put any item on the agenda to be addressed as long as it is done a couple of weeks in advance.

Meeting with elected officials face-to-face is always a good way to be persuasive, especially if one brings many constituents to the meeting. Many lawmakers are just not available, so the chief staffer or environmental staffer ends up meeting with the group. However, if the office of the legislator is not available for a meeting and insists on email communication, an unannounced visit with constituents will do. That works from City Council all the way to a U.S. senator.

C. Getting the community engaged while empowering it

A great example of community engagement and empowerment is what a Canadian city did at the pump.* Warning labels (similar to those on cigarette boxes) were placed right at the fueling station pump, so consumers could see the message while pumping gas. The objective was to encourage drivers to think about being more energy efficient when they drive as well as the kind of fuel they use at the pump.

D. Calling for direct action

Challenging power relationships is how many issues have gotten off the ground and successfully found solutions. By empowering the community members and altering

---

* http://www.pri.org/stories/2016-07-22/canadian-city-putting-warning-labels-gas-pumps.

the relation of power, many realize how far they can go to get targets to move, whether an elected official or a corporation.

While direct action is a tactic, when backed by a solid strategy, it has a very good likelihood of succeeding. In the case of renewable energy, tactics may include showcasing the ties of an elected official to the fossil fuel industry to give him or her a choice of changing positions or facing more aggressive direct actions, such as campaigning for an opponent who supports renewables, if reelection is upcoming.

E. *Activism works; technology amplifies it*

Challenging the oppressors and their power by organizing has been working for a long time (Figure 3.1). From the labor movement to the civil rights movement, organizing and empowering people has worked. While it has not come easy, and the big corporations have tried to sabotage most of the efforts, organizing works. Many techniques remain the same, but it is undeniable that technology and social media have made it easier to organize. It enabled the Arab Spring and has taken politicians down, and for now it has leveled the playing field while big corporations fight net neutrality.

Organizers and activists alike tend to be more informed because of their deep involvement and stake in the issue. If there is a roadblock, they do not hesitate to share information or get more (such as through the freedom of information act) to get to the bottom of an issue.

Coalition building creates power and helps organizations with limited resources fight big corporations with seemingly unlimited resources. A good recent example is the coalition to push for 100% electric buses formed by the Sierra Club, Food & Water Watch, Earthjustice, Environment California, IBEW Local 11, South Bay 350, Union of Concerned Scientists, Communities for a Better Environment, East Yard Communities for Environmental Justice, American Lung Association, Bike San Gabriel, Global Green, Climate Resolve, Jobs to Move America, Los Angeles Community Action Network, and In Defense of Animals. This last organization joined after SoCal Gas (Sempra) decided to start touting methane as a "clean" fuel renaming it "renewable natural gas," which would incentivize CAFOs. At a meeting on March 23, 2017, the Los Angeles County Metropolitan Transportation Authority board heard public comment from most of the coalition members, and Food & Water Watch brought up the fact that the way to solve climate change is not through natural gas, whether fracked or sourced from factory farms—that we do not need to power our buses or power plants with animal abuse. At

FIGURE 3.1   May 14, 2016, Downtown Los Angeles, the people's climate rally "break free" from fossil fuels.

that point, some of the board members flinched, and it hit them: "renewable natural gas" is not clean or humane. The procurement for the buses is still under review, but a few weeks ago it seemed like the purchase of gas buses was almost a done deal.

As far as activism in extreme extraction, Food & Water Watch was the first group to call for an outright ban on fracking while other environmental groups called for "strict regulation," which has proven futile. Very few believed a ban was possible, and today we have a ban in three states: New York, Vermont, and Maryland. The bans are the result of coalition work and politicians who understood what was at stake. It is worth mentioning that extreme extraction is not a partisan issue; in the case of Maryland, Larry Hogan, a Republican governor, called on his legislators to approve the bill and send it to him for his signature.

Celebrities who care about the environment also have made an impact using their social media platforms to reach out to their base and beyond. Mark Ruffalo did it when he took the "toxic tour" offered by Communities for a Better Environment in Los Angeles and ended up being featured on NBC Nightly News. Jane Fonda did the same when she announced she was withdrawing her money from Wells Fargo because of their stake in the Dakota Access Pipeline and called everyone to divest from the bank.

2. Early adoption and beyond
A. *Consumers do bear some responsibility*

While we still depend on legislators to set rules and rebates for early adopters and the ones right after them, there is a point when a consumer must think about the long-term investment and/or consequences of the purchase at hand. In the case of a brand new car, with very few exceptions, everyone should be getting electric cars. They are cheaper to maintain and create no emissions. Even when electricity is produced by coal or gas (which is not the best way to power a car or a home), electric vehicles are way more efficient than internal combustion engines, and so is the fuel.

However, the key to drastically lowering emissions is to use public transportation and to have such transportation be electric and powered by clean, renewable energy. Being a conscientious consumer means actively engaging with lawmakers, corporations, friends, and family about the choices we make and the way we live.

Water is a key part of the renewable equation. About 19% of the electricity used in California is used to move water,[*] and renewables use very little water. So saving water saves energy, saving energy saves water, and going with 100% renewables will save even more water.

B. *Not consuming*

There is efficiency and then there is living within our means: not financially (although that helps too), but rather environmentally. With planned and perceived obsolescence and fast fashion, consumers are pressured into buying, buying, buying stuff that will simply not last very long; since most items are relatively cheap, cleaning out a closet and donating the items does not feel like a strain on the budget—but it should when it is added up and when the impact on the environment is factored in.

According to the Environmental Protection Agency (EPA), in 2012 84% of unwanted clothes in the United States went into either a landfill or an incinerator.[†] In other words, all the energy that went into manufacturing the clothes and transporting them is now compounded by the energy that will be needed to dispose of them. The same applies to food waste and all the disposable items consumers buy and dispose of after a few minutes of use.

---

[*] http://blogs.kqed.org/climatewatch/2012/06/10/19-percent-californias-great-water-power-wake-up-call/.
[†] http://www.newsweek.com/2016/09/09/old-clothes-fashion-waste-crisis-494824.html; https://www.epa.gov/smm/advancing-sustainable-materials-management-facts-and-figures-report.

Consumers not only pay to acquire the clothes and/or disposable items, they pay again with their taxes to dispose of them, and since the amount of clothes that Americans disposed of doubled in the last 20 years from 7 million to 14 million tons (about 80 pounds per person per year), that price is increasing.

C. Setbacks

The Trump administration is undeniably a threat to all the progress that has been made thus far in the field of renewable energy, efficiency, and regulating pollution from fossil fuel sources. From selecting a secretary of state who was the former CEO of the largest oil company in the world (currently under investigation for what it knew about climate change, when it knew it, and how it chose to profit from it) to an EPA administrator who does not believe climate change is manmade.

The setback comes with not only a high price for the environment and health, but also for our economy and jobs. Clean energy is the biggest new source of sustainable high-wage employment in the world.[*,†]

Moreover, the Trump administration thrives on "alternative facts," and it did not hesitate to use them to approve the Keystone XL Pipeline (reversing the decision of the Obama administration). President Trump touted the pipeline would generate thousands of jobs, help America with energy reliability, and use steel made in the U.S. to build the pipeline. None of those three facts is correct; the pipeline will generate between 35 and 50 permanent jobs, the oil will be exported, and the steel will be imported, as TransCanada will enjoy a loophole written into the law.[‡]

Adding insult to injury, President Trump claimed neither the Keystone XL Pipeline nor the Dakota Access Pipeline was ever controversial[§] and that he never got a single call about them. He failed to disclose in the same statement that the comment line to the White House had been closed (Figure 3.2). It reopened a week after he made such claim on February 16.

FIGURE 3.2 January 24, 2017, Los Angeles. With less than 4 h notice, hundreds attend a march near the federal building in Los Angeles to protest the approval of the Dakota access pipeline by President Trump.

---

[*] http://www.irena.org/News/Description.aspx?NType=A&mnu=cat&PriMenuID=16&CatID=84&News_ID=1450.

[†] https://thinkprogress.org/america-last-trumps-budget-cedes-millions-of-high-wage-jobs-to-china-655c943546f1.

[‡] http://www.reuters.com/article/us-transcanada-keystone-steel-idUSKBN16A2FC.

[§] http://tv.fusion.net/story/385440/donald-trump-dakota-access-keystone-xl/.

The fight to keep fossil fuels in the ground will be waged locally. People in Nebraska will fight to keep the Keystone XL Pipeline from being built. For most pipelines, including the Dakota Access Pipeline, the fight will be waged starting with the financial institutions that enable them with the money from citizens.[*]

One would turn to California or New York for greener pastures and the opportunity to fight climate change, but the grass is not necessarily greener. In California, the moderate Democrats are in the pockets of the fossil fuel industry, which is also the biggest and most powerful lobby in the state.[†] Time and time again, the moderate Democrats join forces with the Republicans to advance the fossil fuel industry interests. For nearly seven years, the fossil fuel industry has had the fiercest ally as a governor. Jerry Brown has advocated for lax regulations and business-friendly practices, to the point that he fired the top oil regulators to ensure the industry could appoint someone with whom they could operate more freely and comfortably.[‡] Governor Brown wants to be labeled as an environmentalist, yet he received nearly $10 million from fossil fuel corporations and has returned favors accordingly.[§] A legislation to improve roads would tax nonemitting electric vehicles, making them even more expensive and exempt diesel trucks from air regulations thanks to Governor Brown and his affinity for the fossil fuel industry.[¶]

Many Democrats have demanded that Jerry Brown stop shilling for the fossil fuel industry; they have even booed him at the DNC convention, but to no avail.[**] Even as Porter Ranch saw the biggest methane disaster in the United States, and the community, as well as environmental groups, asked Brown to take action and close the dangerous Aliso Canyon gas storage facility, Brown remains mute. Some speculate it was the contributions from Sempra Energy (parent company of SoCal Gas) or the fact that Jerry Brown's sister, Kathleen Brown, sits on the Sempra board of directors and commands a high salary.[††]

California remains a challenging state for banning oil and gas extraction as there are so many techniques used (Figures 3.3 and 3.4); legislation needs to be written to address them all and avoid any possible loopholes. When a rare opportunity appears, it is usually gutted by a moderate Democrat with a poison pill, or if it passes, like

FIGURE 3.3   July 1, 2016, Los Angeles City Hall. Protest to pressure the city council to shut down the dangerous leaking gas storage facility.

[*] http://www.businessinsider.com/trump-boycott-dapl-banks-standing-rock-2017-3.

[†] http://www.counterpunch.org/2017/03/31/big-oils-revolving-door-in-california/.

[‡] http://www.huffingtonpost.com/2015/07/02/western-states-petroleum-association_n_7715584.html.

[§] http://www.consumerwatchdog.org/newsrelease/report-finds-big-energy-companies-gave-big-and-got-big-favors-governor-brown-dollars-and.

[¶] http://www.latimes.com/opinion/editorials/la-ed-road-funding-20170401-story.html.

[**] http://www.sfgate.com/science/article/Fracking-exposes-rift-between-Jerry-Brown-5305168.php#photo-6001982.

[††] http://www.dailynews.com/government-and-politics/20151218/sister-of-california-gov-jerry-brown-on-socalgas-board.

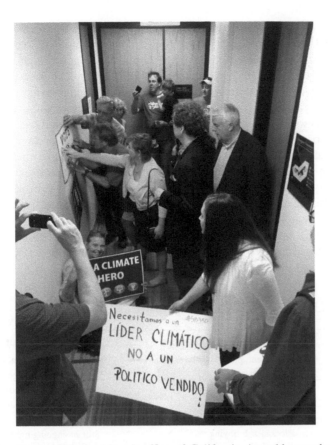

FIGURE 3.4 September 4, 2015, the (closed) office of California Assembly member Sebastian Ridley-Thomas. A group of constituents and environmentalists went to pay him a visit to urge a yes vote on SB350 and Ridley-Thomas unexpectedly closed his office. He proceeded to block everyone who contacted him on social media about the emissions legislation. Ridley-Thomas is a member of the moderate Democrats who has received significant contributions from the fossil fuel industry.

Measure Z in Monterey County, oil companies sue while they try and buy more politicians who will introduce legislation to stall proceedings.

Currently, there is pending legislation that will attack community choice aggregation programs (SB618) in California thanks to Senator Steven Bradford, a moderate Democrat who is heavily financed by the fossil fuel industry and utilities.

New York proved to be a little less challenging, being one of three states that banned fracking. It happened not because Governor Cuomo was ready to do it, but because the community along with a coalition of environmental organizations exerted enough pressure to force his hand, and he signed off on it.

There are also national groups fighting to get out the word about the voting record and leadership on climate issues for legislators. The Courage campaign puts out a report card on many progressive issues, one of them being the environment. But an emerging Super-PAC that looks at the legislator's full record is Climate Hawk Vote. They are well known for their well-planned endorsements and event disruptions at political conventions.

Without a question, the fight will be from the ground up. There are way too many politicians ready to sell out their constituents. Winona LaDuke said it best, "if you are not sitting at the table, you are on the menu." We can no longer trust the politicians and the corporations who put them there; we need a seat at the table to ensure our survival.

In conclusion, there is no question that renewables are being adopted by the United States and other countries, but they need to replace the fossil fuel infrastructure at a much faster pace, preferably by 2030 or sooner. To ensure that happens, we need political will, community engagement, and leaders who can pave the way as well as overcome every obstacle and roadblock the fossil fuel industry has in store.

Electrifying transportation is key, not only for efficiency, but also because the internal combustion engine is killing us. We need to stop burning fuels; whether gasoline, methane, or diesel, they all have to go.* When we show the costs of fuels, we need to show all aspects of them, and when we do, it will be clear that spending any amount extracting fossil fuels is too much. Among the externalities usually omitted by the fossil fuel industry are government subsidies (much higher than renewables), human health, climate change costs, water footprint, and military costs, which are about $80 billion per year according to a study by RAND.†

The bridge to 100% renewable energy is efficiency and conservation. We do not need any fossil fuel as a bridge; nor do we need a pay-to-pollute scheme like a carbon tax or cap-and-trade. We need to act on climate, and we need to do it now!

SUMMARY TABLE

Technology at Stake	Activists' Response	Outcome
Oil Leaks and Pipelines	Divestment, protests, lobbying, valve turning	Still an ongoing fight, plummeting oil/gas prices make practice less popular. Televised leaks and unorthodox security practices have helped get public support against corporations.
Shale Oil	Blockades, lobbying	Still an ongoing fight, plummeting oil/gas prices make practice less popular
Oil Shale	—	—
Bitumen	Blockades, protests, lobbying	Still an ongoing fight, plummeting oil/gas prices make practice less popular
Fracking	Blockades, protests, lobbying, screenings of documentaries, global frackdown	Still an ongoing fight, plummeting oil/gas prices make practice less popular. Many documentaries and reports have made this practice unpopular with most Americans.
Nuclear Power	Protests, lobbying, public comment	On its way out
Cap and Trade/Carbon Tax	Public comment, forums	Big ongoing fight

3.4   A BRIEF HISTORY OF ENERGY-RELATED ACTIVISM IN THE US

3.4.1   CIVIL RIGHTS ROOTS OF ENERGY-RELATED ACTIVISM

From its birth, the fossil fuel industry has capitalized on human rights abuses whether perpetrated by governments or overlooked by them. The abuses go from appropriation of land and forced relocation to violent and even deadly suppression of critics.

Oil and gas production leads to repression, as this resource thwarts democratization by enabling countries to fund internal security. Gender equality is also negatively impacted as the industry is predominantly male driven, which gives women less opportunity and political influence.

---

* http://www.autoblog.com/2014/10/30/the-internal-combustion-engine-is-killing-us-literally/.
† http://www.rand.org/pubs/monographs/MG838.html.

The United States even has a combat training school for Latin American soldiers, located at Fort Benning, Georgia, called the Western Hemisphere Institute for Security Cooperation (WHINSEC) formerly known as the School of the Americas. This institution was initially established in Panama in 1946 and expelled from that country in 1984. Hundreds of thousands of Latin Americans have been tortured, raped, assassinated, "disappeared," massacred, and forced into refuge by those trained at WHINSEC. SOA Watch, a nonprofit organization that exposes abuses by WHINSEC, details one of their operations in Colombia against an activist who fought privatization of the state oil company.

Fossil fuel corporations are notorious for getting away with many transgressions in the United States, where we are supposed to have "law and order." But when it comes to other countries, all bets are off. A good example is Texaco's negligence in the Ecuadorian Amazon that created a public health crisis.

Fighting pay-to-pollute schemes is another way to fight for people who live in disadvantaged communities in the United States and where corporations have chosen to offset carbons. A good example is the program called REDD under cap and trade. Environmental organizations that understand how cap and trade adversely affects communities are vehemently opposed to REDD, the program that has allowed corporations to buy offsets in underdeveloped countries by privatizing forests and evicting long-term residents. That ensures the polluters can keep doing business as usual and polluting at the source, while they keep harming local communities; with cap and trade, they also harm the communities from which they get their offsets. It is noteworthy to mention that the concept of cap and trade was invented by Goldman Sachs and Enron, two corporations that are well known for not putting people before profits.

The fight always starts at a local level, with the neighborhoods and their local representatives. A case in point is STAND-L.A., an environmental coalition of community groups that seeks to end neighborhood drilling to protect the health and safety of Angelenos on the frontlines of urban oil extraction. Most of the urban drilling happening in Los Angeles is in low-income communities where predominantly people of color and immigrants live. It is environmental racism. Corporations have been taking advantage of minorities and the poor, and now the communities are organizing and getting advocates. Drilling happens in rural areas as well, including next to agricultural fields. That means civil rights activists have to fight not only those who want to use toxic pesticides and fertilizers in our food, but also those who want to extract oil right next to them. The continuous exposure to dirty air and now thanks to the deal with agribusiness and oil companies, toxic oil wastewater used in crops, may result in severe illnesses and even death. Most of the farmers and their families who are exposed to all these toxins and emissions are usually low income, people of color, and/or immigrants. That is why we have groups like the Center on Race, Poverty & the Environment that are part of California Environmental Justice Alliance (CEJA).

Environmental racism is not perpetuated by of corporations or government officials; in South Los Angeles even the Catholic Church is involved in such a practice. For years, the Los Angeles Archdiocese has leased two properties to oil corporations for unconventional drilling. The oil wells are in a residential area that is mostly low income. Many have reported getting sick and hearing noise and vibrations at all hours of the day and night, but Archbishop Jose Gomez refuses to cancel the leases. Univision reached out to Gomez in November 2014 when they did a report, but the archbishop refused to go on camera and issued a statement saying that he cares about the health and welfare of the community and is working with the well operators to ensure they are responsible neighbors. Gomez did not have anything to say when confronted about what Pope Francis preaches, to take care of the poor and the environment and live humbly.*†‡

---

* http://america.aljazeera.com/watch/shows/america-tonight/articles/2015/8/14/los-angeles-church-oil-wells-climate.html.

† http://grist.org/cities/the-sad-sickening-truth-about-south-l-a-s-oil-wells/.

‡ http://www.huffingtonpost.com/mark-ruffalo/why-is-la-toxic_b_9910640.html.

4 The Constructal Design Applied to Renewable Energy Systems

Elizaldo D. dos Santos and Liércio A. Isoldi
Federal University of Rio Grande

Mateus das N. Gomes
Federal University of Rio Grande and Federal Institute of Paraná

Luiz A. O. Rocha
University of Vale do Rio dos Sinos

CONTENTS

4.1 INTRODUCTION

Constructal theory is the view that the generation of images of design (pattern, rhythm) in nature is a phenomenon of physics and that this phenomenon is covered by a principle. This principle is the constructal law of design and evolution in nature: "for a finite-size flow system to persist in time (to live) it must evolve such that it provides greater and greater access to the currents that flow through it" [1]. Evolution means design change and how this design change spreads in time, i.e., the flow of knowledge [2].

Constructal law has been applied in engineering by using the constructal design method, and it has been documented in several important references [3–7]. Constructal design is the philosophy of evolutionary design in engineering applications; it is also known as design with constructal theory (DCT) [7,8].

According to Bejan [7], "there is no 'best' in evolutionary design. There is 'better' today, which turns out to be not as good tomorrow." DCT and constructal design are not mathematical optimization methods. This makes it hard to apply them when the system has many degrees of freedom (DOFs). An alternative to engineering project design is to use the constructal design method in association with some optimization methods, e.g., exhaustive search or genetic algorithm [9]. This approach makes it possible to study complex systems, i.e., systems with a larger number of DOFs.

Society is increasing its energy needs: everybody wants to have a better life [2]. Better means education, more travel, power, leisure, etc. However, these needs are followed by the search for clean and renewable sources of energy. Can constructal design methods be used to improve the configuration of devices that produce power by means of renewable energy?

This chapter shows, step by step, how to apply constructal design, associated with an exhaustive search method, to discover a better configuration in the following devices: solar chimney power plant (SCPP), oscillating water column (OWC), and earth–air heat exchanger (EAHE). The design of SCPP is shown, explaining didactically all the details of the method. Later, the architectures that improve the performance of OWC and EAHE are studied to reinforce the application of the method.

4.2 SOLAR CHIMNEY POWER PLANT

The greenhouse effect caused by the burning of fossil fuels associated with energetic resource scarcity has led to big concerns about the generation and use of new sources of energy. Moreover, financial and social problems have been caused by greenhouse effect. For instance, Stern [10] estimated that financial losses caused by global heating can exceed 20% of the world gross domestic product. As a consequence, several countries studied measures to reduce gases emissions and comply with the energy requirements. One of these measures is the incentive to use renewable sources of energy. Solar energy is among the most promising alternatives. According to Li et al. [11], solar energy is free from carbon emission, inexhaustible, free from costs (in its basic form), and practically unlimited. One possibility for converting solar energy into electricity is the SCPP.

The SCPP is composed of three main components: collector, chimney, and turbine [12,13]. More precisely, the chimney is a long cylindrical structure usually placed in the center of a greenhouse collector (made of transparent glass or plastic film) and wind turbines, which are generally inserted at the base of the chimney to improve the power generation [12,13]. Figure 4.1 illustrates the SCPP's main components. The main physical principle of the device is based on the heating of an air layer held inside the collector. The heating of air is caused by the incident solar radiation transmitted by the collector. The transmitted solar radiation strikes the ground surface, and a part of the energy is absorbed, while another part is reflected back to the roof. The multiple reflection results in a higher fraction of energy absorbed, by the ground and heating of the ground surface, which transfers thermal energy to the surrounding air. First, the air is driven by natural convection

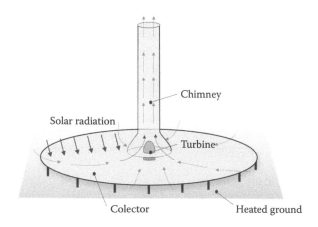

FIGURE 4.1   Illustration of the operational principle of an SCPP.

toward the top region of the chimney. Afterward, suction from the chimney draws in more hot air, and the cooler air from outside the collector enters the chimney to replace the hot air by mixed convection [11] and the airflows as illustrated in Figure 4.1. The airflow expands through a wind turbine in the chimney base converting the kinetic energy of air into mechanical and electrical energy.

Several studies have been performed to improve the knowledge about the SCPP. It is worth mentioning the pilot SCPP of real scale [14,15]. After this was built, evaluations were performed to improve knowledge about the physics of the device and energy production capacity of the prototype [16]. Many works have since been performed in an experimental and numerical framework to understand some physics aspects, e.g., the distribution of fluid dynamic and thermal fields in devices [13,17–24]. Additionally, several other interesting works have been done in order to obtain recommendations about the design of SCPP, which improves the fluid dynamic and thermal performance of the device. Chief examples can be seen in Refs. [22,15–32]. More details about these studies can be seen in Refs. [12,13,33].

In the constructal design framework, Koonsrisuk et al. [34] evaluated analytically the influence of one DOF (the ratio between the height of the chimney and the collector radius) over the power production of SCPP. In this pioneer and important work, it was verified that the power generated per unit of land area is proportional to the length scale of the power plant. Moreover, intermediate ratios between chimney height and the square of the radius maximize the device power. Afterward, Vieira et al. [33] performed a numerical study about SCPP submitted to two constraints areas (chimney and collector areas) and three DOFs: R/H (ratio between the collector radius and its inlet the height), R_1/H_2 (ratio between the radius and height of the chimney), and H_1/H (ratio between the collector outlet height and collector inlet height), which was constant in that study ($H_1/H = 10.0$). The used DOFs allowed taking into account cylindrical, divergent, and convergent chimneys (as illustrated in Figure 4.2) with a sloping collector, studying in combined way the radius of collector, chimney height, and diameter of chimney. This evaluation was also performed for different ground temperatures, which mimics the influence of different solar incidence conditions over the SCPP.

In the present work, the application of constructal design associated with exhaustive search for the SCPP studied in Vieira et al. [33] is detailed with the aim of improving the comprehension of the steps employed for geometrical evaluation of this kind of problem.

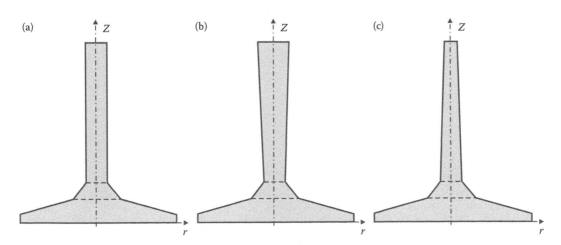

FIGURE 4.2 Different kinds of geometries for a chimney device: (a) Cylindrical, (b) Divergent, (c) Convergent.

4.2.1 PROBLEM DESCRIPTION

The studied problem is idealized; it simulates the main operational principle of SCPP. The following simplifications are taken into account:

1. Thermal radiation is not considered for computation of thermal field.
2. The effect of air heating is performed by the imposition of a prescribed temperature in the ground.
3. The collector and chimney surfaces are thermally insulated.
4. The turbine is not employed.
5. It is considered an axisymmetric domain.
6. Turbulent incompressible flow is considered at the steady state.

Concerning the mathematical modeling of the problem, the time-averaged balance of mass, momentum, and energy, are solved numerically with the finite volume method (FVM) [35,36]. Moreover, the air is treated as an ideal gas. Then, it requires the solution of an additional equation for density given by the ideal gas law. For the treatment of turbulent flow, it employs the standard $k-\varepsilon$ model [37,38]. The details of mathematical and numerical modeling used in this problem can be seen in Vieira et al. [33] and for the sake of brevity will not be repeated here.

The axisymmetric computational domain is illustrated in Figure 4.3. It is considered a gravitational field in the negative direction of azimuthal coordinate (z) and an atmospheric reference pressure ($p_{op}=1.0$ atm). For the boundary conditions, it is considered that a highest temperature (T_S) in the ground surface generates the initial motion of the fluid. Moreover, imposed in this surface are the impermeability and no-slip conditions for fluid dynamic fields. A prescribed pressure ($p=1.0$ atm) and temperature ($T_\infty=303$ K) are imposed at the collector inlet, as well as in the exit of the chimney. For the dashed surfaces in Figure 4.3, no-slip and impermeability boundary conditions are also imposed for momentum, and an adiabatic condition is used for the thermal field. In the dashed

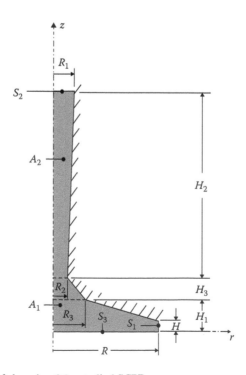

FIGURE 4.3 Computational domain of the studied SCPP.

point line is imposed an axis boundary condition. In the study of Ref. [33], four different ground temperatures are evaluated (T_S=308, 313, 318, and 323 K) in order to evaluate the effect of different solar incidences on the device.

4.2.2 APPLICATION OF CONSTRUCTAL DESIGN FOR SCPP

For the application of constructal design in association with exhaustive search (simulating all geometrical possibilities), the main steps are defined and exemplified for the problem of SCPP. The steps used in the present work are defined as follows:

Step 1—Definition of the flow system

In this step, it is important to define properly the system domain, problem restrictions (geometric, physical, manufacturing limitations, and others), and the DOFs that will be evaluated. In the SCPP studied problem, the collector, turbine, and chimney areas are the problem constraints. The same turbine is used in every case, and the dimensions R_2, R_3, and H_3 are considered fixed. In this sense, for the definition of variable dimensions, the SCPP problem is submitted to the two remaining constraint areas: the collector and chimney (A_1 and A_2):

$$A_1 = (R_3 H_1) + \frac{(H_1 + H)}{2}(R - R_3) \qquad (4.1)$$

$$A_2 = \frac{(R_1 + R_2)}{2} H_2 \qquad (4.2)$$

The problem is defined with eight geometric variables (R, H, H_1, H_3, R_2, R_3, H_2, R_1), three being fixed in the turbine region (H_3, R_2, R_3). There are two equations and five variables to be determined due to the imposed constraints given by Equations 4.1 and 4.2. Then, DOFs are required for closure of the equations system and definition of search space. In this study, the following DOFs are defined: R/H, R_1/H_2, and H_1/H.

Step 2—Identification of flow and objectives

In this step, it is necessary to define what is flowing through the system and the objective(s) of the problem. For the study of SCPP, the system is benefited when the momentum of flow is intensified in the device, mainly in the turbine region allowing a higher conversion of wind energy into an electrical one. The momentum is dependent on the magnitude of convection heat transfer, which acts as the driven force of the flow. Then, the main objective is the maximization of the available power in the solar chimney, which is given by:

$$P = \frac{1}{2}\rho A_t V^3 \qquad (4.3)$$

where A_t is the cross-sectional area of the turbine ($A_t = \pi R_2^2$), and V is the spatial-averaged velocity in this surface (m/s).

Step 3—Evaluation of the first DOF

In this step, one DOF is evaluated keeping the other DOFs and parameters as constraints and physical dimensionless groups fixed. Here it is important to evaluate how the variation in the DOF affects the objective(s) function(s) and internal currents that flow through the system. In the SCPP study, this step consists of the variation of the ratio R/H, keeping fixed the DOFs R_1/H_2 and H/H_1 and the temperature of ground (which is also represented by the Grashof number). The highest magnitude of available power is the once-maximized power, P_m, and the corresponding ratio R/H is the once-optimized ratio, $(R/H)_o$. An illustration of the geometric evaluation performed in this step can be seen in dark lines of Figure 4.4. The gray dashed lines represent the other geometric possibilities, which will be evaluated in the next steps. It is worth mentioning that for each evaluated geometry, the numerical simulation must be performed again. Another important observation is that constructal design is not an optimization method, but rather a method for evaluation of design in flow systems. For optimization of the evaluated problem, an association with an optimization method is required. In the study of SCPP, the exhaustive search was employed.

Step 4—Evaluation of the second DOF

In this step, Step 3 is repeated for other ratios of a second DOF. In this evaluation, it is possible to investigate the influence of this DOF over the effect of the first DOF on the system's performance. In the study of SCPP, this step consists of the repetition of the same process performed in Step 3 for several values of the ratio R_1/H_2 and keeping fixed the remaining parameters (here H/H_1 and Grashof number). Figure 4.5 illustrates the optimization process performed at this level. Dark lines represent the number of simulations performed in this step, while gray dashed lines represent the simulations that will be performed in the next steps. The highest magnitude of P is the twice-maximized available power (P_{mm}), and

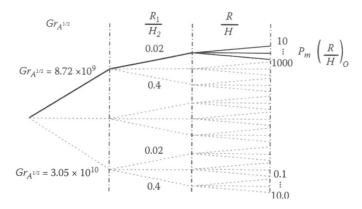

FIGURE 4.4 Flowchart of simulations performed for the evaluation of first DOF.

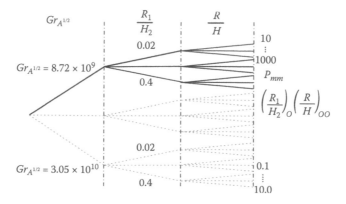

FIGURE 4.5 Flowchart of simulations performed for the evaluation of two DOFs.

the corresponding optimal geometries are the once-optimized ratio of R_1/H_2, $(R_1/H_2)_o$, and the twice-optimized ratio of R/H, $(R/H)_{oo}$.

Step 5—Evaluation of the other DOFs and parameters
In this step, Steps 3 and 4 are repeated for a third DOF. This process continues until the end of DOFs, obtaining the n-times maximized or minimized objective and n-times optimized geometry according to the DOFs established for the problem. Another possibility is the evaluation of other parameters of the flow system. For the SCPP problem, the geometric evaluation was performed for different ground temperatures (or Grashof numbers, as properly defined in Ref. [33]). The complete flowchart performed can be viewed in the reference and for the sake of brevity will not be repeated here.

4.2.3 OPTIMIZATION RESULTS

With the purpose of illustrating the application of constructal design in association with the exhaustive search for geometric evaluation of SCPP, the effect of the first two DOFs (R/H and R_1/H_2) on the available power (P) for $T_S = 323$ K is presented. These results represent an illustration of Steps 3 and 4 described in the previous section. First, Figure 4.6 presents the effect of the ratio R/H on the available power (P) for two different ratios of R_1/H_2 ($R_1/H_2 = 0.04$ and 0.2). As can be seen, for $R_1/H_2 = 0.04$, which represents a chimney with a large height, the highest collector radius led to the best device performance. For this case, the optimal ratio, $(R/H)_o = 1000$, led to a power 4.6 times higher than the worst performance, $R/H = 10$. It was also noticed that a strong variation of power was obtained for the lowest ratios of R/H, while the gain is significantly smoothed for $R/H > 400$. Results also showed that for the highest magnitude of $R_1/H_2 = 0.2$, which represents a chimney with a more divergent shape, the effect of the ratio R/H on the available power behaved in a different way and the optimal shape is achieved not for the highest magnitude of the ratio R/H, but for an intermediate ratio. In general, the results showed that it is important to evaluate the shape of chimney and collector combined.

The best results obtained for different ratios of R_1/H_2 (performed in Step 3) illustrated in Figure 4.6 are compiled in Figure 4.7.

Figure 4.7 shows the effect of the ratio R_1/H_2 on the once-maximized available power (P_m) and the once-optimized ratio of R/H, $(R/H)_o$. Results showed that there is an intermediate ratio of R_1/H_2 that relates to the best performance of a device, i.e., for the twice-maximized available power (P_{mm}). Therefore, the design of chimneys with high convergent or divergent profiles led to the worst performance. Concerning the effect of the ratio R_1/H_2 on $(R/H)_o$, it can be observed that the highest

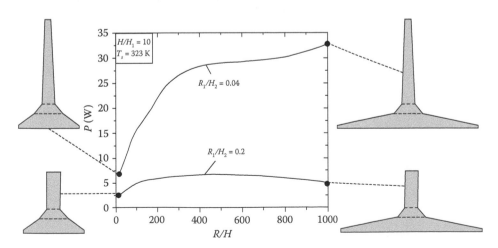

FIGURE 4.6 Effect of the ratio R/H on the chimney power for two different ratios of R_1/H_2 with $T_S = 323$ K.

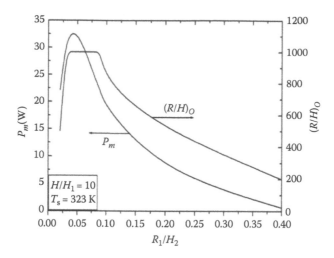

FIGURE 4.7   Effect of the ratio R_1/H_2 on the once-maximized power (P_m) and the once-optimized ratio of R/H, $(R/H)_o$, for $T_S = 323$ K.

magnitudes of $(R/H)_o$ are obtained in the optimal region of the ratio R_1/H_2. Then, the radius of a collector can be augmented only in situations where the chimney has a condition to drive the flow with high intensity generated in the collector. Results also showed that the designs of collector and chimney must be evaluated in a combined way.

In general, results showed that the application of constructal design is important for rationalization of energy conversion in an SCPP device.

4.3   OSCILLATING WATER COLUMN

The OWC was the first concept to be developed for wave energy conversion and is still the favorite technology among a large part of the wave energy conversion community. It can be employed in isolated shoreline or nearshore situations, integrated into a breakwater, or used in single- and multi-OWC floating plants. From the mechanical viewpoint, the OWC is particularly simple and reliable: the only moving part is the rotor of an air turbine, located above seawater, directly driving a conventional electrical generator [39].

OWC devices are, basically, hollow structures partially submerged, with openings to the sea below the water-free surface, as can be seen in Figure 4.8. In accordance with Cruz and Sarmento

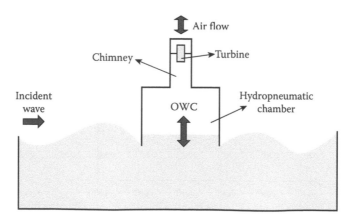

FIGURE 4.8   OWC system.

[40], the electricity generation process has two stages: When the wave reaches the structure, its internal air is forced to pass through a turbine placed in its chimney, as a direct consequence of the augmentation of pressure inside the hydropneumatic chamber. When the wave returns to the ocean, the air again passes by the turbine, but now being sucked from the external atmosphere, due to the chamber's internal pressure decreasing. So, to use these opposite air movements usually the Wells turbine is employed. This type of turbine has the property of maintaining the rotation direction irrespective of the flow direction. The set turbine/generator is responsible for the electrical energy production.

4.3.1 Constructal Design Applied in OWC Devices

Constructal design and exhaustive search are employed to achieve geometric optimal shapes in flow systems. According to these methods, the flow is malleable and the geometry is obtained from a principle of global performance maximization. In addition, geometry must be subject to global constraints and varied according to its DOFs [3]. To apply these methods for the geometric optimization of a physical system requires an objective function (a quantity that will be improved), DOFs (geometric parameters that may vary during the optimization process), and geometric constraints (parameters that are kept constant during the optimization process).

In Figure 4.9, it is possible to verify some DOFs analyzed in an OWC converter. The following DOFs may be considered: H_1/L (ratio between the height and the length of the chamber), H_3 (lip submergence), and H_2/l (ratio between the height and the the length of the chimney).

Two global constraints were considered: the OWC input volume (V_E) and the OWC total volume (V_T), which are defined as:

$$V_E = H_1 L L_1 \tag{4.4}$$

$$V_T = V_E + H_2 l L_1 \tag{4.5}$$

where L_1 is the out-of-plane dimension, which is constant and equal to one, so that a bidimensional analysis can be assumed. From Equation 4.4, one can obtain expressions to determine L and H_1, respectively:

$$L = \left[\frac{V_E}{(H_1/L)L_1} \right]^{1/2} \tag{4.6}$$

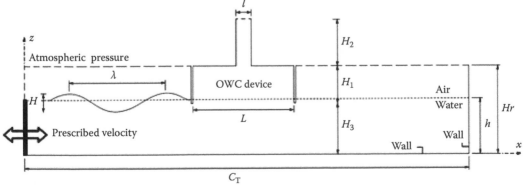

FIGURE 4.9 Schematic representation of the computational domain.

$$H_1 = L\left(\frac{H_1}{L}\right) \tag{4.7}$$

while from Equation 4.5, it is possible to define l and H_2 dimensions, respectively, as:

$$l = \left[\frac{V_T - V_E}{(H_2/l)L_1}\right]^{1/2} \tag{4.8}$$

$$H_2 = l\left(\frac{H_2}{l}\right) \tag{4.9}$$

Besides, using as a reference the wave climate, it was considered that $L=\lambda$ and $H_1=H$, allowing for the value of the constraint V_E. In turn, to determine the value of the constraint V_T, it was assumed that V_E has a value equal to 70% of V_T. Among the considered objective functions are mass flow rate, pressure, and hydropneumatic power.

4.3.2 Constructal Design Applied to OWC Converter in a Laboratory Scale

Constructal design has been applied with the objective of analyzing geometry of OWC converter and thus maximizing energy conversion; some works can be highlighted in the OWC realm [41–46].

Several numerical studies have been developed about wave energy converters. One possibility of numerically analyzing these converter devices is through the volume of fluid (VOF) method, proposed by Hirt and Nichols [47]. The VOF method is a multiphase numerical model that can be used to treat the interaction between water and air. To solve the mathematical model, the commercial code FLUENT, based on the FVM, is used [48].

The tendency is that more and more researchers use computational modeling to analyze the device's wave energy converters. Currently, numerical simulation is an important tool in engineering projects, and it is possible to verify that these researchers have used the VOF model in the numerical simulations of wave energy converters [45,49–53].

The proposal is to apply constructal design to optimize the two-dimensional geometry of an OWC converter device for maximum conversion of wave energy. The goal is to optimize DOFs: H_1/L (ratio between the height and the length of the chamber), H_3 (lip submergence), and H_2/l (ratio between the height and the length of the chimney).

The chamber input volume of the OWC device (V_E), Equation 4.4, and the total volume of the OWC device (V_T), Equation 4.5, are also kept constant and are the geometric constraints of the problem. The geometric variables of the computational domain can be observed in Figure 4.9.

In the present work, a wave is used in a laboratory scale. The wave and the numerical wave tank that are considered have the following characteristics: period $T=0.8\,\text{s}$, height $H=0.14\,\text{m}$, length $\lambda=1.0\,\text{m}$, depth $h=0.5\,\text{m}$, numerical wave tank length $C_T=5.0\,\text{m}$, and numerical wave tank height $H_T=0.8\,\text{m}$.

In all cases, the DOF H_3 is varied considering the following measures: $H_3=0.535\,\text{m}$, $H_3=0.500\,\text{m}$, $H_3=0.465\,\text{m}$, $H_3=0.430\,\text{m}$, $H_3=0.395\,\text{m}$, and $H_3=0.360\,\text{m}$, respectively. The following values were adopted for the constants of the problem: $V_E=0.14$, $V_T=0.20$, and $H_2/l=3.0$. Thus, by knowing the DOF H_1/L, it is possible to calculate the dimensions of the OWC converter devices of Equations 4.6 through 4.9.

Figure 4.10a shows the behavior of RMS average hydropneumatic power when the DOFs H_1/L and H_3 are considered. One can note in Figure 4.10 that the maximum values of the RMS hydropneumatic power were achieved when $0.56 < H_1/L < 0.84$. This range comprises the cases in which the hydropneumatic chamber length (L) is approximately half of the wavelength (λ), i.e., $L \approx \lambda/2$, and the hydropneumatic chamber height (H_1) is around double the wave height (H), i.e., $H_1 \approx 2H$.

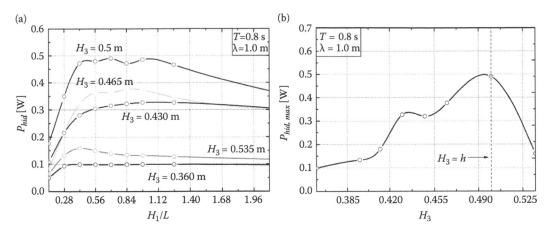

FIGURE 4.10 RMS average behavior: (a) of the hydropneumatic power with DOFs H_1/L and H_3 and (b) of the maximum hydropneumatic power for each H_1/L in respect to H_3 for T=0.8 s.

Another DOF analyzed is the submergence depth of the OWC converter, H_3. The results indicate that when H_3 decreases or increases in relation to h, the performance decreases, as it is possible to verify in Figure 4.10b. Also, in analyzing Figure 4.10b, it is necessary to consider other effects, e.g., the tidal effect and the depth variation.

It is possible to verify in Figure 4.10 that the best case is when H_3=0.5 m and $(H_1/L)_o$=0.7. In this situation, the average RMS for hydropneumatic power is equal to 0.4796 W and the efficiency is equal to 69.11%. The worst case occurs when H_3−0.535 m and H_1/L−0.14, when the average RMS for hydropneumatic power is equal to 0.0643 W and the efficiency is equal to 19.00%.

It should be noted that a redistribution of the geometry based on constructal design can provide a better performance of the OWC device. Besides, the dimensions of the converter were related to the wave climate. Therefore, the results of this computational model applying constructal method can be used to supply theoretical information for the construction of the OWC prototype to take advantage of the wave energy potential at any location with appropriate wave climate.

4.4 EARTH–AIR HEAT EXCHANGERS

Taking into account that the amount of solar energy received per minute by the earth's surface is larger than the energy consumption of the entire global population in 1 year [54], it is evident that this energy must be harnessed. A way to take advantage of the energy from the sun is through the EAHE device. The EAHE harnesses the ground source heat, since the earth's shallow subsurface can be regarded as a huge storage heater stoked by solar energy [55]. Its operational principle consists, basically, of the installation of one or more buried ducts in the underground through which the air insufflated by a fan exchanges heat with the surrounding soil. This air will be cooled or warmed depending on the season, since the subsoil is used, respectively, as a heat sink or source [56]. Thus, the outgoing air of the EAHE can be used to improve the thermal condition of built environments, allowing a considerable reduction of electrical energy consumption of the air-conditioning equipment [57]. There are several published studies about EAHE regarding experimental [58–60], analytical [61], and numerical [62–65] approaches.

Moreover, it is well known that the constructal design method has been extensively used for geometry evaluation in fluid mechanics and heat transfer engineering applications. However, even with the EAHE operating principle based on these two areas, only a few works employ the constructal design to analyze EAHE performance. We believe that the first published work about this subject is Isoldi et al. [66]. Ever since, the constructal design method has been applied to improve EAHE performance, as presented in the works published by Rodrigues et al. [67], Brum et al. [68], and Ramalho et al. [69]. In these studies, constructal design allows the use of different geometric configurations for arrangements with one, two, three, and four straight and parallel ducts, as indicated in Figure 4.11.

To guarantee an adequate thermal performance comparison among the six geometries of Figure 4.11, it was necessary to define the air volume fraction (ϕ) and installation volume fraction (ψ) parameters, which are given, respectively, by:

$$\phi = \frac{V_a}{V_s} = \frac{n\pi d_i^2 L}{4 W_s H_s L} = \frac{n\pi d_i^2}{4 W_s H_s} t \tag{4.10}$$

$$\psi = \frac{V_i}{V_s} = \frac{A_i L}{W_s H_s L} = \frac{A_i}{W_s H_s} \tag{4.11}$$

with V_a, the ducts' internal volume of EAHE installation (representing the volume where the air flows); V_s, the soil total volume; V_i, the EAHE installation volume; n, the number of ducts of the installation; π, the mathematical constant; d_i, the duct diameter of the installation; L, the soil (and duct) length; W_s, the soil width; H_s, the soil height; and A_i, the area occupied by the EAHE

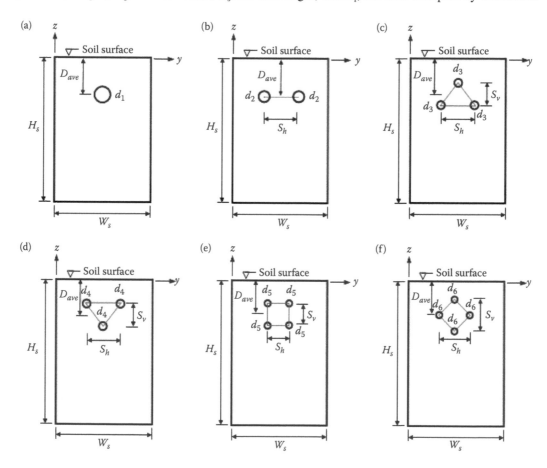

FIGURE 4.11 EAHE installations proposed by constructal design: (a) I1, (b) I2, (c) I3, (d) I4, (e) I5, and (f) I6.

installation (grey lines in Figure 4.11c–f). For the installations I3 (Figure 4.11c), I4 (Figure 4.11d), and I6 (Figure 4.11f), A_i is defined as:

$$A_i = \frac{S_h S_v}{2} \tag{4.12}$$

while for the installation I5 (Figure 4.11e) A_i is given by:

$$A_i = S_h S_v \tag{4.13}$$

where S_h and S_v are, respectively, the horizontal and the vertical spacing between ducts (see Figure 4.11). It is important to highlight that for the installations I1 (Figure 4.11a) and I2 (Figure 4.11b), because of their geometric configurations, the ψ parameter is not valid.

The ϕ and ψ parameters are constraints for the constructal design application, as well as the following relations: $S_v > d_i$, $S_h > d_i$, $S_v < (2D_{ave} - d_i)$, $S_h < (W_s - d_i)$, and $D_{ave} > d_i/2$. In addition, the DOF S_v/S_h (ratio between the vertical and horizontal spacing of the ducts) is adopted for each installation, having as an objective function the EAHE thermal performance maximization.

The transient and tridimensional numerical simulations were carried out by means of the verified and validated computational model presented in Brum et al. [64]. The clay soil dimensions ($L = 26$ m in x-direction, $W_s = 10$ m in y-direction and $H_s = 15$ m in z-direction), reproducing the soil type of the Viamão city, which is located in southern Brazil, and the average installation depth ($D_{ave} = 3$ m) were based on those used in Vaz et al. [57] and Brum et al. [64]. Moreover, the ducts' diameters for the installations I1–I6 are, respectively, $d_1 = 220$ mm, $d_2 = 55$ mm, $d_3 = 127$ mm, $d_4 = 127$ mm, $d_5 = 110$ mm, and $d_6 = 110$ mm, as in Rodrigues et al. [67] and Brum et al. [68], aiming to obtain a constant value for the air volume fraction of $\phi = 2.534 \times 10^{-4}$, which ensures that all EAHE installations have the same mass flow rate. An installation area of $A_i = 1.5$ m^2 and hence an installation volume fraction of $\psi = 0.010$ were adopted. For the sake of brevity, more details about the computational modeling, as verification and validation procedures, boundary conditions, initial conditions, and material properties, can be obtained in Vaz et al. [57], Brum et al. [64], and Rodrigues et al. [67].

The thermal performance of the several proposed EAHE configurations was analyzed by means of its thermal potential (TP), which represents the average monthly temperature difference between the inlet air and the outlet air of EAHE.

Therefore, to summarize, the application of the constructal design method was performed considering $\phi = 2.534 \times 10^{-4}$, $\psi = 0.010$, six ducts arrangements (see Figure 4.11), and several values of S_v/S_h for each installation type, aiming to maximize the EAHE TP during the year simulated. In this way, it was possible to determine the best geometric configuration, i.e., the geometry for each installation type that conducts with superior thermal performance. So, the maximized TP values, represented by TP_m, for heating and cooling, respectively, for the months of June and December are presented in Figure 4.12.

One can observe in Figure 4.12 that as the number of ducts increases, there is an augmentation in TP_m value. Hence, among all studied ducts arrangements, the superior thermal performance is achieved with installation I6 (diamond type) when $S_v/S_h = 0.05$, as illustrated in Figure 4.13.

Figure 4.13 shows the temperature distribution of installation I6 with $S_v/S_h = 0.05$ in the regions of inlet (air temperature around 30°C) and outlet (air temperature around 22°C) of the EAHE. Therefore, these results indicate that with the EAHE device, it is possible to promote a significant improvement of the thermal condition of a built environment.

In addition, it is evident that the EAHE has a cooling TP larger than the heating TP, making its use more effective in the summer season. In this case, from Figure 4.13, the installation I6 is better than installations I1, I2, I3, I4, and I5, respectively, 83.18%, 27.61%, 14.88%, 10.72%, and 3.54%.

Based on the presented results, it is possible to prove the effectiveness of the constructal design method employment to evaluate the geometry influence in the EAHE thermal performance.

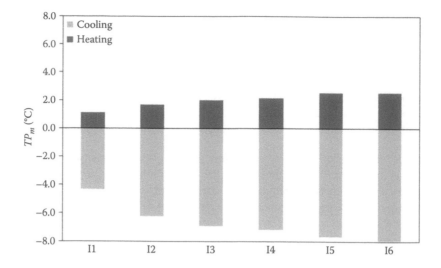

FIGURE 4.12 Maximized *TP* for heating and cooling of each EAHE installation type.

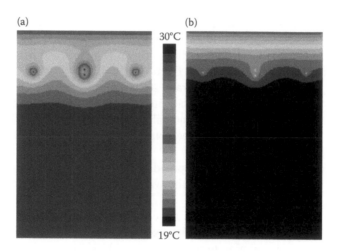

FIGURE 4.13 EAHE installation I6 with $S_v/S_h = 0.05$: (a) inlet air with 30°C and (b) outlet air with 22°C.

Furthermore, if an optimization technique is used, as the exhaustive search (which analyzes all geometric configuration proposed by the constructal design), it is also possible to obtain the optimized EAHE shape that conducts to the superior thermal performance.

The message that emerges is that geometry matters. Architecture, configuration, shape, and structure can be found by using a useful tool: constructal design. This method based on a physical law, constructal law, and combined with the enormous amount of current optimization methods should give more power to engineers to improve everyday designs.

ACKNOWLEDGMENTS

The authors E.D. dos Santos, L.A. Isoldi, and L.A.O. Rocha thank CNPq (*Conselho Nacional de Desenvolvimento Científico e Tecnológico-Brasília*, DF, Brazil) for the research grant and for financial support in Universal projects (Process: 445095/2014-8 and Process: 445558/2014-8).

REFERENCES

1. A. Bejan and S. Lorente, The constructal law of design and evolution in nature, *Philosophical Transactions of The Royal Society B*, vol. 365, no. 1545, pp. 1335–1347, 2010.
2. A. Bejan, *The Physics of Life: The Evolution of Everything*, St. Martin's Press, New York, 2016.
3. A. Bejan, *Shape and Structure: From Engineering to Nature*, Cambridge University Press, Cambridge, 2000.
4. A. Bejan and S. Lorente, Constructal theory of generation of configuration in nature and engineering, *Journal of Applied Physics*, vol. 100, p. 041301, 2006.
5. A.H. Reis, Constructal theory: From engineering to physics, and how flow systems develop shape and structure, *Applied Mechanics Reviews*, vol. 59, pp. 269–281, 2006.
6. A. Bejan and S. Lorente, Constructal theory of generation of configuration in nature and engineering, *Journal of Applied Physics*, vol. 100, pp. 041301, 2006.
7. A. Bejan and S. Lorente, *Design with Constructal Theory*, Wiley, Hoboken, NJ, 2008.
8. A. Bejan, interviewed by A.W. Kosner, Freedom is good for design, how to use constructal theory to liberate any flow system, *Forbes*, March 18, 2012.
9. G. Lorenzini, C. Biserni, E.S.D. Estrada, E.D. dos Santos, L.A. Isoldi and L.A.O. Rocha, Genetic algorithm applied to geometric optimization of isothermal Y-shaped cavities, *Journal of Electronic Packaging*, vol. 136, pp. 031011-1–031011-8, 2014.
10. N. Stern, *The Economics of Climate Change: The Stern Review*, Cambridge University Press, Cambridge, 2007.
11. W. Li, P. Wei and X. Zhou, A cost-benefit analysis of power generation from commercial reinforced concrete solar chimney power plant, *Energy Conversion and Management*, vol. 79, pp. 104–113, 2014.
12. A. Dhahri and A. Omri, A review of solar chimney power generation technology, *International Journal of Engineering Advanced and Technology*, vol. 2, pp. 1–17, 2013.
13. X. Zhou, F. Wang and R.M. Ochieng, A review of solar chimney power technology, *Renewable and Sustainable Energy Reviews*, vol. 14, pp. 2315–2338, 2010.
14. W. Haaf, K. Friedrich, G. Mayr and J. Schlaich, Solar chimneys: Part I: Principle and construction of the pilot plant in Manzanares, *International Journal of Solar Energy*, vol. 2, pp. 3–20, 1983.
15. W. Haaf, Solar chimneys: Part II: Preliminary test results from the Manzanares pilot plant, *International Journal of Solar Energy*, vol. 2, pp. 141–161, 1984.
16. J. Schlaich, R. Bergermann, W. Schiel and G. Weinrebe, Design of commercial solar updraft tower systems—Utilization of solar induced convective flows for power generation, *Journal of Solar Energy Engineering*, vol. 127, pp. 117–124, 2005.
17. X. Zhou, J. Yang, B. Xiao and G. Hou, Experimental study of temperature field in a solar chimney power setup, *Applied Thermal Engineering*, vol. 27, pp. 2044–2050, 2007.
18. M. Ghalamchi, A. Kasaeian and M. Ghalamchi, Experimental study of geometrical and climate effects on the performance of a small solar chimney, *Renewable and Sustainable Energy Reviews*, vol. 43, pp. 425–431, 2015.
19. S. Lal, S.C. Kaushik and R. Hans, Experimental investigation and CFD simulation studies of a laboratory scale solar chimney for power generation, *Sustainable Energy Technology Assessment*, vol. 13, pp. 13–22, 2016.
20. A.R. Shahreza and H. Imani, Experimental and numerical investigation on an innovative solar chimney, *Energy Conversion and Management*, vol. 95, pp. 446–452, 2015.
21. A.B. Kasaeian, E. Heidari and S.N. Vatan, Experimental investigation of climatic effects on the efficiency of a solar chimney pilot power plant, *Renewable and Sustainable Energy Reviews*, vol. 15, pp. 5202–5206, 2011.
22. H. Pastohr, O. Kornadt and K. Gürlebeck, Numerical and analytical calculations of the temperature and flow field in the upwind power plant, *International Journal of Energy Research*, vol. 28, pp. 495–510, 2004.
23. R. Sangi, M. Amidpour and B. Hosseinizadeh, Modeling and numerical simulation of solar chimney power plants, *Solar Energy*, vol. 85, pp. 829–838, 2011.
24. P. Guo, J. Li and Y. Wang, Numerical simulations of solar chimney power plant with radiation model, *Renewable Energy*, vol. 62, pp. 24–30, 2014.
25. M.A. dos S. Bernardes, R.M. Valle and M.F.-B. Cortez, Numerical analysis of natural laminar convection in a radial solar heater, *International Journal of Thermal Science*, vol. 38, pp. 42–50, 1999.
26. A. Koonsrisuk and T. Chitsomboon, Effect of tower area change on the potential of solar tower, *2nd International Conference Sustainable Energy Environment*, Bangkok, Thailand, pp. 1–6, 2006.

27. S.K. Patel, D. Prasad and M.R. Ahmed, Computational studies on the effect of geometric parameters on the performance of a solar chimney power plant, *Energy Conversion and Management*, vol. 77, pp. 424–431, 2014.
28. E. Gholamalizadeh and M.-H. Kim, Thermo-economic triple-objective optimization of a solar chimney power plant using genetic algorithms, *Energy*, vol. 70, pp. 204–211, 2014.
29. S. Dehghani and A.H. Mohammadi, Optimum dimension of geometric parameters of solar chimney power plants—A multi-objective optimization approach, *Solar Energy*, vol. 105, pp. 603–612, 2014.
30. N. Pasumarthi and S.A. Sherif, Experimental and theoretical performance of a demonstration solar chimney model—Part I: Mathematical model development, *International Journal of Energy Research*, vol. 22, pp. 277–288, 1998.
31. A.J. Gannon and T.W. von Backström, Solar chimney cycle analysis with system loss and solar collector performance, *Journal of Solar Energy Engineering*, vol. 122, pp. 133–137, 2000.
32. J.P. Pretorius and D.G. Kröger, Critical evaluation of solar chimney power plant performance, *Solar Energy*, vol. 80, pp. 535–544, 2006.
33. R.S. Vieira, A.P. Petry, L.A.O. Rocha, L.A. Isoldi and E.D. dos Santos, Numerical evaluation of a solar chimney geometry for different ground temperatures by means of constructal design, *Renewable Energy*, vol. 109, pp. 222–234, 2017.
34. A. Koonsrisuk, S. Lorente and A. Bejan, Constructal solar chimney configuration, *International Journal of Heat and Mass Transfer*, vol. 53, pp. 327–333, 2010.
35. S.V. Patankar, *Numerical Heat Transfer and Fluid Flow*, McGraw-Hill, New York, 1980.
36. H.K. Versteeg and W. Malalasekera, *An Introduction to Computational Fluid Dynamics: The Finite Volume Method*, Pearson, London, 2007.
37. D.C. Wilcox, *Turbulence Modeling for CFD*, 2nd ed., DCW Industries, La Canada, CA, 2002.
38. B.E. Launder and D.B. Spalding, *Lectures in Mathematical Models of Turbulence*, Academic Press, London, 1972.
39. A.F.O. Falcão and J.C.C. Henriques, Oscillating water column wave energy converters anda ir turbines: A review, *Renewable Energy*, vol. 85, pp. 1391–1424, 2016.
40. J.M.B.P. Cruz and A.J.N.A. Sarmento, *Energia das Ondas: Introdução aos Aspectos Tecnológicos, Económicos e Ambientais*, Ed. Instituto do Ambiente, Amadora, Portugal, 61 p, 2004.
41. N. Lopes, F.S.P. Sant'Anna, M.N. Gomes, J.A. Souza, P.R.F. Teixeira, E.D. Santos, L.A. Isoldi and L.A.O. Rocha, Constructal design optimization of the geometry of an oscillating water column wave energy converter (OWC-WEC), *Proceedings of Constructal Law Conference*, UFRGS, Porto Alegre, Brazil, 2011.
42. M.N. Gomes, C.D. Nascimento, B.L. Bonafini, E.D. Santos, L.A. Isoldi and L.A.O. Rocha, Two-dimensional geometric optimization of an oscillating water column converter in laboratory scale, *Engenharia Térmica*, vol. 11, pp. 30–36, 2012.
43. M.N. Gomes, E.D. dos Santos, L.A. Isoldi and L.A.O. Rocha, Two-dimensional geometric optimization of an oscillating water column converter of real scale, *Proceedings of 22nd International Congress of Mechanical Engineering* (COBEM 2013), Ribeirão Preto, Brazil, November 3–7, 2013.
44. L.A.O. Rocha, S. Lorente and A. Bejan, *Constructal Law and the Unifying the Principle of Design*, 1st ed., Springer-Verlag, New York, 2013.
45. M. das N. Gomes, Constructal Design de Dispositivos Conversores de Energia das Ondas do Mar em Energia Elétrica do Tipo Coluna de Água Oscilante, Doctoral Thesis, PROMEC-UFRGS, 2014.
46. M. das N. Gomes, M.F.E. Lara, S.L.P. Iahnke, B.N. Machado, M.M. Goulart, F.M. Seibt, E.D. dos Santos, L.A. Isoldi and L.A.O. Rocha, Numerical approach of the main physical operational principle of several wave energy converters: Oscillating water column, overtopping and submerged plate, *Defect and Diffusion Forum*, vol. 362, pp. 115–171, 2015.
47. C.W. Hirt and B.D. Nichols, Volume of fluid (VOF) method for the dynamics of free boundaries, *Journal of Computational Physics*, vol. 39, no. 1, pp. 201–225, 1981.
48. H.K. Versteeg and W. Malalasekera, *An Introduction to Computational Fluid Dynamics*, Longman, Malaysia, 257 p, 1999.
49. M. Horko, CFD optimisation of an oscillating water column energy converter, Masters Thesis, School of Mechanical Engineering, University of Western Australia, 145 p, 2007.
50. J.M.P. Conde and L.M.C. Gato, Numerical study of the air-flow in an oscillating water column wave energy converter, *Renewable Energy*, vol. 33, pp. 2637–2644, 2008.
51. M. das N. Gomes, C.R. Olinto, L.A.O. Rocha, J.A. Souza and L.A. Isoldi, Computational modeling of a regular wave tank, *Engenharia Térmica*, vol. 8, pp. 44–50, 2009.

52. Z. Liu, B. Hyun and K. Hong, Numerical study of air chamber for oscillating water column wave energy convertor, *China Ocean Engineering*, vol. 25, pp. 169–178, 2011.
53. P.R.F. Teixeira, D.P. Davyt, E. Didier and R. Ramalhais, Numerical simulation of an oscillating water column device using a code based on Navier-Stokes equations, *Energy*, vol. 61, pp. 513–530, 2013.
54. Z. Sen, *Solar Energy Fundamentals and Modeling Techniques: Atmosphere, Environment, Climate Change and Renewable Energy*, Springer, London, 276 p, 2008.
55. D. Banks, *An Introduction to Thermogeology: Ground Source Heating and Cooling*, John Wiley & Sons, Chichester, UK, 339 p, 2008.
56. F. Kreith and D.Y. Goswami, *Handbook of Energy Efficiency and Renewable Energy*, CRC Press, Boca Raton, FL, 1510 p, 2007.
57. J. Vaz, M.A. Sattler, E.D. dos Santos and L.A. Isoldi, Experimental and numerical analysis of an earth-air heat exchanger, *Energy and Buildings*, vol. 43, pp. 2476–2482, 2011.
58. P. Hollmuller and B. Lachal, Cooling and preheating with buried pipe systems: Monitoring, simulation and economic aspects, *Energy and Buildings*, vol. 33, pp. 509–518, 2001.
59. J. Pfafferott, Evaluation of earth-to-air heat exchangers with a standardised method to calculate energy efficiency, *Energy and Buildings*, vol. 35, pp. 971–983, 2003.
60. J. Vaz, M. Sattler, R.S. Brum, E.D. Santos and L.A. Isoldi, An experimental study on the use of earth-air heat exchangers (EAHE), *Energy and Buildings*, vol. 72, pp. 122–131, 2014.
61. M. De Paepe and A. Janssens, Thermo-hydraulic design of earth-air heat exchangers, *Energy and Buildings*, vol. 35, pp. 389–397, 2003.
62. G. Mihalakakou, M. Santamouris and D. Asimakopoulos, Modelling the thermal performance of earth-to-air heat exchangers, *Solar Energy*, vol. 53, pp. 301–305, 1994.
63. H. Wu, S. Wang and D. Zhu, Modelling and evaluation of cooling capacity of earth-air-pipe systems, *Energy Conversion and Management*, vol. 48, pp. 1462–1471, 2007.
64. R.S. Brum, L.A.O. Rocha, J. Vaz, E.D. dos Santos and L.A. Isoldi, Development of simplified numerical model for evaluation of the influence of soil-air heat exchanger installation depth over its thermal potential, *International Journal of Advanced Renewable Energy Research*, vol. 1, pp. 505–514, 2012.
65. R.S. Brum, J. Vaz, L.A.O. Rocha, E.D. Santos and L.A. Isoldi, A new computational modeling to predict behavior of earth-air heat exchangers, *Energy and Buildings*, vol. 64, 2013, pp. 395–402.
66. L.A. Isoldi, R.S. Brum, M.K. Rodrigues, J.V. Ramalho, J. Vaz, J.A. Souza, E.D. dos Santos and L.A.O. Rocha, Constructal design of earth-air heat exchangers, *Proceedings of Constructal Law Conference*, Nanjing University of Science and Technology, Nanjing, 2013, pp. 88–96.
67. M.K. Rodrigues, R.S. Brum, J. Vaz, L.A.O. Rocha, E.D. dos Santos and L.A. Isoldi, Numerical investigation about the improvement of the thermal potential of an earth-air heat exchanger (EAHE) employing the constructal design method, *Renewable Energy*, vol. 80, pp. 538–551, 2015.
68. R.S. Brum, M.K. Rodrigues, J.V.A. Ramalho, L.A.O. Rocha, L.A. Isoldi and E.D. dos Santos, On the design of two EAHE assemblies with four ducts, *Defect and Diffusion Forum*, vol. 372, pp. 31–39, 2017.
69. J.V.A. Ramalho, R.S. Brum, L.A.O. Rocha, L.A. Isoldi, E.D. Dos Santos and M. Sulzbacher, Fitting new constructal models for the thermal potential of earth-air heat exchangers, *Acta Scientiarum. Technology*, 2017 vol. 40, 2018.

5 Integration of Renewable Energy Technologies in Buildings and Cities[*]

Andy Walker
National Renewable Energy Laboratory

CONTENTS

5.1 INTRODUCTION

Sustainable Energy Technologies, such as the solar, wind, and waterpower systems described in other chapters of this book, need to be integrated into our energy systems to expand access to energy while avoiding harmful effects on the environment. The existing utility system or "grid" is a network that consists of components for generation, transmission, distribution, and control of electric power from producers (also called "resources") to consumers (also called "loads" or "demand"). Public safety necessitates that the grid be reliable, and public policy requires that grid power be affordable and broadly accessible. Much of the existing power grid was constructed before the advent of modern information technology for controls and automation. The existing grid is facing new challenges related to increases in power demand, complexity in managing more diverse and distributed resources, power quality issues introduced by "non-linear" loads (that have parasitic inductance and capacitance like LED lights or computers), physical and cybersecurity threats, and challenges to the existing economic model of costs and revenues. Increases in power demand include not only growth in population and per-capita electricity use, but also call for electricity to meet the need for heating and transportation as well. Electric vehicles are a rapidly growing example of this electrification of previously fossil-fueled uses. [Your author's Chevrolet Volt vehicle uses about the same amount of

---

[*] This chapter was prepared as an account of work sponsored by an agency of the United States government. The National Renewable Energy Laboratory (NREL) was not asked to endorse the findings of this chapter nor should any implied endorsement by NREL be assumed. The views and opinions of authors expressed herein do not necessarily state or reflect those of the United States government or any agency thereof.

electricity as his house, around 300 kWh/month. Fortunately, both are served with solar photovolta-ics (PVs)]. Therefore, there is an immediate need for improvement toward a reliable, self-regulating, efficient, and economical grid system that will allow the integration of renewable distributed power generation.

Energy System Integration (ESI) addresses how increasing amounts of intermittent renewable energy (RE) generation can be integrated into the larger utility grid to realize energy cost savings and improve system reliability. Challenges and solutions related to ESI occur at all levels (Walker et al., 2016):

- At the level of the building and its distribution feeder
- At the utility substation
- At the level of the transmission grid
- At the level of the other generator plants on an electric system
- At a level affecting utility business operations and finance

Years ago, grid integration issues were not critical because RE projects were small and often located in off-grid applications. Around 2008, we started to see large RE systems in grid-connected appli-cations, many benefiting from utility policies such as "net metering" and "feed-in tariffs," which require significant interaction of an RE generator with the grid. Challenges surfaced as buildings and campuses strived to integrate intermittent resources while relying on the grid to meet demand at times when RE resources were not producing. Problems were first notable in small grids, such as islands or remote communities, that might have only one generator plant and lack a diversity of loads and resources to smooth out fluctuations in both. These isolated grids are harbingers for prob-lems that will emerge in larger utility systems too.

In this chapter, we present challenges that project developers or building owners face when they implement a high penetration of RE on utility circuits. Challenges become greater as the renewable system gets larger, gets farther away from a substation, and gets farther away from a local voltage regulation device. The implications for the utility are that distributed generation increases cycling of the equipment that regulates voltage, undermines previous voltage regulation and overcurrent protection schemes, and changes the power factor required of the utility (reactive power). The inter-mittent nature of RE generation also increases the amount of spinning reserve required. For each challenge, we will explore solutions. We will also discuss some of the cost and efficiency issues the utility encounters. Solving these problems will require significant creativity and investment in technology:

- Expansion of transmission and reconfiguration of existing distribution systems
- Forecasting of both resources and loads
- Control of interruptible loads
- Energy storage, including batteries and nonelectric storage such as chilled water
- Sophisticated and interconnected controls to harmonize operation of all the components in the energy system
- New regulatory approaches to allocating the costs of building and operating the utility system, as well as programming cost recovery for utility companies.

5.2 ENERGY ISSUES FACING CITIES

Energy is central to the social, technological, environmental, economic, and political challenges faced by cities across the globe. Population growth, transportation, and industrial development are all driving energy use in cities, which is in turn driving concerns about climate change and local air pollution. Trends to address these concerns include RE, green buildings, outdoor lighting and night-time design, and mass transit and city walkability as alternatives to our love affair with our personal cars. An integrated approach is required to bring diverse stakeholders such as utilities, architects, engineers, investors, lobbyists, regulators, technology providers, and policy makers together to craft workable solutions. By coalescing expertise in all disciplines involved in urban design, cities can become more sustainable, more pleasant to live in, more sociable, safer, more secure, healthier, and easier to get around.

Cities are already energy intensive, and many of the approaches to reduce fossil fuel use involve electrification of loads that were previously based on fuels. Use of efficient heat pumps and PVs for heating and cooling could supplant the use of fuel oil or natural gas. As another example, mass transit and electric vehicles would both require electric infrastructure that does not currently exist in many cities. These increasing electric uses will cause us to redesign and rebuild our electrical infrastructure in cities. Energy use in New York City varies from less than 50 kWh/year per m^2 of land area in some regions to almost 5,000 kWh/year per m^2 of land area at the highest density, plus a maximum of about 1,500 kWh/year per m^2 of land area in fuel energy (Sherpa, 2016). The solar resource in that area is on the order of 1500 kWh/m^2 of land area. So while some buildings such as single-family residential or low-rise multifamily residential buildings might be able to achieve net-zero energy use with onsite solar systems, most city buildings will rely on an interconnected utility distribution system.

5.3   INCREASING PENETRATION OF INTERMITTENT RE

An estimated 147 gigawatts (GW) of renewable electric generation capacity were added globally, in 2015, and solar met more than 1% of electricity demand in 22 countries, with far higher fractions in Italy (7.8%), Greece (6.5%), and Germany (6.4%, REN21, 2016). While these percentages are currently low compared to other energy sources, RE technologies such as wind and solar are the fastest growing part of capacity additions. For example, 50 GW of solar PV were added in 2015, an increase of 28% to the 177 GW cumulative capacity of the previous year. As the amount of renewable generation in a system increases, system operation must be altered in order to accommodate the intermittent renewable power. These changes to the operation generally reduce the efficiency and cost effectiveness of the conventional resources, although many advocates of RE point out that distributed RE might forestall the need for huge investments in central power generation.

5.4   STRATEGIES FOR ENERGY SYSTEMS INTEGRATION

Radial utility power systems are based on one-way power flow—real power always flows from the central generation plant, through substations, and on to loads. In contrast, distributed RE systems are deployed in a network and inject power into the grid from multiple locations. Design of the overall system has to anticipate how the power will get from the distributed resources to all the loads, but since the distributed generation might occur at or closer to the distributed loads, losses in the system could be lessened.

The variability and intermittency of PV or wind power means that the overall system must be designed to accommodate rapid increases in RE power when it occurs and, likewise, provide for continued load when RE output drops off suddenly. For this reason, information must flow in two directions, so that available resources can be dispatched in advance of the need and interruptible loads can be controlled should resources be insufficient.

Challenges can occur for both the energy customer and the utility on the following scales when installing a large RE system:

1. *Building and Electric Distribution System*: limits to current-carrying capacity of utility circuit; location and setting of overcurrent protection (breakers, fuses); voltage out of limits
2. *Substation*: routing of power from circuit with excess RE to load on other circuits; voltage fluctuations

3. *Generation Plant*: increase in spinning reserve requirements; limits to sudden changes imposed by ramp-rates of generators
4. *Utility Operations*: allocation of costs associated with RE integration among utility customers; equitable distribution of operating cost among ratepayers
5. *Investors*: current model of a centralized utility in which guaranteed wholesale power rates return long-term investments (such as with baseload coal or nuclear plants) is challenged by onsite generation; wholesale prices are lowered by RE generation based on its low marginal cost compared to the fuel cost of fossil plants.

5.5   BUILDING LEVEL CHALLENGES AND SOLUTIONS

There are five areas of consideration related to the effects of high penetration of RE on the electrical service to individual buildings.

1. Configuration of Electrical Service to Each Building
   Many cities have "network" power systems rather than purely "radial" systems. It is more problematic to integrate distributed RE generation into a network system than it is into a radial system. This is because in a network system, each large building, or each transformer vault serving multiple smaller buildings, is fed by more than one feeder. This network configuration increases reliability, improves voltage stability, and reduces the need for reserve capacity on each circuit. Each feeder is provided with a "network protector" to open in the event of reverse power flow because if there is reverse power flow, it indicates a fault with another feeder. This prevents distributed generation from exceeding the load at a site or even an approved fraction of the load, and contributing to the larger city load. A time-delay feature is included to accommodate some reverse power from regenerative elevators in tall buildings. A solution to allow the distributed generator to contribute some power to the larger system would be a current-supervision function that allows the network protector to stay closed when the network current is lower than the expected level of reverse-power flow from the distributed generator (IEEE, 2000). The net effect is a two-level reverse-power function: time delayed for low levels of reverse power but instantaneous at higher levels, as occur during faults on the primary feeder (Behnke et al., 2005). One solution is to install a minimum import relay or a reverse power relay. The minimum import relay disconnects the PV system if the powers flowing from the utility drop below a set value, and the reverse-power relay would open if the power flows the wrong way (Jordan, 2014). A more sophisticated control on the network protector would be able to:
   - Adjust the trip level according to the maximum amount of solar power expected as a function of time of day, day of year and latitude calculated with a "clear-sky model" (Meinel and Meinel, 1976, initially, and others more recently)
   - Monitor power at the main feeder and send a control signal to a dynamically controlled inverter, which initiates a reduction in generated power, if required (Jordan, 2014)
   As an alternative to curtailing the output of the RE generator, energy storage (batteries) would be able to absorb energy in excess of the load, preventing a back flow of power, and then deliver that power at a later time preserving the ability of the distributed generator to offset energy (kWh) consumption of the building.
2. Capacity of Electrical Service to Each Building
   The electrical service to a building is designed to accommodate the peak load expected of the building plus a safety factor for future load growth. For example, a building with a 1 MW load might have a 2 MW service. If, as many projects are now doing, the building seeks to achieve "net-zero" and offset 24 MWh/day of energy with on-site solar energy, which occurs only over 6 h per day, a 4 MW PV system would be installed and the electrical

service upgraded from 2 MW to at least 3 MW to accommodate the flow of solar power out of the building during the day (4 MW − 1 MW = 3 MW).

3. Overcurrent Protection on Electrical Service to a Building with an RE Generator

    Coordination of overcurrent protection is an important consideration in the addition of RE generation. Some type of fuse, fusible link, or circuit breaker limits the current coming from a transformer to a building to a value less than that for which the line is rated. An RE generator at some midpoint on that line could add additional current and result in a maximum current (sum of current from transformer and distributed generation) greater than the line is capable of carrying. During initial distribution system planning, a protection coordination study informed the location and settings of fuses, breakers, and protective relays and synchronized other protective equipment to isolate electrical faults in a planned sequence. The addition of an RE generator invalidates previous protection studies because it changes the amount of current available for conductors to carry and changes the location and setting of breakers to prevent excessive current flow. New studies may be required in order to determine how the maximum current from an RE generator increases currents flowing through the electrical system when a line becomes overloaded or when a fault (short-circuit) occurs. Results of this study include any required changes to the protective system used to detect and interrupt current and isolate faults. The protective system may need to be reset or redesigned to properly set the coordination and timing of devices such as protective relays, reclosers, circuit breakers, and fuses in a system to detect and isolate faults. Measures might include replacing or resetting overcurrent limits. The sequence in which breakers trip is important: we want the fault to be localized by having local breakers trip before those upstream on the distribution system or at the substation. In extreme cases, it may be necessary to upgrade power lines to accommodate larger amounts of current than they were originally installed to handle.

4. Voltage Regulation of Electric Service to Building with RE

    A new source of energy, such as distributed PV systems, in the distribution system increases voltage (volts) at the point power is injected. This can mislead control systems that are based on load on the substation rather than actual voltage measured at the building. Voltage regulation is often cited as the primary concern of utilities when considering interconnection of an on-site RE generator. Utilities are required to keep these voltages within a relatively narrow operating range, for example +/−6% of the nominal voltage value. Since voltage drops along the powerline as the load goes up, current technology changes the voltage at the transformer based on the load, rather than a measurement of voltage at the building. This is called "line drop compensation." PV generation on a circuit will reduce the load seen at the transformer, the line drop compensation will lower the voltage at the transformer, and voltage will drop further as it moves along the line to the building. Loads served by the PV system may then see a voltage lower than the specified limits. Planning and modifications are needed to avoid deviations in voltage at various locations on the distribution system as PV inverters inject power at their locations. Also, at the substation, the utility will study the need to install new types of voltage regulation equipment (based on measured voltage rather than load). The intermittent operation of the RE generator may cause increased cycling on the new or existing voltage regulation equipment. Power-flow based software tools can determine cycle counts on voltage regulation equipment during changes in variable generation power output. If it is found to be excessive, one strategy to reduce cycling is increasing the delay time of the voltage regulators. Again, as with the control of network protectors, a more sophisticated control and communications system is needed to communicate the voltage at the load to the voltage adjustment at the transformer or substation.

5. Sourcing of Reactive Power

The utility generally provides customers with both real power and reactive power. The cosine of the phase angle between the two is the power factor. Reactive power is wasteful, and the utility will impose charges and experience generator problems if the power factor is much different from unity (1.0). Currently inverters are limited by standards to provide only real power, so the proportion of reactive power the utility must provide could increase with the addition of a PV system on the load. Inductive loads such as electric motors induce a lagging power factor (current lags voltage), but nonlinear loads such as LED lights and computers might also induce a leading power factor. Equipment used to actively provide reactive compensation for power factor correction can also cycle excessively due to the high penetration of intermittent RE generators. The reason for this is that most installed PV systems with static inverters supply power at unity power factor (power factor = 1). This means that these PV systems can supply real power to loads but not reactive power, which must still be supplied by the substation.

The reactive power supplied by the utility must increase to accommodate each additional kW of PV as it varies widely and suddenly as the RE resource changes. One solution is to specify inverters that provide reactive power if needed, as will soon be available under new inverter standards (revision to IEEE1547). Here again, new controls and communications would be required to send a signal to the inverter to tell it what power factor to operate at and to "source" or "sink" reactive power.

5.6 SUBSTATION-LEVEL CHALLENGES AND SOLUTIONS

Distributed generation on buildings has effects on the utility substations serving those buildings.

1. Effects of RE on the Local Substation

An electrical substation consists of a transformer by which voltage is converted from transmission to distribution voltage and a bus with circuit breakers, current transducers, and isolating switches on each circuit coming off the bus. Configurations including transfer buses, double buses, and ring buses are available, but because they are simple and inexpensive, single bus configuration is common. This single bus may serve as a means to accept power from circuits with excess RE and deliver it back out to circuits with excess load. However, in configurations other than a single bus, sectionalizing switches or isolation switches must be positioned for this exchange of power to occur, and new legs between circuits may need to be added. Analysis of expected RE generation and loads may inform required reconfigurations to the buses and circuits. The intent of this reconfiguration would be to allow power to flow from circuits with excess RE generations to other circuits with load to serve. A substation is not designed to conduct this transfer if it was designed for power to flow only out to circuits.

2. Voltage Stability of Electric Grid

Voltage stability refers to the ability of the grid to resume nominal voltage levels after disturbances such as sudden increase in load or sudden decrease in generator output or demands for reactive power that cannot be met. Inverter-based RE generators have voltage set points, which could trip offline during abnormal voltage conditions and exacerbate a low-voltage problem. New standards for static inverters out on a grid improve voltage stability on large, complex power systems; they can currently be deployed with mutual agreement of the utility and the system owner (IEEE, 1547). New inverter standards will allow inverters to contribute to recovery from such a disturbance by "riding through" low voltage or low frequency conditions. These new inverter standards involve a nonunity power factor, low voltage ride through, and low frequency ride through. Again, a sophisticated control system would be required to control such an inverter.

5.7   UTILITIES' ELECTRICAL GENERATION SYSTEMS

1. Ramp-Rates and Spinning Reserve

Building owners installing large renewable systems, especially on small utility grids such as on islands, should realize that a kWh delivered does not generally equal a kWh saved at the generation plant. It takes time to start a generator and increase its power output, especially a big heavy coal or nuclear plant. The heat-engine portion of a generator is always operated with some extra reserve capacity to make up for variations in the electrical demand, and in a modular plant, an additional generator may run with minimal loads. This operational overhead is called "spinning reserve." Without RE generation, the amount of spinning reserve must cover anticipated changes in the load. But with the addition of RE to a system, the amount of spinning reserve must cover both sudden increases in the load and sudden decreases in RE output at the same time. Spinning reserve is undesirable since it consumes some fuel use and incurs run time on the generator machine. Much maintenance is planned based on run hours rather than power output. Several strategies have been identified to reduce, but not eliminate, the amount of waste associated with spinning reserve:

- Utilities' controls may implement sophisticated program logic that looks at the instantaneous contribution of RE and calculates the appropriate amount of spinning reserve.
- Demand side management (DSM) is another alternative to spinning reserve; it controls the electrical loads on the system rather than aiming to meet all loads at any given time. However, DSM calls for at least some participation by the energy consumers by allowing the utility system operator to delay noncritical loads such as air conditioners and water heaters for several minutes at a time.
- Forecasting the output of an RE generator minutes, hours, or days in advance also can reduce the need for spinning reserve. Technical developments in forecasting software have allowed operators of utility transmission and distribution systems to predict in advance solar system or wind output.
- Storage alternatives to spinning reserve such as batteries, flywheels, or pumped hydro storage may be more cost effective than spinning reserve.

2. Synchronization of Generation Equipment on the Electric Grid

Rotor angle stability refers to the ability of the generators in a power system to maintain synchronism after disturbances, such as a drop in frequency caused by the addition of a large load or drop in RE output due to natural fluctuation in the RE resource. Output of inverter-based RE generators can change very quickly, in turn affecting generator dynamics. With high-penetration levels of inverter-based RE, effects on synchronization of all of the equipment on an electric grid may be significant. RE project developers working with utilities can hire consultants to use commercial software packages in order to simulate transient events, such as output variations due to clouds or sudden loss of output from an RE generator. These model the details of generator machines, controls, detailed electrical circuit, and load models in order to determine whether the system will return to stability following a transient event, such as a sudden change in the output of a specific RE generator.

5.8   ALLOCATING COSTS AND INVESTMENT

1. Net Metering

Under a net metering policy, power from an RE system in excess of the local load can be absorbed by the larger utility system and credited to the load's utility bill, with the load still able to consume power from the grid when the RE is not generating. A building with enough PV might offset enough energy to bring its electric bill, which is based on consumption, to nearly zero. Yet the building is still using the utility on a daily basis even though it does not pay for that service. So we perceive net metering as an incentive for distributed generation that is offered at the expense of other ratepayers. Many utilities have net metering policies, but with limits for each site or for the system as a whole, and with the realization that there are both technical and economic reasons that not every utility customer can net meter. Some jurisdictions have already rescinded such net metering policies and added charges a building owner would pay for the service offered by the utility, even if over the course of a month or a year they produce a surplus of energy.

2. Utility Revenue Models

Under most net metering policies and tariff structures, RE reduces utility revenue that is based on consumption (c/kWh). Utilities incur costs as they implement the measures described in the previous sections and provide the ongoing spinning reserve in order to accommodate distributed RE generation. Utilities argue that RE generation on their system saves them only some portion of the fuel they would otherwise use and that all debt service, O&M expenses, and administrative costs remain the same. At the same time, the building owners that install large RE systems seek to minimize their financial payment to the utility. RE is characterized as a high capital cost but a low operating cost; the marginal cost of solar and wind power is essentially zero, compared to substantial fuel costs for the fossil alternatives. This has the effect of suppressing wholesale prices paid for fossil energy when RE is generating. This affects the current investment model for utilities and creates a problem for investors expecting a high rate of return on large coal and nuclear plants. Large fossil-fueled power plants entail billions of dollars of investment, with the prospectus that they will recover an agreed-upon rate of return by selling power at a higher price during the day than at night and a higher price in the summer than in the winter (because loads are higher at such times). Distributed generation based on solar PV carves into this revenue stream, threatening these investments with a nonviable financial prospectus. Utility rate changes that may be expected in the future include:

- More fixed charges on utility bills that are paid regardless of energy consumption
- Higher demand charges based on peak utilization (kW) rather than energy consumption
- Stand by charges that are paid on a per kW basis even if the demand does not occur
- Charges for itemized services and system upgrades provided by utilities

There is a necessity for a new business model that rewards low-carbon energy production, covers the cost of reliability, equitably assigns costs, and involves investors directly investing in required grid upgrades. Rather than recovering their costs based on how much energy you consume, utilities of the future might charge for the services, such as reliability of power and power quality, that you actually do use even when you generate your own on-site RE. These changes to the operation generally reduce the efficiency and cost-effectiveness of the conventional resources, although many advocates of RE point out that distributed RE might forestall the need for huge investment in central generation.

3. Integrated Resource Planning

Although primarily a utility issue, energy consumers and RE technology providers participate in "integrated resource planning" with other stakeholders to determine what resources should be deployed to meet the system load. These resources are "dispatched" according to their cost effectiveness and availability. For instance, power plant operators can optimize fuel efficiency by loading generators near their maximum capacity with other generators following the demand and variability of the intermittent RE generation. Integrated resource planning occurs in the hearing rooms of public utility commissions, or other regulatory bodies, in each state. Clean energy advocates must participate in those proceedings to ensure that their interests are represented.

5.9 ENERGY STORAGE

Many of the challenges in this chapter can be solved with energy storage. Energy storage may be deployed at any scale in the system: from individual buildings to problematic feeders at substations to the generation plant (for example pumped hydro storage is on this scale). Battery technology is rapidly evolving in response to this need for storage. New material systems offer improvement in toxic and hazardous materials, energy density, cycle life, and cost effectiveness. Economy of scale in the energy storage market also promises lower cost. However, batteries remain an expensive and maintenance-intensive component of a system, and there are strategies to reduce the amount of battery storage required including:

- *Demand control*: turning off an interruptible load, such as a 4 kW water heater, for 15 minutes and then turning it on when generation resources are sufficient has the same effect as 1 kWh of battery storage at a small fraction of the cost.
- *Forecasting*: advance knowledge of when the load will demand energy and when the resources will be available can optimize operation of storage and thus minimize the amount of storage required.
- *Nonelectric storage*: many buildings have large year-round cooling loads, such as data centers. Since cooling is provided by electricity, storage of ice or chilled water can have the same effect as electrical storage (freeze ice when solar PV is generating, melt it when cooling is needed) at a fraction of the cost of battery storage. Similarly, heating hot water tanks or subcooling freezer warehouses would be an effective way to store energy. All of these nonelectric storage alternatives should be considered in order to minimize the amount of battery storage required.

5.10 FORECASTING

Forecasting of both loads and resources has been indicated as a solution to some of the problems described above (Walker, 2013).

- *Conventional weather forecasts*: Forecasting of building loads can be accomplished with building energy models informed by future weather forecasts related to temperature, humidity, and solar heat gains. Models of RE generation based on solar radiation, temperature, and wind speed can use conventional weather forecasts to predict output. However, such forecasts are available hours and days in advance when often control must be accomplished on a scale of minutes, leading to a need for sky observations and cloud-formation models.
- *Sky observations*: Sky observations use a digital camera to observe the sky over a PV system. Image processing and software assign a vector to each cloud edge with a speed and direction that can be used to estimate when a cloud will cover the PV system minutes in advance.
- *Atmospheric models*: Models of the thermodynamics and transport phenomenon in the atmosphere can also be used with other information to predict the formation of clouds in an area and provide advance notice required to dispatch conventional generation resources.

5.11 SMART GRID

The term "smart grid" is used to describe the combination of advanced controls and hardware indicated in many of the solutions above. Smart grid principles can allow greater grid penetration of distributed generation, reduced losses in transmission and distribution, effective demand-side management, utilization of forecasts for loads and variable resources, and hierarchical control for grid security and reliability. Four features of smart grids are integration, control, communication, and metering. Integration refers to the connection of heterogeneous types of energy sources with the grid using appropriate converters. Controls in smart grids are made intelligent to extract the maximum power from distributed and variable RE resources, operational scheduling of energy sources, avoidance of overloads, control of transients, and provision of both real and reactive power. For effective operation of the diverse smart grid, communication between various control nodes is necessary. Communication standards for smart grids usually are set by protocols, and most of them involve the interconnection of a secure communication line to the main control unit by the local area network and wide area network. The interconnection should be accompanied by a firewall at various levels for the cybersecurity of the smart grid. Smart meters are also employed in smart grids to provide more information than just cumulative production, such as voltage needed to inform voltage regulation controls described above. Demand control hardware is also employed in smart grids. All the devices measure the energy parameters of the load remotely, transfer the data through the communication network, and return vital control signals to the devices so that the grid can simultaneously satisfy requirements related to the amount of power, power quality, and reliability (Reddy et al., 2014).

5.12   SUMMARY

Let's conclude with a recap of strategies to increase the amount of intermittent RE generation that can be added to the energy system in buildings and cities.

- Utility circuits can be reconfigured to allow power to get from a generator to a load, including locations and settings of network protectors, voltage regulation, and overcurrent protection devices.
- If PV is distributed over a large area rather than installed all in one place, there is less chance that a cloud could affect the output and the generation is smoothed out over a partly cloudy day.
- Fixed mounts facing south would all deliver their maximum power at noon, but by using tracking mounts and different PV orientations we can smooth out the daily delivery: less at noon but more in morning and evening hours.
- The new IEEE 1547a standard will allow for more advanced inverter features. These include low voltage ride-through, low frequency ride-through, and power-factor correction. Another capability is the ability of an inverter to curtail RE output in response to a signal from the utility or grid conditions.
- Forecasting of RE plant output would provide a utility system operator with enough advance notice to ramp up a fossil-fueled generator rather than maintain such a large amount of spinning reserve. Even a few minutes' warning could help out tremendously in this regard.
- Controlling the instantaneous demand of noncritical loads is also a good idea: it is less wasteful to turn off an appliance for the 15 minutes it takes to ramp up a generator than it is to keep that generator running as spinning reserve.
- Energy storage is the ultimate solution to almost all of these problems, and battery technology and cost are improving all the time in response to this future need for a large amount of energy storage.
- Changes in how utilities collect revenue from purely commodity pricing based on how much energy (kWh) you consume toward charging for services that support your RE generator and your load, such as fixed charges and stand by charges that charge for the ability to serve your load should your RE generator fail to do so.

This chapter has described some of the most important hardware and operational measures that can be taken to resolve common challenges and achieve a higher level of clean energy and distributed generation in buildings and cities. Meeting our clean energy goals is possible but requires collaborating early with utilities and other stakeholders in the deployment of systems and in the regulatory process steering the evolution of the overall energy system.

REFERENCES

Behnke, M., W. Erdman, S. Horgan, D. Dawson, W. Feero, F. Soudi, D. Smith, C. Whitaker, B. Kroposki. 2005. Secondary network distribution systems background and issues related to the interconnection of distributed resources, Report Number NREL/TP-560-38079, July 2005.

IEEE 2000, Standard C57.12.44-2000, IEEE Standard Requirements for Network Protectors.

IEEE 2003, IEEE Standard 1547-2003, IEEE Standard for Interconnecting Distributed Resources with Electric Power Systems.

Jordan, N. 2014, Integration of network protector relays on downtown distribution networks with penetration of renewable energy, Thesis, Louisiana State University and Agricultural and Mechanical College, May 2014.

Meinel, A. B. and M. P. Meinel. 1976. *Applied Solar Energy: An Introduction,* 2nd ed., Addison-Wesley Publishing Co., Reading, MA, ISBN-13: 978-0201047196.

Reddy, K. S., M. Kumar, T. K. Mallick, H. Sharon, and S. Lokeswaran. 2014. A review of Integration, Control, Communication and Metering (ICCM) of renewable energy based smart grid. *Renewable and Sustainable Energy Reviews* 38 (October): 180–92.

REN21. 2016. Renewable Energy Policy Network for the 21st century (REN21) Renewables 2016 Global Status Report. www.ren21.net/gsr.

Sherpa. 2016. Estimated total annual building energy consumption at the block and lot level for NYC. http://qsel.columbia.edu/nycenergy/index.html, Accessed 22 April, 2017.

Walker, A. 2013. *Solar Energy: Technologies and Project Delivery for Buildings*, John Wiley & Sons, Inc., Hoboken, NJ, September 16, 2013; ISBN 978-1-118-13924-0. 298 p.

Walker, A., J. Scheib, C. Turchi, R. Robichaud, G. Tomberlin, K. Burman, M. Hillesheim, B. Kroposki, and M. Qu. 2016. *Integration of Renewable Energy Systems* ASME Technologies for Sustainable Life-Concise Monograph Series, ASME Press, New York, ISBN 9780791861240.

6 Next Developments in Solar Thermal Technologies

Valeriano Ruiz-Hernández
Advanced Technology Centre for Renewable Energies (CTAER)

CONTENTS

6.1 INTRODUCTION: THE GLOBAL ENERGY SYSTEM

The global energy system is changing in decisive ways:

- Primary energy consumption has more than doubled in the last 41 years (1973–2014). It has gone from 6.1 to 13.7 Gtep (573.6 EJ) with data from the International Energy Agency (IEA) current methodology "based on efficiency" (IEA, 2017).
- Oil remains the predominant primary energy source, although its relative importance decreased from 46.2% in 1973 to 31.3% in 2014. Coal is second, but it has increased from 24.5% in 1973 to 28.6% in 2014, with the corresponding negative impact on climate change. Natural gas has also increased its share of primary energies, and biofuels (mainly firewood) maintain their percentage at around 10%. All these data assure us that human beings continue to increase greenhouse gas emissions, as shown by the increase in radiative forcing ($1.6\,W/m^2$ in 2005 and $2.3\,W/m^2$ in 2011), another indicator of the environmentally regressive process in which humanity is committing suicide.
- Total electricity generation rose from 6131 TWh in 1973 to 23,816 TWh in 2014; that is to say, it multiplied almost fourfold in 41 years; this fact shows the electrifying tendency of the system. Coal remains the main source of electricity generation, followed by natural gas, hydro, and nuclear. The "curious" point is that the electricity of hydraulic origin (3983 TWh in 2014) is greater than that generated by nuclear fission (2533 TWh in 2014), but when transferred to primary energy, their ranks are reversed: hydroelectric is valued at 2.4% and nuclear at 4.8%. Mysteries of IEA statistics!

For the purposes of this chapter, it is interesting to note that "other" sources (wind, solar, geothermal, biomass) in IEA electricity data represented 6.3% (1500.4 TWh) in 2014 as compared to 0.6% in 1973 (36.8 TWh); that is to say, in 41 years it grew by a factor of 41.

As can be seen, we are walking, albeit slowly, in the correct path of change in the energy system (especially the electrical subsystem) toward a greater contribution by renewable energies. Let us now see, in this brief review of the world energy situation, the evolution of renewable energy in recent years.

According to the last REN21 Global Status Report released in June 2017, the total capacity for electricity generation from renewable sources was 2017 GW in 2016, of which the bulk (1096 GW) was hydroelectric; followed by wind power with 487 GW; photovoltaic (PV) solar power with 303 GW; bioenergy, 112 GW; geothermal, 13.5 GW; and solar thermal (concentrating solar), with 4.8 GW, of which—to date—2.3 GW are installed in Spain, (Figure 6.1). On June 8, 2017, at 11:40 am this technology was generating 2.191 GW in Spain, according to the Spanish Electric Grid System of Information (REE, 2017).

For comparative purposes, the total installed electrical capacity in the world is in the order of 5000 GW; so, renewable electricity in terms of installed power represents around 40%. The solar heating of water grew in 2016 to 456 GW of heating capacity, with 651 million square meters of installed thermal solar collectors. Biofuels (bioethanol and biodiesel) produced in 2016 were 129.4 billion liters.

It is worth expounding the energy consumed by the hundreds of millions of people who do not have modern energy (electricity and fossil fuels). We are referring mainly to wood for cooking. According to data from the IEA, it is about 10% of the world's primary energy consumption, in the order of 1370 Mtoe, not accounted for by several agencies. As shown in Figure 6.2, biomass remains an essential energy form for many human beings.

In any case, what is worth highlighting is the speed with which the incorporation of renewable energies in the world energy system is taking place.

The solar thermal began to soar in 1985 with 14 MW; in 1991, there were 350 MW, and in 2015 we had 4800 MW, maintaining the growth rate in a practically continuous way in recent years.

PVs also grow at a sustained pace: from 5.1 GW in 2005 to 303 GW in 2016, with an annual growth of around 30 GW/year in the last 5 years. Wind power also has exceptional increases, from 59 GW in 2005 to more than 500 GW in the first quarter of 2017, with a growth rate of 39 GW/year in the last 5 years. Solar water heating also shows very significant annual growth with 40 GW_{th} added in 2015.

In short, renewable energies are being introduced at an accelerated pace in recent years, which is a hopeful sign of a more reasonable energy system in the future. However, the effort is not sufficient if the use of coal, petroleum products, and natural gas is not restricted. Nuclear energy is insignificant from the point of view of the amount of energy contributed to the system, although there are those who insist the opposite. The important thing about nuclear energy is that there are few mineral resources, and they are exhaustible; so, the future of the system cannot be based on that energy source, in addition to the problem of radioactive waste that we leave in inheritance to our descendants.

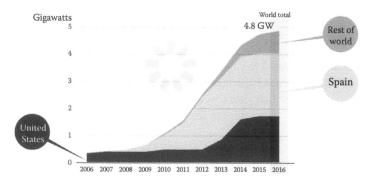

FIGURE 6.1 Concentrating solar thermal power global capacity, by country and region, according to REN 21. (From REN21, *Renewables 2017 Global Status Report*, 2017, Figure 20.)

FIGURE 6.2 Firewood on rooftops in Nepal. (Photo by Donald Macauley—Flickr: IMG_4133, CC BY-SA 2.0, https://commons.wikimedia.org/w/index.php?curid=15290825.)

6.2 PERSPECTIVES OF SOLAR TECHNOLOGIES

Solar radiation is the most abundant, safe, and clean energy source available to meet our needs. Apart from that, it is the origin of almost all of the other energy sources. Specifically, it activates photosynthesis, the origin of life and of biomass, an abundant energy resource that will play a very important role in the future energy system, especially in hybridization with solar energy.

Solar radiation is transformed into thermal energy when it is absorbed by any material and in electricity when the material has special characteristics that are achieved through highly specialized technological treatment. We give a review of the two forms of transformation: thermal and PV. The corresponding devices are very diverse and quite complex. We only give a brief review of thermal energy of medium and high temperature attained by systems of concentration of the direct component of solar radiation.

6.2.1 PV TECHNOLOGY

PV technologies allow for the direct transformation of solar radiation into electricity. Their basic devices are photovoltaic cells of which there is a great variety with very different characteristics. These cells are connected in series and, in turn, in parallel to give the voltage and intensity required for each application. The most common cells are based on crystalline silicon (mono and poly); they cover more than 90% of the market, equally divided between the two types. The rest are thin film technologies and the emerging perovskites-based PV cells. It is important to note that these technologies have a brilliant and wide future, given their adaptability to consumption due to their great modularity. To date, there are large facilities with a large number of modules up to several MWs, increasing every day. There are already 500 MW installations, and others that are even bigger are being built as we speak.

It is in the small- and medium-size facilities that the most important market niche is found because the future electrical system must be distributed, with the generation of electricity at the consumption point. A very specific and desirable case is that each consumer's home generates a significant part of the electricity needed. This modality is becoming fashionable in Europe with legislation that facilitates this option; consumers are able to share generated electricity with the neighbors.

PV technologies can improve the level and quality of life of many human beings around the planet. In addition to isolated installations in rural homes, the so-called micro grids in small human groups are making significant progress in many parts of the world.

6.2.2 Solar Thermal Power Technologies

In these technologies the beam component of solar radiation is transformed into medium and high temperature thermal energy (above 100°C) that can be used directly in thermal processes in diverse applications: thermal processes in agrifood industries, air conditioning, solar cooling and heating, and, from thermal energy obtained at the highest possible temperature, electric power can be obtained through the appropriate thermodynamic cycles (Rankine and Brayton in particular). In the latter case—electric power generation—there is a very interesting option: utilizing thermal storage that allows electricity generation at times when there is insufficient solar radiation, as shown below. Solar thermal facilities are increasingly hybridized with another energy source that produces thermal energy. The incorporation of the two possibilities (storage and hybridization) allows for complete manageability of solar thermal installations. Figure 6.3 shows the Gemasolar CSP installed in Andalucia, Spain. It was the first commercial-scale plant in the world to apply central tower receiver technology and thermal storage in molten salts. The relevance of this plant resides in its technological singularity, since it paved the way to new thermosolar electric generation technologies. With a nominal capacity of 19.9 MW, it has 2650 heliostats and a storage tank of hot salts that allows an autonomy of electric generation of up to 15 h without solar input.

Since many publications (including some of my own) already exist, describing the four main typologies (parabolic channel, Fresnel, central tower receiver, and parabolic disk) of photo thermal concentration technologies, and many commercial installations of around 5 GW exist, I shall limit myself to a review of the main challenges facing the more or less immediate future of each. Now is the time for technology leaps in solar thermal hybrids to catch up with the already very competitive

FIGURE 6.3 Gemasolar solar thermal concentration plant in Spain. (Credit: Torresol Energy, Wikimedia Commons.)

PV technologies. Investors rightly see in these technologies the future of the energy system whose keys are collaboration with technology centers and the race for competitive advantages among companies.

6.3 R+D+I ISSUES OF SOLAR THERMAL TECHNOLOGIES

After about 40 years working on these topics, I still find it interesting to learn about recent advances in thermal solar research, development, and innovation (R+D+i). First, designs are improving in relation to reflecting surfaces and tracking of the Sun's trajectory. One of the main limits to the optical efficiencies of the systems is the cosine-factor that strongly limits the concentrated radiation reaching the corresponding receivers, especially when the reflectors are fixed to the ground or have to be adjusted to horizontal surfaces. One way to reduce these losses is to move the device tracking the Sun. The case of the parabolic disks is paradigmatic because they are designed so that the direct incident radiation is always normal to the plane of opening of the device so that the cosine-factor is always 1. But in the other typologies, this is not the case, so there is an interesting possibility to improve the optical behavior of the devices. In experimental developments of "variable geometry" in the desert of Tabernas, Andalucia, we have involved both the tower and the parabolic channels (CTAER, 2017) (Figure 6.4).

Another issue is the durability of the reflective characteristics of mirrors, which is dependent on their quality. It is an aspect of the technologies that is not yet sufficiently taken into account and has a great influence on the profitability of the facilities. The interception factor of all typologies is also important. Most of the time, the percentage of the reflected radiation that "enters" the receiver is unknown. I do not think it is less than 10%–15%. Correcting this effect also greatly influences profitability since the performance of the installations is directly related to those avoidable optical losses with a good design of the device. The most obvious solution is a second stage of concentration coupled with the receiver.

The receiver also requires some attention. If the reflected radiation intercepted by a plane perpendicular to it is a circle, then the receivers in the tower centers need to have a circular entry section, not a square one as most often done up till now! Thus, we have designed a hybrid solar-fuel gaseous receiver in the CTAER that is to be constructed and tested. We are convinced that it will outperform those currently in operation.

Other aspects of importance—in the thermal subsystem—are the working and storage fluids.

For the fluids involved in the transformation of radiation into thermal energy, transport to the storage, and devices transforming thermal energy into mechanical energy, it is necessary to keep in mind not only the thermodynamic and mechanical aspects of fluids, but especially the environmental ones. In particular, a problem that may become fundamental is accident prevention, especially

FIGURE 6.4 R & D facilities in thermosolar technologies of variable geometry (a) tower and (b) experimental heliostat (From CTAER, *Solar Research Areas in the CTAER,* in Spanish, https://ctaer.com/es/areas-de-investigacion/solar, 2017.)

seismic breakage of the absorber tubes of the large parabolic trough plants, which could spill large quantities of mineral oils onto ground and groundwater.

As a matter of fact, there are already spills due to faulty connections between modules. In that respect, the tower units have lesser problems because the working fluid is more constrained to the surroundings of the tower, and the amount of fluid involved in the process is much smaller. Plants based on the Fresnel typology working at a lower temperature use water at medium pressure (20 bar at 200°C), so that the possible environmental impact is practically nonexistent.

At very high temperatures, a gas—even air—can be used as the working fluid, so there is no problem, or it is minor. Regarding storage, the main issue to be resolved is the volume of the corresponding container in direct relation to the specific volume and the physical state of the substance being used. For the moment, when using sensible heat and temperatures in the surroundings of 400°C, the volume is elevated. It is in the order of 16,000 m^3 for a 50 MW plant with storage of 7.5 h, which has become almost standard in Spanish plants. In that case, the fluid used is molten (liquid phase) salts. A competitive solution has to be found in phase-change materials or in reversible chemical reactions.

In terms of efficiency, direct thermal applications (e.g., industrial processes or air conditioning), rather than electricity generation, will have maximum yield and, therefore, good economic profitability. This is an ideal field of application for the Fresnel and parabolic channel typologies. If electricity is to be obtained, the highest possible temperatures have to be sought, in order to have maximum thermodynamic performance. In that sense, the typologies of concentration at one point have an advantage. In fact, large tower centers are an advantage. But small-scale, rooftop, parabolic disks should be developed with a view to a specific market niche.

Another issue of considerable importance in power plants is the necessary cooling of the power cycles, especially considering that the best way to get rid of residual heat is with water and that this is usually scarce in areas where there are higher levels of direct radiation. If we add that the mirrors need to be cleaned from time to time to maintain adequate reflectivity, decreasing water needs is another transcendent challenge.

In a nutshell, there are lots of research, development, and innovation challenges, and interested solar companies have to boost their involvement in these processes if they want to win the competitive battles that are being presented.

6.4　CONCLUSION

To conclude this brief update of energy problems affecting the upcoming sustainable energy systems, my conclusion is that solar resource growth at the current pace has to be achieved in a sustainable energy system without major environmental problems. Solar energy is by far the largest source of clean energy, and we must use it wisely so as not to damage the environment while covering all our present and future energy needs.

REFERENCES

CTAER. 2017. *Solar Research Areas in the CTAER* (in Spanish). https://ctaer.com/es/areas-de-investigacion/solar.
IEA. 2017. *World Energy Balance 2014*. https://www.iea.org/Sankey/#?c=World&s=Balance.
REN21. 2017. *Renewables 2017 Global Status Report*. Paris: REN21 Secretariat.
Red eléctrica de España. 2017. *Real-Time Demand of Electricity, Generation Structure, and CO$_2$ Emissions* (in Spanish). https://demanda.ree.es/demanda.html.

7 Advances in Solar-Assisted Air Conditioning Systems for Tropical-Humid Locations

Jorge E. González
City College of New York

CONTENTS

7.1 INTRODUCTION

Solar-assisted air conditioning systems have been proven economically feasible. This chapter reviews analysis and design strategies for solar-assisted air conditioning systems in hot and humid climates. The general strategy consists of presenting three different options: (1) an absorption unit coupled with an array of flat-plate collectors, (2) an absorption unit coupled with an array of flat-plate collectors and a liquid desiccant system, and (3) an absorption unit coupled with an array of flat-plate collectors and a solid desiccant system. Specifically:

A. Independent mathematical models describing the thermal performance of absorption, desiccant, and solar collectors systems will be presented. Combinations of the optimized independent models will be introduced to find the optimum hybrid system for different geographical regions and its optimal operational conditions.
B. Two types of flat-plate collectors were considered: conventional flat-plate collectors and evacuated tubes. Solar collectors are the major cost and system component of the proposed system. Under normal conditions, higher temperatures can be reached with evacuated tubes. However, their performance has not been verified in low latitudes. Therefore, results from an experiment to determine the performance of both types of flat-plate collectors under similar conditions are presented.
C. The cost effectiveness of the systems considered was fully explored using a *life cycle cost* analysis for different cities across the Caribbean. Variables considered in the economic analysis included interest and inflation rates, cost of local electricity, initial cost of components, maintenance and operational costs, tax incentives, and environmental impact costs.

7.2 BACKGROUND ON SOLAR-ASSISTED AIR CONDITIONING SYSTEMS

The objective of this section of the chapter is to provide the reader with enough technical and informational background to allow him/her to become familiar with the technologies considered in this chapter. The section will also summarize the operation of closed absorption machines and liquid and solid desiccants, when coupled with arrays of solar collectors. A brief historical background on the development of solar-assisted air conditioning and dehumidification technologies is also provided.

7.2.1 Principles of Absorption Refrigeration Machines

Cooling buildings with solar energy is an attractive idea, as solar radiation is usually in phase with the cooling load of a building. A closed absorption machine coupled with an array of solar collectors has been the most widely used application. The closed absorption system is a heat-operated unit using a refrigerant that is alternatively absorbed and liberated at certain temperatures and pressures. The main components of this system are absorber, generator, condenser, and evaporator. Figure 7.1 shows the principal features of these systems. The basic operation is as follows.

In the evaporator, the refrigerant is evaporated under low pressures as it is sprayed on the tubes through which chilled water flows. The chilled water is then delivered to the heat exchanger coil; however, the air is cooled and discharged as needed to the room space for air conditioning. In the absorber, the concentrated solution absorbs the vaporized refrigerant from the evaporator at the same low pressure. As the solution is sprayed on the cooling tubes within the absorber container, it is cooled down and the pressure is lowered to be capable of absorbing more vapor. Any absorption heat produced is suppressed by cooling water. The absorbent solution is diluted due to the absorption of the vaporized refrigerant, being reduced to a weak solution with less capacity of absorbing water vapor. In the regenerator, the diluted solution pumped from the absorber is heated and concentrated by dispelling some refrigerant vapor, so that the solution again is capable of absorbing refrigerant

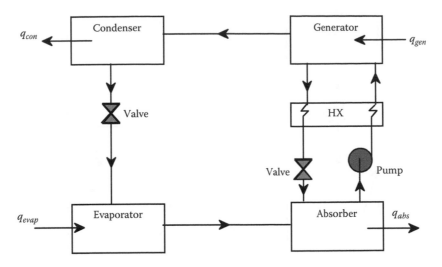

FIGURE 7.1   Closed absorption single-effect cooling cycle.

vapor. This separation occurs because the concentration of the refrigerant in the absorber decreases with temperature and is called regeneration. The heat must be given from outside and may come from gas, oil, solar energy, or combinations of the three. The heat source will depend on the boiling temperature of the refrigerant in the weak solution. The strong solution at the high temperature is cooled though a heat exchanger with the weak solution from the absorber before being brought into the absorber. This process serves to preheat the weak solution, which will be eventually brought to much higher temperatures at the regenerator. In the condenser, the high temperature vapor produced in the regenerator is condensed on the surface of the cooling tubes through the cooling tower water flow. The liquid refrigerant then enters into the evaporator by a throttling process, being evaporated to take heat out of the chilled water. The cycle is thus completed.

The absorbent solution is usually a salt with strong power to absorb moisture. The desirable properties for the refrigerant are high heat of vaporization, low specific heat capacity, and good thermal stability, while for the absorbent, chemical stability, high boiling point, and low heat capacity are advantageous. The most common refrigerant/absorbent pairs are water/lithium-bromide (H_2O/LiBr) and ammonia/water (NH_3/H_2O).

The performance of closed-absorption machines is usually expressed by the coefficient of performance (*COP*). The *COP* of an absorption chiller is defined as the ratio of the room cooling effect (Q_s) to the input energy to the regenerator (Q_g):

$$COP = \frac{Q_s}{Q_g} \tag{7.1}$$

Typical values of *COPs* for commercially available absorption chillers range between 0.5 and 1.5. Commercial units are available for the H_2O/LiBr absorbent/refrigerant pair. Companies such as Carrier Corporation, U.S.A.; Arkla, U.S.A.; and Yazaki, Japan; build chillers operated by gas burners and have modified these units to be operated by hot water from solar collectors. Units of NH_3/H_2O are not commonly found for domestic applications due to the toxicity of NH_3. Arkla-Servel and Bryant, U.S.A. have marketed aqua-ammonia air conditioning units that can be used outside of commercial or industrial buildings.

7.2.2   Double-Effect Absorption Machines

Since the advent of closed absorption machines more than 30 years ago, improvements in the simple absorption cycle have taken several forms, the most significant being the incorporation of multiple

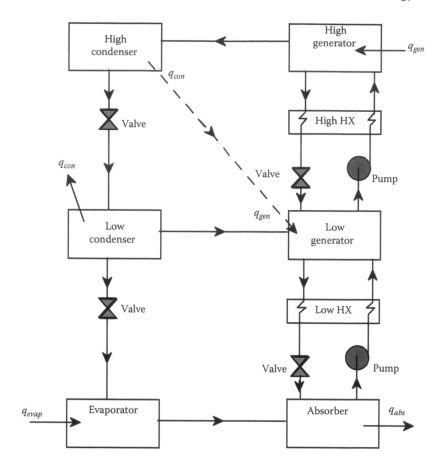

FIGURE 7.2 Double-effect closed absorption cooling cycle.

effects generators. Some years ago an absorption chiller that operated with two regenerators was proposed. In this chiller, most of the refrigerant is desorbed in a high-temperature regenerator. This high-temperature refrigerant exchanges heat with the weak solution at a lower pressure, and the weak solution is further regenerated. Figure 7.2 shows a typical two-stage absorption chiller.

7.2.3 COMPARISON BETWEEN ABSORPTION MACHINES AND VAPOR-COMPRESSION (VC) SYSTEMS

Most air conditioning systems operate with the conventional (*VC*) cycle. The VC cycle is shown in Figure 7.3; it consists of four basic processes, described as follows. The cold gas/liquid of refrigerant absorbs heat from the room at the evaporator and changes the phase to full vapor. The saturated vapor is then compressed to a higher pressure in the compressor, increasing its temperature. The vapor refrigerant at high temperature and pressure is then condensed by exchanging heat with cold water and, later, is expanded through a throttling process, thus completing the cycle. The major difference between the VC cycle and the absorption cycle relies on the heat input required to drive the systems. The VC cycle uses work to drive the compressor, while the absorption cycle uses heat at the regenerator. Work is a high quality form of energy and is not easily available. Most VC cycles use electricity to drive the systems, which is a form of work. On the other hand, heat is a low quality form of energy and easily available. Heat can be obtained as a waste of primary processes, from cogenerations processes, from geothermal sources, from directly burning fossil oils, or from solar

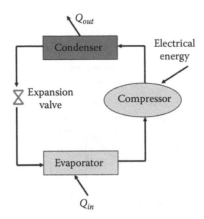

FIGURE 7.3 Vapor compression cycle.

radiation. Note that heat was used once to generate electricity and is now converted back into heat at the VC machine. Therefore, it is expected that electricity will carry deficiencies in conversion efficiency and thus higher costs. It is not surprising, thus, that *COPs* of VC machines are generally higher than those of absorption systems, typically three times higher. This indicates that more heat can be removed from the room per unit of energy input. However, the cost of the heat used by absorption machines is far lower than the cost of the work used in VC systems, and in many cases the heat can be obtained free as is the case in solar applications.

7.2.4 DESICCANT SYSTEMS

In tropical climates, annual average humidity is relatively high, and dehumidification is necessary in many cases in order to achieve room comfort. Dehumidification can be achieved in several forms. Excessive temperature reductions in refrigeration systems will lead to dehumidification by condensation if the temperature is below the dew point of the moist air. Air can also be dehumidified, however, without cooling by use of hygroscopic materials that absorb or adsorb part of the water present in humid air. These hygroscopic materials are known as desiccants and may be liquids or solids.

If used for room comfort, the dehumidification process must be associated with an air-cooling process. The dehumidification in desiccant systems is usually associated with heat release and thus temperature increase of the air. The cooling of the air can be achieved by means of a VC or absorption system. Cooling of the air can also be achieved by addition of moisture by means of heat and mass exchange between air and atomized water or evaporative cooling. The use of desiccant systems for both dehumidification and cooling has been suggested for use with solar energy. However, the feasibility of this proposal is still pending. We therefore will focus the discussion on desiccant systems for dehumidification purposes only in this report.

The basic operation of a solid desiccant is quite simple. The moist air is brought into contact with the solid drying material, which absorbs water. The air experiences a temperature increase during the absorption process passing thus to a sensible cooling process. The solid desiccant material must be regenerated by means of heating to desorb the moisture gained. The necessary heat for the regeneration may come from several sources including solar energy. Figure 7.4 shows the basic components of a rotary-bed solid desiccant system. The desiccant material can be a silica gel and zeolite (molecular sieve.) The stream of air to be dehumidified is supplied to one of the circular faces of the cylinder and withdrawn from the other face after passing through the desiccant. Moisture is adsorbed in the silica gel, or any other desiccant, and the air is heated due to the heat of adsorption. In order for the process to be continuous, the cylinder slowly rotates (one to three revolutions per minute), so that the desiccant material that has adsorbed moisture moves into another stream of air

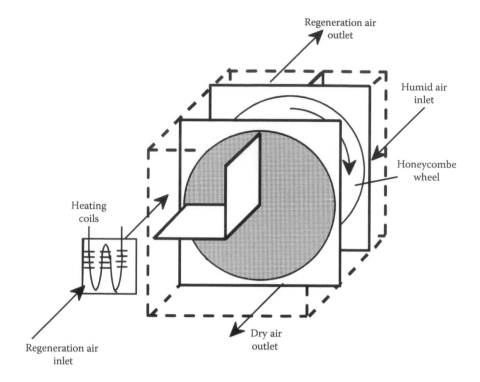

FIGURE 7.4   Solid-desiccant wheel.

that has been previously heated. As the hot air stream passes through the desiccant solution, moisture is evaporated from it into the air and is discarded to the atmosphere. The dry desiccant slowly rotates again into the stream of air being dehumidified, and the process continues. Typical regeneration solid desiccant temperatures are near 150°C. This temperature is difficult to achieve with solar panels only and may require additional heating from gas or electricity to operate solid desiccants with solar energy. Companies such as Desi/Air, Bry-Air, and Cargoaire, U.S.A. manufacture solid desiccant wheels for industrial applications.

Dehumidification can also be achieved by absorbing water vapor by a hygroscopic liquid. The important requirements of the liquid are its miscibility (i.e., capacity for dilution) with water and a water vapor pressure in the solution significantly lower than the partial pressure of water in moist air. The boiling point of the liquid must be much higher than that of water in order to be recycled and reused. Liquids that have been commercially used as liquid desiccants include calcium chloride, lithium chloride, and triethylene glycol. Kathabar, U.S.A. manufactures a line of liquid desiccant dehumidifiers for industrial applications. Figure 7.5 shows the basic processes of a liquid desiccant dehumidifier system. Room air is brought in contact with the warm liquid desiccant. The air goes through a sensible or evaporative cooling process, while the dilute solution of liquid desiccant goes to the regenerator, which exchanges heat to desorb the moisture gained, which is then released into the environment. A heat exchanger is placed between the absorption and the desorption towers to preheat the liquid desiccant prior to the regeneration. Typical regeneration temperatures for liquid desiccant range from 60°C to 80°C, which are suitable for solar heating applications.

7.2.5   OPERATING EXPERIENCES OF SOLAR-ASSISTED AIR CONDITIONING SYSTEMS

Solar-assisted air conditioning systems have received wide attention for more than 40 years. Commercial demonstration projects 25 years old can be encountered in several countries, including

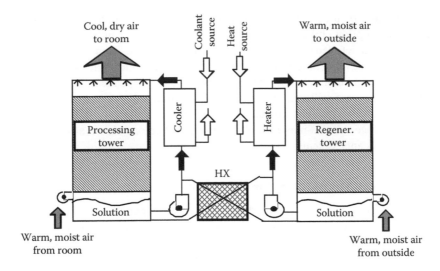

FIGURE 7.5   Basic liquid-desiccant machine.

the former Soviet Union, Japan, Australia, India, and the United States. Kakabaev et al. [2] installed two open absorption systems in the former Soviet Union during the mid-1970s; they produced 7 and 180 kW of cooling, and they were driven by 60 and 180 m² of collector area, respectively. Both systems used lithium-chloride and water as working substances. Other demonstration projects with open absorption systems during the 1970s included a solar powered liquid desiccant in India [3] and a glycol-desiccant system in South Africa used to cool a large laboratory [4]. More recently, the Colorado State University (CSU) Experimental House III used a 10.5 kW closed absorption system to cool this two-story house with a floor area of about 130 m² [5]. The chiller used LiBr/H₂O as the absorbent/refrigerant pair and required about 5 m² of collector area per kW of cooling load. Measured solar fractions for the CSU Solar House were below 30%. In 1987, Bong et al. [6] installed and operated a LiBr/H₂O absorption chiller of 7 kW of cooling capacity in Singapore. The system was driven by 64 m² of heat-pipe evacuated collectors, rather than conventional solar collectors, and a solar fraction of about 40% was reported.

A solar-assisted absorption system using lithium-bromide/water as the absorber and refrigerant fluids, respectively, was installed in the City of Mayagüez, Puerto Rico, during the early 1980s [7]. Parabolic trough concentrators were used on that system. The high diffuse insolation available in Mayagüez [8] led to low concentrator performance, which, along with operational difficulties, contributed to a subsequent shutdown of the system. However, the long-term operation of a solar-assisted air conditioning system using flat-plate collectors has been achieved with great success elsewhere [9]. That system has been operating in Sacramento, California, for more than 20 years providing a full solar cooling load of 49 kW to a commercial building during the summer months. The Sacramento system uses a ratio of 4.5 m² per kW of cooling load and has reported a payback period of less than 8 years. A pilot solar-assisted air cooling and dehumidification system was tested in Cabo Rojo, Puerto Rico. The system consisted of a 35-kW LiBr-H₂O absorption chiller, a 113 m² array of selective surface flat-plate collectors, a 5700-L hot water storage tank, an 84-kW cooling tower, a 1800-L per hour air handling unit, and a 50-kW auxiliary heater [10].

7.3   MATHEMATICAL MODEL OF SOLAR-ASSISTED AIR CONDITIONING SYSTEMS

The objective of this section is to describe the thermal performance of solar-assisted air conditioning and dehumidification systems for possible operation in the Caribbean. The strategy used in describing these systems consists of formulating mathematical models of the independent components, i.e., solar collectors, absorption chillers, and desiccant systems. The formulation of the mathematical models was based on a detailed analysis of each system component. Then, the resulting set of equations was arranged to develop time-efficient modular computer codes. Each component model was verified against experimental data whenever possible. Once independent models were described, combinations of these were explored in order to find the optimum arrangement for a given climate and application. The following subsections are divided in the mathematical description for the heat and mass transfer processes that may take place in solar collectors, closed absorption chillers, and liquid and solid desiccants systems.

7.3.1 Mathematical Model for Solar Collectors

Due to the highly diffuse nature of solar radiation in the Caribbean [8], beam radiation concentrating collectors are a good choice for providing energy to the absorption system. Flat plates are best suited for that purpose because they make use of both the diffuse and beam components of solar radiation and represent a low cost alternative to accumulate solar energy.

The first step in the modeling of a flat-plate collector array was based on the estimation of hourly tilted radiation, including both its beam and diffuse components, for a typical day in the month. The Perez model [11] was found to be a good estimator of solar energy for tropical areas like the one prevailing in the Caribbean, which has a high diffuse component of radiation. Therefore, the monthly average hourly tilted radiation was given by:

$$\bar{I}_T = \bar{I}_b \bar{R}_b + \bar{I}_d(1-F_1)\left(\frac{1+\cos\beta}{2}\right) + \bar{I}_d F_1 \frac{a}{b} + \bar{I}_d F_2 \sin\beta + \bar{I}\rho_g\left(\frac{1-\cos\beta}{2}\right) \quad (7.2)$$

where the terms F_1, F_2, a, and b, are functions of the radiation angles of incidence and other parameters that describe the sky conditions [11]. The remaining variables consider the geometric relationships between the receiving surface and the incoming solar radiation. The ratio of hourly diffuse radiation to daily diffuse radiation was obtained from [12]:

$$r_d = \frac{\pi}{24} \frac{\cos\omega - \cos\omega_s}{\sin\omega_s - \frac{\pi\omega_s}{180}\cos\omega_s}. \quad (7.3)$$

The daily horizontal diffuse radiation for Puerto Rico was predicted from a correlation developed by González [13], suitable for tropical regions:

$$\bar{K}_D = \frac{\bar{H}_d}{\bar{H}} = \exp(-0.355 + 0.0711\bar{K}_T - 2.51\bar{K}_T^2) \quad (7.4)$$

where the monthly average clearness index is:

$$\bar{K}_T = \frac{\bar{H}}{\bar{H}_o}. \quad (7.5)$$

The second step in the modeling was the evaluation of absorbed solar radiation and useful energy gain from the collectors. A heat transfer analysis, which takes the whole flat-plate collector as the system, is explained in [14]. This analysis considers convection and radiation heat losses to ambient and conduction heat transfer through the plate and insulation. From this analysis, the outlet fluid temperature could be obtained by:

$$\frac{T_o - T_a - S/U_L}{T_i - T_a - S/U_L} = \exp(-A_c U_L F'/\dot{m}c_P) \quad (7.6)$$

where U_L is the overall heat transfer coefficient, and the collector efficiency factor is:

$$F' = \frac{1/U_L}{W\left[\dfrac{1}{U_L[D+(W-D)F]}\right]} \tag{7.7}$$

$$F = \frac{\tanh[m(W-D)/2]}{m(W-D)/2} \tag{7.8a}$$

$$m^2 = \frac{U_L}{k\delta}. \tag{7.8b}$$

The actual useful energy gain from a collector is calculated from:

$$Q_u = A_c F_R[S - U_L(T_i - T_a)] \tag{7.9}$$

where the collector heat removal factor is given by:

$$F_R = \frac{\dot{m}c_P}{A_c U_L}\left[1 - \exp\left(-\frac{A_c U_L F'}{\dot{m}c_P}\right)\right]. \tag{7.10}$$

Also, the absorbed solar radiation is expressed as:

$$S = F_{dust} F_{shading} (\tau\alpha)_e \bar{I}_T \tag{7.11}$$

where the effective transmittance–absorptance product accounts for the optical properties of the collector:

$$(\tau\alpha)_e = (\tau\alpha) + (1 - \tau_a)\frac{U_t}{U_{c-a}}. \tag{7.12}$$

A hot water storage tank allows the meeting of cooling loads at any time when solar energy is not available. Simple modeling for such tank considered the assumptions of no thermal stratification and uniform heat loss through its walls [14]. Therefore, a uniform water temperature inside the tank was estimated for given conditions of input useful energy gain from the collectors and total regeneration load, L_S, from the air conditioning systems:

$$T_s(t+\Delta t) = T_s(t) + \frac{\Delta t}{(mc_P)_s}[Q_u - L_S - (UA)_s(T_s - T_a')]. \tag{7.13}$$

An auxiliary heating input was used to compensate for low output water temperatures from the storage tank.

7.3.2 ABSORPTION COOLING SYSTEM

In the closed absorption refrigeration system, the refrigerant vapor is absorbed by a salt at the evaporator pressure, pumped to the condenser pressure, and then separated from the salt, which returns again to the absorber as shown in Figure 7.1. This absorption system is referred to as closed. Thermal energy is used in the generator, where heat is added to drive the refrigerant out of the refrigerant-salt solution. The overall performance of the absorption system, in terms of refrigeration effect per unit of energy input, is generally low; however, cost-free energy obtained from solar radiation can be used to achieve better overall energy conservation.

Computer models for different arrangements of closed absorption systems have been developed for steady-state conditions [15, 16]. These models treat each internal device of the systems as a black box, considering it to be at a fixed temperature during operation. Mass and energy balances for the absorber, condenser, evaporator, generator, and heat exchanger, as separate control volumes, lead to the following relations describing the performance of the absorption system.

$$\dot{m}_8 = \frac{\dot{q}_{evap}}{h_8 - h_7} \tag{7.14}$$

$$\dot{m}_5 = \dot{m}_8 \tag{7.15}$$

$$X = \frac{\dot{m}_2}{\dot{m}_5} = \frac{x_3}{x_3 - x_2} \tag{7.16}$$

$$\dot{m}_2 = X \times \dot{m}_5 \tag{7.17}$$

$$\dot{m}_3 = \dot{m}_2 - \dot{m}_5 \tag{7.18}$$

$$T_2 = T_1 + \varepsilon_{HX} \frac{(\dot{m}c_P)_{min}(T_3 - T_1)}{\dot{m}_1 c_{P1-2}} \tag{7.19}$$

$$\dot{q}_{gen} = \dot{m}_3 h_3 + \dot{m}_5 h_5 - \dot{m}_2 h_2 \tag{7.20}$$

$$COP = \frac{\dot{q}_{evap}}{\dot{q}_{gen}}. \tag{7.21}$$

Lithium bromide was used as the absorbing salt and water as the refrigerant. It was assumed, for simplicity, that pure refrigerant flows in the condenser and evaporator. Then the concentrations are $x_5 = x_6 = x_7 = x_8 = 0$. Also, assuming mass transfer equilibrium implied that $P_{cond} = P_{gen}$ and $P_{evap} = P_{abs}$. These pressures related the refrigerant temperature to the solution temperature and concentration.

A simple model of a cooling coil [17] was also utilized to take into account the sensible and latent heat removal capacity of the absorption system. In this model, the inlet air and water conditions, as well as the coil surface characteristics NTU_o and NTU_i, needed to be known first hand. An exit temperature for the water in the coil was assumed, and an average specific heat for saturated moist air was calculated from:

$$c_{P,a} = \frac{h_{i,w} - h_{o,w}}{T_{i,w} - T_{o,w}}. \tag{7.22}$$

The outlet air enthalpy from the cooling coil was expressed as:

$$h_{o,c} = \frac{h_{i,w}\left(1 - e^{-(1-C_1)C_2}\right) + h_{i,c}\left(1 - C_1\right)e^{-(1-C_1)C_2}}{1 - C_1 e^{-(1-C_1)C_2}} \tag{7.23}$$

where:

$$C_1 = \frac{\dot{m}_c c_{P,a}}{\dot{m}_w c_{P,w}} \tag{7.24}$$

$$C_2 = \frac{NTU_o NTU_i}{NTU_i + NTU_o C_1}. \tag{7.25}$$

Finally, an effective surface enthalpy was obtained from:

$$h_s = h_{i,c} - \frac{h_{i,c} - h_{o,c}}{1 - e^{-NTU_o}}. \tag{7.26}$$

With the corresponding effective surface temperature, the outlet air temperature was calculated as:

$$T_{o,c} = T_s + e^{-NTU_o}\left(T_{i,c} - T_s\right). \tag{7.27}$$

A new value for the outlet water enthalpy was calculated using an overall energy balance. The convergence was accomplished when the difference between the calculated and assumed values of outlet water temperatures was within certain specified criteria.

7.3.3 Liquid-Desiccant Dehumidification System

In open absorption systems, flows of heat and mass take place to and from the system. Among these, dehumidification systems take the air in, either from outside or from the building, and absorb its moisture in a solid or liquid desiccant, cooling it also by exchange of sensible heat. The desiccant is then regenerated by using low-grade heat, such as that provided by solar collectors. Stevens et al. [18] have described a computationally efficient model of liquid-desiccant heat and mass exchangers derived from an effectiveness model of a cooling tower. The inlet solution and water conditions and the system NTU were needed, as well as an assumption of the exit conditions for the solution. A specific heat for saturated moist air was obtained from:

$$c_{P,sat} = \frac{dh_{T_s,sat}}{dT_s}. \tag{7.28}$$

Then, an effectiveness was calculated as:

$$\varepsilon = \frac{1 - e^{-NTU(1-m^*)}}{1 - m^* e^{-NTU(1-m^*)}} \tag{7.29}$$

where a capacitance ratio was defined by:

$$m^* = \frac{\dot{m}_a c_{sat}}{\dot{m}_{s,i} c_{P,s}}. \tag{7.30}$$

The outlet air enthalpy was expressed as:

$$h_{a,o} = h_{a,i} + \varepsilon(h_{T_s,sat} - h_{a,i}). \tag{7.31}$$

Using an energy balance, the outlet solution enthalpy could be obtained. An effective saturation enthalpy is defined as:

$$h_{T_s,sat,eff} = h_{a,i} + \frac{h_{a,o} - h_{a,i}}{1 - e^{-NTU}}. \tag{7.32}$$

With the corresponding effective humidity ratio, the outlet air humidity ratio can be calculated with:

$$\omega_{a,o} = \omega_{T_s,sat,eff} + \left(\omega_{a,i} - \omega_{T_s,sat,eff}\right)e^{-NTU}. \tag{7.33}$$

Mass balances and the known states were used to calculate the outlet air temperature, the outlet solution mass flow rate, and new values for its concentration and temperature. Therefore, the previous procedure called for an iterative scheme. This approach was used to model the solar-assisted hybrid air conditioning system shown in Figure 7.5. First, an absorber inlet solution temperature was obtained such that the air leaving the absorber has the desired humidity for the supply air. Then, a regenerator inlet solution temperature was found to ensure that the moisture released in the regenerator was equal to the moisture added in the absorber.

7.3.4 THERMODYNAMIC PROPERTIES

For both closed and open absorption systems, many solutions have been tested to perform as absorbent-refrigerant pairs. Among them, water–ammonia, lithium bromide–water, and lithium chloride–water, are the most commonly used pairs. To develop computer models for absorption systems, there is need to have reliable correlations predicting the thermodynamic properties of these types of solutions. Thermodynamic properties were evaluated for both lithium bromide–water and lithium chloride–water solutions using the equations given in the References, as follows:

   I. Enthalpies
      Lithium bromide–water solution [19]
      Lithium chloride–water solution [20]
  II. Specific heats
      Lithium bromide–water solution [21]
      Lithium chloride–water solution [22]
 III. Concentrations
      Lithium bromide–water solution [19]
      Lithium chloride–water solution [22]

7.4 TYPICAL RESULTS FOR TROPICAL CLIMATES

In this section, the use of mathematical modeling describing the thermal performance of each system component of a solar-assisted air conditioning system (SAACS) will be demonstrated for practical applications. The mathematical model of each system component was described in detail in the previous section. The described equations were assembled in a computer code to simulate the operation of SAACS under any weather conditions. Three tropical cities were chosen as base cases for this study: Mayagüez and Ponce in Puerto Rico and Piarco in Trinidad, which represent the typical weather spectrum of the Caribbean. Year-round simulations were carried out for the absorption and the desiccant/absorption hybrid machines, with both systems being driven by arrays of flat-plate collectors and operating under weather conditions typical of the chosen cities. Results for the performance of SAACS are shown for the three cities, including variable thermal load, optimum number of collectors, thermal storage capacity, and daily and monthly solar energy contributions.

7.4.1 WEATHER DATA

The performance of the absorption and hybrid systems are demonstrated here under the variable climatic conditions of three cities across the Caribbean: Mayagüez and Ponce in Puerto Rico, and Piarco in Trinidad. These cities represent typical weather scenarios across the Caribbean, which include high-latitude humid (Mayagüez), high-latitude semi-dry (Ponce), and low-latitude humid (Piarco). Figure 7.6 presents a map of the Eastern Caribbean region.

Detailed weather data are essential as input variables to describe the performance of the systems if operating in these cities. Figures 7.7 through 7.9 show typical monthly weather data for the three

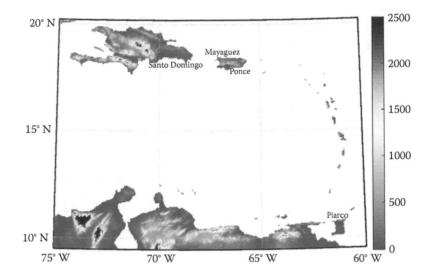

FIGURE 7.6   The Eastern Caribbean region.

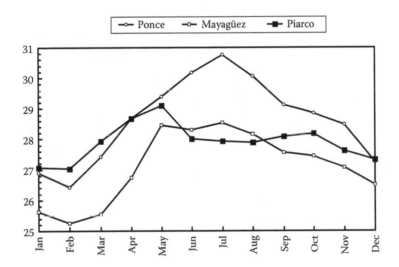

FIGURE 7.7   Monthly average ambient temperature variation for the cities considered.

cities including ambient temperature, humidity, and solar radiation. It can be noted that the city of Ponce has the highest solar radiation and the lowest humidity ratio values, whereas Mayagüez has the lowest radiation values and Piarco the highest relative humidity values.

7.4.2   BUILDING THERMAL LOAD

When detailed building load data is not available, simple relationships can be generated to simulate both sensible and latent loads. In many places, and particularly in the Caribbean, building loads are in phase with the incident solar radiation. The approach used in this study was to estimate the room sensible and latent heat loads by assuming a linear relationship between the sensible heat and the monthly average hourly horizontal solar radiation. The latent load was assumed to be a fraction

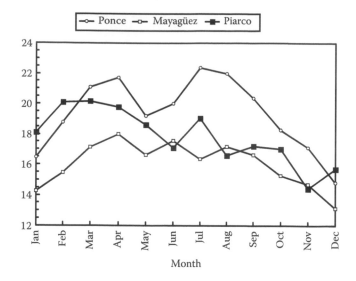

FIGURE 7.8   Monthly average daily horizontal total radiation variation for the cities considered.

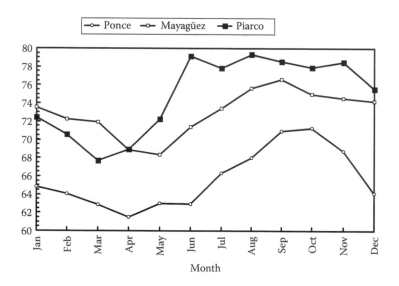

FIGURE 7.9   Monthly average relative humidity variation for the cities considered.

of the sensible load, f_{RLH}. In this study, fractions of 20% and 50% were considered. These linear relationships were as follows:

$$RSH = 3517\left(RH_{min} + RH_{mid}\, \frac{\overline{I} - \overline{I}_{min}}{I_{max} - \overline{I}_{min}} \right) \tag{7.34}$$

$$RLH = f_{RLH} \times RSH \tag{7.35}$$

where

$$RH_{mid} = \frac{RH_{max}}{f_{RLH} + 1} - RH_{min}. \tag{7.36}$$

7.4.3 OPTIMIZATION OF SOLAR COLLECTOR/STORAGE SUBSYSTEM

7.4.3.1 Solar Collectors

Before we proceed to determine the performance of the solar-assisted air conditioning systems under consideration, optimization of the solar collector array and thermal storage system should be performed. The main variables to be considered in a solar collector array should be the mass flow rate and the inclination angle. The optimum volume to store energy must also be found.

Two solar collector systems were considered for driving the system: conventional flat plate with selective surface and evacuated tube heat pipes flat-plate collectors (ETHPC). Both of these collector systems are capable of reaching moderate temperatures (70°–100°C) under high diffuse skies as those of the Caribbean region. The main difference between the conventional flat-plate collector and the ETHPC is the way in which energy is transferred to the working fluid. In conventional flat-plate collectors, the energy is directly supplied to the working fluid while in the ETHPC a secondary fluid that changes phase is used to transfer the solar energy to the working fluid. Figure 7.10 shows a schematic of this collector. In a parallel effort to this project, both collectors were tested following ASHRAE (American Society of Heating, Refrigerating and Air-Conditioning Engineers) Standards under outside conditions in the City of Cabo Rojo, Puerto Rico, located in the southwestern corner of the Island [23]. *Figure 7.11 shows the summarized results for the thermal efficiency of both collectors and clearly indicates that high performance flat-plate collectors are more efficient than ETHPC in low-latitude locations.*

7.4.3.2 Optimum Mass Flow Rate

The optimum mass flow rate through flat-plate collectors should correspond to the one that produces the best possible balance between outlet temperature and solar energy collection efficiency. Several simple simulations were executed to determine this optimum value of mass flow rate for a constant solar radiation of $1000\,W/m^2$, constant inlet temperature of 50°C, and constant collector length of 2 m resulting in an optimum water flow rate of 0.02 kg/s per area of collector in most cases. This value was also verified with experimental data and coincides with recommended values by ASHRAE. *The value of 0.02 kg/s per area of collector is then recommended and used in all subsequent simulations.*

7.4.3.3 Optimum Tilt Angle

The amount of maximum solar radiation a flat-plate collector can receive is highly dependent on the geographical location and angle from the horizontal at which the collector is inclined. The earth's annual relative motion in the north-south axis is ±23.45°, commonly referred to as declination motion. This motion causes seasonal changes and will affect the position of the local solar noon. Therefore, collectors should be inclined to compensate for this offset of the solar noon to maximize incident radiation. The ideal practice will be to continuously rotate the collector such that

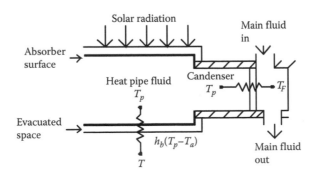

FIGURE 7.10 Schematic of ETHP.

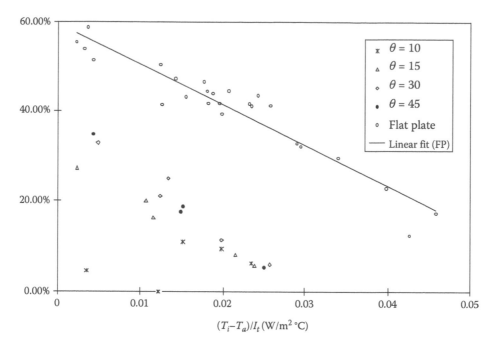

FIGURE 7.11 Comparison of thermal efficiencies for a conventional flat-plate collector and an ETHPC.

it always faces the sun at normal positions. However, this practice becomes cumbersome in actual applications. Common practice is to incline flat-plate collectors at the local latitude (i.e., collectors placed in Ponce, Puerto Rico, should be inclined for 18°). This practice is in general valid for normal annual sky conditions and moderate clouds. In the Caribbean, the sky is in general clear during the springtime and heavily cloudy in the fall (due to the rainy season). Therefore, a revision of the common practice of inclining collectors at the local latitude is mandatory for any application. We verified this practice for the cities considered by finding the optimum angles that generated maximum solar radiation at a given collector angle for each month and averaging all monthly angles. Figure 7.12 shows an example of this optimization procedure for Ponce, Puerto Rico. The average optimum

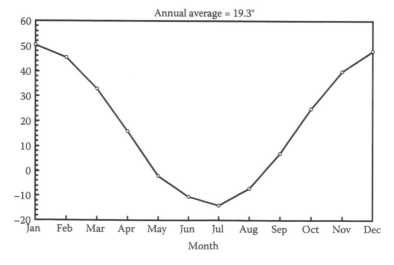

FIGURE 7.12 Collector angle optimization.

angle found was about 19.3°. A similar procedure was carried out for Mayagüez and Piarco, and optimum inclination angles were found to be 19° and 10.5°, respectively. Therefore, we conclude that collectors can be inclined at local latitudes with enough confidence and minimum error for any given city in the Caribbean region.

7.4.3.4 Thermal Storage Volume

The next parameter to optimize in any solar collector/thermal storage system is the ratio of thermal storage volume to collector area. The thermal storage serves to offset transient effects in solar radiation as well as to store any excess energy the collectors can provide. This parameter is of major importance for large thermal loads such as those of SAACS. An optimization procedure was carried out to determine the optimum storage tank volume under constant thermal loads for all cities considered. Figures 7.13 and 7.14 show results from this optimization process for the city of Ponce, Puerto Rico, for constant thermal loads of 35 and 88 kW, respectively. The figures suggest the use of a storage volume to collector area ratio of 40 L/m²; this is the recommended value for any given load. Values beyond this

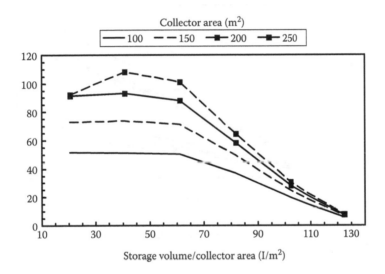

FIGURE 7.13   Storage volume optimization for various collectors areas for Ponce, Puerto Rico, and constant load of 35 kW.

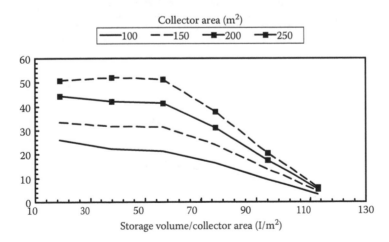

FIGURE 7.14   Storage volume optimization for various collectors areas for Ponce, Puerto Rico, and constant load of 88 kW.

will tend to decrease the solar energy contribution to meet the required load due to cooling of the water and increase thermal losses caused by increase in surface area. Similar trends were found for the cities of Mayagüez and Piarco, respectively. Thus, the value of 40L/m² is used in subsequent calculations.

7.4.4 OPTIMIZATION OF SOLAR COLLECTOR AREA

The optimization criterion for the sizing of the collector array was to meet all the air conditioning and dehumidification load for a typical day in the hottest month of the year with the highest possible solar fraction. Based on this criterion, the optimum numbers of collectors for variable loads in the desired load range are shown in Figures 7.15 and 7.16 for both absorption and absorption-liquid desiccant systems.

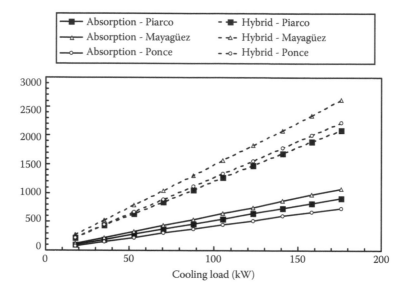

FIGURE 7.15 Area optimization for the solar-assisted absorption and hybrid air conditioning systems—$F_{rlh}=0.2$.

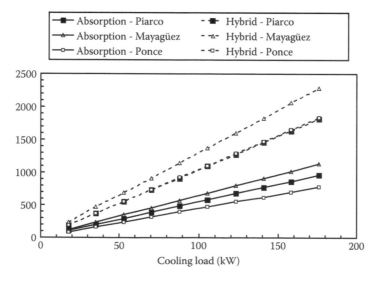

FIGURE 7.16 Area optimization for the solar-assisted absorption and hybrid air conditioning systems—$F_{rlh}=0.5$.

Results are shown for room latent heat to room sensible heat ratios of 20% and 50%, respectively. As seen in the figures, a higher latent load demands a greater collector area in the case of the solar-assisted absorption system and a smaller area in the case of the hybrid system. For the simulations presented here, banks of conventional flat plate, single cover collectors arranged in parallel, were used inclined at the optimum angles mentioned earlier with water flow rates of 0.02 kg/s per collector area.

7.4.5 ESTIMATES OF SOLAR FRACTIONS

With the required collector area being optimized for a given load, a comparison of the solar fractions, or the ratio of solar energy contribution to the total required load, which result from the solar-assisted absorption and hybrid systems simulations proceeds. Results are shown in Figures 7.17 through 7.20.

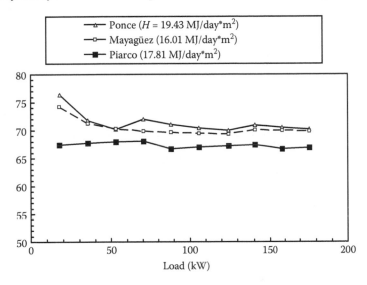

FIGURE 7.17   Annual solar fractions for the solar-assisted absorption air conditioning system—$F_{rlh}=0.2$.

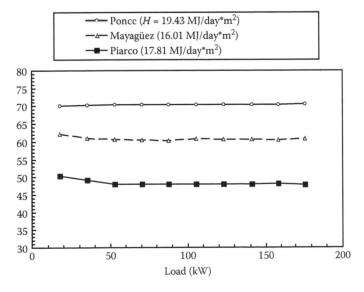

FIGURE 7.18   Annual solar fractions for the solar-assisted absorption/liquid-desiccant hybrid air conditioning system—$F_{rlh}=0.2$.

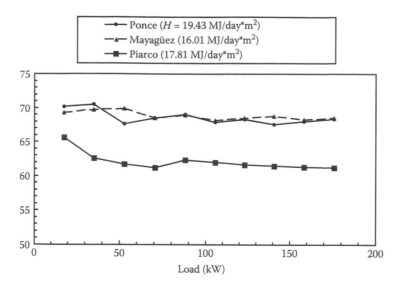

FIGURE 7.19 Annual solar fractions for the solar-assisted absorption air conditioning system—F_{rlh}=0.5.

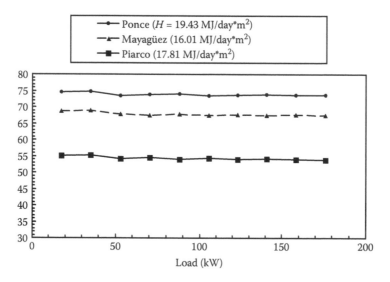

FIGURE 7.20 Annual solar fractions for the solar-assisted absorption/liquid-desiccant hybrid air conditioning system—F_{rlh}=0.5.

These results were generated for the variable load ranges mentioned earlier and for a room latent heat to room sensible heat ratio of 20% and 50%. In these figures, the solar fractions in Ponce are higher than in the other cities considered. This was expected because, as discussed previously, Ponce is the location with the highest solar radiation among the three cities. The hybrid system yielded poor solar fractions in the cities of Mayagüez and Piarco because of the low solar radiation prevailing in the former, and the low relative humidity of the latter (see Figures 7.8 and 7.9). On the average, the solar-assisted chiller system showed to be the more effective system when working with solar energy.

7.4.6 LIMITING CASES FOR HIGH HUMIDITY CONDITIONS

In view of the previous results, an explanation of the better performance of the solar-assisted absorption system is in order. Figure 7.21 compares the resulting *COP* for the case of variable

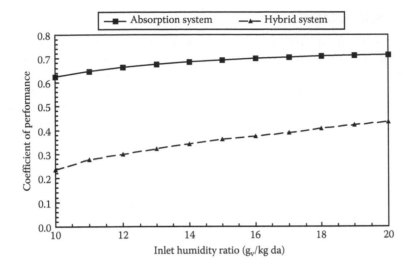

FIGURE 7.21 *COP* variation with inlet process air humidity ratio.

humidity ratio and uniform temperature in process inlet air. It must be specified that these results were obtained by keeping the outlet air conditions at a dry-bulb temperature of 15.6°C, a humidity ratio of $9\,g_v/kg\,da$, and a mass flow rate of $1\,kg/s$. Regeneration inlet air was maintained at a dry-bulb temperature of 27.5°C, a humidity ratio of $10\,g_v/kg\,da$, and a mass flow rate of $1.3\,kg/s$. It is seen that, although the hybrid system *COP* improves at higher humidity ratios, the solar-assisted absorption system still yields the highest *COPs*. As seen in Figure 7.22, the hybrid liquid desiccant-based system requires lower water temperatures at higher humidity ratios. However, for the previously described simulations, typical process inlet air humidity ratios are near $12\,g_v/kg\,da$, where the differences in temperature and *COP* justify the better performance of the absorption systems over the hybrid ones.

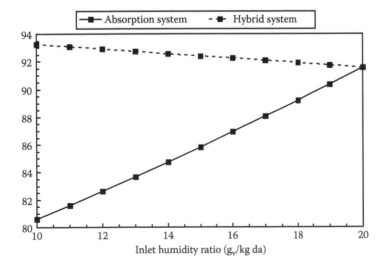

FIGURE 7.22 Maximum required water temperature variation with inlet process air humidity ratio.

7.5   ECONOMIC ANALYSIS

This chapter concerns the determination of the economic potential of SAACS when operating in the Caribbean region. The economic potential of SAACS is shown through a simple Life Cycle Cost analysis, which considers initial installation costs, actual cost of electricity, solar energy contribution, parasitic energy usage, tax incentives, and standard economic figures of merit. The analysis presented here does not quantify the mitigation in environmental impact, which the use of SAACS may reduce. Dollar quantification due to greenhouse gas reductions has not been standardized yet.

7.5.1   ECONOMIC MODEL

The economic model used in this report is based on the present worth factor (*PWF*), which is the net value of savings or losses in the future brought to the present time. The life-cycle cost is the sum of all of the present worths. The *PWF* is mathematically defined as:

$$PWF(N,i,d) = \begin{cases} \dfrac{1}{(d-i)}\left[1-\left(\dfrac{1+i}{1+d}\right)^{N}\right] & \text{if } i \neq d \\[4mm] \dfrac{N}{(1+i)} & \text{if } i = d \end{cases} \tag{7.37}$$

where
   N = number of years considered
   i = inflation rate
   d = market discount rate (interest rate)
   The payback time is defined as the time needed for the cumulative savings to equal the total initial investment. This value can then be obtained using the *PWF* as:

$$(Net\ Savings)*PWF(N_p,\ i,\ d) = Total\ Initial\ Investment. \tag{7.38}$$

Here, the net savings (*NS*) are given by:

$$NS = E_S + O_S \tag{7.39}$$

where,
   E_S are the electrical energy savings for using a solar energy system, rather than a conventional air conditioning unit, to meet the thermal building load, and O_S are the operating costs savings for using SAACS. The net initial investment (*NII*) is the cost difference between installing a vapor compression system and a SAACS. The payback years N_p can be estimated from Equations 7.37 and 7.38 as:

$$N_p = \frac{\ln\left[\dfrac{NII*(i-d)}{NS}+1\right]}{\ln\left(\dfrac{1+i}{1+d}\right)}. \tag{7.40}$$

Payback times are computed here for several base scenarios. Variables considered in this economic analysis include initial investment, net cooling load, variable solar energy contribution to meet the load, fuel cost, inflation rate, and discount rate.

7.5.2 Installation Costs

Installation costs will be parametrized in terms of costs per unit of cooling load installed ($/kW). Typical total installation costs of an absorption chiller driven by flat-plate collectors range between $1000 and $3000/kW in Puerto Rico and may increase in other Caribbean islands. Installation costs for a conventional vapor compression system range between $300 and $600/kW. Installation costs should include cost of the absorption machines, cost of the collectors, cost of the storage tanks and auxiliary heater, piping costs, air handling units, A/C ducts, and laboring. Furthermore, most governments provide a tax credit to users of solar-based systems, which consists of 30% of the total cost of collectors. Here, a flat 30% reduction of the total cost of the system will be assumed as tax credit in order to present a significantly different scenario when compared with the no tax credit case.

7.5.3 Annual Energy and Operating Costs Savings

Annual energy savings are calculated on the annual cooling load demand, electricity costs, solar energy contribution, and typical performance of a vapor compression system (VCS). Annual energy savings (E_s) are estimated as:

$$E_s = C_e * \left(TCL * SF/COP_{VCS} - S_e \right) \tag{7.41}$$

where C_e is the cost of the electricity in $/kWH, TCL is the annual total cooling load in kWh/year, SF is the annual solar fraction, and S_e is the annual energy usage by the solar system in kWh/year. Energy consumption by the solar system is significantly lower than that of the VCS and is due to the usage of pumps and controls. Normally, this value accounts for about 5%–10% of the total cooling load delivered by the solar system. A conservative value of 10% has been used in this study. The coefficient of performance of the vapor compression system (COP_{VCS}) is assumed here to be 2.5. The TCL is estimated from the annual cooling load distribution of the building. For simplicity of demonstrating the economic potential of SAACS, this load is assumed to be constant of 10 h of continuous operation per day. Electricity costs are high and extremely variable across the Caribbean basin ranging from $0.12/kWh to about $0.30/kWh.

Annual maintenance costs for SAACS should be significantly lower than those of VCSs. The former involves service and repairing of piping systems, while the latter involves replacement of compressors and expensive maintenance resulting from the use of refrigerants. Here, we have assumed a value of $10/kW for the SAACS and $30/kW for the conventional one. Therefore, a rate of annual operating costs savings of $20/kW is used in this chapter.

7.5.4 Inflation and Discount Rates

Typical general inflation rates in Puerto Rico are about 3% with fuel tending to increase at relatively higher rates. Therefore, a conservative figure of 5% annual fuel inflation rate has been used in this study. The interest and the discount rates are assumed to be the same in this study. For the last 3 years, these interest rates have fluctuated between 5% and 10%. Therefore, both scenarios are explored, as well as the typical 8% case.

7.5.5 Results

Results are presented here for the case of 12 operating hours, 360 days/year, which is the case for many government and service offices. Results are shown for two base cooling loads, 35 and 175 kW. The sensitivity of the payback period is investigated as a function of the annual solar fraction, the cost of electricity, tax credits, and total installation costs. Inflation and discount rates used are 5% and 8%, respectively. Parasitic costs are considered to represent 10% of the total cooling load and

TABLE 7.1
Assumed Constant Parameters in Economic Analysis

Parameter	35 kW-Unit	175 kW-Unit
Fuel inflation rate (%)	5	5
General inflation rate (%)	5	5
Discount rate (%)	8	8
Daily operating hours	12	12
Total annual operating hours (h/year)	4320	4320
COP vapor compression system	2.5	2.5
V/C installation costs ($/kW)	500	500
Solar system maintenance costs ($)	300	1000
V/C system maintenance costs ($)	1000	2000
Fraction of electric energy usage by solar system (%)	10	10

annual maintenance costs to be $300 in the 35 kW and $1000 in the 175 kW unit, respectively. The economic performance of the solar-assisted air conditioning system is compared with one corresponding to a VCS, which operates with a *COP* of 2.5 and has installation costs of $500/kW and annual maintenance costs of $1000 for the 35 kW unit and $2000 for the 175 kW unit. A summary of other assumed fixed parameters in the economic analysis is shown in Table 7.1.

Results for the 35 kW unit are shown in Figures 7.23 and 7.24 for variable electricity costs and annual solar fractions, respectively. The figures show results for the cases with no tax credit for the installation costs and for the case of a tax credit of 30%. Several conclusions can be drawn from these figures. It can be immediately noticed that the payback period is highly sensitive to the initial installation costs, electricity cost, tax credits given, and annual solar fractions. For conditions expected to prevail in Puerto Rico, solar fraction of 80%, electricity cost of $0.12/kWh, and installation costs ranging from $1000–$1300/kWh, the payback period will be less than 6 years with or without tax credit. For cities such as Piarco, installation costs are similar but annual solar fractions

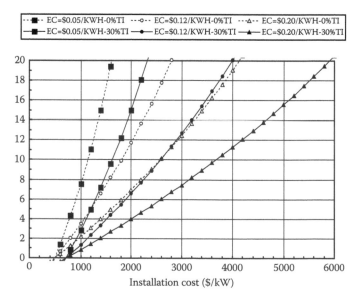

FIGURE 7.23 Payback for a 35 kW unit system for several electricity costs (ec) and tax credits: annual solar fraction = 80%.

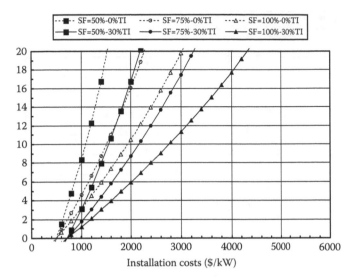

FIGURE 7.24   Payback for a 35 kW unit system for several annual solar fractions and tax credits: electricity cost = $0.12/kWh.

are in the range of 50%. However, fuel costs may be higher than in Puerto Rico, and therefore payback periods can be expected to be approximately 6 years as well. Figures 7.25 and 7.26 show very similar trends for a 175 kW unit as those for a 35 kW unit. Therefore, results presented here should be independent of the size of the unit.

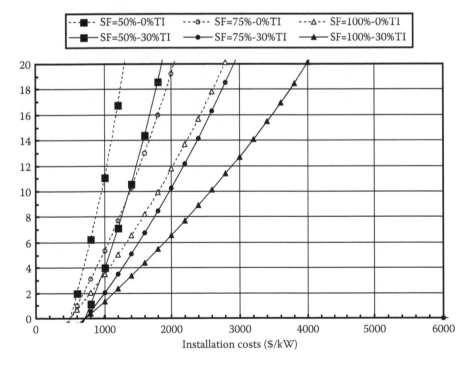

FIGURE 7.25   Payback for a 175 kW unit system for several electricity costs (ec) and tax credits: annual solar fraction = 80%.

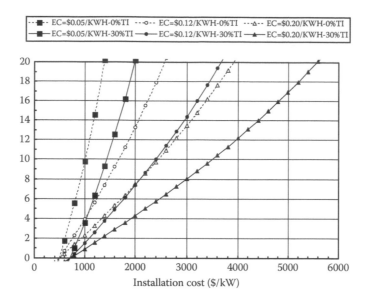

FIGURE 7.26   Payback for a 175 kW unit system for several annual solar fractions and tax credits: electricity cost = $0.12/kWh.

7.6   CONCLUSIONS AND RECOMMENDATIONS

The main objective of this chapter was to present analysis and design tools for solar-assisted air conditioning systems for applications in humid tropics such as the Caribbean. Several technologies were considered including closed and open absorption machines. Combinations of these machines were investigated in detail such as closed absorption/liquid-desiccant (open absorption). The closed absorption/solid-desiccant was not further considered due to the high operating temperatures required for the humid tropics. These hybrid combinations of sensible cooling mechanisms (closed absorption) and dehumidification systems (desiccant based systems) were considered keeping in mind the hot and humid climate of the Caribbean basin. Furthermore, all systems required thermal energy, which is abundant in these regions in the form of solar radiation. High performance flat-plate collectors were used in the study in order to maximize energy collection of the high diffuse solar radiation of the region.

The main tasks of this chapter were the development of detailed mathematical models of the systems considered; the generation of extensive computational simulations to determine the most optimum system for applications under Caribbean climates; the investigation of the cost effectiveness of the optimum systems; and the generation of a complete design using the tools developed for the possible installations of these systems. The major findings can be summarized as follows:

- Two types of flat-plate collectors were studied experimentally, namely, a high performance conventional flat-plate collector, and an ETHP solar collector. The experiment consisted of measuring the thermal performance of both collectors simultaneously while changing the inlet temperature and the tilt angle. Experimental observations demonstrated that conventional flat plates with selective surface coating perform better than ETHPs in low latitude conditions and high temperature applications. ETHPs, however, represent a potential technology for high temperature applications under cloudy skies.
- Three cities in the Caribbean were considered in a parametric study to investigate the performance of the technologies, namely, Mayagüez and Ponce in Puerto Rico, and Piarco in Trinidad. These cities were chosen to represent the entire climatic pattern of the Caribbean

including high latitude-humid (Mayagüez), high latitude-dry (Ponce), and low latitude-humid (Piarco). Variable cooling loads in the range of 10–180 kW were investigated in all three cities representing low- to medium-size cooling loads in commercial applications. The cooling load was varied linearly as a function of ambient parameters. Variables optimized in the parametric study included collector tilt angle, thermal storage capacity, number of collectors, and solar fraction. Optimum flat-plate collectors' tilt angles were found to be close to the latitude, and optimum thermal storage volume to collector area ratio was found to be about 40 L/m². Results for all cities showed that solar-assisted closed absorption systems averaged higher coefficients of performance and solar fractions than hybrid ones. This finding is justified on the basis of the relatively medium moisture level found in commercial applications. Results also show that in high-moisture removal applications, such as in supermarkets and "clean" rooms, the hybrid absorption/liquid desiccant system seems to be a highly promising technology. Solid desiccant-based systems were found to require extremely large collector areas to operate. The city that showed the highest solar fraction in all cases was Ponce, which had the lowest humidity level and the highest incident solar radiation. Furthermore, annual solar fractions ranged between 50% and 85% for all cases considered.

- The general results described above concerning the technical feasibility of SAACS in three different cities across the Caribbean should be valid for other tropical islands as long as their climate is fairly well represented by any of the cities investigated in this study. A similar statement cannot be made at this time about feasibility of SAACS in subtropical regions such as continental U.S. as the climate of continents is, in general, different from islands.
- An economic analysis to determine the cost-effectiveness of the proposed technologies was also carried out in this study. This economic analysis used the life cycle cost savings approach with the payback period (PB) as the main economic indicator. PBs are defined as the time (in years) in which cumulative energy savings equal the initial investment of the system considering discount and inflation rates. PBs were investigated as a function of the installation cost, in $/kW installed, solar fraction, cost of electricity, and tax credits. It was shown that typical installation costs for a solar-absorption system should range between $1000 and $1500/kW and that payback periods for these costs are less than six years independent of tax credits. Payback periods are also highly sensitive to solar fractions provided by the system, and it was shown that the systems proposed here are not economically feasible for cases in which the solar fraction is less than 75%.

This chapter represents a major step forward in promoting solar-assisted air conditioning systems in the humid tropics. Several suggestions will therefore be made, which may smooth the commercialization process for the proposed technologies.

- The installation of demonstration projects is essential for several reasons: (1) it will provide an actual operational scenario in which to calibrate the design tools developed under the present effort; (2) it will provide an actual cost and installation scenario of the entire system; and (3) it will be useful in stimulating confidence of potential investors, manufacturers, and users.
- The establishment of an infrastructure to supply major system components across the Caribbean region, such as the absorption chillers and high temperature flat-plate collectors, will result in the reduction of installation costs. This infrastructure may consist of the availability of local manufacturers and/or direct distribution centers of the required technologies.
- Efforts should be made to focus on conducting practical research in adapting or developing solar thermal technologies capable of achieving higher temperatures than those of existing

flat-plate collectors under cloudy skies. Technologies suggested for investigation should include compound parabolic concentrators, evacuated tube based technologies, including the heat pipes investigated in this study, and technologies that make use of phase change materials such as those reported by Alva et al. [24].

- Additional technology developments may consider air-cooled absorption systems to reduce the use of water-based cooling towers [25] and lower temperature heat driven air conditioning units such as the emergent adsorption machines [26].
- The success of any technology commercialization process is highly dependent on aggressive and effective promotional campaigns. Therefore, an aggressive campaign across the Caribbean should be implemented as soon as the feasibility of the technology is demonstrated through the demonstration project suggested above. This campaign should include government subsidies and tax incentives and be supported by local utility companies as a mechanism to offset the present oil dependency on most of the Islands.

NOMENCLATURE

A	storage tank surface area (m^2)
A_c	area of one collector (m^2)
a, b	terms accounting for radiation incidence angles
COP	coefficient of performance
c_P	fluid specific heat (kJ/kg·K)
D	tube diameter (m)
F	fin efficiency
F_1, F_2	circumsolar and horizon brightness coefficients
F_{dust}	dust effect factor
$F_{shading}$	shading effect factor
f_{RLH}	ratio of room latent heat to room sensible heat
$\overline{H}$	monthly average daily horizontal radiation (kJ/day·m^2)
HX	heat exchanger
H	specific enthalpy (kJ/kg)
$\overline{I}$	monthly average hourly horizontal radiation (kJ/h·m^2)
K	plate thermal conductivity (W/m·°C)
L_S	rate of removal of energy from the tank to the load (kJ/h)
m	defined variable used to solve for fin efficiency (m-1)
$\dot{m}$	total collector flow rate (kg/s)
$m*$	capacitance ratio (kJ/kg·K)
NTU	number of transfer units
P	pressure (kPa)
$\dot{q}$	heat transfer rate (kW)
$\overline{R}_b$	geometric factor
RH	room heat (tons)
RLH	room latent heat (W)
RSH	room sensible heat (W)
T	temperature (°C)
T_a	ambient temperature (°C)
T_a'	ambient temperature around the storage tank (°C)
T	time (s)
U	storage tank overall heat transfer coefficient (W/m^2·°C)
U_L	overall heat transfer coefficient (W/m^2·°C)
U_t	top loss heat transfer coefficient (W/m^2·°C)

U_{c-a} heat transfer coefficient between cover and ambient (W/m$^2 \cdot$°C)

W tube spacing (m)

X mass flow rate of solution from absorber per unit mass flow rate of refrigerant [(kg/s solution)/(kg/s water)]

X Weight-based concentration of salt in solution (kg salt/kg solution)

GREEK SYMBOLS

β collector tilt (slope) (°)

δ plate thickness (m)

ε effectiveness (%)

η collector efficiency

ρ_g diffuse ground reflectance

τ_α beam radiation transmittance

$(\tau\alpha)$ transmittance–absorptance product

ϕ relative humidity (%)

ω hour angle (°), humidity ratio (gv/kg da)

ω_s sunset hour angle (°)

SUBSCRIPTS

1, 2, …   state points

A   air

abs   absorber

B   beam

C   coil

D   diffuse

cond   condenser

eff   effective

evap   evaporator

gen   generator

HX   heat exchanger

I   inlet, inner

max   maximum

mid   intermediate

min   minimum

O   extraterrestrial, outer, outlet, outside

S   surface, storage tank

sat   saturated

T   tilted

w   water

REFERENCES

1. Alva, L., C.J. Gonzalez, J. Hertz (2005). Impact of construction materials in the energy consumption in homes in the Caribbean. *ASME 2005 International Solar Energy Conference*, pp. 113–121, August 6–12, Orlando, FL. doi: 10.1115/ISEC2005-76188.
2. Kakabaev, A., O. Klyshchaeva, A. Khandurdiev (1972). Refrigeration capacity of an absorption solar refrigeration plant with flat glazed solution regenerator. *Geliotekhnika*, 8, pp. 60–67.
3. Ghanhidasan, P. (1983). Testing of an open solar regenerator. *Proceedings of the 18th IECEC Meeting*, Vol. 4, pp. 1946–1951, 1983, Orlando, FL.

4. Johannsen, A. (1984). Performance simulation of a solar air-conditioning system with liquid desiccant. *International Journal of Ambient Energy*, 5, pp. 59–88. doi: 10.1080/01430750.1984.9675412.
5. Karaki, S., T.E. Brisbane, G.O.G. Löf (1984). Colorado State University Report SAN-11927-15 to U.S. Department of Energy. Performance of the Solar Heating and Cooling System for CSU House III—Summer Season 1983 and Winter Season 1983–1984.
6. Bong, T.Y., K.C. Ng, H. Bao (1993). Thermal performance of a flat-plate heat pipe collector array. *Solar Energy*, 50, pp. 491–498.
7. Bonnet, J.A., G. Pytlinski, K.G. Soderstrom (1982). Solar Energy Storage for Cooling Systems in the Caribbean. *Presented at the International Workshop on Solar Energy Storage*, Jeddah, Saudi Arabia.
8. López, A.M. and K.G. Soderstrom (1983). Insolation in Puerto Rico. *ASME Solar Engineering*, 1983, pp. 70–75.
9. Bergquam, J.B. 1983. Final Report to U.S. National Renewable Energy Laboratory Subcontract No. AR-2-12138-1. *Solar Absorption Air Conditioning/Sacramento Municipal Utility District.*
10. Meza J.I., A.Y. Khan, J. E. González (1998). Experimental assessment of a solar-assisted air conditioning system for applications in the Caribbean. *Solar Engineering*, 1998, pp. 149–154.
11. Perez, R., R. Seals, P. Ineichen, R. Stewart, D. Menicucci. A new simplified version of the Perez diffuse irradiance model for tilted surfaces. *Solar Energy*, 39, pp. 221–231.
12. Liu, B.Y.H. and R.C. Jordan (1960). The interrelationship and characteristic distribution of direct, diffuse and total solar radiation. *Solar Energy*, 4, pp. 1–19.
13. González, J.E. (1989). Modeling of the thermal performance of shallow solar ponds for Puerto Rico and the Caribbean. M.S. Thesis, University of Puerto Rico at Mayagüez, 1989.
14. Duffie, J.A. and W. A. Beckman (1993). *Solar Engineering of Thermal Processes.* 2nd ed. New York: John Wiley & Sons.
15. Grossman, G. and E. Michelson (1985). A modular computer simulation of absorption systems. *ASHRAE Transactions*, 91(2b): pp. 1808–1827.
16. Siddiqui, M.A. (1991). Optimal cooling and heating performance coefficients of four biogas powered absorption systems. *Energy Conversion and Management*, 31(1): pp. 33–49.
17. Khan, A.Y. (1994). Heat and mass transfer performance analysis of cooling coils at part-load operating conditions. *ASHRAE Transactions*, 100(1a): pp. 54–62.
18. Stevens, D.I., J.E. Braun, S.A. Klein (1989). An effectiveness model of liquid-desiccant system heat/mass exchangers. *Solar Energy*, 42: pp. 449–455.
19. ASHRAE (2001). *ASHRAE Handbook of Fundamentals.* Atlanta: American Society of Heating, Refrigerating and Air-Conditioning Engineers.
20. GRI Annual Report. Gas Research Institute, October, 1991.
21. Jeter, S.M., J.P. Moran, A.S. Teja. Properties of lithium bromide–water solutions at high temperatures and concentrations—Part III: Specific heat. ASHRAE Transactions, 98(1): pp. 137–149.
22. Uemura, T. (1967). Studies of the lithium chloride–water absorption refrigerating machines. *Technical Report*, Kansai University, Japan.
23. Rosa F., J.E. González, A.Y. Khan (1997). Performance comparison of evacuated tube heat pipe collectors versus flat Plate collectors for tropical climates. *Solar Engineering*, 1997, pp. 319–325.
24. Alva, L.H., J.E. González, N. Dukham (2006). Initial analysis of PCM integrated solar collectors. *Journal of Solar Energy Engineering*, 128, pp. 173–177.
25. Alva, L.H. and J.E. González (2002). Simulation of an air-cooled solar-assisted absorption air conditioning system. *Journal of Solar Energy Engineering*, 124, pp. 276–282.
26. Ratnanandan, R. and J.E. González (2012). A system modeling approach for active solar heating and cooling system with phase change material (PCM) for small buildings. *Proceedings of the ASME 2012 International Mechanical Engineering Congress and Exposition*, Houston, TX, November 9–15, 2012, Paper No. IMECE2012-93038, pp. 207–215; doi:10.1115/IMECE2012-93038.

8 Absorption Thermodynamic Cycles
Advanced Cycles Based on Ammonia/Salt

Antonio Lecuona-Neumann, Pedro A. Rodríguez-Aumente, Mathieu Legrand, and Antonio Famiglietti
Carlos III University

CONTENTS

8.1 INTRODUCTION

Absorption is a technology many years old. Since the French scientist Ferdinand Carré built the first absorption machine in 1858, it has been intensively developed, and it now offers many possibilities of use. An absorption machine is able to transfer heat from a cold source to a hot sink consuming

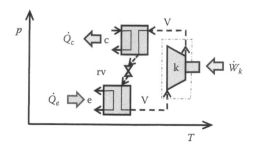

FIGURE 8.1 Basic layout of the Inverse Rankine Cycle using a mechanical compressor. position of elements indicates the pressure and temperature.

heat from an external third source, working as a heat-driven *heat pump*. Although many applications have reached a commercial stage and a lot of innovative systems are being investigated, absorption machines are not so common now, owing to a higher cost and bulk than the work-driven technologies based on the *inverse Rankine cycle*, which uses a mechanical compressor (k), Figure 8.1, usually consuming electricity. In both kinds of cycles, the phase change of a volatile substance, called *refrigerant*, is used internally as a working fluid. It produces cold when evaporating in the evaporator (e) and heat when condensing at higher pressure p in the condenser (c) and consequently at a higher temperature as the refrigerant usually is a pure substance. Since absorption cycles consume heat instead of work, many different energy sources can be used as input, going from convectional gas combustion to waste heat and renewable sources. This positions them in premium places in the energy scenario, as heat is frequently available at temperatures not high enough for an efficient heat to work conversion.

Absorption machine performances and applications depend on the working fluids used inside the machine as well as on the thermodynamic cycles. Among others, the combination ammonia/salt is used as a working fluid in absorption cycles, and it is considered in this chapter. Ammonia is a natural refrigerant, thus with no greenhouse effect or stratospheric ozone depleting potential, as many organic refrigerants incorporate. It offers classical and innovative possibilities for energy conversion into heat or cold and even into mechanical power, consuming renewable energies with high resilience and autonomy.

New developments of advanced thermodynamic cycles offer interesting possibilities of integrating absorption cycles into sustainable energy strategies based on distributed generation. Hybrid cycles that combine thermal energy and power have been also developed. They allow complementing time-variable renewable energy sources with grid electricity consumption or stored biomass burning to satisfy the user demand. The variant of combined cycles produces not only thermal energy but also work and the moderate and even low temperature input heat makes them suitable for waste energy valorization and solar thermal conversion to electricity.

This chapter presents in a progressive way more recent but classical cycles using ammonia absorbed in salts as a working fluid, thus forming an introduction to some concepts and acting as a background for more advanced and newly developed cycles currently materialized on prototypes or even commercialized. Abundant references are provided for further consult, creating a kind of information center.

8.2   FUNDAMENTALS

Absorption has been a well-known technology for decades but is still under development in some areas. In Niebergall (1981), Herold et al. (1996), ASHRAE (2013), and Ochoa Villa et al. (2011), the fundamentals are detailed. The absorption machines were initially intended for producing cold, although they can be configured to produce heat at a different temperature of the heat consumption. Absorption technology has evolved into many varieties of thermodynamic cycles including heating heat pumps (with the variant of thermal converters), (work) power production, and power consumption (Srikhirin et al., 2001).

The key processes of absorption technology involve two basic processes of the refrigerant vapor (V):

- *Sorption* in the absorber (a), releasing heat; actually the sum of the heat of going from gas phase to liquid phase, condensation, plus a smaller amount of separating the solver molecules to give room to the refrigerant
- *Desorption*, in the vapor generator (g) consuming heat; actually the sum of boiling heat plus the heat to approach the solvent molecules

Owing to the usually large phase change enthalpy, the *solution heat* generally is of smaller amount. This makes the heat transferred in the absorber similar to the generator heat.

8.2.1 BASIC ABSORPTION CYCLE LAYOUT

The basic absorption cycle is similar to the inverse Rankine cycle except for the way the vapor refrigerant is compressed. The mechanical vapor compressor of the inverse Rankine cycle, Figure 8.1, is substituted for a thermochemical compressor, tchk in Figure 8.2, named *compression stage*, as several stages can be stacked either in series or in parallel. tchk is basically composed of *absorber* (a), *vapor generator* (g) and *solution heat exchanger* (shx). The refrigerant vapor coming from the *evaporator* is absorbed by the liquid solution in the absorber at T_a. The solution is pumped to the vapor generator where external heat at a temperature above ambient T_g is supplied to evaporate the refrigerant, which then goes to the *condenser*. The solution, poor in refrigerant, goes back to the absorber, preheating the solution moving to the generator through the counterflow heat exchanger. Since the solution is pumped in a liquid state, much less work is consumed than with a vapor compressor.

Both liquids, refrigerant and solution (poor in refrigerant) are expanded by *valves* (rv and sv, respectively) to the lower pressure part of the machine, closing the cycle. Depending on the operating temperatures, the cycle can work as chiller, heat pump, or *heat transformer*. For cold production both absorber and condenser reject heat $\dot{Q}_a + \dot{Q}_c = \dot{Q}_{ac}$ to ambient at T_{amb} and $\dot{Q}_e$ is the useful heat

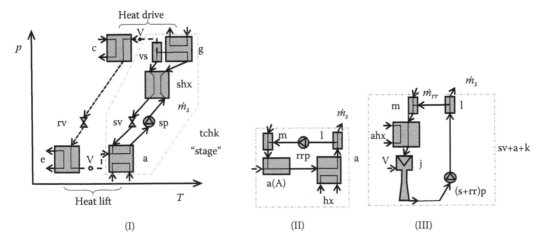

FIGURE 8.2 (I) Basic layout of the single effect absorption continuous cycle with diabatic absorber. position of elements indicates the pressure and temperature. (II) adiabatic absorber (a) forming a diabatic absorber with an additional pump and heat exchanger. (iii) adiabatic absorber (a) forming a diabatic absorber with an enlarged solution pump and an ejector. Dash lines indicate refrigerant; short dash is for liquid. V indicates when vapor is assumed. Figure 8.5 indicates external fluids heat transfers.

Sustainable Energy Technologies

extraction as $T_e < T_{amb}$, Figure 8.5. For heat production as a heat pump $T_e \approx T_{amb}$, $\dot{Q}_{ac}$ is the useful production and $T_g > T_{ac}$. A heat transformer is obtained when $T_c, T_a > T_g$ and $\dot{Q}_{ac}$ is the useful production (Wu et al., 2014a).

Figure 8.2 shows the basic layout of the closed absorption cycle working in a continuous fashion (Sarbu and Sebarchievici, 2013). A comparative study of different cooling cycles can be found in (Gupta et al., 2008).

The solution desorbs vapor at high pressure when heat is applied to the generator (g in Figure 8.2). It absorbs the vapor when put into contact with the refrigerant vapor at lower pressure and temperature in the absorber. This releases heat that has to be evacuated. This way the thermochemical compressor consumes heat instead of the conventional mechanical work consumption. The moderate temperatures needed for this make the tchk attractive for consuming waste heat and thermal renewables.

The *solution heat exchanger* (shx) Figure 8.2(I), recovers thermal energy from the outlet of the generator to preheat the solution inlet, thus reducing $\dot{Q}_g$ that increases the efficiency of the cycle, called *Coefficient of Performance* (*COP*), Equation 8.1 and reducing the heat load on the absorber $\dot{Q}_a$. The liquid solution can be substituted by the *adsorption* of the vapor into a porous solid that is alternatively connected to low and to high pressure in order to adsorb and desorb respectively (Meunier, 2013). This kind of sorption machine is not the main scope of this chapter, although some works are presented. A practical application is Zhai et al. (2008). Many variants of the basic cycle can be achieved, as discussed in the following sections of this chapter.

What is done with the vapor at high pressure determines the purpose of the absorption machine. If it is condensed and downstream evaporated at lower pressure, it forms a generic heat pump as in Figure 8.2. If it is expanded in a turbine, electric power can be obtained.

The heat that is released condensing the refrigerant in the condenser (c) can be used to separate more vapor from the solution at a lower pressure generator, increasing the production of refrigerant in what is called a double effect (DE) machine in contrast with the single effect cycle (SE) so far described, where this heat is useful. Triple effect (TE) is also possible. Many variants of the basic cycle have been proposed. In Herold et al. (1996) and Srikhirin et al. (2001) there are good presentations of the main types. The variants, including hybrid cycles and power producing cycles, are constructed by adding and combining the basic elements shown in Figure 8.3. The use of semipermeable membranes to separate the refrigerant from the solution (Srikhirin et al., 2001) is under research; its analysis is out of the scope of this chapter.

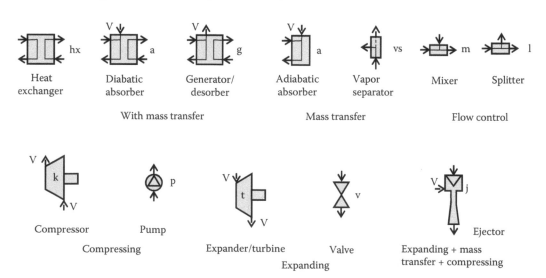

FIGURE 8.3 Components for building absorption cycles. V indicates when vapor is assumed.

8.2.2 SOME ADVANCED ELEMENTS

Adiabatic absorbers (Aa), proposed some time ago (Ryan, 1993), are especially indicated for viscous solutions such as $NH_3/LiNO_3$, e.g., (Sarbu and Sebarchievici, 2013). Their absorption capability is limited by the absorption self-heating, so that a *recirculation circuit* (rr) is necessary for cooling and repeating absorption, Figure 8.2(II). The *absorber heat exchanger* (ahx) is now of single-phase internal flow, thus of standard design (plate type in favor of compactness) and is more effective. Its compactness contributes to a lower NH_3 inventory, consequently increasing safety and lowering cost. An experimental work shows its feasibility using sprayed $NH_3/LiNO_3$ solution in the adiabatic absorber (Ventas et al., 2012).

The *ejector* (j in Figure 8.2) is a component where the kinetic energy of a free expanding jet is transferred to another flow at lower pressure, entering in a plenum chamber, mixing with it in a wider straight tube named *nozzle* because of a converging entrance. The mixture is compressed in the nozzle but can be further compressed in a downstream *diffuser,* and if the expanded flow is poor in refrigerant solution it can absorb refrigerant very efficiently. This way the ejector can perform three functions: solution expanding valve (sv), adiabatic absorber (Aa), and compressor (k) with no external energy consumption, Figure 8.2(III) (Vereda et al., 2012). As the absorber generally is the biggest element, this innovative device can lead to a machine-size reduction.

There is another kind of absorption or adsorption machine. It operates in an *open cycle* to atmosphere, as part of its duty is to reduce the humidity of the supply air to a building, in addition to controlling supply temperature (e.g., Sarbu and Sebarchievici, 2013). Thus, they use a solution where the refrigerant is water and the solvent is a *desiccant,* such as BrLi, ClLi, silica gel, zeolites, etc.

8.2.3 SOLUTION COMPOSITION

Many substances have been chosen and studied to form the solution, e.g., (Srikhirin et al., 2001). The refrigerant is the initial choice, as water (H_2O) and ammonia (NH_3) are the preferred choices for natural refrigerants with high phase change enthalpy. Hydrocarbons and industrial refrigerants have been used, dissolved in organic or synthetic but less volatile substances, but they exhibit either GWP, ODP, they are inflammable or do not exhibit much affinity with the solvent. With LiBr or LiCl as solvents, H_2O has been much used in absorption machines for its many advantages, only shadowed for its freezing at 0°C. NH_3 has been used as refrigerant mainly in large machines, professionally operated owing to the risks of the toxic leaks. It allows cold production down to −70°C. H_2O and NH_3 and many other refrigerants can produce not only heat and cold (Wu et al., 2014b). Mechanical power can be produced by expansion if the pressure and the density of the vapor at the operating temperatures are large enough, thus excluding H_2O. Nowadays the use of NH_3 in absorption machines continues, using H_2O as a solvent. Removing H_2O from the vapor generated consumes heat in a rectifying tower that removes H_2O, lowering the efficiency of the chiller. Thus, and aiming at reducing corrosion, the use of inorganic salts as solvents/absorbents for NH_3 is under intensive research and development since its proposal (Gensch, 1937).

8.2.4 REPRESENTATIVE VALUES OF *COP*

COP accounts for the efficiency of the cycle, being the ration between the benefit obtained and the expenditure. With the basic cycle working as chiller the COP_{coo} is defined as follows:

$$COP_{coo} = \frac{\dot{Q}_e}{\dot{Q}_g} \qquad (8.1)$$

For fixed external fluids temperatures, averaged between inlet and outlet T_e and $T_a = T_c = T_{ac}$, the representative COP_{coo} of absorption chillers is shown in Figure 8.4. The COP_{coo} for adsorption

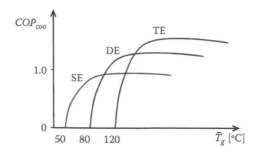

FIGURE 8.4   Representative evolution of COP_{coo} of basic cycle absorption machines versus average external generator temperature $\bar{T}_g$ with fixed $\bar{T}_e$ and $\bar{T}_{ac}$. Power is varying, Section 8.10.

chillers can be around 1/3 to 2/3 of this value, especially because these chillers have to operate in batch mode.

With the basic cycle layout, the *COP* of a heating heat pump COP_{hea} can be estimated, although the temperatures are not the same, as:

$$
\left.
\begin{aligned}
COP_{hea} &= \frac{\dot{Q}_a + \dot{Q}_c}{\dot{Q}_g} \\[6pt]
\text{Energy balance}: \dot{Q}_a + \dot{Q}_c &= \dot{Q}_g + \dot{Q}_e
\end{aligned}
\right\} COP_{hea} = 1 + COP_{coo} \tag{8.2}
$$

Typical values for COP_{coo} are indicated in Figure 8.4. A very simple model for the consideration of the effect of the three average temperatures on performances is presented in Section 8.10, allowing to describe curves such as in Figure 8.4.

8.3   AMMONIA/SALT AS A SOLUTION

This chapter addresses recent advances in absorption machines that use the combination of NH$_3$/salt as an example of an innovative solution to introduce new tendencies and concepts, but conventional concepts are the same as with conventional solvents. These salts are mainly lithium nitrate (LiNO$_3$) and sodium thiocyanate (NaSCN), although ternary mixtures have been explored (e.g., Farshi, Infante Ferreira et al., 2014). One of them is simply adding water to the previous pairs, up to around 30% to increase heat conductivity and decrease viscosity of the solution (Libotean et al., 2008; Cuenca et al., 2014). Limiting this proportion avoids water evaporation and to some extent avoids internal corrosion.

8.4   SOLAR HEATING/COOLING AND ADVANCES CYCLES

Using solar thermal collectors to activate/drive absorption chillers is an established technology but is still under development (Srikhirin et al., 2001; Kim and Infante Ferreira, 2008; Balaras et al., 2007, among others). The investment can be considered still high, especially in front of PV-based mechanical compression machines, although the easy addition of thermal/chemical storage is a non-negligible advantage, as well as the capability of *thermal energy storage* (TES). For solar-driven absorption, the maintenance becomes complex as there is a thermal fluid loop that brings heat from the solar field to the machine, requiring some power for pumps and devices to avoid boiling or freezing (see Figure 8.5). Simplifying this scheme seems a research objective for advanced cycles. Section 8.8.1 of this chapter brings a recent proposal in this line of work.

When solar energy is not enough for the user demand, a separate mechanical compression cycle is needed for increasing production, with the result of an increased overall complexity and cost. Moreover, when the solar/absorption system is put out of the loop, some solar energy is wasted.

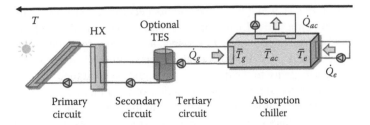

FIGURE 8.5 Generic solar cooling circuit layout with an absorption chiller and hot water storage tank.

It seems reasonable to explore hybridizing the absorption cycle with some sort of mechanical compression to *boost* the absorption cycle, helping in the lift of the pumped heat, when solar power is limited. A single machine will result in a simpler and more cost effective facility. A compressor in an absorption cycle brings the difficulty of lubrication, with associated cost and possible contamination of the solution. A recent proposal that circumvents this issue is presented as part of Section 8.6.1.2.

The introduction of the electricity vector into absorption technology opens a new opportunity for energy integration. For building acclimatization in moderate climates, there could be seasons when neither heat nor cold is required. In industrial solar/absorption facilities during weekends or non-working days the same happens, but solar energy or residual heat can be still there. Producing and selling electricity during these periods could reduce the amortizing period of the asset, one of the main financial parameters considered. This chapter introduces this topic within the framework of extending the use of absorption as an element helping to pave the way to fully distributed and sustainable energy integration (see Figure 8.8).

It is much claimed that a high integration of renewables will be favored if cost effective energy storage is realized. This will reduce the need for backup electric power, typically expensive and/or fossil fueled. Local redirection of the power produced (in a polygeneration scheme) to the local demand (cold/heat/electricity) as a result of the flexibility of energy conversion machines reduces the need for the storage, backup, and capacity of energy transport (see Figure 8.8).

8.5 RECENT WORKS

This section introduces the most canonical cycles and then shifts into advanced cycles, starting with hybrid cycles and following with the variants that offer power (electricity) production and combined thermal/power production. A general presentation of the main absorption technologies can be found in Srikhirin et al. (2001), excepting the power production cycles.

8.5.1 SINGLE EFFECT SINGLE PRODUCTION CYCLES

Single stage SE cycles act as heat pumps for cold or heat production. In Best et al. (1991) the basics of using $NH_3/LiNO_3$ in absorption cycles are presented, although previous studies on this solution were performed, such as in Infante Ferreira (1984). In Hernández-Magallanes et al. (2014) a steady-state operation of a prototype SE absorption chiller is analyzed, using $NH_3/LiNO_3$. The chiller showed promising performances for air conditioning or refrigeration, registering a $COP_{coo} = 0.45 - 0.70$ with external fluids generating temperatures of 85°C–105°C, rejecting/recooling temperatures $(\bar{T}_a = \bar{T}_c = \bar{T}_{ac})$ of 18°C–36°C and T_e down to 1°C.

Air cooling of a chilling machine means rejecting/recooling directly to the atmosphere the heat from absorber and condenser $\dot{Q}_{ac}$, this way eliminating its cost and bulk and eliminating the infection risk by *legionella* bacteria associated with wet cooling towers. Dry heat rejection into the air signifies a larger heat exchange and larger temperature difference/jump with atmosphere, reducing the *COP*. *Direct air cooling* refers to a finned absorber and condenser, as in Kim and

Infante Ferreira (2009). In Cai et al. (2014), the authors perform a thermodynamic analysis of a SE refrigerating cycle using either $NH_3/LiNO_3$ or $NH_3/NaSCN$ with an absorber of the diabatic type. They propose a solution recirculation loop to enhance absorption, rejecting heat (similar to Figure 8.2(III)), although the results presented ignore this loop characteristic of adiabatic absorbers. In Ventas et al. (2012) the sensitivity to the degree of recirculation is studied.

Cai et al. (2014) assume the simplest approximation of a temperature jump with external heat interchange flows for the generator and condenser. Pumps and fan powers were not taken into account. An external exergy analysis of the cycle is performed. The results show that the generating temperature must be limited because of the closeness of solution crystallization conditions at the inlet of the absorber, the lowest temperature and refrigerant mass fraction x in the cycle. From this viewpoint, $NH_3/LiNO_3$ is a better choice than $NH_3/NaSCN$, but the latter gives higher exergetic efficiencies and starts to produce cold at slightly lower generating temperatures. In this context a higher exergetic efficiency stands for the exergy delivered over the exergy supplied. Exergy is the capacity of a system to produce useful work. One interesting issue is that a prescribed absorption efficiency is applied to the model as $x / x_{sat} < 1$ at the absorber outlet indicating not reaching saturated conditions. The study indicates that the growth of this parameter improves the exergetic efficiency of the cycle. In the second part of that paper, an internal component exergy destruction analysis shows that 80% of the total can be attributed to generator, absorber, and solution heat exchanges.

8.5.2 Two Stages Double Effect Cycle

A higher COP for absorption chillers is possible if heat sources above 100°C are available using multiple effect cycles (see Figure 8.4). With NH_3 as refrigerant, pressures will be very high for multiple effect cycles, in the order of 50–70 bar for DE, making construction difficult. In Herold et al. (1996), the two compression stages cycle is presented for NH_3/H_2O, with the same maximum pressure as an SE cycle, but the high heat requirement for *rectification* reduces the COP dramatically. In Ventas et al. (2016), the two-stage (in parallel)/double effect (TSDE) cycle is proposed using $NH_3/LiNO_3$ with a maximum pressure of an SE cycle ~ 20 bar. With this cycle, the second effect is produced transferring heat from the higher temperature absorber to the lower temperature generator. The performances are studied thanks to a numerical model, and the limitation imposed by crystallization of the $NH_3/LiNO_3$ is determined for both adiabatic and diabatic absorbers. Both variants offer similar COP_{coo}; however, the adiabatic variant shows a larger margin against solution crystallization thus allowing a higher T_g or/and a lower T_{ac}. This cycle can produce cold for external inlet evaporator temperatures down to −10°C without crystallization. The maximum COP_{coo} can be 1.25 for an external circuit inlet generator temperature of 100°C, a lower temperature than required for DE cycles yielding this COP_{coo}. Comparison with $H_2O/LiBr$ and NH_3/H_2O DE cycles shows advantages regarding COP, evaporation and condensation temperatures, and crystallization.

8.6 HYBRID CYCLES

Hybrid cycles a) use at least two different sources of energy or/and b) produce at least two different forms of energy, in which case it is called combined cycles or *polygeneration cycles*. The main idea behind hybrid/combined cycles is to adapt the production to the user energies' demand. Using two driving energies joins efforts, giving the temperature lift to heat (see Figure 8.2(I)). This results in an increase in efficiency and enhances flexibility, as exposed in Subsection 8.6.1.

8.6.1 Cycles Consuming Heat and Work

8.6.1.1 Mechanically Boosted Cycles

Hybridizing the thermochemical compressor that the solution circuit forms, according to Figure 8.2(I), with a mechanical compressor has been explored somehow. It can be performed implementing the mechanical compressor (k) parallel to the tchk, such as in Ayala et al. (1997), so that both compressors add their refrigerant mass flow with the same *compression ratio* $CR = p_g/p_a$; $p_g = p_c$; $p_a = p_e$.

An alternative is locating the compressor in-series inside the refrigerant line, allowing it to have arbitrary lower CR (Xie at al. 2012). This way the cycle incorporates three pressure levels. The compressor can be located either downstream the generator (closed point V in Figure 8.2(I), $p_c / p_g = CR > 1$, $p_e = p_a$) or upstream the absorber (open point V in Figure 8.2(I), $p_a / p_e = CR > 1$, $p_g = p_c$). In an equivalent way, they both reduce the temperature lift the thermochemical compressor has to perform. The first choice seems less attractive for producing cold, owing to the higher enthalpy at the inlet of the compressor, meaning there is a higher consumption of specific work w for the same CR and polytropic efficiency η_{pk}:

$$w = \frac{\dot{W}_k}{\dot{m}_V} = h_{ou} - h_{in} \cong h_{in} \left(CR^{\frac{\gamma-1}{\gamma\eta_{pk}}} - 1 \right) \tag{8.3}$$

The basic figure of merit for booster compressors is a kind of electrical *COP* for its implementation: the increase of cooling capacity over the electrical compressor power consumed:

$$ECOP_{coo} = \frac{\dot{Q}_{e,with} - \dot{Q}_{e,without}}{\dot{W}_k} \tag{8.4}$$

From a primary energy consumption viewpoint $ECOP_{coo}$ should be higher than the COP_{coo} of the same compressor acting in an inverse Rankine cooling cycle. As the CR of the compressor can be easily varied, it substantially increases flexibility of the absorption cycle for increasing its capacity, lowering the generation temperature, increasing the heat rejection temperature, lowering the evaporator temperature, or a combination, so that some optimum can be reached.

Pressure boosting is not only favorable for chillers, it is also good for absorption heat pump heaters, as some works show. Wu et al. (2013) presents a thermodynamic study of an absorption heat pump cycle using NH_3 as vapor refrigerant and $LiNO_3$ or $NaSCN$ as solvents for producing acclimatization heat. The crystallization limit is extended by using a compressor between the evaporator and the absorber. In their study, $LiNO_3$ exhibits superior performances, in terms of both COP_{coo} and crystallization resistance. Generation temperatures varied between 100°C and 150°C, while evaporation temperatures varied between −30°C and 10°C and the condensing temperatures ranged between 30°C and 65°C.

The most used working pair in commercial chillers is NH_3/H_2O in SE configuration, as DE configuration has not raised enough interest. Under difficult operating conditions, mechanical boosting can significantly extend its applicability. In Wu et al. (2015), a theoretical study demonstrates the flexibility and high energy efficiency of an NH_3/H_2O SE cycle working as a heat pump during the cold season at different Chinese cities, counting on the benefits of a compressor booster, with up to $CR = 3.0$, increasing the absorber pressure.

Ventas, Lecuona, Zacarías et al. (2010) present a thermodynamic study of the benefits of using a compressor to raise the absorption pressure above the evaporator one in an SE cooling cycle using $NH_3/LiNO_3$. This cycle, driven with hot water at 67°C and with a compressor $CR = 2.0$ yields the same capacity of an SE absorption cycle at 94°C. The $ECOP_{coo}$ was found to be higher than when the same compressor was applied to an ammonia vapor compression cycle for comprehensive working conditions. In Angrisani et al. (2016), a partial materialization of such a cycle with NH_3/H_2O is presented, being fed by a reciprocating internal combustion (IC) engine.

8.6.1.2 Ejector Booster and Mechanically Boosted Ejector Cycles

The most common application of ejectors is to use a high-pressure flow after it is accelerated by a free expansion to transfer kinetic energy to a lower pressure flow. In Eames et al. (1995) and Witte (1965), details about this device and its use can be found. Ejectors have been used to completely substitute for the compressor in inverse Rankine heat pumps with a gas-gas ejector (Eames et al., 1995), thus consuming heat to boil the liquid refrigerant instead of compressor work, only consuming a little of work for a pump. Many configurations are possible (e.g., the one performed in Zeyghami et al., 2015).

With NH_3 as a refrigerant in absorption SE cycles, the pressure difference between generator and absorber (4–10 bar) is high enough to drive a liquid-gas ejector for substituting both the solution expansion valve and the booster compressor, therefore compressing NH_3 vapor with a jet of solution. This way the additional work consumption by the booster compressor is eliminated (e.g., Wang et al., 1998; Vereda et al., 2012; Farshi, Mosaffa et al., 2014, among others). The improvements in the cycle performances are a lower requirement for heat drive with a fixed lift (e.g., lower temperature for vapor generation/activation *ceteris paribus*) and a higher *COP* in spite of the low efficiency of the compressing process in an ejector. In the theoretical study of Vereda et al. (2012), it is supposed that no absorption takes place in the ejector nozzle and diffuser in order to obtain conservative performances, in spite of the high contact surface vapor bubbles solution that occurs inside the ejector, located just upstream of the inlet of the conventional diabatic absorber. Ejectors seem of high potential, but they tend to degrade performances when out of the design operating point, thus variable geometry seems an appropriate solution. The ejector in that paper incorporates the only easily implemented variable geometry.

In a later article (Vereda et al., 2014), a new idea is proposed and analyzed thanks to the experimental evidence that adiabatic absorption takes place effectively in the mixing tube and diffuser of the ejector because of the formation of tiny bubbles in a much turbulent mixing flow of NH_3 vapor and $NH_3/LiNO_3$ solution. In that article, it is shown that the ejector can additionally act as both an effective adiabatic absorber and a compressor booster. This is achieved by increasing the flow in the recirculation circuit by augmenting the power supplied to the larger capacity solution pump, $\dot{W}_{(s+rr)p}$ as indicated in Figure 8.2(III). Results of the numerical simulation show that this innovation behaves like a compressor boosted hybrid cycle without carrying the complexity of a compressor. Thus, this cycle can be named mechanically boosted ejector cycle.

In Vereda (2015), it is experimentally demonstrated that the mechanically boosted ejector combination produces a very compact component. One added singular phenomenon is the absence of desorption inside the ejector plenum chamber when the solution results superheated after expansion in the convergent nozzle. The hereby-described (Ejector Adiabatic Absorber or EAA) cycle allows for decreasing the activation temperature about 15°C for a recirculation ratio of $RR = \dfrac{\dot{m}_{rr}}{\dot{m}_s} = 3.0$ and increasing the cooling capacity about 100% for a generation temperature of 80°C. The $COP_{E,coo} = \dot{Q}_e / \dot{W}_{(s+rr)p}$ that results from the simulation shows higher values than for the mechanical compression (inverse Rankine) cycle for the same heat temperature lift (Figure 8.2) T_e to T_{amb}. One additional advantage is that for high enough generator temperatures, so that the SE bare cycle shows the characteristic slow down increase in *COP* when T_g increases (Figure 8.4), the $COP_{E,coo}$ of the EAA cycle reaches values between 8 and 16, and increasing with T_g what might be interpreted as an exergetic efficiency increase. The advantages and drawbacks of recirculation were previously described by Ventas, Lecuona, Legrand et al. (2010).

Both the mechanically boosted cycles and the ejector cycles open the possibility of increasing capacity of either/both heat/cold production by consuming work when the driving heat source is scarce (low $\dot{Q}_g$ or T_g).

8.7 POWER CYCLES

The moderate and low temperatures that allow producing vapor in absorption cycles make them first-order candidates for low temperature residual/waste heat valorization and solar thermal conversion to electricity. A general presentation can be found in Ibrahim and Klein (1996) and Brückner et al. (2015). The Organic Rankine cycle (ORC) is a competitive technology using a synthetic refrigerant instead of H_2O (e.g., Chen et al., 2016). It can be sophisticated for superior efficiencies (e.g., He et al., 2012 and Lecompte et al., 2015). But ORC does not allow the heat pumping possibilities that some absorption cycles offer as cycle inversion looks difficult. Although the variants are numerous in both cases, one can say that in terms of efficiency, there is no clear winner.

In this section, the classical absorption power cycles are briefly presented as a background for more advanced proposals. In Section 8.7.1, the focus will be put on cycles that can simultaneously or alternatively produce cold or electricity. Ayou et al. (2013) is a detailed review of combined power/cold cycles. No booster ejector compressor of any kind is considered in this section.

8.7.1   BASIC POWER CYCLES

In Kalina (1983), a breakthrough was established: an irreversibility reduction was offered in the heat transfer from external enthalpy sources by using the *Kalina cycle*. Instead of using as a working fluid a pure substance, two miscible liquids with different but not too dissimilar volatilities, typically NH_3/H_2O were used. The resulting vapor generation evolution of temperature shows a *temperature glide*; this means a temperature increase corresponding to saturation when the liquid is enriched on the less volatile fluid during the heat addition, as in the absorption machines up to now described. If the thermodynamic properties of the working fluid and mass flow rates are appropriate, this leads to a fairly constant temperature difference with a counter-current external flow that transfers its enthalpy to the cycle, allowing the extraction of more heat than if the boiling occurred with a constant-temperature pure liquid. A similar situation happens when the two-component vapor is absorbed at lower pressure, rejecting heat to an external counter-current flow that in turns can be heated by a higher amount because of the glide. The larger pressure for heat rejection is inconvenient, so that the vapor solution from a distillation unit is mixed with a bypass basic solution, according to Ayou et al. (2013). The combined stream is then cooled and condensed in a second condenser, which adds extra degrees of freedom to the system and allows the distillation unit to operate at a lower pressure than the boiler pressure. Working with medium to low temperature heat carrying flows shows high potential. For solar thermal activation, the advantage of glide is not the same, but allows reducing the heat transport fluid mass flow rate. The use of NH_3/H_2O means natural substances for which there is much experience. The expansion does not show problems of wetness in the working fluid, even considering that NH_3 is a *wet vapor* (Quoilin et al., 2013). The high H_2O content of the vapor precludes using this cycle for cooling. Kalina cycles have been proposed and sometimes commercialized under many variants (e.g., Ayou et al., 2013). In that paper, several figures of merit are discussed to fairly evaluate the efficiency of a combined cycle with a similar aim to the $ECOP_{coo}$ defined in Equation 8.4.

Excluding the solution $H_2O/LiBr$ as working fluid for its extremely high specific volume of the H_2O vapor, H_2O/NH_3 does not seem attractive for circulating inside a DE cycle for its high pressure (similar to $NH_3/LiNO_3$) but also for the high H_2O content of the generated vapor. This last difficulty is not present with $NH_3/LiNO_3$. The high saturation pressure is circumvented by producing the second effect.

8.7.2   COMBINED POWER AND COLD PRODUCTION

The well-documented Goswami Cycle combines power/cold production (Goswami, 1998). It was initially based on an SE cycle where the entire NH_3/H_2O vapor produced is expanded in a turbine without rectification. Its vapor outlet could be cooler than ambient temperature thus providing cold from its sensible thermal energy in a heat exchanger named *precooler*. Tamm et al. (2004) present a theoretical study where the Goswami Cycle is optimized either from the energy and exergy points of view aiming at its use for the combined production of power and cold in a flexible way varying the cycle temperatures.

Some experimental developments of small size show moderate first law efficiencies (e.g., Mendoza et al., 2014) although excellent for the small size and moderate driving temperatures. The main reason for them can be the moderate efficiency of the purposely adapted small size scroll expander. In this cycle, $NH_3/LiNO_3$ and $NH_3/NaSCN$ were used as working pairs, showing not much efficiency difference. In addition, NH_3/H_2O was tested, adding rectification over the Goswami Cycle, recovering the heat evacuated from the rectification tower for preheating the solution. Rectification allows producing cold in an evaporator. A splitter allows choosing the amount of vapor toward the expander, but the split ratio was fixed by the constant rotational speed of the volumetric expander used.

Adsorption has also been proposed for combined power/cold production. In Wang et al. (2013) the authors offer the description of a readsorption cycle with two modes of operation and several metallic salts as adsorbers. One is a conventional readsorption cycle where one chemisorption compressor composed by two solid beds is connected to an identical chemisorption inverted compressor, thus forming a chemisorption expander, this way offering a cooling machine where there is no liquid-phase ammonia, in favor of safety. It operates with high COP_{coo} up to 0.77, according to the canonical model offered. In the other mode of operation, the vapor chemically pumped is expanded in an isentropic turbine, producing power, although no energy efficiency is given. Instead, the exergy efficiency ε obtained reaches high values, above the Goswami Cycle, 0.69 for electricity and 0.29 for refrigeration, although not much information is given for the cooling temperatures. The expanded vapor is cold enough to produce a small quantity of cold in a precooler upstream from the absorber with a $COP_{coo} = 0.05$ maximum. There is the necessity of switching between modes what carries some losses to reach the operating regime.

In Ventas et al. (2017) a TSDE cycle using $NH_3/LiNO_3$ and producing cold with a high COP_{coo} is complemented with an expander/turbine to produce power on demand in a parallel location between the two generators and the two absorbers. This can be a retrofit for an existing chiller. A numerical model based on a thermodynamic cycle indicates that wetness at the exit of the expander/turbine can be satisfactorily low without superheating. When producing only power, the efficiency is located in the range 14%–19%.

8.8   SOLAR HEATING AND COOLING

Solar heating can be produced directly by a solar thermal collector, but either heat pumps or heat converters using sorption cycles can be applied to adapt the production to the temperature needs (see Figure 8.8). Solar cooling can also be produced with sorption machines, SE for flat plate or vacuum tube collectors and DE with sun concentrating collectors, typically being concentrating parabolic stationary collectors, parabolic troughs, and Fresnel type linear collectors (Li and Sumathy, 2000; Fan et al., 2007, among others). In Bataineh and Taamneh (2016), a review of some installations is offered. Today those concentrating collectors that operate in the 150°C–250°C range, called medium temperature (MT) type are receiving special attention. They are especially indicated for DE machines. They inherit part of the technology of high temperature models for solar thermal power plants, reaching about 350°C–400°C. A new development is the Photovoltaic/Thermal (PVT) collector (Sharaf and Orhan, 2015). In them, the majority of solar light that is not converted to electricity by the PV cell is used to heat a flow for transporting heat to the user. There are contradictory needs: on one hand, the PV cell does not endure high temperature; they diminish the conversion efficiency and can even be destroyed. On the other hand, higher temperatures for the collected heat mean higher efficiency of the thermal machine. The dramatic price decrease of PV panels poses high competition as electricity can be managed more easily than thermal energy and the $COP_{coo,R} = \dfrac{\dot{Q}_e}{\dot{W}_k}$ (Figure 8.1) of practical inverse Rankine cycles is about 3–5, what compensates a lower commercial panel efficiencies from solar to electricity $\eta_{PV} = \dfrac{\dot{W}_{PV}}{G_T A_a}$ of around 0.1–0.15 before thermal solar collectors $\eta_{TH} = \dfrac{\dot{Q}_{TH}}{G_T A_a}$ in the range of 0.5–0.7 (Allouhi et al., 2015). G_T is the total tilted irradiance normal to the aperture area A_a of the solar collector.

8.8.1 Direct Solar Vapor Production

In thermally activated layouts there is one or several loops of thermal fluids that bring the driving heat to the generator heat exchanger (e.g., Figure 8.5). If these loops can be eliminated, drawing the working fluid of the cycle directly through the heat recovering exchanger, a more compact and efficient system will be constructed if the loop extension can be minimized. Moreover, these thermal fluids can be expensive and polluting as a residue, as specialty oils are often required, which also pose fire risks. When using steam, a nonnegligible installation cost will have to be amortized with the durable cost.

Implementing the vapor generator and separator into a solar collector is an innovation that seems of interest in favor of compactness and simplification. Direct steam production in solar power plants is now of high interest (e.g., Kalogirou et al., 1997; Roldán et al., 2013; Biencinto et al., 2016), for reasons similar to the ones presented below.

As Figure 8.5 shows, the current layout of a solar/absorption facility includes an outdoor primary circuit of a heat transfer fluid frequently including antifreeze capability; a secondary indoor circuit, typically carrying water, eventually including a heat storage tank (TES); and a tertiary circuit heated water for driving the chiller (see Kalogirou, 2004; Lecuona et al., 2009; Bermejo et al., 2010 for more detail). A hydronic heat rejection circuit is also needed when the machine is not directly air cooled. All the pumps indicated contribute to the parasitic energy consumption.

The end result is an increase in the *levelized cost of energy* (*LCOE*). All that induces some reluctance to invest, reinforced by the tailored character of every contemporary facility. The Carbon Thrust confirms this appreciation: *"The lack of an optimized solution is the most critical barrier for rural or off-grid application of CSP"* (2013).

In Lecuona-Neumann, Ventas-Garzón et al. (2016), an air-cooled SE absorption machine is proposed (see Figure 8.6). It is directly connected to at least one *MT concentrating solar collector* (Jradi and Riffat, 2012). The $NH_3/LiNO_3$ solution slides internally in a parabolic inclined receiving tube through solar collectors under a stratified two-phase gravity-driven flow regime (Wallis 1969), being collected at its lower end and sent back to the machine. The vapor produced by boiling/evaporation $\dot{m}_v$ raises to the upper side of the tube, separated from the liquid, and flows in the opposite direction from the solution. It is collected at the upper end of the tube. Low velocities are involved $v \sim 1\,m/s$ so that an almost flat interphase is formed.

This machine must be installed next to the collectors, thus outdoors. This implies fewer requirements for safety against NH_3 leaks. The solution and NH_3 are the only working fluids, this way no freezing or crystallization problems arise when the machine stops. Only the produced energy is delivered to the building, thus not consuming valuable machine room space or penetrating the building envelope with the machine heat transfer piping.

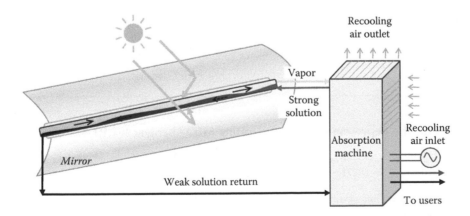

FIGURE 8.6 Scheme of an elementary solar cooling layout with stratified flow in the receiver tube and attached air cooled absorption machine. Liquid waviness exaggerated.

Large receiver tube inclinations angles β favor the collection of solar energy at medium to large latitudes for North-South collector orientation. With this embodiment (see Figure 8.6), solution mass flow rate $\dot{m}_l$ can be varied widely, producing a change in wet (liquid l) and dry (vapor V) cross section perimeters P and liquid central angle φ; this allows an exit temperature control. This layout could lead to perimetral temperature differences (Lecuona-Neumann et al., 2016).

Figure 8.7 shows some numerical results for an MT parabolic trough with an aperture width L_{ap} and a length L, for simplicity assuming liquid inlet temperature at saturated condition, which is imposed all along the axial length z from inlet. For inlet condition, the condenser temperature chosen is $T_c = T_{amb} + 10$ K. This corresponds to $p = 15.5$ bar inside the collector. Choosing an inlet temperature $T_{in} = 100\,°C$ it corresponds to a vapor mass fraction $x_{in} = 0.428$. The results validate the concept, as well as the ongoing test campaigns. For a tilted beam irradiance $G_{Tb} = 800\,\mathrm{Wm}^{-2}$ the production of ammonia vapor mass fraction for this case is $\Delta x = 8.41\%$, similar to a nonsolar generator. This means a crystallization temperature of about 15°C. The counterflow layout allows one to minimize vapor superheating, reducing the load in the condenser (Table 8.1).

The higher driving temperature with MT collectors allows the application of DE and other advanced cycles (Ventas et al., 2016). This way, several opportunities appear to reduce *LCOE* and increase coverage of the user demand, boosting cooling capacity by (1) consuming electricity with

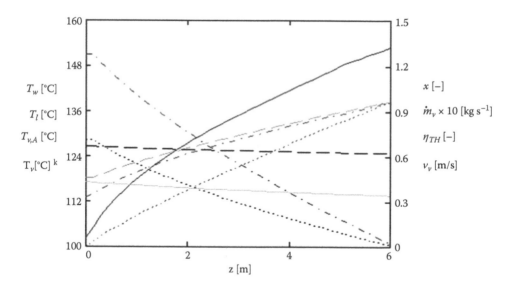

FIGURE 8.7 Evolution along z of wall W, liquid l and vapor V temperatures with $L = 6$ m; inner tube diameter $D = 3$ cm; $\dot{m}_l = 0.0833$ kg s^{-1}; $\beta = 15°$; tube inner roughness $\varepsilon = 0.1$ mm; $\varphi = 101°$; $v_l = 0.97\,\mathrm{m\,s}^{-1}$; $x_{ou} = 0.339$; $T_{amb} = 30°C$. $T_{V,A}$ stands for vapor adiabatic flow. A is for adiabatic; η_{TH} is for thermal efficiency.

TABLE 8.1
Basic Data for the Representative Concentrating MTC Abengoa-IST PT-1

$L_a{}^a$ (m)	L^a (m)	$a_0 \equiv F'(\tau\alpha)_{en}$ [–]	a_1 $\left[\mathrm{W\,m^{-2}\,K^{-1}}\right]$	a_2 $\left[\mathrm{W\,m^{-2}\,K^{-2}}\right]$	**Original Fluid**	T_{max} [°C]
2.2	6.0	0.71	0.3581	0.0019	water	250

Note: Effects of wind, sky infrared radiation and incidence are ignored.

[a] For reported net aperture surface $A_a = L_a L$. Normalized Collector Efficiency: $\eta_{TH}\langle T_w\rangle = \dfrac{a_0 - a_1(T_w - T_{amb}) - a_2(T_w - T_{amb})^2}{G_{Tb}}$ modified considering azimuthal average wall temperature $T_w \approx T_{water}$.

the same machine (Vereda et al., 2014), (2) direct or pumped heat production in winter, (3) increased *COP* using two stages/double effect TSDE cycle (Ventas et al., 2016) and (4) even allowing electricity production in periods when neither heat nor cold is needed.

8.9   ABSORPTION INTEGRATED WITH ELECTRICITY

The booster cycles signify a step toward the integration of absorption into the electrical grid and eventually into smart grids. Until now, adding electricity production with hybrid absorption cycles has not been extensively studied as a further step toward this aim. Another requirement is flexibility, meaning the capability of modulating between electricity production/consumption and more interestingly, the capability of modulating the cold/heat production consuming the available renewable/residual heat and complementing it with grid consumption. Another possibility can be opened by generating electricity in a flexible way when the demand of heat/cold is lower than what is possible from renewable/residual heat. Finally, if PV is one of the renewable sources, splitting between direct electricity consumption/grid injection and generating heat/cold is another possibility through an electric controller. Figure 8.8 illustrates such a highly integrated energy system with absorption technology and PV.

Some materializations of absorption machines aim at research studies on energy integration. One attempt is Angrisani et al. (2016). Although still using fossil gas for feeding a reciprocating IC engine, both the residual heat and the generated electricity drive a compressor-boosted SE machine for producing cold; the engine remnant residual heat can also be a product.

8.10   CHARACTERISTIC TEMPERATURE DIFFERENCE MODEL

The change in heat power consumed and capacity of an absorption machine when operating at different external temperatures is of high interest. Moreover, the solution pumps are usually operating at constant solution flow rate that facilitates generalization. Sometimes the available data from the manufacturer is scarce, so a predictive model helps in this direction by predicting the machine performance at different operating conditions than the given ones.

This practical model is based on comparing the external circuit's heat temperature drive, defined as the difference of the average driving temperature and ambient, thus the thermal potential: $\bar{T}_g - \bar{T}_c$ (somehow exergy content) Figure 8.2, with the heat temperature lift $\bar{T}_a - \bar{T}_e$ (exergy delivered). In Puig-Arnavat et al. (2010) and Lecuona et al. (2009) practical applications of this model track back to its origin. Assuming that typically absorber and condenser are tied together in parallel $\bar{T}_a = \bar{T}_c = \bar{T}_{ac}$,

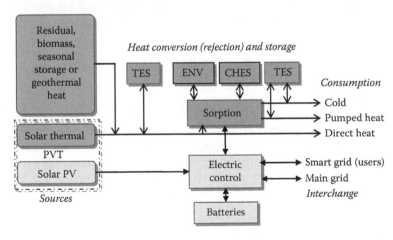

FIGURE 8.8   Integration of sorption machines in the energy services system. ENV indicates heat that is exchanged with the environment, either the atmosphere, water reservoirs or geothermal sources/sinks.

rejecting heat for producing cold, the basic variable determining the exchanged powers of the machine is the characteristic temperature difference $\Delta\Delta T$, under many simplifying assumptions. For the basic cycle layout, see Figure 8.2:

$$\bar{T} = \frac{T_{in} + T_{out}}{2}; \; \Delta\Delta T = \overbrace{\left(\bar{T}_g - \bar{T}_{ac}\right)}^{\text{Drive}} - \beta\overbrace{\left(\bar{T}_{ac} - \bar{T}_e\right)}^{\text{Lift}} = \bar{T}_g - d_{ac}\bar{T}_{ac} + d_e\bar{T}_e \, f0; \beta, \, d_{ac}, d_e > 1 \qquad (8.5)$$

A linear dependence is commonly accepted:

$$\frac{\dot{Q}_g}{\dot{Q}_{e,nom}} = g_0 + g_1 \Delta\Delta T > 0; g_0, g_1 > 0; \frac{\dot{Q}_e}{\dot{Q}_{e,nom}} = e_0 + e_1 \Delta\Delta T > 0; e_0, e_1 > 0 \qquad (8.6)$$

Here $\dot{Q}_{e,nom}$ is the nominal cooling capacity of the chiller. This model considers a linear dependence of the external average temperatures $\bar{T}$. It serves for partial load operation and for greater-than-normal production, as the absorption machines are thermally driven. The coefficients d_{ac}, d_e, g_0, g_1, e_o, and e_1 can be totally or partially calculated by the least squares fitting to the manufacturer's data. This model somehow resembles a differently grounded model (Gordon and Ng, 2000).

8.11 COMMENTS

8.11.1 PRIMARY ENERGY CONSUMPTION

Absorption heat pumps seem to have a modest COP (see Figure 8.4) in front of mechanically compression inverse Rankine ones, exhibiting practical $COP_{coo,R} \approx 3-5$. In terms of primary energy consumption, this is not the case, as this example shows.

Let's consider primary energy at the wall of the power station. Let's assume that the average thermal power plant efficiency is $\eta_{pp} = 40\%$ (this means a combination of low efficiency power plants: coal, nuclear; medium efficiency: gas turbine, Diesel; and high efficiency, such as combined cycle).

Let's assume a typical 8% loss of electricity transport and distribution down to consumption point, thus $\eta_{line} = 92\%$.

The final conversion into cold using a mechanical compressor chiller delivers for 1 unit of primary energy: $\eta_{pp}\eta_{line}COP_{coo,R} = 0.4 \times 0.92(3 \text{ to } 5) = 1.1 \text{ to } 1.84$.

An absorption SE chiller operating at appropriate temperatures yields a representative $COP_{coo} = 0.7$ and with DE around $COP_{coo} = 1.2$ (Figure 8.4). Thus, only the double effect machine is competitive from this standpoint. In Section 8.8 the use of renewables is considered.

8.11.2 PRACTICAL USE OF CHARACTERISTIC TEMPERATURE DIFFERENCE MODEL

For a small-scale SE chiller the fitted coefficient values were found for $\bar{T}\,[°C]$: $d_{ac} = 2.5$; $d_e = 1.8$; $g_0 = 0.2$; $g_1 = \frac{0.051}{°C}$; $e_0 = 0.09$; $e_1 = \frac{0.042}{°C}$. With this model it is straightforward to obtain $\dot{Q}$ and COP versus the three temperatures. It reveals the negative effect of increasing $\bar{T}_{ac} < \bar{T}_g$ and decreasing $\bar{T}_e < \bar{T}_{ac}$ and the positive effect of increasing $\bar{T}_g > \bar{T}_{ac}$. There are limit values beyond which operation is not possible, and the model cannot predict this (e.g., crystallization, power limitations, etc.). The model correctly manifests that increasing $\bar{T}_g$ above the minimum value that makes $\dot{Q}_e > 0$ COP_{coo} raises at a high pace, but slows down reaching an asymptote, somehow resembling Figure 8.4. Not predicting the COP_{coo} decrease for high $\Delta\Delta T$ is not crucial as this zone is not used frequently. Feeding an absorption machine with $\bar{T}_g$ so high that the COP_{coo} plateau has been reached is not convenient from the exergy point of view, although power can be high.

BIBLIOGRAPHY

Allouhi, A., T. Kousksou, A. Jamil, P. Bruel, Y. Mourad, and Y. Zeraouli. Solar driven cooling systems: An updated review. *Renewable and Sustainable Energy Reviews* 44 (2015): 159–181.

Angrisani, G., M. Canelli, C. Roselli, A. Russo, M. Sasso, and F. Tariello. A small scale polygeneration system based on compression/absorption heat pump. *Applied Thermal Engineering* 114 (2017): 1393–1402.

ASHRAE. *ASHRAE Handbook—Fundamentals.* Atlanta, GA: American Society of Heating, Refrigerating and Air-conditioning Engineers, 2013.

Ayala, R., C.L. Heard, and F.A. Holland. Ammonia/lithium nitrate absorption/compression refrigeration cycle. Part I. Simulation. *Applied Thermal Engineering* 17, no. 3 (1997): 223–233.

Ayou, D.S., J.C. Bruno, R. Saravanan, and A. Coronas. An overview of combined absorption power and cooling cycles. *Renewable and Sustainable Energy Reviews* 21 (2013): 728–748.

Balaras, C.A. et al. Solar air conditioning in Europe—An overview. *Renewable and Sustainable Energy Reviews* 11, no. 2 (2007): 299–314.

Bataineh, K., and Y. Taamneh. Review and recent improvements of solar sorption cooling systems. *Energy and Buildings* 128 (2016): 22–37.

Bermejo, P., F.J. Pino, and F. Rosa. Solar absorption cooling in Seville. *Solar Energy* 84, no. 8 (2010): 1503–1512.

Best, R., L. Porras, and F.A. Holland. Thermodynamic design data for absorption heat pump systems operating on ammonia-lithium nitrate-part one. Cooling. *Heat Recovery Systems and CHP* 11, no. 1 (1991): 49–61.

Biencinto, M., L. González, and L. Valenzuela. A quasi-dynamic simulation model for direct steam generation in parabolic troughs using TRNSYS. *Applied Energy* 161 (2016): 133–142.

Brückner, S., S. Liu, L. Miró, M. Radspieler, L.F. Cabeza, and E. Lävemann. Industrial waste heat recovery technologies: An economic analysis of heat transformation technologies. *Applied Energy* 151, no. 1 (2015): 157–167.

Cai, D., G. He, Q. Tian, and W. Tang. Exergy analysis of a novel air-cooled non-adiabatic absorption refrigeration cycle with NH3-NaSCN and NH3-LiNO3 refrigerant solution. *Energy Conversion and Management* (2014): 66–78.

Chen, Q., G.P. Hammond, and J.B. Norman. Energy efficiency potentials: Contrasting thermodynamic, technical and economic limits for organic Rankine cycles within UK industry. *Applied Energy* 164, no. 15 (2016): 984–990.

Cuenca, Y., D. Salavera, A. Vernet, A.S. Teja, and M. Vallès. Thermal conductivity of ammonia+lithium nitrate and ammonia+lithium nitrate+water solutions over a wide range of concentrations and temperatures. *International Journal of Refrigeration* 38 (2014): 333–340.

Eames, I.W., S. Aphornratana, and D.W. Sun. The jet-pump cycle-A low cost refrigerator option powered by waste heat. *Heat Recovery Syslems and CHP* 15, no. 1 (1995): 711–721.

Fan, Y., L. Luo, and B. Souyri. Review of solar sorption refrigeration technologies: Development and applications. *Renewable and Sustainable Energy Reviews* 11, no. 8 (2007): 1758–1775.

Farshi, L.G., A.H. Mosaffa, C.A. Infante Ferreira, and M.A. Rosen. Thermodynamic analysis and comparison of combined ejector–absorption and single effect absorption refrigeration systems. *Applied Energy* 133 (2014): 335–346.

Farshi, L.G., C.A. Infante Ferreira, S.M.S. Mahmoudi, and M.A. Rosen. First and second law analysis of ammonia/salt absorption refrigeration systems. *International Journal of Refrigeration* (Elsevier) 40 (2014): 111–121.

Gensch, K. Lithiumnitratammoniakat als absorptionsflüssigkeit für kältemaschinen.*Z. für Die Gesam. Kälte Indus.* 2 (1937): 24–30.

Gordon, J.M., and K.C. Ng. *Cool Thermodynamics.* Cambridge: Cambridge International Science Publishing. Viva Books, 2000.

Goswami, D.Y. Solar thermal technology: Present status and ideas for the future. *Energy Sources* 20 (1998): 137–145.

Gupta, Y., L. Metchop, A. Frantzis, and P.E. Phelan. Comparative analysis of thermally activated, environmentally friendly cooling systems. *Energy Conversion and Management* 49 (2008): 1091–1097.

He, Y.L., D.H. Mei, W.Q. Tao, W.W. Yang, and H.L. Liu. Simulation of the parabolic trough solar energy generation system with Organic Rankine Cycle. *Applied Energy* 97 (2012): 630–641.

Hernández-Magallanes, J.A., L.A. Domínguez-Inzunza, G. Gutiérrez-Uruetab, P. Soto, C. Jiménez, and W. Rivera. Experimental assessment of an absorption cooling system operating with the ammonia/lithium nitrate mixture. *Energy* 78 (2014): 685–692.

Herold, K.E., R. Radermacher, and S.A. Klein. *Absorption Chillers and Heat Pumps.* Boca Raton, FL and California: CRC Press, 1996.

Ibrahim, O.M., and S.A. Klein. Absorption power cycles. *Energy* 21, no. 1 (1996): 21–27.

Infante Ferreira, C.A. Thermodynamic and physical property equations for ammonia/lithium nitrate and ammonia-sodium thiocyanate solutions. *Solar Energy* 32 (1984): 231–236.

Jradi, Muhyiddine, and Saffa Riffat. Medium temperature concentrators for solar thermal applications. *International Journal of Low-Carbon Technologies* (Oxford University Press) 9 (2014): 214–224.

Kalina, A.I. Combined cycle and waste-heat recovery power systems based on a novel thermodynamic energy cycle utilising low temperature heat for power generation. *ASME Paper; 83-JPGC-GT-3* (ASME), 1983.

Kalogirou, S., S. Lloyd, and J. Ward. Modelling, optimisation and performance evaluation of a parabolic trough solar collector steam generation system. *Solar Energy* (Elsevier) 60, no. 1 (1997): 49–59.

Kalogirou, S.A. Solar thermal collectors and applications. *Progress in Energy and Combustion Science* 30 (2004): 231–295.

Kim, D.S., and C.A. Infante Ferreira. Air-cooled LiBr–water absorption chillers for solar air conditioning in extremely hot weathers. *Energy Conversion and Management* 50, no. 4 (2009): 1018–1025.

Kim, D.S., and C.A. Infante Ferreira. Solar refrigeration options—A state-of-the-art review. *International Journal of Refrigeration* 31, no. 1 (2008): 3–15.

Lecompte, S., H. Huisseune, M. van den Broek, B. Vanslambrouck, and M. De Paepe. Review of organic Rankine cycle (ORC) architectures for waste heat recovery. *Renewable and Sustainable Energy Reviews* 47 (2015): 448–461.

Lecuona, A., R. Ventas, M. Venegas, A. Zacarías, and R. Salgado. Optimum hot water temperature for absorption solar cooling. *Applied Energy* 83, no. 10 (2009): 1806–1814.

Lecuona-Neumann, A., M. Rosner, and R. Ventas-Garzón. Transversal temperature profiles of two-phase stratified flow in the receiver tube of a solar linear concentrator. Simplified analysis. *EUROSON 2016.* Palma de Mallorca, Spain: ISES, 2016.

Lecuona-Neumann, A., R. Ventas-Garzón, V. Vereda-Ortiz, and M. Legrand. Linear tube solar receiver as stratified flow vapor generator/separator for absorption machines using NH3/LiNO3. *11th ISES EuroSun Conference EUROSUN 2016.* Palma de Mallorca, Spain: ISES, 2016.

Li, Z., and K. Sumathy. Technology development in solar absorption air conditioning systems. *Renewable and Sustainable Energy Reviews* 4, no. 3 (2000): 267–293.

Libotean, S., A. Martín, D. Salavera, M. Valles, X. Esteve, and CoronasA. Densities, Viscosities, and Heat Capacities of Ammonia+Lithium Nitrate and Ammonia+Lithium Nitrate+Water Solutions between (293.15 and 353.15) K. *Journal of Chemical and Engineering Data* 53, no. 10 (2008): 2383–2388.

Mendoza, L.C., D.S. Ayou, J. Navarro-Esbrí, J.C. Bruno, and A. Coronas. Small capacity absorption systems for cooling and power with a scroll expander and ammonia based working fluids. *Applied Thermal Engineering* 72 (2014): 258–265.

Meunier, F. Adsorption heat powered heat pumps. *Applied Thermal Engineering* 61, no. 2 (2013): 830–836.

Niebergall, W. *Sorptions-Kältemaschinen. Handbuch die Kältemaschinen, Band 7.* Berlin: Springer Verlag, 1981.

Ochoa Villa, A.A., J.C. Charamba Dutra, and J.C. Henríquez Gerrero. *Introduçao a análise de sistemas de refrigeraçao por absorçao.* Recife, Brazil: Editorial Universitária de UFPE, 2011.

Puig-Arnavat, M., J. Lopez Villada, J.C. Bruno, and A. Coronas. Analysis and parameter identification for characteristic equations of single- and double-effect absorption chillers by means of multivariable regression. *International Journal of Refrigeration* 33 (2010): 70–78.

Quoilin, S., M. Van Der Broek, S. Declaye, P. Dewallef, and V. Lemort. Techno-economic survey of Organic Rankine Cycle (ORC) systems. *Renewable and Sustainable Energy Reviews* 22 (2013): 168–186.

Roldán, M.I., L. Valenzuela, and E. Zarza. Thermal analysis of solar receiver pipes with superheated steam. *Applied Energy* 103 (2013): 73–84.

Ryan, W.A. Water absorption in an adiabatic spray of aqueous lithium bromide solution. *International Absorption Heat Pump Conference, AES.* 1993. 155–162.

Sarbu, I., and C. Sebarchievici. Review of solar refrigeration and cooling systems. *Energy and Buildings* 67 (2013): 286–297.

Sharaf, O.Z., and M.F. Orhan. Concentrated photovoltaic thermal (CPVT) solar collector systems: Part I—Fundamentals, design considerations and current technologies. *Renewable and Sustainable Energy Reviews* 50 (2015): 1500–1565.

Srikhirin, P., S. Aphornratana, and S. Chungpaibulpatana. A review of absorption refrigeration technologies. *Renewable and Sustainable Energy Reviews* 5, no. 4 (2001): 343–372.

Tamm, G., D.Y. Goswami, S. Lu, and A.A. Hasan. Theoretical and experimental investigation of an ammonia–water power and refrigeration thermodynamic cycle. *Solar Energy* 76 (2004): 217–228.

The Carbon Thrust. *Small-scale Concentrated Solar Power. A review of current activity and potential to accelerate deployment.* London: The Carbon Thrust, 2013.

Ventas, R., A. Lecuona, A. Zacarías, and M. Venegas. Ammonia–lithium nitrate absorption chiller with an integrated low-pressure compression booster cycle for low driving temperatures. *Applied Thermal Engineering* 30 (2010): 351–360.

Ventas, R., A. Lecuona, C. Vereda, and M. Legrand. Two-stage double-effect ammonia/lithium nitrate absorption cycle. *Applied Thermal Engineering* (Elsevier) 94 (2016): 228–237.

Ventas, R., A. Lecuona, C. Vereda, and M.C. Rodriguez-Hidalgo. Performance analysis of an absorption double-effect cycle for power and cold generation using ammonia/lithium nitrate. *Applied Thermal Engineering* 115 (2017): 256–266.

Ventas, R., A. Lecuona, M. Legrand, and M.C. Rodríguez-Hidalgo. On the recirculation of ammonia-lithium nitrate in adiabatic absorbers for chillers. *Applied Thermal Engineering* 30 (2010): 2770–2777.

Ventas, R., C. Vereda, A. Lecuona, M. Venegas, and M.C. Rodríguez-Hidalgo. Effect of the NH_3–$LiNO_3$ concentration and pressure in a fog-jet spray adiabatic absorber. *Applied Thermal Engineering* 37 (2012): 430–437.

Vereda, C. *Eyector-absorbedor adiabático como potenciador de un ciclo híbrido para refrigeración por absorción basado en la disolución amoniaco-nitrato de litio (PhD Thesis).* Leganés, Madrid, Spain: Universidad Carlos III de Madrid, 2015.

Vereda, C., R. Ventas, A. Lecuona, and M. Venegas. Study of an ejector-absorption refrigeration cycle with an adaptable ejector nozzle for different working conditions. *Applied Energy* 97 (2012): 305–312.

Vereda, C., R. Ventas, A. Lecuona, and R. López. Single-effect absorption refrigeration cycle boosted with an ejector-adiabatic absorber using a single solution pump. *International Journal of Refrigeration* (Elsevier) 38 (2014): 22–29.

Wallis, G.B. *One-Dimensional Two-phase Flow.* New York: McGraw-Hill Book Co., 1969.

Wang, J., G. Chen, and H. Jiang. Study of a solar driven ejection absorption refrigeration cycle. *International Journal of Energy Research* 22, no. 8 (1998): 733–739.

Wang, L., F. Ziegler, A.P. Roskilly, R. Wang, and Y. Wang. A resorption cycle for the cogeneration of electricity and refrigeration. *Applied Energy* 106 (2013): 56–64.

Witte, J.H. Efficiency and design of liquid-gas ejectors. *British Chemical Engineering* 10, no. 9 (1965): 602–607.

Wu, W., B. Wang, W. Shi, and X. Li. Crystallization Analysis and Control of Ammonia-Based Air Source Absorption Heat Pump in Cold Regions. *Advances in Mechanical Engineering* (Hindawi Pub. Corp.) 2013 (2013).

Wu, W., B. Wang, W. Shi, and X. Li. Absorption heating technologies: A review and perspective. *Applied Energy* 130 (2014a): 51–71.

Wu, W., B. Wang, W. Shi, and X. Li. An overview of ammonia-based absorption chillers and heat pumps. *Renewable and Sustainable Energy Reviews* 31 (2014b): 681–707.

Wu, W., W. Shi, B. Wang, and X. Li. Annual performance investigation and economic analysis of heating systems with a compression-assisted air source absorption heat pump. *Energy Conversion and Management* 98 (2015): 290–302.

Xie, G., Q. Wu, X. Fa, L. Zhang, and P. Bansal. A novel lithium bromide absorption chiller with enhanced absorption pressure. *Applied Thermal Engineering* 38 (2012): 1–6.

Zeyghami, M., D.Y. Goswami, and E. Stefanakos. A review of solar thermo-mechanical refrigeration and cooling methods. *Renewable and Sustainable Energy Reviews* 51 (2015): 1428–1445.

Zhai, X.Q., R.Z. Wang, J.Y. Wu, Y.J. Dai, and Q. Ma. Design and performance of a solar-powered air-conditioning system in a green building. *Applied Energy* 85, no. 5 (2008): 297–311.

9 Hybrid Solar and Geothermal Heating and Cooling

Bernd Weber
Mexico State University

CONTENTS

9.1 INTRODUCTION

In energy applications, a design engineer generally selects a hybrid system to obtain a process that has some benefits over a conventional uncombined system. The benefits of combined systems in fulfilling heating and cooling purposes are generally found in the form of energy efficiency. As shown in previous chapters, energy efficiency is a key factor in (environmental) sustainability and economy, the latter of which can be used as a measurement of how well individual technical solutions can compete with other alternatives. Thus, thermal solar energy systems, characterized by the capability to provide energy at a high density though with intermittent availability, and geothermal systems, having constant availability but mostly present at low temperature levels, are emerging as replacements for fossil-fuel based systems. Without a doubt, harvesting solar energy has been integrated into human activities since prehistoric times and geothermal energy usage was first reported at least a few centuries ago; however, both lost prominence in covering human energy needs with the onset of the industrial revolution. The exception has been during the last few decades as renewable energies have gained market share, a trend mostly driven by the scarcity of fossil fuels.

9.2 SOLAR THERMAL ENERGY APPLICATIONS IN THE RESIDENTIAL SECTOR

In the past, the use of thermal solar energy systems was limited to the integration of small solar collectors into existing fossil fuel-based heating systems. The share of energy provided by solar energy systems of the total thermal energy demand of a building is defined as the fractional energy savings (F_{sav}) (Jordan and Vajen, 2001). The seasonal discrepancy between solar availability and building thermal energy demand leads to low fractional energy savings in most cases. However, the use of better thermal insulation and the implementation of heat recovery units in air ventilation systems have now allowed for the reduction of thermal energy demands in residences to such a degree that in some demonstration projects solar thermal fractional energy savings rise to nearly 100% (Oliva et al., 2015). The challenge of reaching thermal autonomy in buildings necessitates not only solar collector

TABLE 9.1

Performance of Solar Thermal Energy Systems

Type of Application	F_{sav} (%)	Storage Volume (M³)	Storage Type	Reference/Additional Information
Kindergarten, Ge	40	13	Vacuum isolation	(Bauer and Drück, 2016) Simulation Study
Solar Active House, Ge	80	16	Vacuum isolation	(Bauer and Drück, 2016) Simulation Study
Calabria, It	28	91		(Novo et al., 2010)
Lykovrissi, Gr	500	70		(Novo et al., 2010)
Ottrupgaard, Dk	16	1,500	Polyuretane	(Novo et al., 2010)
Hamburg Bramfeld, Ge	56	4,500	Concrete Water/Rock Storage	(Benner et al., 2003) Central Solar System For 124 Residential Buildings
Friedrichshafen-Wiggenhausen, Ge	35	12,000	Concrete Water/Rock Storage	(Benner et al., 2003) Central Solar System
Neckarsulm, Ge	39	63,360	Multiple Borehole Ground Storage	(Benner et al., 2003)

installation, but also the use of recent developments in thermal energy storage systems. Studies have shown that vacuum-based thermal insulation with a global heat transfer coefficient (U) of about 0.05 W m^{-2} K^{-1}, as compared to traditional insulation based on polyurethane with a u of 0.2 W m^{-2} K^{-1}, can increase fractional energy savings considerably (Bauer and Drück, 2016). The performance of solar thermal energy systems in residential settings has been monitored in various studies that revealed that in practice most harvested energy is lost during seasonal storage (Table 9.1).

The success of these technical achievements has given rise to new building guidelines. For example, the European guideline for edification (EPBD) specifies that over the yearly average on-site energy harvesting must be equal to at least the energy demand of the building, which must be met by 2019 for public buildings and 2021 for residential buildings.

9.3   BASICS OF HEAT PUMPS FOR GEOTHERMAL APPLICATIONS

The objective of this study is the evaluation of ground source heat pump systems in combination with solar energy. The geothermal heat in ground source heat pump systems is collected via boreholes, earth collectors, and thermoactive foundations or as open systems gaining benefit from groundwater resources. In addition, wastewater streams in sewers are considered anthropogenic resources of geothermal energy, and recent innovative developments have shown the feasibility of using such systems (Dbu, 2012).

Geothermal systems can first be classified by the depth of their ground heat exchangers. Systems deeper than 400 m working at higher temperature levels are intended for power generation applications. Alternatively, "near ground" geothermal systems that typically have applications in space heating are run at temperature levels in which the available heat must be transferred to higher temperatures by heat pumps. Thermodynamic laws explain the exergy input required to transfer thermal energy from a lower temperature level to a higher temperature level. Thus, the efficiency of heat pumps is evaluated by the coefficient of performance (COP), the relation of thermal energy provided and electric energy needed (exergy input) to transfer thermal energy between the two heat exchangers of the heat pump (Equation 9.1). Experimentally, the COP is measured in specialized laboratories under standardized conditions according to American or European Guidelines (Ahri Standard 340/360 (I-P)-2015 and En 14511).

$$COP = \frac{\dot{Q}_{Output}}{P_{el,\,heat\,pump}} \tag{9.1}$$

The *COP* of a heat pump running in the reversible mode can be described alternatively as the inverse of the Carnot efficiency

$$COP_{Rev} = \frac{1}{\eta_{Carnot}} = \frac{T_{Sink}}{T_{Source} - T_{Sink}} \tag{9.2}$$

The *COP* of a system operating in reversible mode is of particular interest because the relation between the *COP* of the heat pump and the *COP* of the reversible heat pump represents how far the system is running from the thermodynamic optimum. In fact, the relation between the *COP* of a heat pump and the reversible heat pump determines the exergy efficiency of the process.

$$\varepsilon = \frac{COP}{COP_{Rev}} = \frac{\eta_{Carnot} \cdot \dot{Q}_{Output}}{P_{el,\,heat\,pump}} \tag{9.3}$$

In an ideal case, the reversible heat pump runs with ambient air temperature as t_{source} and room temperature as t_{sink}. For common situations with a room temperature of 20°C and an outside air temperature of −15°C, −5°C, and 5°C the *COP* of a reversible heat pump is 8.4, 11.8, and 19.7, respectively. Thus, from a thermodynamic point of view, heating a building is a simple task because the thermal energy provided is mainly composed of anergy. Knowing that heat pump systems for residential heating purposes have a *COP* 2 to 5 times smaller, it can be concluded that the performance of technical systems is far from the benchmark set by a reversible heat pump, a result of the irreversibilities caused by temperature gradients in the heat exchangers, gas compression, and pressure reduction by a throttling valve.

Thus, in practice the great challenge when designing heat pump systems is to find the highest source temperature available and implement heating systems running with low temperature differences, as observed in floor heating systems. Mathematically, this situation is described by equation 9.2. The temperature variations between source and sink that naturally occur over longer periods are the reason that instead of using the *COP* a better evaluation of the heat pump can be determined by the seasonal or annual performance factor (*SPF* and *APF*, respectively).

$$SPF = \frac{Q_{Output}}{E_{el,\,heat\,pump} + E_{Aux}} \tag{9.4}$$

This equation also considers the energy needed for running auxiliary components like vans, pumps for brine circulation, defrosting functions, and back-up heaters. It is worth mentioning that the system boundary does not include pumps for heat circulation in the heating circuit of the building to allow for benchmarking with alternative heating systems. However, today's energy policies are looking to the primary energy demand and consequently to the comparison of primary energy demands in a heat pump system with other solutions. Thus, the primary energy factor must be taken into account in Equation 9.4 as shown in Equation 9.5.

$$Q_{Prim} = \frac{Q_{Output}}{SPF} \cdot f_P \tag{9.5}$$

The primary energy factor for electricity production is influenced by the energy mix of a country and by how that mix changes over time. For example, in Germany, as a result of the introduction of renewable energy into the energy mix, the primary energy factor for electricity was reduced from 2.6 to 1.8 over the last two decades (Esser and Sensfuss, 2016). Thus, heat pump systems are becoming more sustainable as the share of renewable energy in the power generation sector increases.

A simple calculation can determine the minimum *SPF* of a heat pump system necessary for the system to be more efficient in the consumption of primary energy than a condensing boiler using gas (assumed efficiency of the combustion system based on the net calorific value of 100%).

$$SPF = f_{P,\,el} \cdot f_{P,\,Gas} = 1.8 \cdot 1.2 = 2.16 \tag{9.6}$$

Even under unappropriated conditions in which heat pumps run the evaporator with ambient air, monitoring reports have shown that heat pump systems under real conditions reach the *SPF* calculated in Equation 9.6 (Miara et al., 2011).

In general, the heat pumps employed are based on gas compression technology, where the compressor is driven by an electric engine. Depending on the nationwide energy mix, gas compressors driven by an internal combustion engine may have a better primary energy demand. The author has shown that heat pumps using an internal combustion engine are a competitive alternative to cogeneration systems (Weber et al., 2014). Aside from gas compression technology, some heat pumps are based on absorption and adsorption technology, which are evaluated with the annual heating energy ratio (*AHER*) instead of the *SPF*. Although the alternatives to gas compression technology mentioned have demonstrated better performance in most applications, the following analysis is restricted to gas compression technology, specifically because such systems represent the largest share of the market.

In contrast to geothermal sources, heat pumps can take thermal heat from ambient air or waste heat from industrial applications. Heat pumps using ambient air as a heat source work throughout the year with a wider temperature interval than ground source heat pumps. Unfortunately, the heat demand pattern of a building is in opposition to the temperature pattern (the lowest ambient temperature results in the highest heat demand). In a monitoring survey for the period of 2007–2010 realized by the Fraunhofer Institute ISE, the average *SPF* of 18 air source heat pumps monitored was determined to be 2.89. The average *SPF* of the systems analyzed decreased from 3.03 to 2.87 for the years 2008–2010, which was mainly attributed to the decease of ambient temperatures by about 0.6 K based on the annual average, thus indicating a strong dependency on ambient temperature. In the same study, a total of 56 ground source heat pump systems were analyzed and exhibited an average *SPF* of 3.88. Contrary to the air source heat pump data, the ground source heat pump *SPFs* did not show dependence on the ambient air temperature (Miara et al., 2011).

9.3.1 Borehole Heat Exchangers

Generally, heat transfer from the ground to the heat carrier medium can be realized with vertical boreholes, although the literature also reports boreholes with a certain tilt angle in geothermal applications. Most boreholes are drilled to a depth of about 20–100 m and have a diameter of 100–150 mm. The heat carrier medium is conducted through a coaxial pipe or u-pipe. Frequently a double u-pipe is chosen because heat transfer from the surrounding earth through the borehole fill material into the pipes is adequate and borehole functionality is not completely lost in the case of damage to one pipe. Alternatively, coaxial pipes have the advantage that during maintenance the inner pipe may be pulled out. Another advantage of coaxial pipes is that heat transfer to the heat carrier medium can be optimized by thermal insulation of the inner pipe in the upper part of the borehole because in that part of the borehole the heat carrier medium is already heated up and the main task is to avoid cooling of the recently introduced medium.

The energy transferred from the earth to the borehole is by reason of a temperature reduction in the earth close to the borehole causing the heat transfer rate to be reduced. Consequently, the energy extracted during a heating season must be regenerated by the surrounding earth throughout the rest of the year, otherwise longtime stability of the boreholes cannot be guaranteed. The specific heat transfer capacity of a borehole depends on the thermal properties of the earth, the temperature distribution of the soil, and the components employed in the borehole.

Analytically, the temperature profile around the borehole can be described by Kelvin's line source theory (Ingersoll and Plass, 1948). The following equation, which describes the temperature difference at a specific point as a function of time and distance to the center of the borehole, is produced by taking into account boundary conditions and some simplifications.

$$\Delta T = T_\infty - T_{(r)} = \frac{0.1833 \cdot q_H}{k} \cdot log_{10}\left(\frac{a \cdot t}{r^2} + 0.106 \cdot \frac{r^2}{a \cdot t} + 0.351\right) \tag{9.7}$$

Equation 9.7 includes two thermodynamic soil properties: the thermal diffusivity (a) and the thermal conductivity (k). In Equation 9.7 the symbol q_H stands for the specific energy flow extracted per meter of borehole depth. The maximum admissible value of q_H depends on the duration of the heating period and the number of geothermal boreholes drilled. It is evident that the use of Equation 9.7, which is based on radial heat transfer, is limited when analyzing the regeneration of internal energy in the surrounding soil because of vertical heat transfer from both the surface and deeper soil.

The depth of the borehole plays an important role in the share of recharge energy coming from the surface and the share coming from deeper soil. Huber and Pahud showed that with increasing time, the regeneration share from the surface rises to values of over 80% (Huber and Pahud, 1999). The study applied the EWS program, which is based on g-functions developed by Eskilson (1987). The g-function is a dimensionless temperature response to heat extracted from a borehole. An example of such a temperature response is that of a heat pump put into operation in a vertical borehole. The g-function is described as follows and can be determined for simple boreholes as well as for systems of boreholes using analytical methods or numerical simulations (Haller et al., 2012):

$$g\left(Es, \frac{r_b}{H}\right) = \frac{\Delta T \cdot 2 \cdot \pi \cdot k_{Soil}}{\dot{q}} \tag{9.8}$$

where Es is dimensionless time and is defined as:

$$Es = \frac{t \cdot 9 \cdot a}{H^2} \tag{9.9}$$

The time constant of a borehole, t, is also of special interest:

$$t = \frac{H^2}{9 \cdot a} \tag{9.10}$$

For soils with a thermal diffusivity (a) varying from $0.35 \cdot 10^{-6}$ to $1.2 \cdot 10^{-6} m^2 \, s^{-1}$ a borehole with a depth of about 100 m results in a time constant of 30–100 years, and thus, a thermal equilibrium is reached within decades. It is important to note that the interaction of various boreholes is not negligible. The corresponding g-functions for multiple boreholes demonstrate that the radius of influence is in the range of the borehole depth. Thus, typical residential applications are likely to experience lower source temperature since the density of such systems increases in residential areas (Huber and Pahud, 1999).

Given the above, geothermal heating systems using vertical boreholes may not necessarily be considered renewable energy since the thermal equilibrium of such systems is reached within decades and the energy potential is not significant enough to provide for villages, towns, or cities. Therefore, the use of vertical boreholes for heating and cooling purposes in combination with active thermal energy recharge with solar thermal systems represent a viable solution for a more sustainable use of this technology as will be shown below.

9.3.2 Ground Heat Collectors

Ground heat collectors are defined as drawing geothermal heat from surface soils, up to approximately five meters in depth. Different heat exchanger geometries are available on the market including helixes or serpentines formed by pe pipes with a diameter in the range of 15–32 mm. Because the temperature of the soil near the ground is influenced by the ambient temperature and its hydrologic balance, the soil energy potential is generally regenerated during the spring and summertime, which represent the most significant energy-harvesting period for geothermal collectors.

These regeneration periods can be characterized by the quantity of heat extracted; for example, the VDI 4640 guideline recommends a maximum specific heat extraction of q_H per square meter of surface and the total of thermal energy extractable during a heating period. These data are tabulated as a function of the duration of a heating period, climatic conditions, and soil type. As a rough approximation, the specific heat extraction capacity is between 15 W m^{-2} and 40 W m^{-2} and the specific energy potential between 30 kWh m^{-2} y^{-1} and 70 kWh m^{-2} a^{-1}. If the latter values are compared with the average geothermic release of the earth, 0.6 kWh m^{-2} y^{-1}, it is obvious that the local environment plays an important role in the energy balance of such a system. Specifically, capacity is significantly reduced when installations are located above the ground collector. In such an aboveground arrangement, the collectors are classified as thermoactive foundations, dependent on other energy sources like solar energy, since recharging by air convection and direct solar irradiation of the earth's surface is no longer given.

9.3.3 Thermoactive Foundations

In thermoactive foundation systems, heat exchanger pipes are built directly into the foundation. Such solutions can be designed as plate foundations as well as pile foundations. Pile foundations are particularly apt to significantly reduce the cost of a ground source heat pump system since borehole drilling is the principal cost in ground source heat pump systems. However, the thermal efficiency of such systems is not usually optimized in such combinations given that the design of the foundation is based on structural considerations as opposed to maximizing heat transfer considerations. Nevertheless, the high thermal conductivity of the concrete used in foundations facilitates heat transfer to the surrounding ground, and recently implemented applications show that thermoactive foundations in residential and public constructions are effective in covering heating and cooling needs (Kwag and Krati, 2013). Since simple calculation methods for thermoactive foundations have not yet been developed, planning of such systems is frequently based on transient three- or two-dimensional modeling. Finally, the accuracy of heat transfer models in soil can be enhanced by exhaustive validation. Thus, various investigations have been completed or are in progress to systematically measure heat transfer in completed projects.

The thermal capacity of a thermoactive foundation and the surrounding earth lead to annual temperature variation of 4°C–20°C. In comparison, the natural soil temperature below the soil surface in Germany, for example, exhibits much smaller variation throughout the year and remains between 8°C and 12°C depending on location. At such temperature levels, a direct cooling operation during the summertime is sufficient and over the heating season, heat pump systems can reach an *SPF* close to 4. If the direct cooling operation option is not applicable, some heat pumps on the market can switch their operation mode. The above temperature interval is only given when cooling demand and heating demand are nearly balanced. The climate in Germany demands such a balance for office buildings, but for residential buildings, the balance is shifted to a higher heating demand. This is because there are usually lower inner thermal loads present in residences, and more effective natural or forced ventilation is realized during the nighttime. A solution for the cases where the heating demand prevails, the integration of solar thermal collectors, would be constructive in maintaining soil temperature at an adequate temperature range.

9.4 PASSIVE GEOTHERMAL COOLING

The soil close to a building has a more stable temperature than the outside temperature. Though the soil temperature close to the surface is generally influenced by the average season temperature, the soil at depths of 15 m or more (depending on thermodynamic properties of the soil) is closer to the average yearly temperature. Grigull and Sander developed Equation 9.7, which describes the soil temperature as a function of time and depth (1990). The model is based on the heat conduction theorem, average heat transfer coefficient at the surface h_{surf}, and thermodynamic parameters of the soil. Although this model does not consider underground water flows, freezing of water in soil, or strong radiation at the surface, it can provide soil temperature information if the soil has potential to act as cooling storage.

$$T_{Soil}(z, t) = \bar{T}_{average} \cdot \frac{\Delta T_{a, amp.} \cdot e^{-\varphi}}{\sqrt{1 + 2 \cdot \beta + 2\beta^2}} \cdot \cos\left(\frac{2 \cdot \pi \cdot t}{t_o} - \omega_o - \varphi - \varepsilon\right) \tag{9.11}$$

where:

$$\varphi = z \cdot \sqrt{\frac{\pi}{a \cdot t_o}}; \quad \beta = \frac{k}{h_{Surf.}} \cdot \sqrt{\frac{\pi}{a \cdot t_o}}; \quad \varepsilon = \arctan\left(\frac{\beta}{1 + \beta}\right) \tag{9.12}$$

Given Equation 9.11, in desert regions with large variations between daytime and nighttime temperatures, the heat exchanger should be placed close to the surface. Alternatively, in regions where seasonal temperature variation is large, better performance is reached when the ground heat exchanger is placed deeper. As an example, for a typical residence located in central Europe, the thermal analysis of the building results in a required cooling capacity of 6.85 kW; about 50% of the heating capacity is during the winter. Helix type ground collectors from the Rehau Company located at a depth of 1.5–4 m each transfer about 230 W from the heat carrier medium to the ground. Temperature differences registered during daytime cooling reach 3.5°C in the upper part of the soil (2.4 m depth) and 4.4°C in the deeper part of the coil (4.4 m depth), which are completely regenerated overnight. A third temperature measuring point placed at a depth of 0.8 m directly over the coil (representing a distance of 0.7 m to the upper part of the coil) does not show a significant temperature difference from uninfluenced soil temperature at the same depth. In summary, soil close to the surface is an attractive temperature sink to discharge waste heat in passive and active cooling applications.

Finally, passive cooling is also possible through the use of air-soil heat exchangers particularly when cooling incoming air in ventilation systems. Such systems are an additional component of central heating, ventilation, and air conditioning (HVAC) systems. In air-soil heat exchangers, energy savings are obtained since cooling capacity is in most cases higher than the additional energy needed for forced air circulation in the underground tubes. Studies have shown that with a 30 m long underground tube, incoming air at 32°C can be cooled down to 20°C (Dibowski, 2005). Furthermore, history and nature have shown that such systems can also be run with natural venting. In nature, termite hills are cooled with natural air circulation in vertical earth tunnels connected to an aquifer (Tributsch, 1999). Historically, buildings in Iran were constructed with ground air-soil heat exchangers based on natural venting (Bansal et al., 1994; Chohfi, 1988). Thus, recent conceptual designs have been based on such historical developments (Amara et al., 2011).

9.5 HYBRID SOLAR AND GEOTHERMAL APPLICATIONS

The integration of solar thermal systems into geothermal applications can be realized in different ways. The most common are:

1. Heating a buffer tank
2. Heating brine between a borehole and heat pump
3. Recharging a ground source

Since geothermal systems use a buffer tank between the heat pump and consumer, solar energy can be used to heat up the buffer tank with the effect that the heat pump operation hours are reduced. The main limitation for such an arrangement is that a reasonable compromise must be made between the buffer tank size designed for a heat pump and that designed for a thermal collector.

If the solar collector receives lower irradiance, its temperature level may be too low to feed the buffer tank; however, heating the borehole heat carrier is still possible. Such an alternative is only available when the heat pump is in operation. For example, the *COP* of a heat pump running at a source temperature of 0°C and 50°C is about 2.4, and when the brine inlet temperature is shifted to 10°C using solar energy the *COP* rises to about 3 (Pärisch et al., 2011). A study from Marx realized with a low cost uninsulated solar absorber showed that the annual performance for such an arrangement was increased (Marx, 2013). Only a few hydraulic modifications are necessary to achieve ground source recharge using such an arrangement (Figure 9.1). Specifically, unidirectional heat flow from the ground surrounding the heat carrier pipes to the boreholes must be considered. This behavior does not change at a sufficient distance from the borehole when using solar recharge. Thus, as opposed to the solar storage systems described technically in Table 9.1 as an alternative to the hybrid system, no solar energy is lost. However, heat conduction inside the earth represents an exergy destruction, which has to be replaced by the heat pump when heating residences.

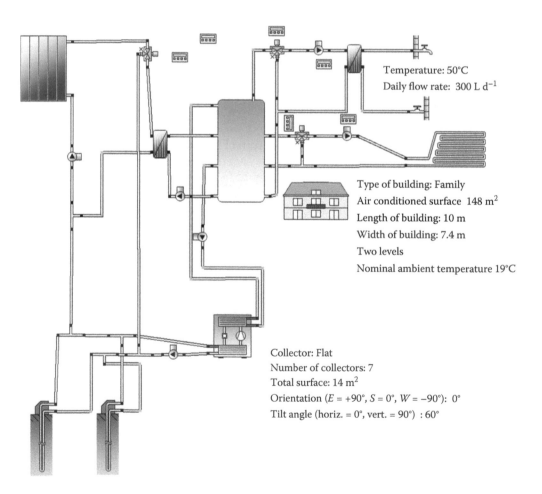

Temperature: 50°C
Daily flow rate: 300 L d^{-1}

Type of building: Family
Air conditioned surface 148 m^2
Length of building: 10 m
Width of building: 7.4 m
Two levels
Nominal ambient temperature 19°C

Collector: Flat
Number of collectors: 7
Total surface: 14 m^2
Orientation (E = +90°, S = 0°, W = −90°): 0°
Tilt angle (horiz. = 0°, vert. = 90°) : 60°

FIGURE 9.1 Hydraulic scheme of hybrid solar and ground source heat pump system generated for simulation with polysun4.

Experimental studies have shown that *SPF* is increased significantly by borehole recharge with solar energy. For example, through the simulation of a case study, Bertram found that for a residential application the *SPF* rises from 3 to 5 with the implementation of $15\,m^2$ collector area (2013). Commercial software like the Polysun 4, EWS, EED, and Trnsbm have program packages to do simulations of configurations similar to those shown in Figure 9.1.

As an example, in a simple set-up for a residential building with a heated surface of $120\,m^2$, a moderate energy demand for space heating of $70\,kWh\,m^{-1}\,a^{-1}$, and a hot water demand of $3000\,kWh\,a^{-1}$, the combination of a heat pump with thermal solar can be parameterized. The VDI 4640 guideline recommends not exceeding the extraction rate of $49\,W\,m^{-1}$ for a soil with a thermal conduction of $2.6\,W\,m^{-1}\,K^{-1}$ when implementing a heat pump running 1800h per year at full capacity. With an average *SPF* of 3, the four boreholes should be 70 m deep to fulfill the energy requirements of the building. Given a constant heat carrier medium temperature of 45°C and the g-functions for the four boreholes, both the temperature in the borehole and the temperature of the heat carrier at the source can be calculated. Under such conditions, the *SPF* of the heat pump can be assessed and is shown as a function of the borehole depth as seen in the reference system in Figure 9.2. The same figure displays how the conditions change when solar thermal collectors cover energy demand during the summertime and recharge excess energy into the boreholes. While without solar support the *SPF* remains below 3.5, the *SPF* for solar supported systems increases to over 6.5. This increase is mainly a result of the increased share of solar energy in the Q_{output} (Equation 9.4).

Figure 9.2 also shows the iso-lines of the relative investment of a heat pump system with four 70 m deep boreholes without solar thermal support. By choosing the specified alternative, one can reduce the borehole depth to 45 m and install $12\,m^2$ of solar thermal collectors; thus, with the same investment the *SPF* of 3 for the reference system can be increased to 4.3. Consequently, the alternative is a more ecoefficient solution for the heating needs of the reference building.

The share of solar energy in Equation 9.4, with almost no added electricity needs, increases the *SPF* in heat pump systems using ground heat collectors as a source in a similar way. However, the heat transfer in soil close to the surface is made more complicated by additional recharging with

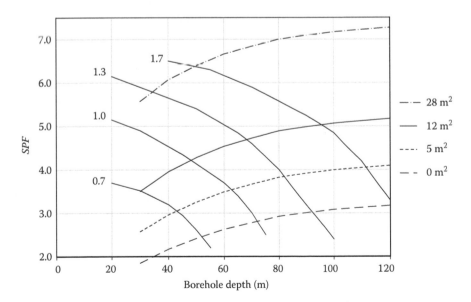

FIGURE 9.2 Seasonal performance *SPF* of a ground source heat pump in function of the borehole depth (4 boreholes) combined with different area of solar thermal collectors (0, 5, 12 and $28\,m^2$). The figure also contains iso-lines for a relative investment of a heat pump system and borehole depth of 70 m without solar thermal support.

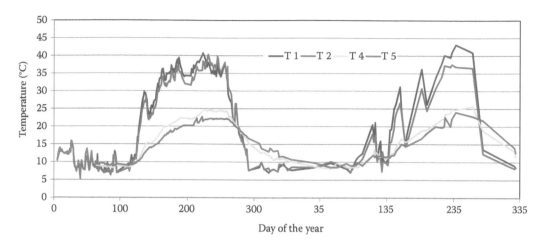

FIGURE 9.3 Pattern of manually taken temperatures in tank (T1, T2) and soil (T4, T5) at 0.5 and 1.0 m distance from tank shell.

solar energy, which requires that simulation models for such arrangements should be developed and validated for completed projects.

While experience has shown that seasonal thermal energy storage adheres to higher energy losses, the author investigated the effectiveness of an uninsulated buried tank for seasonal thermal heat storage (Weber et al., 2016). With such an arrangement, the challenge is to reach a high temperature level in the liquid medium and to recover heat losses through the tank wall when storage capacity of the tank is exhausted and consequently heat transfer is reversed. The latter stage corresponds to the energy recovery from boreholes when recharging is realized with solar excess heat. The relation between collector area and storage tank capacity is about one square meter per cubic meter. This relation is similar to storage systems for small communities where maximum temperatures of about 70°C are reached (Benner, 2003). In the alternative of an uninsulated buried tank, presented in Figure 9.3, maximum temperatures of 45°C are reached.

9.6   CONCLUSION

Technical and economic limitations are large factors in the divergence of today's heat pump technologies from the thermodynamic maximum. All heat exchangers employed in heat pumps have to operate with a reasonable temperature gradient because the surface cannot be extended ad infinitum, which would correspond to zero exergy losses in this step. Recent trends have focused on looking for energy, more precisely anergy, sources nearer to the required process temperature to overcome the heat exchange barrier in heat pump systems. One such energy source is found in the soil, which usually has a higher temperature than ambient air during the heating season. Another way to increase heat pump performance or to reduce heat pump operating hours is the integration of solar energy, and when combined with ground source heat pumps a proper solution for the seasonal energy storage is found. Monitoring programs realized to determine the performance of heat pump systems have revealed that systems based on ambient air as an energy source are already competitive with fossil fuel-based heating systems with respect to primary energy consumption. It is worth mentioning that the introduction of renewable energy continuously decreases the primary energy factor of electricity, thus, making heat pump systems even more attractive. Taking into account improvements to optimize the energy sources presented in this manuscript and further technical improvements of heat pumps, it is evident that heat pump systems will play an important role in future residential heating and cooling applications.

BIBLIOGRAPHY

Amara, S.; Nordell, B.; Benyoucef, B. (2011) Using Fouggara for Heating and Cooling Buildings in Sahara. *Energy Procedia* 6: 55–64.

Bansal, K.N.; Hauser, G.; Minke, G. (1994) *Passive Building Design, a Handbook of Natural Climatic Control*, Elsevier Science Verlag, Amsterdam.

Bauer, D. and Drück, H. 2016. Entwicklung Grossvolumiger, Preiswerter Warmwasserspeicher Mit Hocheffizienter Dämmung Zur Aussenaufstellung. Final Report 0325992a+B to the Federal Ministry for Economic Affairs and Energy (Bmwi), Stuttgart, Germany.

Benner, M.; Bodmann, M.; Mangold, D.; Nussbicker J.; Raab, S.; Schmidt, T.; Seiwald, H. (2003) Solar Unterstützte Nahwärmeversorgung Mit Und Ohne Langzeit-Wärmespeicher. Final Report 0329606 S to the Federal Ministry for Economic Affairs and Energy (Bmwi), Stuttgart, Germany.

Bertram, E. (2013) Solar Assisted Heat Pump Systems with Ground Heat Exchanger Simulation Studies. *Energy Procedia* 48: 505–514.

Chohfi, R.E. (1988) Computer Programme to Simulate the Performance of Earth Cooling Tubes, Graduate School of Architecture and Urban Planning, University of California, *13th National Passive Solar Conference*, Cambridge, MA, USA, ASES.

Dibowski, H. G. (2005). Luft-Erdwärmetauscher L-Ewt Planungsleitfaden Teil 1: Einleitung. Rapport Ag Solar, Köln.

Dbu (2012) Nahwärmeversorgung Mit Wärmepumpe Und Abwasserkanal Wärmetauscher Terrot Areal Stuttgart-Bad Cannstatt. Dbu-Abschlussbericht-Az-27080.

Eskilson, P. (1987) Thermal Analysis of Heat Extraction Boreholes. PhD Dissertation, Department of Mathematical Physics. University Of Lund, Sweden.

Esser, A. and Sensfuss, F. (2016) Evaluation of Primary Energy Factor Calculation Options for Electricity (Final Report). Specific Contract Ener/C3/2013-484/02/Fv2014-558 under the Multiple Framework Service Contract Ener/C3/2013-484. Https://Ec.Europa.Eu/Energy/Sites/Ener/Files/Documents/Final_Report_Pef_Eed.Pdf (Accessed December 28, 2016).

Grigull, U. and Sandner, H. (1990) *Wärmeleitung*, 2nd Ed, Berlin Heidelberg New York: Springer-Verlag.

Huber, A. and Pahud, D. (1999) Untiefe Geothermie – Woher Kommt Die Energie? Bundesamt Für Energie. Schlussbericht (33206).

Haller, Y.M.; Bertram, E.; Dott, R.; Hadorn, J.-C. (2012) Review of Component Models for the Simulation of Combined Solar and Heat Pump Heating Systems. *Energy Procedia* 30: 611–622.

Ingersoll, L.R. and Plass, H.J. 1948. Theory of the Ground Pipe Source for the Heat Pump. *Ashve Trans.* 54: 339–348.

Jordan, U. and Vajen, K. 2001. Influence of the DHW Load Profile on the Fractional Energy Savings: A Case Study of a Solar Combi-System with Trnsys Simulations. *Solar Energy* 69: 197–208.

Kwag, B. and Krati M. 2013. Performance of Thermoactive Foundations for Commercial Buildings. *J. Sol. Energy Eng.* 135. doi: 10.1115/1.4025587.

Marx, R.; Bauer, D.; Drueck, H. 2013. Energy Efficient Integration of Heat Pumps into Solar District Heating Systems with Seasonal Thermal Energy Storage. *Energy Procedia* 57: 2706–2715.

Miara, M.; Günther, D.; Kramer, T.; Oltersdorf, T.; Wapler, J. 2011. Heat Pump: Efficiency: Analysis and Evaluation of Heat Pump Efficiency in Real Life Conditions: Project Funded by the Federal Ministry of Economics and Technology (Reference 0327401a). Https://Wp-Monitoring.Ise.Fraunhofer.De/Wp-Effizienz/Download/Final_Report_Wp_Effizienz_En.Pdf (Accessed December 28, 2016).

Novo, A.V.; Bayon, J.R.; Castro-Fresno, D.; Rodriguez-Hernandez, J. (2010) Review of Seasonal Heat Storage in Large Basins: Water Tanks and Gravel–Water Pits. *Appl. Energy* 87: 390–397.

Oliva, A.; Stryi-Hipp, G.; Kobelt, S.; Bestenlehner, D.; Drück, H.; Dash, G. 2015. Solar-Active-Houses – Dynamic System Simulations to Analyze Building Concepts with High Fractions of Solar Thermal Energy. *Energy Procedia* 70: 652–660.

Pärisch, P.; Kirchner, A.; Wetzel, W.; Voss, S.; Tepe, R. (2011) Test System for the Investigation of the Synergy Potential of Solar Collectors and Borehole Heat Exchangers in Heat Pump Systems. Paper Presented at the Solar World Congress, September/Oktober, Kassel, Germany.

Tributsch, H. (1999) *Solartechnische Pionierleistungen Aus Naturvorbildern Und Tradioneller Architektur In 'Das Bauzentrum 14/99'*, Verlag Das Beispiel, Darmstadt.

Weber, B.; Cerro, E.; Martínez, I.G.; Rincón, E.A.; Durán, M.D. (2014) Efficient Heat Generation for Resorts. *Energy Procedia* 57: 2666–2675.

Weber, B.; Solís, E.; Durán, M.D.; Martínez, I.G.; Rincón-Mejía, E. (2016) The Utilization of Ground Surroundings for Seasonal Solar Energy Storage. *ASME 2016 10th International Conference on Energy Sustainability*, 26–30 June 2016, Charlotte, NC. doi:10.1115/Es2016-59663.

Yang, H.; Fang, Z.; Cui, P. (2010) Vertical-Borehole Ground-Coupled Heat Pumps: A Review of Models and Systems. *Appl. Energy* 81: 16–27.

10 Thermal Energy Storage Systems for Solar Applications

Aran Solé
Jaume I University

Alvaro de Gracia
Rovira i Virgili University

Luisa F. Cabeza
University of Lleida

CONTENTS

10.1 INTRODUCTION

According to the IPCC Special Report on Renewable Energy, solar energy is abundant and offers significant potential for near-term and long-term climate change mitigation (Arvizu et al., 2011). Solar thermal can be used for a wide variety of applications, such as for domestic hot water (DHW), comfort heating of buildings, and industrial process heat. At the same time, solar energy typically has a variable production profile with some degree of unpredictability that must be managed. Thermal energy storage (TES) can provide such management capabilities.

TES, also commonly called heat and cold storage, allows the storage of heat or cold to be used later (Mehling and Cabeza, 2008). Advantages of using TES in an energy system are the increase of overall efficiency and better reliability, but it can also lead to better economics, reduced investment and running costs, less pollution of the environment, and fewer CO_2 emissions (de Gracia and Cabeza, 2015).

Three technologies can be used for TES (Figure 10.1). The first one is sensible heat storage, where heat (or cold) is stored, raising the temperature of the storage material; the amount of heat stored can be calculated as follows:

$$Q_{sensible} = \int_{T_1}^{T_2} C_p \cdot dT$$

where C_p is the heat capacity of the storage medium. Sensible heat storage is by far the most common method for heat storage.

The second technology is latent heat storage, where heat is stored during a phase change; usually, the solid–liquid change is used. For latent heat storage, the amount of heat is calculated taking into account both the sensible and latent heat stored:

$$Q_{latent} = \int_{T_1}^{T_{pc}} C_{p,s} \cdot dT + \Delta H_{pc} + \int_{T_{pc}}^{T_2} C_{p,s} \cdot dT$$

where ΔH_{pc} is the enthalpy difference between the solid and liquid phase, called phase change enthalpy, melting enthalpy, or heat of fusion. Latent heat storage allows higher energy storage than the sensible one, especially if the temperature change in the process is small. Another advantage of latent heat storage is the delivery of the heat at a constant temperature during the phase change.

Finally, the third technology is thermochemical energy storage, where heat is stored using a sorption process (with zeolites, silica gel, or salts) and chemical reactions. This technology uses the heat of reaction to store energy. Any chemical reaction with high heat of reaction can be used for TES if the products of the reaction can be stored and if the heat stored during the reaction can be released when the reverse reaction takes place.

The three TES technologies are based on reversible processes, which allow the recovery of the stored heat and consequent use for the final purpose. All of them can be implemented in different

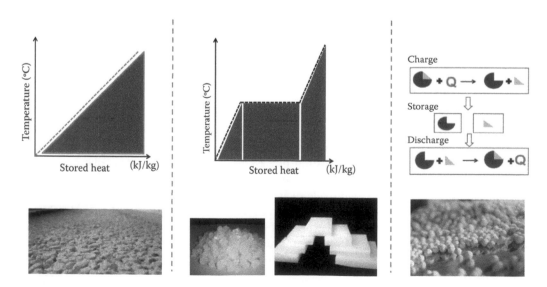

FIGURE 10.1 TES technologies. From left to right: Sensible heat storage (solid materials as example), latent heat storage (paraffins as example), and thermochemical energy storage—sorption and chemical storage (zeolites as example).

technologies, and these technologies can be classified according to the level of temperature of the storage system:

- Low (−18°C–120°C): passive and active systems for building comfort and DHW, freezers, cooling devices (medical transportation, electronics, etc.), and others
- Medium (120°C–200°C): solar cooling, waste heat recovery, etc.
- High (201°C–1100°C): CSP for electricity generation, waste heat recovery, etc.

Another classification of the TES technologies is according to the application itself; solar applications are the focus of this chapter, such as building applications (thermal comfort), electricity generation by concentrating solar power plants (CSP), drying processes, etc. (see Section 10.3).

The deployment of these technologies in the market shows that the status of research between them differs depending on the storage mechanism and application. For instance, sensible materials are widely deployed as hot water tanks, but research is needed for high-temperature storage because its main barriers are cost and low capacity. In the case of phase change material (PCM), they show a high market implementation of cold storage (ice), while for high-temperature applications, the implementation is very low. The last technology, thermochemical materials (TCM), is still under research although prototypes are being designed and tested for active systems for building comfort (Solé et al., 2015a) and high-temperature applications, such as CSP (Prieto et al., 2016).

This chapter presents the recent research on TES systems for solar applications. First of all, due to the importance of the maturity of the storage media in each TES system, a state-of-the-art report of the recent developments of TES materials (sensible, latent, and thermochemical) is provided. Therefore, the materials research section gives an overview of the newest published research toward the advancement on the materials side for solar TES applications. This section is followed by the TES systems for solar applications section, the core of the chapter. There, the TES systems that are implemented for each application are reviewed and detailed.

10.2 MATERIALS RESEARCH

There is a huge field of research regarding all three mechanisms (sensible, latent, and thermochemical) from the material behavior point of view. Different ways are being explored to improve materials' properties and/or overcome some material barriers, such as thermal conductivity and specific heat, avoiding corrosion with the container material, improving the gas transport within the material in the case of TCM, etc. The most recent developments are reviewed in this section.

10.2.1 Sensible Heat Storage Materials

Sensible heat storage is one of the most developed technologies for thermal storage and has been used for many years in both the domestic and industrial sectors. The storage medium can be liquid or solid. Some of the most common liquid sensible heat storage materials are water, mineral oil, molten salts, metals, and alloys. The most representative materials for solid sensible heat storage are rocks, concrete, sand, and bricks (Alva et al., 2017).

Besides physical properties (such as density and specific heat of the storage materials) and operational temperatures, there are properties that are important to take into account for high efficient systems: thermal conductivity and diffusivity, vapor pressure, compatibility among materials and their stability, heat losses as a function of the surface area-to-volume ratio, and cost of the materials and systems.

From the literature, it can be gathered that although the molten salt TES technology is relevant for current CSP, the nitrate mixture is limited to about 600°C. Therefore, it is not adapted to the next generation of CSP power tower working at higher temperatures, and efforts are being made to find alternative TES materials with high availability, low cost, and preferably advantageous sustainable character. For instance, some of the solid candidates are recycled ceramic materials obtained from vitrified asbestos containing wastes (Py et al., 2011), silicate oxide-based refractory materials, municipal solid wastes, fly ashes (Meffre et al., 2015), pellets of iron, and iron oxide (Kousksou et al., 2014).

Moreover, nanomaterials are attracting attention when dispersed in common liquid sensible energy storage materials. Nanomaterials can provide higher thermal conductivity and/or specific heat capacity (C_p) of the final material. Nanofluid can be thought of as the storage medium itself or as the heat transfer fluid (HTF). It is obtained by dispersing nano-sized particles, fibers, and tubes in the conventional base fluids such as water, EG, engine oil, etc. (Ahmed et al., 2017). On one hand, nanomaterials such as copper oxide (CuO), titanium oxide (TiO_2), and alumina (Al_2O_3) have been used to enhance water thermal conductivity showing enhancements between 7% and 60% depending on the concentration and size, among other parameters, while multi-walled carbon nanotubes (MWCNT) and graphite are dispersed in ionic liquids and oils, respectively, to enhance their thermal conductivity between 6% and 36%. On the other hand, silica (SiO_2) and alumina (Al_2O_3) nanoparticles are used to enhance specific heat of the solar salt (a binary salt consisting of 60% of $NaNO_3$ and 40% of KNO_3, which melts at 221°C) (Arthur and Karim, 2016). It has been experimentally proven on a laboratory scale that 1 wt.% is the optimal concentration of nanoparticles giving around 30% of specific heat enhancement in the liquid phase (Hentschke, 2016).

More research is needed in this field to study the long-term stability of nanomaterial, corrosion effects, and modelling to predict heat storage capacity enhancement as a function of temperature, size, and concentration of the nanoparticle.

10.2.2 LATENT HEAT STORAGE MATERIALS

Latent heat storage materials, also called PCM, are usually classified according to the material origin; thus, they can be either organic or inorganic. Paraffin, fatty acids, and sugar alcohols are the most representative families of organic PCM, whereas salts and metals are the most used as inorganic PCM. Some of the PCM are commercially available. In addition, eutectics of binary (also tertiary, quaternary, etc.) mixtures are synthesized to behave as PCM at the desired temperature. These mixtures can be organic–organic, inorganic–organic, or inorganic–inorganic. As can be seen in Figure 10.2, there is a wide range of PCM melting temperatures and therefore of applications, but at high temperatures, basically salts are found.

Properties of the available PCM are well known, studied, and published. In general, paraffins and fatty acids present no corrosion and low thermal conductivity, while salt hydrates usually are more corrosive, depending on the PCM metal or polymer used; they may present subcooling and phase separation and have high thermal conductivity (Chandel and Agarwal, 2017). To overcome some of these material-side barriers, research with PCM has been focused on encapsulation and composite PCM development.

Macro- and microencapsulation experimental and numerical studies are found in the literature as heat transfer enhancement techniques to improve PCM rates of charging and discharging of energy (Agyenim et al., 2010). Nevertheless, unless the matrix encapsulating the PCM has a high thermal conductivity, the microencapsulation system suffers from low heat transfer rate (Zhang et al., 2016). In addition, microencapsulation and shape-stabilization technology make PCM possible to integrate into a building material as long as the shell or supporting material used to encapsulate PCMs is compatible with the building materials (Khadiran et al., 2016).

Experimental studies showed that with the optimum amount of nanoparticles, PCM has the potential of providing a more efficient means of storage. Various high thermal conductive fillers (e.g., metallic fins, ceramic powder fillers, graphitic carbon fibers, carbon nanofibers, graphite particles, and exfoliated graphite) were reported to improve the effective thermal conductivity of paraffin.

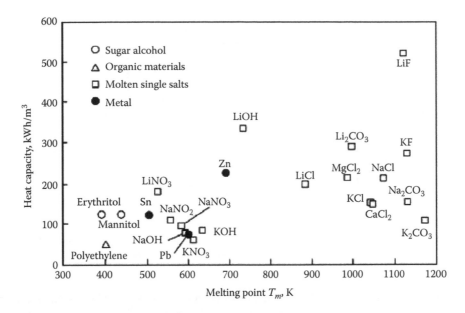

FIGURE 10.2 Heat capacity and melting temperature of some PCM. (From Hoshi et al., *Solar Energy*, 79, 332–339, 2005.)

The heat storage units of new concentrated solar plants could benefit from a HTF that incorporates encapsulated particles of PCM (Zhang et al., 2016). The heat transfer capacities would be improved by the high conductivity and the high energy density of the PCM. In comparison with a conventional thermal fluid, the specific heat will be increased by a fraction of 3 while the energy density will increase by about 5.

Up to now, from the literature review, there is no commercial high-temperature (>800°C) TES module implementing PCM storage; only experimental modules have been constructed and assessed.

10.2.3 THERMOCHEMICAL MATERIALS

One of the most novel technologies for storing thermal energy or transforming excess renewable power into heat is reversible reactions, using TCM (Fernandes et al., 2012). A large number of reversible chemical reactions can be used for thermochemical energy storage, each within a given range of equilibrium temperatures (60°C–1000°C) and heats of reaction (up to 1700 kJ/kg) (Khadiran et al., 2016).

Recent developments toward TCM enhancement, for building applications (<120°C), are based on developing composites to improve mass and heat transfer in the reactors, seeking higher permeability and thermal conductivity, respectively. There is a compromise between thermal conductivity and permeability of the TCM, both properties being contrary depending on the porosity (see Figure 10.3). Contrastingly, porosity also has an effect on the density of the salt and at the end on the storage density (Solé, 2015; Fopah-Lele et al., 2016). In this sense, different host matrices or additives are suggested: vermiculite, silica gel, zeolite, activated carbon, expanded natural graphite, etc. Several studies with expanded graphite show improvement of vapor diffusivity during the hydration process (Aydin et al., 2015).

Since the majority of the TCM for low-temperature applications are salt hydrates, corrosion studies under the operating conditions between the TCM and the vessel material are mandatory. Polymeric materials as vessel containers and/or heat exchangers and coatings are being researched to overcome corrosion (Solé et al., 2015b, 2016).

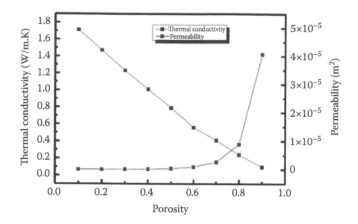

FIGURE 10.3 The heat and mass transfer dilemma in function of the bed porosity for the magnesium chloride system. (From Fopah-Lele et al., *Applied Thermal Engineering*, 101, 669–677, 2016.)

The most representative TCM for high-temperature applications (>400°C), such as CSP plants, are metal oxides, hydroxides, sulfates, and carbonates (André et al., 2016). Meanwhile, metal oxides reactions involve oxygen (O_2) as a reactant, and carbonates require carbon dioxide (CO_2). Metal oxide reactions are most favorable, since air can be the reactant of the reverse reaction.

New concepts to overcome the loss of reversibility in high-temperature applications are emerging. For instance, 10 wt.% of nonreactive oxides were added to Co_3O_4, and the results showed an improvement in the reversibility of the Co_3O_4/CoO redox reaction (Tescari et al., 2014). Also, Cabeza et al. (2017) published a proof of concept of TCM based on consecutive reactions instead of reversible reactions. By changing from reversible endothermic/exothermic reactions toward consecutive reactions, a higher chemical reaction conversion can be achieved; this would also contribute in a better material stability, leading to a longer material life and thus longer operational plant periods.

10.3 TES IN SOLAR APPLICATIONS

10.3.1 BUILDING SECTOR

TES is identified as one of the most promising technologies to enhance the use of solar energy and other renewable sources in the building sector. The integration of the TES system to store solar energy can be in the form of active (Navarro et al., 2016a) or passive (Navarro et al., 2016b) systems.

Building designs have traditionally taken advantage of sensible TES capacity of materials such as stone, earth, brick, and concrete. However, these materials require a lot of space to provide significant TES capacity for the building envelope, and recently the use of latent heat materials has been explored to store energy at a desired thermal range limiting the amount of required material (Castell et al., 2010). Other than high thermal mass materials, solar walls are popular building systems and integrate TES. These systems offer a feasible technique for the exploitation of directional flow of heat in buildings. According to Saadatian et al. (2012), solar walls are classified as standard Trombe wall, solar water wall, solar transwall, composite solar wall, or fluidized solar wall; only the Trombe wall and solar water wall include storage in their design. Trombe walls consist of a high thermal mass covered by glass, leaving an air channel between the layers. Trombe walls can be designed without ventilation, with circulation to provide heating supply during winter or free cooling in summer (Figure 10.4). Moreover, solar water walls follow the same working principle as solar Trombe walls but replace the thermal mass element (concrete, brick, or stone) with water containers.

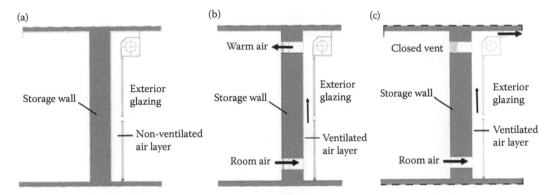

FIGURE 10.4 Trombe wall configurations: (a) without ventilation, (b) winter mode with air circulation, and (c) summer mode with cross ventilation. (From Stazi et al., *Renewable Energy*, 75, 519–523, 2012.)

Even though the use of water increases the thermal storage capacity of the system, it can present some drawbacks due to condensation.

Other than its use in solar building passive systems, TES has been applied in active building systems. These TES systems are charged and/or discharged mechanically by the use of pumps, compressors, or fans, being able to adapt these processes to the demand and/or the production schedules of the energy fluxes. The use of appropriate control strategies is mandatory in buildings' active systems to maximize energy benefits and justify their high investment costs, thus making them attractive to architects and engineers, as well as final users (de Gracia et al., 2015).

Another aspect of high relevance is the integration of these active systems in building without occupying high volume. Active solar TES systems have been integrated into the building design for walls, external solar facades, ceilings, floors, suspended surfaces, or in ventilation systems. In these systems, solar energy is usually stored by means of high thermal mass sensible materials or PCMs to provide heating to the building after the sunny hours. For instance, Chen et al. (2010) developed and tested a ventilated concrete slab, which is actively charged by airflow from a collector located in the roof. The heat stored is passively discharged to cover the heating demand of the building. A similar design was proposed by Navarro et al. (2016c); however, in this technology PCM was placed inside the concrete hollows to enhance the TES capacity of the system and discharge the stored heat passively or actively depending on the heating demand. This system was tested at prototype scale (Figure 10.5) demonstrating a potential of energy reduction up to 60% in comparison with a standard air-to-air heat pump.

Furthermore, the implementation of solar TES in external facades and walls of buildings has been explored. Within this context, Fraisse et al. (2006) studied a solar air collector connected to a concrete wall cavity inside the building, in which the heat can be passively discharged from the wall surfaces to the indoors or actively discharged using an airflow. In addition, de Gracia et al. (2012) experimentally tested the energy performance of a ventilated double skin facade with PCM in its air chamber. In this system, the facade is closed during sunny hours allowing the PCM to be melted by the greenhouse effect in the cavity. The heat stored in the PCM is mechanically (fans) or naturally (solar chimney effect) discharged to create a heating supply to the indoor environment through the solidification process of PCM (at 22°C). The system was able to reduce 20% of the energy consumption of the whole building, limited by heat losses during the storage period.

Thermal storage water tanks have a crucial role in the final efficiency of solar systems used for DHW. The most critical aspects of these systems are the design of the tank and connections to enhance the thermal stratification (Han et al., 2009) and its integration in the building.

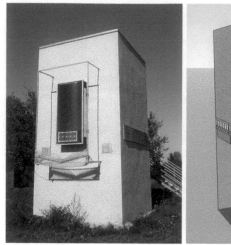 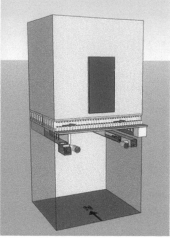

FIGURE 10.5 Prototype of an active concrete slab with PCM connected to a solar air collector. (From Navarro et al., *Renewable Energy*, 85, 1334–1356, 2016c.)

10.3.2 UNDERGROUND SEASONAL TES

Underground seasonal TES systems are used to store thermal energy produced by solar collectors during summer and use it during winter for heating and DHW (Pinel et al., 2011). The underground TES can be based on water tanks, boreholes, or aquifers. Table 10.1 presents a main description of the different systems used for underground seasonal TES.

10.3.3 CONCENTRATED SOLAR POWER PLANTS

Although CSP technology dates back to 1970, most of the current plants were installed this decade (2010–2020). In 2015, the CSP market had a total capacity of 5840 MWe worldwide, among which 4800 MWe is operational and 1040 MWe is under construction (Liu et al., 2016).

CSP plants can be based on parabolic trough collectors (PTC), linear Fresnel reflectors (LFR), solar power tower (SPT), and parabolic dish systems (PDS). Among these CSP technologies, PTC accounts for 95% of the global CSP installations.

CSP provides electricity from solar energy, a renewable resource, when the sun is available. To be able to have a continuous production, at night and on cloudy days, other technologies or resources are required. CSP can be hybridized with another resource (renewable or nonrenewable) like biomass, gas, etc., or can be designed with a TES technology.

Thermal energy for CSP systems can be stored in sensible heat storage using either solid or liquid storage media, latent heat storage by using PCMs and thermochemical storage through reversible chemical reactions (see Section 10.2).

The most deployed (and only mature one today at an industrial scale) TES system for CSP plants is based on sensible energy storage technology, specifically the two-tank storage of molten salts (mainly Hitec/Hitec XL or Solar Salt). The two molten salt tanks (Figure 10.6) can be implemented in solar trough-based CSP or solar towers (Meffre et al., 2015). In some of the installed CSPs, for instance, in the Andasol (Granada, Spain) standard-like plant, 28,000 tons of solar salt (60% $NaNO_3$+40% KNO_3 by weight) are thermally cycled between 292°C and 386°C to supply 7.5 h of TES capacity. In the Gemasolar plant of 20 MWe, a similar thermal storage of 15 h in capacity with a variation in salt temperature from 292°C to 565°C allows the production of electrical power 24 h a day. A larger similar plant of 110 MWe is already under construction in the U.S.

TABLE 10.1
Underground Seasonal TES Systems Overview

Tank TES

- Suitable geological conditions: tank construction can be built almost independently from geological conditions, as much as possible avoiding groundwater
- Depth: from 5 to 15 m
- Heat storage capacity: between 60 and 80 kWh/m^3
- Tank's characteristics: **Structure** made of concrete, stainless steel, or fiber-reinforced polymer. A coating of polymer or stainless steel covers the inside tank surface. The outside surface has an **insulation layer** of foam glass gravel for the bottom part and expanded glass granules in the membrane sheeting for walls and top.

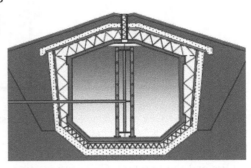

Pit TES (PTES)

- Suitable geological conditions: almost independent from geological conditions, as much as possible avoiding groundwater
- Depth: from 5 to 15 m
- PTES filled with water or gravel-water mixture (gravel fraction 60%–70%)
- Heat storage capacity with gravel-water mixture: between 30 and 50 kWh/m^3 (equivalent to 0.5–0.77 m^3 of water)

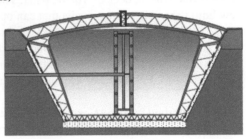

Borehole TES

- Suitable geological formations: rock or water-saturated soils with no or only very low natural groundwater flow. The ground should have high thermal capacity and impermeability.
- Depth: from 30 to 100 m
- Heat directly stored in the water-saturated soil: u-pipes, also called ducts, are inserted into vertical boreholes to build a huge heat exchanger.
- Heat storage capacity of the ground: between 15 and 30 kWh/m^3

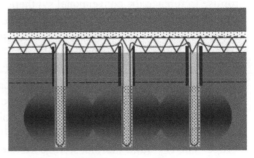

Aquifer TES

- Suitable geological formations: aquifer with high porosity, ground water, and high hydraulic conductivity (kf > 10–4 m/s), small flow rate, up and down enclosed with leak-proof layers.
- Aquifers defined as naturally occurring self-contained layers of ground water are used for heat storage.
- Heat storage capacity: between 30 and 40 kWh/m^3

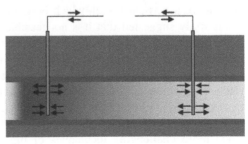

Source: Schmidt et al., *Solar Energy*, 76, 165–174, 2004.

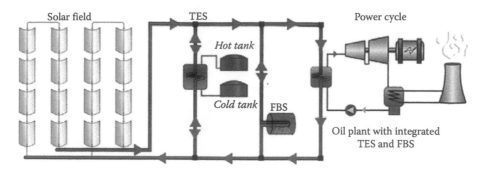

FIGURE 10.6 CSP plant diagram with molten salt tanks as TES technology. (From Boukelia, T.E. et al., *Energy*, 88, 292–303, 2015.)

Beyond the two-tank CSP configuration, research in this field is seeking other, more efficient systems that will reduce costs of the plant. The research is mainly focused on (1) the development of new liquid sensible storage media that can operate over a wider range of temperatures than molten salts, (2) the development of cheap solid heat storage media, (3) research on PCM and highly efficient latent energy storage systems, and (4) identification of TCM suitable for this operating range (Liu et al., 2016). Latest developments concerning TES materials (points 1, 2, and 4) have already been discussed. Published developments focused on the CSP application itself are provided in this section.

One important part of a CSP is the HTF. Research is being done as it is a novelty and offers major opportunities to enhance SPT commercial potential. The use of particle suspensions to transfer solar heat from the receiver to the energy conversion process offers major advantages in comparison with water/steam, thermal fluids, or molten salts (Zhang et al., 2016). In addition, ionic liquids (ILs) promise to replace the conventional synthetic oil/molten salts as the working fluid for the parabolic trough CSP plants (Moens et al., 2003). Other developments, like the addition of nanoparticles to molten salts, have been discussed in Section 10.2.1.

Besides being implemented directly as storage media, solid sensible TESs can be employed as filler materials in nitrate salts—thermocline storage systems to prevent convective mixing and reduce the amount of salts required. The commonly investigated system is the packed bed/dual-media thermocline system using rocks or other ceramic materials (Kuravi et al., 2013; Yang et al., 2014). A report from the Abengoa Solar Company (Kelly, 2010) analyzed and compared supercritical receiver designs using a packed bed thermocline and two-tank system for thermal storage. The two-tank thermal storage system consists of a cold salt tank, operating at a temperature of 288°C, and a hot salt tank, operating at a temperature of 565°C. The thermocline system consists of a vertical vessel, which is filled with a granular ceramic material, such as quartzite. Their study suggested that it is not feasible to use supercritical steam as the HTF in a thermocline system, but that it can work with supercritical CO_2. Also, the report suggests that two-tank nitrate salt thermal storage systems, are strongly preferred over thermocline systems when using supercritical heat transport fluids. One of the critical problems in thermocline systems is thermal ratcheting, which happens when the tank wall progressively expands from cyclic operation leading to catastrophic failure of the wall material (Liu et al., 2016).

Latent heat TES using high-temperature PCM is not yet popular for commercialized CSP plants since more challenging techniques of high-temperature PCM encapsulation and heat transfer enhancement are in research and development (Strumpf, 1987). Nonetheless, the high melting point latent heat TES approach can reduce energy storage equipment and containment cost, since the energy density is large. Importantly, energy with constant temperature and pressure can be supplied to turbines running at optimal efficiency, unlike the case with sensible heat storage systems, where the temperature drops as storage is discharged. Hoshi et al. (2005) discussed theoretically and numerically several PCMs for compact LFR technology. They concluded that the most promising PCMs are molten salts (i.e., $NaNO_3$), sodium carbonate (Na_2CO_3) for multitower solar array operating with Brayton Cycle turbines, sodium nitrite ($NaNO_2$) for compact LFR plants using low-pressure turbines, and zinc (Zn) for trough systems using higher boiler temperatures. Nevertheless, it is necessary to do more research in design, vessel corrosion, and vessel integrity.

One topic of interest in the development of latent heat thermal storage systems for CSP is heat pipes. Research is being done on how to reduce thermal resistance between the PCM and the HTF and to embed heat pipes as thermal conduits. Few experimental studies are found, especially for high temperatures, but some numerical studies show a capital cost reduction by at least 15% when comparing a PCM storage system with staggered heat pipes for a 50 MWe parabolic trough CSP plant with a conventional two-tank sensible storage system of the same capacity (Roback et al., 2011).

Another TES technology being studied for implementation in CSP is thermochemical energy storage. Thermochemical energy storage for CSP is still at the laboratory scale, far from transfer to a commercial scale. Nevertheless, it is a topic of interest since it is a nearly lossless way of storing energy, and it could provide remarkable energy density of the storage media. Prieto et al. (2016) reviewed this technology for CSP and concluded that calcium carbonate has the potential to be viable and should thus be attempted at the research plant scale once a reactivation cycle can be designed; manganese oxide cycle, while less developed, is fundamental enough to be a suitable application for desert climates over the remaining water-frugal or even water-avoiding cycles. A review published by André et al. (2016) provides a selection with respect to requirements such as suitable range of reaction temperature for CSP application and nontoxicity: $CaCO_3$, $SrCO_3$, $BaCO_3$, $Ca(OH)_2$, $Sr(OH)_2$, $Ba(OH)_2$, Co_3O_4, Fe_2O_3, CuO, BaO_2, and Mn_2O_3. Besides identifying and characterizing suitable material, big efforts are carried out in the line of reactor design (kind of reactor, contact patterns between reactants, etc.). Considerable efforts pursue the identification and characterization of suitable materials; the development of novel reactor designs with enhanced contact patterns between reactants and new locii of thermochemical reactions; as well as scale up and integration of reactors into CSP plants (Ströhle et al., 2016).

Other configurations of storage systems for CSP like cascade PCM, steam accumulators, and combined systems are also being investigated (Kuravi et al., 2013). Corrosion, vessel materials, sensors, and the environmental impact of these systems are topics of great interest (Gasia et al., 2017).

10.3.4 SOLAR COOLING

The use of thermal absorption chillers driven by solar energy can offset a major part of the energy consumed for cooling purposes in buildings and industry applications. In these systems, the refrigerant effect is produced by physical and/or chemical changes between the refrigerant and the sorbent, which requires different amounts of heat at different temperature ranges, depending on the type of vapor absorption system (half/single/double/triple effect) (Chidhambaram et al., 2011). According to the literature, there are approximately 100 solar heating and cooling systems installed in the EU region (SOLARI, 2014), most of them being single-stage absorption with water as HTF and storage media.

The solar intermittency limits the application of this technology unless an adequate TES unit is considered to produce cooling based on the end-user need. TES systems used for solar cooling applications can reduce peak electricity demand, improve the management of energy flows in the electricity market, increase the solar fraction, and ensure stable operation by preventing rapid start-up/slow-down events (Pintaldi et al., 2015). Depending on the solar production and especially on the cooling demand and schedules of cooling load, the storage can be designed to store heat produced by the solar field for its later use to provide cooling by providing heat to the absorption chiller, or to store directly cool from the absorption chiller and distribute this cool to the demand once required.

The type of vapor absorption system and the way the absorption solution is regenerated determine the temperatures of operation of the whole chiller and storage system. In the case of operating with single-effect, hot water-driven absorption chillers, water is used as the storage medium and heat transfer fluid, pumped between 65°C and 95°C for the regeneration of the absorption solution. Multiple-effect absorption chillers have a higher coefficient of performance (COP), which might lead to a reduction of the solar collector area and its associated cost reduction but can operate only at high temperatures. In the case of double effect (heat inlet at 155°C–185°C) and triple effect (heat inlet above 200°C), the chilled water is cooled down two or three times, respectively. These systems are designed to usually operate with pressurized water (35 bars in case of operating at 200°C) or thermal oil, which can be operated at atmospheric pressure and is commonly selected for parabolic trough power plants (Pintaldi et al., 2017). For double- and triple-effect chillers with thermal oil as HTF, solid storage media such as concrete has been evaluated as a potential candidate since it can operate up to 400°C (Laing et al., 2006). However, for high performance of the chiller, the heat has to be supplied within a narrow thermal range, which limits the potential of sensible storage units, as they are not able to provide a large quantity of heat to the regenerator at a given temperature when no solar radiation is available.

Latent heat storage systems have been explored and tested, as they are able to supply heat to the regenerator at an almost constant temperature during long periods. Several designs of latent storage tanks for solar cooling have been investigated on a laboratory scale such as tube and fins (Medrano et al., 2009), shell and tubes (Gil et al., 2013), spherical capsules, and packed bed units (Felix Regin et al., 2009). Moreover, at prototype scale, a 5 Tn storage tank was installed in a real solar cooling plant in Seville (Figure 10.7) using hydroquinone as PCM (Gil et al., 2014). The use of latent heat storage systems allows the implementation of cascaded storage systems for solar cooling. These systems offer vast potential: when designed with PCM at different melting temperatures arranged in series, they can store energy at different temperature levels. The cascaded systems connect the lower temperature tank to the inlet of the solar collector, improving the efficiency in comparison with conventional storage tanks. Moreover, during the discharge period, a combination of the multiple effect and half/single effect based on temperature availability would be the best choice. However, cost and space requirements might limit the implementation of this technology.

As previously stated, there is the possibility of storing the cool energy produced during sunny hours for later use. Cool thermal energy storage (CTES) systems have gained attraction during the last years for industrial refrigeration, food preservation, process cooling, and building air conditioning. These systems are mainly based on chilled water (Karim, 2010), ice storage (Ezan et al., 2010), or PCM tanks (Brancato et al., 2017); use ethylene glycol as HTF; and require high levels of insulation as the energy available in the cool state is expensive.

FIGURE 10.7   Latent heat storage tank in a real solar cooling installation at the University of Sevilla, Spain. (From Gil et al., *International Journal of Refrigeration*, 39, 95–103, 2014.)

The control of the system during its operation such as temperatures, speeds of pumps, position of three-way mixing valves, or backup boilers has a massive impact on its efficiency (Henning, 2007). The control of both solar and storage loops is challenging due to the transient nature of solar radiation as well as different possible demand profiles. Regarding the operation of the storage loop, Bujeto et al. (2011) compared three different control strategies in a real solar air conditioning plant over 3 years. In the first strategy, the mass flow of hot water going to the chiller was fixed; in the second strategy, variable speed pumps were used to maintain maximum COP; in the third strategy, the goal was to maintain the temperature at the bottom of the cold storage at 10°C so that the chiller could operate without interruption. It was proven that the third strategy increases the efficiency of the solar field 12% and tripled the solar fraction of strategy 1. Moreover, in the last years the incorporation of forecasts (weather, occupancy, and/or electricity price) has gained interest and is being implemented in the control strategies of solar cooling plants with TES.

10.3.5 USE OF TES IN OTHER SOLAR APPLICATIONS

Other than building applications and integration in concentrated solar power plants and solar cooling plants, TES units have been integrated in other applications such as desalination (Gnaneswar Gude et al., 2012), greenhouses (Cuce et al., 2016), solar cooking (Sharma et al., 2000), and solar drying (Kant et al., 2016).

10.3.5.1 Solar Cooking

The solar cooker can offer a viable option in the domestic sector to reduce the use of fuel and wood, especially in developing hot climatic countries (Li et al., 2016). The intermittency of solar radiation limits the application of solar cookers in cloudy conditions or evenings unless TES is integrated (Figure 10.8). Both sensible and latent TES units have been investigated for application in solar cookers. Regarding the sensible heat storage systems, different designs and materials have been explored such as pebbles (Schwarzer et al., 2003), used engine oil (Nahar, 2003), and sand (Ramadan et al., 1988). The use of PCM in solar cookers got the attention of the scientific community as it reduces the

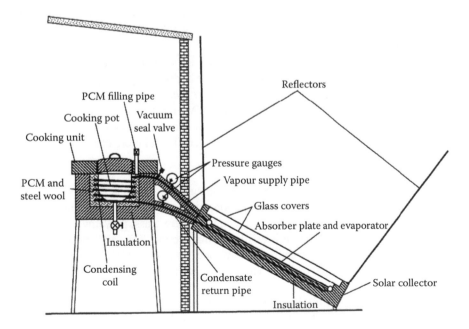

FIGURE 10.8 Cross sectional side view of a solar cooker with PCM. (Hussein et al., *Energy Conversion and Management*, 49, 2237–2246, 2008.)

volume of storage, delivering the stored heat at a given temperature. Sharma et al. (2000) recommend that the melting temperature of the PCM should be between 105°C and 110°C.

10.3.5.2 Solar Drying

Drying processes, reducing moisture from products, are basic in food or agricultural product storage. In industrial dryers, energy is a major operational cost and usually comes from the combustion of fossil fuels (Aghbashlo et al., 2013). One way to move toward a more sustainable drying process is to use renewable energies as the heat source. This shift would also allow cost reduction since the cost of solar drying (per kg) is less in comparison with electric drying. Solar energy and biomass have been investigated as possible alternatives (Tiwari et al., 2016). Biomass is a renewable source, but its combustion still releases greenhouse gases, whereas solar energy is cleaner and more cost competitive.

The viability of drying processes by means of solar energy goes through TES systems because of the intermittence of this renewable source. This combination also leads to a constant temperature of the continuous airflow for drying. In addition, it offers the possibility of shifting from fossil fuels to a major percentage or 100% solar thermal energy instead of electricity and therefore promotes the nonemission of greenhouse gases and the reduction of the costs of these processes.

Until now, sensible and latent heat storage systems have been implemented in solar dryers. Kant et al. (2016) published a review where TES-based solar drying systems are classified in sensible and latent TESs, and both groups are divided into indirect solar dryers and greenhouse dryers. Several studies have been published regarding experimental set ups, modeling and simulation of solar dryers, and optimization of the equipment.

One of the studies, published by Rabha et al. (2017), analyzes the energy and exergy of the drying process of the ghost chili pepper and sliced ginger dried in a newly developed forced convection solar tunnel dryer. A scheme and picture of the dryer are shown in Figure 10.9; it can be seen that a shell-and-tube latent heat storage system is located between the air heater and the drying chamber. However, the authors do not specify the PCM being tested.

The studies that explored sensible materials as TES media mainly worked with sand, granite, or rock. For instance, Ayyappan et al. (2015) conducted an experiment with a natural convection solar greenhouse dryer using different sensible heat storage materials (concrete, sand, and rock bed) in order to study their thermal performance. Sand or rock bed 4″ (10.16 cm) thick provided optimal drying environment during both day and night.

Amongst the available thermal heat storage techniques for solar drying applications, latent heat TES is particularly attractive due to its ability to provide high-energy storage density per unit mass and per unit volume in a more or less isothermal process, i.e., store heat at a constant temperature corresponding to the phase-transition temperature of PCM (Kant et al., 2016). A list of the suitable PCMs for drying purposes is given by these authors. However, Tiwari et al. (2016) suggest that further research is required to find new PCMs that are less costly and more efficient.

10.3.5.3 Greenhouses

Greenhouses are built for plants that require certain features for efficient cultivation. Conventional façade materials utilized in greenhouse constructions currently have poor thermal insulation features. The most common façade materials for greenhouses are glass, polyethylene, semirigid plastic, and plastic film (Cuce et al., 2016). It is estimated that around 20%–40% of the energy loss in a typical greenhouse occurs from the greenhouse envelope (Hee et al., 2015). Of course, temperature, lighting, and humidity vary in each region; therefore, the technology applied in greenhouses in various regions is also different.

It is well known that growing food in greenhouses needs specific conditions, where temperature plays a key role. Depending on the climate, energy costs for heating and cooling, respectively, are high, and depending on the energy source used to provide greenhouse comfort, pollutant gases are released to the atmosphere. The goal is to move toward more sustainable energy sources and

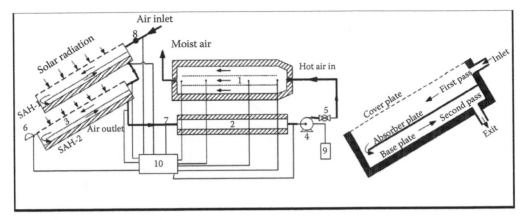

1. Tunnel dryer; 2. shell-and-tube heat exchanger for energy storage; 3. solar air heater; 4. air blower; 5. ball valve;

6. pyranometer; 7. thermocouple; 8. flow meter; 9. energy meter; 10. data acquisition system.

FIGURE 10.9 Experimental setup of the solar dryer developed by Rabha et al. (2017).

more efficient technologies and constructive solutions to be implemented in greenhouses. These technologies are photovoltaic (PV) modules, solar thermal collectors, hybrid PV thermal collectors, TES, heat pumps, preheating and cooling ventilation systems, and alternative façade materials.

Focusing on TES technology, water tanks, rocks, PCMs, and underground (see Section 10.3.2) have been investigated and reviewed by Cuce et al. (2016). From this last publication, it can be gathered that in countries where heating demand is higher than cooling, underground TES systems are preferred. Proper use of seasonal TES via vertical ground heat exchangers can easily meet the heating demand of greenhouses in different climatic regions. For such applications, polyethylene can be a good choice of insulation material with its low thermal conductivity and low cost. Polyethylene surrounding the soil can isolate the uncontrollable heat conduction from heat exchanger to soil.

Regarding PCM, calcium chloride hexahydrate ($CaCl_2 \cdot 6H_2O$), sodium sulfite decahydrate ($Na_2SO_3 \cdot 10H_2O$), and paraffin have been explored (Benli and Durmuş, 2009). A greenhouse implementing a ground heat pump and $CaCl_2 \cdot 6H_2O$ as PCM was designed and tested by Benli (2011) (see Figure 10.10), from where it is obtained that the COPs of the heat pump COP_{HP} and COP_{sys} are found between 2.3 and 3.8 and between 2 and 3.5, respectively, depending on the temperature of the environment in the greenhouse. The ground heat pump produces a heat of 5°C–10°C, and the chemical materials produce an average of 1°C–3°C auxiliary heat. PCM can be utilized in greenhouse applications along with solar thermal collectors to enhance the thermal energy content of the system for reducing the cost of space heating, which is notably high in extreme climatic regions.

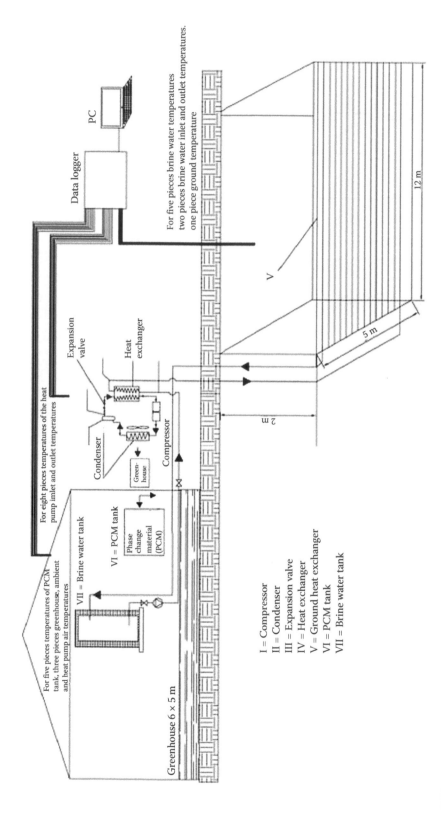

FIGURE 10.10 Schematic view of the greenhouse developed by Benli (2011) where PCM as storage unit is implemented.

REFERENCES

Aghbashlo, M., Mobli, H., Rafiee, S., and A. Madadlou. 2013. A review on exergy analysis of drying processes and systems. *Renewable and Sustainable Energy Reviews* 22, 1–22.

Agyenim, F., Hewitt, N., Eames, P., and M. Smyth. 2010. A review of materials, heat transfer and phase change problem formulation for latent heat thermal energy storage systems (LHTESS). *Renewable and Sustainable Energy Reviews* 14, 615–628.

Ahmed, S.F., Khalid, M., Rashmi, W., Chan, A., and K. Sahbaz. 2017. Recent progress in solar thermal energy storage using nanomaterials. *Renewable and Sustainable Energy Reviews* 67, 450–460.

Alva, G., Liu, L., Huang, X., and G. Fang. 2017. Thermal energy storage materials and systems for solar energy applications. *Renewable and Sustainable Energy Reviews* 68, 693–706.

André, L., Abanades, S., and G. Flamant. 2016. Screening of thermochemical systems based on solid-gas reversible reactions for high temperature solar thermal energy storage. *Renewable and Sustainable Energy Reviews* 64, 703–715.

Arthur, O., and M.A. Karim. 2016. An investigation into the thermophysical and rheological properties of nanofluids for solar thermal applications. *Renewable and Sustainable Energy Reviews* 55, 739–755.

Arvizu, D., Balaya, P., Cabeza, L.F., Hollands, T., Jäger-Waldau, A., Kondo, M., Konseibo, C., Meleshko, V., Stein, W., Tamaura, Y., Xu, H., and R. Zilles. 2011. Direct Solar Energy. In *IPCC Special Report on Renewable Energy Sources and Climate Change Mitigation* [O. Edenhofer, R. Pichs-Madruga, Y. Sokona, K. Seyboth, P. Matschoss, S. Kadner, T. Zwickel, P. Eickemeier, G. Hansen, S. Schlömer, C. von Stechow (eds)], Cambridge University Press, Cambridge and New York.

Aydin, D., Casey, S.P., and S. Riffat. 2015. The latest advancements on thermochemical heat storage systems. *Renewable and Sustainable Energy Reviews* 41, 356–367.

Ayyappan, S., Mayilsamy, K., and V.V. Sreenarayanan. 2015. Performance improvement studies in a solar greenhouse dryer using sensible heat storage materials. *Heat and Mass Transfer* 52, 1–9.

Benli, H. 2011. Energetic performance analysis of a ground-source heat pump system with latent heat storage for a greenhouse heating. *Energy Conversion and Management* 52, 581–589.

Benli, H., and A. Durmuş. 2009. Performance analysis of a latent heat storage system with phase change material for new designed solar collectors in greenhouse heating. *Solar Energy* 83, 2109–2119.

Brancato, V., Frazzica, A., Sapienza, A., and A. Freni. 2017. Identification and characterization of promising phase change materials for solar cooling applications. *Solar Energy and Materials and Solar Cells* 160, 225–232.

Boukelia, T.E., Mecibah, M.S., Kumar, B.N., K.S. Reddy. 2015. Investigation of solar parabolic trough power plants with and without integrated TES (thermal energy storage) and FBS (fuel backup system) using thermic oil and solar salt. *Energy* 88, 292–303.

Bujedo, L.A., Rodríguez, J., and P.J. Martínez. 2011. Experimental results of different control strategies in a solar air-conditioning system at part load. *Solar Energy* 85, 1302–1315.

Cabeza, L.F., Solé, A., Fontanet, X., Barreneche, C., Jové, A., Gallas, M., Prieto, C., and A.I. Fernández. 2017. Thermochemical energy storage by consecutive reactions of higher efficient concentrated solar power plants (CSP): Proof of concept. *Applied Energy* 185, 836–845.

Castell, A., Martorell, I., Medrano, M., Perez, G., and L.F. Cabeza. 2010. Experimental study of using PCM in brick constructive solutions for passive cooling. *Energy and Buildings* 42, 534–540.

Chandel, S.S., and T. Agarwal. 2017. Review of current state of research on energy storage, toxicity, health hazards and commercialization of phase changing materials. *Renewable and Sustainable Energy Reviews* 67, 581–596.

Chen, Y., Galal, K., and A.K. Athienitis. 2010. Modeling, design and thermal performance of a BIPV/T system thermally coupled with a ventilated concrete slab in a low energy solar house: Part 2, ventilated concrete slab. *Solar Energy* 84, 1908–1919.

Chidambaram, L.A., Ramana, A.S., Kamaraj, G., and R. Velraj. 2011. Review of solar cooling methods and thermal storage options. *Renewable and Sustainable Energy Reviews* 15, 3220–3228.

Cuce, E., Harjunowibowo, D., and P.M. Cuce. 2016. Renewable and sustainable energy saving strategies for greenhouse systems: A comprehensive review. *Renewable and Sustainable Energy Reviews* 64, 34–59.

de Gracia, A., and L.F. Cabeza. 2015. Phase change materials and thermal energy storage for buildings. *Energy and Buildings* 103, 414–419.

de Gracia, A., Fernández, C., Castell, A., Mateu, C., and L.F. Cabeza. 2015. Control of a PCM ventilated facade using reinforcement learning techniques. *Energy and Buildings* 106, 234–242.

de Gracia, A., Navarro, L., Castell, A., Ruiz-Pardo, A., Alvarez, S., and L.F. Cabeza. 2012. Experimental study of a ventilated facade with PCM during winter period. *Energy and Buildings* 58, 324–332.

Ezan, M.A., Ozdogan, M., Gunerhan, H., Aytunc, E., and A. Hepbasli. 2010. Energetic and exergetic analysis and assessment of a thermal energy storage (TES) unit for building applications. *Energy and Buildings* 42, 1896–1901.

Felix Regin, A., Solanki, S.C., and J.S. Saini. 2009. An analysis of a packed bed latent heat thermal energy storage system using PCM capsules: Numerical investigation. *Renewable Energy* 34, 1765–1773.

Fernandes, D., Pitié, F., Cáceres, G., and J. Baeyens. 2012. Thermal energy storage: How previous findings determine current research priorities. *Energy* 39, 246–257.

Fopah-Lele, A., Kuznik, F., Osterland, T., and W.K.L. Ruck. 2016. Thermal synthesis of a thermochemical heat storage with heat exchanger optimization. *Applied Thermal Engineering* 101, 669–677.

Fraisse, G., Johannes, K., Trillat-Berdal, V., and G. Achard. 2006. The use of a heavy internal wall with a ventilated air gap to store solar energy and improve summer comfort in timber frame houses. *Energy and Buildings* 38, 293–302.

Gasia, J., Miró, L., and L.F. Cabeza. 2017. Review on system and materials requirements for high temperature thermal energy storage. Part 1: General requirements. *Renewable and Sustainable Energy Reviews*. 75, 1320–1338.

Gil, A., Oró, E., Miró, L., Peiró, G., Ruiz, A., Salmerón, J.M., and L.F. Cabeza. 2014. Experimental analysis of hydroquinone used as phase change material (PCM) to be applied in solar cooling refrigeration. *International Journal of Refrigeration* 39, 95–103.

Gil, A., Oró, E., Peiró, G., Álvarez, S., and L.F. Cabeza. 2013. Material selection and testing for thermal energy storage in solar cooling. *Renewable Energy* 57, 366–371.

Gnaneswar Gude, V., Nirmalakhandan, N., Deng, S., and A. Maganti. 2012. Low temperature deslination using solar collectors augmented by thermal energy storage. *Applied Energy* 91, 466–474.

Han, Y.M., Wang, R.Z., and Y.J. Dai. 2009. Thermal stratification within the water tank. *Renewable and Sustainable Energy Reviews* 13, 1014–1026.

Hee, W.J., Alghoul, M.A., Bakhtyar, B., Elayeb, O., Shameri, M.A., Alrubaih, M.S., and K. Sopian. 2015. The role of window glazing on day lighting and energy saving in buildings. *Renewable and Sustainable Energy Reviews* 42, 323–343.

Henning, H.M. 2007. Solar assisted air conditioning in buildings—An overview. *Applied Thermal Engineering* 27, 1734–1749.

Hentschke, R. 2016. On the specific heat capacity enhancement in nanofluids. *Nanoscale Research Letters*, 11–88.

Hoshi, A., Mills, D.R., Bittar, A., and T.S. Saitoh. 2005. Screening of high melting point phase change materials (PCM) in solar thermal concentrating technology based on CLFR. *Solar Energy* 79, 332–339.

Hussein, H.M.S., El-hetany, H.H., and S.A. Nada. 2008. Experimental investigation of novel indirect solar cooker with indoor PCM thermal storage and cooking unit. *Energy Conversion and Management* 49, 2237–2246.

Kant, K., Shukla, A., Sharma, A., Kumar, A., and A. Jain. 2016. Thermal energy storage based solar drying systems: A review. *Innovative Food Science and Emerging Technologies* 34, 86–99.

Karim, M.A. 2010. Performance evaluation of a stratified chilled-water thermal storage system, world academy of science. *Engineering and Technology Journal: Mechanical and Aerospace Engineering* 5, 18–26.

Kelly, B. 2010. Advanced thermal storage for central receivers with supercritical coolants. Abengoa Solar Inc. 2010. DE-FG-08GO18149.

Khadiran, T., Hussein, M.Z., Zainal, Z., and R. Rusli. 2016. Advanced energy storage materials for building applications and their thermal performance characterization: A review. *Renewable and Sustainable Energy Reviews* 57, 916–928.

Kousksou, T., Bruel, P., Jamil, A., El Rhafiki, T., and Y. Zeraouli. 2014. Energy storage: Applications and challenges. *Solar Energy Materials and Solar Cells* 120, 59–80.

Kuravi, S., Trahan, J., Goswami, D.Y., Rahman, M.M., and E.K. Stefanakos. 2013. Thermal energy storage technologies and systems for concentrating solar power plants. *Progress in Energy and Combustion Science* 39, 285–319.

Laing, D., Steinmann, W.D., Tamme, R., and C. Richter. 2006. Solid media thermal storage for parabolic trough power plants. *Solar Energy* 80, 1283–1289.

Li, G., and X. Zheng. 2016. Thermal energy storage system integration forms for a sustainable future. *Renewable and Sustainable Energy Reviews* 62, 736–757.

Liu, M., Tay, N.H.S., Bell, S., Belusko, M., Jacob, R., Will, G., Saman, W., and F. Bruno. 2016. Review on concentrating solar power plants and new developments in high temperature thermal energy storage technologies. *Renewable and Sustainable Energy Reviews* 53, 1411–1432.

Medrano, M., Yilmaz, M.O., Nogues, M., Martorell, I., Roca, J., and L.F. Cabeza. 2009. Experimental evaluation of commercial heat exchangers for use as PCM thermal storage systems. *Applied Energy* 86, 2047–2055.

Meffre, A., Py, X., Olives, R., Bessada, C., Veron, E., and P. Echegut. 2015. High-temperature sensible heat-based thermal energy storage materials made of vitrified MSWI fly ashes. *Waste Biomass Valor* 6, 1003–1014.

Mehling, H., and L.F. Cabeza. 2008. Heat and Cold Storage with PCM. An Up to Date Introduction into Basics and Applications. Berlin and Heidelberg: Spriner.

Moens, L., Blake, D.M., Rudnicki, D.L., and M.J. Hale. 2003. Advanced thermal storage fluids for solar parabolic trough systems. *Journal of Solar Energy Engineering* 125, 112–116.

Nahar, N.M. 2003. Performance and testing of a hot box storage solar cooker. *Energy Conversion and Management* 44, 1323–1331.

Navarro, L., de Gracia, A., Castell, A., and L.F. Cabeza. 2016a. Experimental study of an active slab with PCM coupled to a solar air collector for heating purposes. *Energy and Buildings* 128, 12–21.

Navarro, L., de Gracia, A., Colclough, S., Browne, M., McCormack, S.J., Griffiths, P., and L.F. Cabeza. 2016b. Thermal energy storage in building integrated thermal systems: A review. Part 1. Active storage systems. *Renewable Energy* 88, 526–547.

Navarro, L., de Gracia, A., Niall, D., Castell, A., Browne, M., McCormack, S.J., Griffiths, P., and L.F. Cabeza. 2016c. Thermal energy storage in building integrated thermal systems: A review. Part 2. Integration as passive system. *Renewable Energy* 85, 1334–1356.

Pinel, P., Cruickshank, C.A., Beausoleil-Morrison, I., and A. Wills. 2011. A review of available methods for seasonal storage of solar thermal energy in residential applications. *Renewable and Sustainable Energy Energy Reviews* 15, 3341–3359.

Pintaldi, A., Perfumo, C., Sethuvenkatraman, S., White, S., and G. Rosengarten. 2015. A review of thermal energy storage technologies and control approaches for solar cooling. *Renewable and Sustainable Energy Reviews* 41, 975–995.

Pintaldi, S., Sethuvenkatraman, S., White, S., and G. Resengarten. 2017. Energetic evaluation of thermal energy storage options for high efficiency solar cooling systems. *Applied Energy* 188, 160–177.

Prieto, C., Cooper, P., Fernández, A.I., and L.F. Cabeza. 2016. Review of technology: Thermochemical energy storage for concentrated solar power plants. *Renewable and Sustainable Energy Reviews* 60, 909–929.

Py, X., Calvet, N., Olives, R., Meffre, A., Echegut, P., Bessada, C., Veron, E., and S. Ory. 2011. Recycled material for sensible heat based thermal energy storage to be used in concentrated solar thermal power plants. *Journal of Solar Energy Engineering* 133, 1–8.

Rabha, D.K., Muthukumar, P., and C. Somayaji. 2017. Energy and exergy analyses of the solar drying processes of ghost chilli pepper and ginger. *Renewable Energy* 105, 764–773.

Ramadan, M., Aboul-Enein, S., and A. El-Sebaii. 1988. A model of an improved low cost indoor. *Solar Wind Technology* 5, 387–393.

Roback, C.W., Bergman, T.L., and A. Faghri. 2011. Economic evaluation of latent heat thermal energy storage using embedded thermosyphons for concentrating solar power applications. *Solar Energy* 85, 2461–2473.

Saadatian, O., Sopian, K., Lim, C.H., Asim, N., and M.Y. Sulaiman. 2012. Trombe walls: A review of opportunities and challenges in research and development. *Renewable and Sustainable Energy Reviews* 16, 6340–6351.

Schmidt, T., Mangold, D., and H. Müller-Steinhagen. 2004. Central solar heating plants with seasonal storage in Germany. *Solar Energy* 76, 165–174.

Schwarzer, K., and M.E.V. Silva. 2003. Solar cooking system with or without heat storage for families and institutions. *Solar Energy* 75, 35–41.

Sharma, S.D., Buddhi, D., Sawhney, R.L., and A. Sharma. 2000. Design development and performance evaluation of a latent heat unit for evening cooking in a solar cooker. *Energy Conversion and Management* 41, 1497–1508.

SOLARI. 2014. Solar consortium, increasing the market implementaiton of solar air conditioning systems for small and medium applications in residential and commercial buildings.

Solé, A. 2015. Phase change materials (PCM) characterization and thermochemical materials (TCM) development and characterization towards reactor design for thermal energy storage. PhD Thesis, University of Lleida, Lleida, Spain.

Solé, A., Barreneche, C., Martorell, I., and L.F. Cabeza. 2016. Corrosion evaluation and prevention of reactor materials to contain thermochemical material for thermal energy storage. *Applied Thermal Engineering* 94, 355–363.

Solé, A., Martorell, I., and L.F. Cabeza. 2015a. State of the art on gas-solid thermochemical energy storage systems and reactors for building applications. *Renewable and Sustainable Energy Reviews* 47, 386–398.

Solé, A., Miró, L., Barreneche, C., Martorell, I., and L.F. Cabeza. 2015b. Corrosion of metals and salt hidrates used for thermochemical energy storage. *Renewable Energy* 75, 519–523.

Stazi, F., Mastrucci, A., and C. di Perna. 2012. The behaviour of solar walls in residential buildings with different insulation levels: An experimental and numerical study. *Energy and Buildings* 47, 217–229.

Ströhle, S., Haselbacher, A., Jovanovic, Z.R., and A. Steinfeld. 2016. The effect of the gas-solid contacting pattern in a high-temperature thermochemical energy storage on the performance of a concentrated solar power plant. *Energy and Environmental Science* 9, 1375–1389.

Strumpf, H.J. 1987. *Advanced Heat Receiver Conceptual Design Study*. Washington, DC. National Aeronautics and Space Administration.

Tescari, S., Agrafiotis, C., Breuer, S., de Oliveira, L., Puttkamer, M.N., Roeb, M., and C. Sattler. 2014. Thermochemical solar energy storage via redox oxides: Materials and reactor/heat exchanger concepts. *Energy Procedia* 49, 1034–1043.

Tiwari, S., Tiwari, G.N., and I.M. Al-Helal. 2016. Development and recent trends in greenhouse dryer: A review. *Renewable and Sustainable Energy Reviews* 65, 1048–1064.

Yang, X., Yang, X., Qin, F.G.F., and R. Jiang. 2014. Experimental investigation of a molten salt thermocline storage tank. *International Journal of Sustainable Energy* 35, 1–9.

Zhang, H., Bayens, J., Cáceres, G., Degrève, J., and Y. Lv. 2016. Thermal energy storage: Recent developments and practical aspects. *Progress in Energy and Combustion Science* 53, 1–40.

11 Solar Photocatalytic Energy

Lourdes Isabel Cabrera-Lara
Independent Researcher

CONTENTS

11.1 INTRODUCTION

Exhaustible energy sources are being exploited at increasing rates, while societies increase their energy requirements. The implacable consequence is that today's fossil fuels cannot sustain demand. The solution to such a problem is to develop renewable energy sources [1]. They can be understood as "energy obtained from naturally repetitive persistent flows of energy occurring in the local environment" [2]. Among these, the extraordinary flow of solar energy reaching Earth stands out: it can supply the increasing demand. It is low cost, abundant, and inexhaustible (at least for several billion years) [1].

Thermal and light solar energy can be harvested by converting it into heat, electrical energy, and chemical fuels by using devices that allow such transformation. These final energies have encountered a significant problem: storage. Advances have been made in this matter. However, among these three energies, chemical fuels offer a great advantage. In this chapter, we are interested in the devices that researchers have designed and built to transform light energy into chemical fuels.

Chemical fuels store energy in covalent bonds [3]. Nature has found a way to perform such transformation by a process called photosynthesis, where raw materials such as CO_2 and H_2O produce carbohydrates (glucose) and O_2 by means of sunlight [4]. Electrochemists and material engineers have tried to mimic this process since 1972, when Fujishima and Honda reported the generation of H_2 from water reduction in an electrochemical system using TiO_2 as photoanode and Pt as cathode [5]. For decades, photoelectrochemical (PEC) cells have been developed to convert solar light energy into fuels like H_2 or MeOH (from CO_2 reduction) [6], due to the abundance of the required reagents, water and sunlight [6].

Hydrogen is regarded as an ideal fuel for meeting future needs [1,7,8]. Hydrogen is nontoxic, stable, can be produced from water using renewable energy sources, does not pollute the atmosphere (when burning with oxygen it produces water), and has high storage capacity [1,8]. Hence, its life cycle is clean and sustainable [8]. However, today's hydrogen synthesis is achieved by using fossil fuels, coal, natural gas, biomass, or nuclear energy. The process most commonly used for commercial and industrial production of hydrogen is steam reforming of natural gas (methane) with attendant water gas shift (WGS) reactions [9].

This process involves the conversion of hydrocarbons in the presence of water steam into hydrogen by means of a multiple-step process [9,10]. The reaction products are known as synthesis gas or syngas, a mixture of H_2, CO, CO_2, and H_2O (the two latter ones in lower quantities [9,10], which are the result of reactions (11.1) and (11.2)):

$$CH_4 + H_2O \leftrightarrow CO + 3H_2 \quad \Delta H°_{298} = 206.2 \text{ kJ/mol} \tag{11.1}$$

$$CH_4 + 2H_2O \leftrightarrow CO_2 + 4H_2 \quad \Delta H°_{298} = 165.0 \text{ kJ/mol} \tag{11.2}$$

Both reactions are endothermic, where reaction conditions involve high temperatures (around 800°C and 1000°C), pressures of 14–20 atm, and a nickel-based catalyst [9]. The syngas is then transferred to a WGS reactor, where a third somewhat exothermic reaction (11.3) takes place [9]:

$$CO + H_2O \leftrightarrow CO_2 + H_2 \quad \Delta H°_{298} = -412.0 \text{ kJ/mol} \tag{11.3}$$

allowing the use of low temperatures [9].

Coal can also be used as a source of hydrogen employing this method. Moreover, coal can be transformed to methanol, which can be used as a hydrogen precursor [9].

Syngas can be generated by biomass. In order to achieve it, two main procedures are used [10]:

1. Fluidized bed gasification followed by steam reforming ($T \approx 900°C$)
2. Entrained flow gasification ($T \approx 1300°C$), which involves a long process of pretreatment. Hence, the challenge is to design a system that allows the use of sunlight photons to reduce water and produce hydrogen at low temperatures and low pressure values [7,11]. The objective is to achieve industrial synthesis of hydrogen fuel with high efficiency under mild conditions and low energy costs [6,11].

An electrolysis configuration can drive water splitting when the generated photovoltage, V_{photo} is greater than the water electrolysis potential, V_{H_2O} [12]. The thermodynamic potential $E_{H_2}^o$ for the water-splitting reaction is given by reaction (11.4):

$$H_2O \rightarrow H_2 + 1/2O_2 \tag{11.4}$$

$$E_{H_2O}^° = E_{O_2}^° - E_{H_2}^°$$

$$E_{H_2O}^° (25°C) = 1.229 \text{ eV}$$

The free energy change for the conversion of one H_2O molecule into H_2 and $1/2O_2$ under standard conditions is $\Delta G = +237.2 \text{ kJ mol}^{-1}$, which, according to the Nernst equation, corresponds to $\Delta E_{H_2O}^° = 1.229 \text{ eV} (25°C)$ per electron transferred. In order to drive this reaction, the photoactive material (generally a semiconductor) must absorb radiant light to make its electrode potential higher than 1.23 eV (ca. 1.6–2.4 eV) [13]. If the material is irradiated by light that has energy greater than the band gap of the photoactive material, then the electrons of the valence band (VB) will be excited into the conduction band (CB) while the holes remain in the VB. In other words, to accomplish water splitting, the CB edge and VB edge should extend the reduction/oxidation potentials of water splitting [13]. Figure 11.1 shows the hydrogen evolution reaction (HER) and oxygen evolution reaction (OER) using electrons/holes generated under illumination [14,15]:

To carry out one or both reactions without recombination, photo-induced electrons and holes in the semiconductor must flow to the solution junction. The excess energy required to generate the electron–hole pair accounts for losses due to the concentration and kinetic overpotentials of the hydrogen and oxygen evolution reactions [13–15].

With the purpose of determining the incident photon-to-current conversion efficiency of photoanode materials, Equation 11.1 is used:

$$IPCE(\%) = \frac{1240\,J}{\lambda \times I} \times 100\% \tag{11.1}$$

where J represents the photocurrent density (mA cm^{-2}), λ is the wavelength of incident light (nm), and I is the intensity of incident light (mW cm^{-2}) [15].

The overall efficiency of solar energy conversion (η) is given by Equation 11.2 [15]:

$$\eta = 100\% \times \frac{J\left(E_{redox}^{\circ} - E\right)}{I_{light}} \tag{11.2}$$

where E is the bias voltage applied against the counter electrode ($E=0$ V for the photoanode), J is the photocurrent density (mA cm^{-2}) at the applied bias E, I_{light} is the irradiance intensity ($I_{light}=100$ mW cm^{-2} for AM 1.5 G), and $E_{redox}^{\circ}=1.229$ V (pH 0) for the water splitting reaction.

Finally, the solar-to-hydrogen (STH) efficiency for a photoelectric system constructed by using photoelectrodes with sunlight and water as inputs can be expressed by Equation 11.3 [15]:

$$\eta = 100\% \times J \times \frac{1.229}{I_{light}} \tag{11.3}$$

For example, in order to start to estimate the efficiency of a photoelectrode to drive the OER or HER, overpotentials of 400 mV at 10 mA cm^{-2} for water oxidation and 50 mV at 10 mA cm^{-2} for H$_2$ evolution at a semiconductor/liquid junction may be important values [15]. If the requirements are

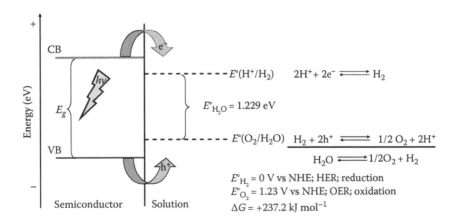

FIGURE 11.1   HER and OER of water using electrons and holes when the semiconductor is illuminated.

not fulfilled, water splitting can be very slow or remote when the VB potential of a semiconductor is not sufficiently positive for water oxidation or the CB potential is not sufficiently negative for proton reduction [15].

Therefore, the challenge is to find and use a semiconductor system with a band gap at a combined maximum power point voltage tuned to V_{H_2O} in a system also providing high photo efficiency [12].

Nowadays, researchers have developed a number of techniques to carry out water splitting by using the photovoltaic (PV) principle. These techniques can be categorized as:

- Coupled PV- electrolysis system [16]
- PEC water splitting
- Photocatalytic water splitting

Direct solar-to-chemical energy conversion in PEC cell is considered a more practical and efficient and less expensive method of H_2 production compared to electrolysis of water using PV-generated electricity, because the former integrates light energy collection and water splitting in one device [17].

PEC and photocatalytic water splitting differ in the fact that PEC takes place on an electrode. On the other hand, photocatalytic water splitting consists of colloidal systems. PEC cells using macroscale electrodes have a general advantage over colloidal systems. An external bias across an electrode can easily be applied to give control over redox potentials of the semiconductor. This is coupled with an efficient venue for dimensional electron–hole pair separation. Colloidal systems, however, have a much greater surface area to volume ratio, are simpler to produce, and have a greater potential to be scaled into larger systems [18].

In both systems, the limiting factor affecting efficiency is recombination of charge carriers. After photogeneration of electron–hole pairs in the semiconductor in a timescale of nanoseconds (or shorter), the hole and electron may migrate to the surface on the picosecond timescale to then participate in oxidation/reduction reactions. If the carriers recombine faster than they react, the STH efficiency is quite moderate. Consequently, electron–hole recombination is a major loss pathway that limits the quantum efficiency of photosynthetic systems. Thus, for optimal efficiency the carriers should be separated as far as possible, on the surface of a photocatalyst [19].

11.2 PHOTOELECTROCHEMICAL PRODUCTION OF HYDROGEN

In these systems, solar conversion of luminescent energy to hydrogen is performed at PEC devices. In a PEC water-splitting device, an external or self-bias voltage suppresses the recombination of photogenerated charge carriers, which improves their separation and transfer [5]. These devices require semiconducting photoelectrodes that should be stable in direct contact with an electrolytic aqueous solution.

The photoelectrode must also fulfill several requirements for it to be employed [5,13,16,20,21]:

1. High stability and resistivity to corrosion and photocorrosion
2. Absorption of photons related to the band gap, preferentially in the visible spectrum
3. CB minimum value (E_C), above the H_2O/H_2 electrochemical level of water reduction ($E_C > E_{O_2/H_2}$)
4. Suitable band edge positions VB maximum value (E_V) below the O_2/H_2O electrochemical level of water oxidation ($E_V < E_{O_2/H_2}$)
5. Efficient transport of photogenerated charge carriers
6. Generates H_2 or O_2 at its surface
7. Low overpotential
8. Low cost and toxicity and availability

The performance of a PEC water-splitting system is dominated by the properties of the semiconductor photocatalysts. However, electrolytes are essential in the PEC system for charge transfers, which usually are NaOH, Na_2SO_4, etc. Some sulfides and organic agents can also serve as electrolytes [5], since in these substances there is no oxygen evolution, as water oxidation potential is more positive than the oxidation potential of these compounds [5].

Semiconducting materials found in nature are not suitable for the demands established for photoelectrodes. Various effective ultraviolet (UV) light-responsive photocatalysts suffer from photocorrosion and are not active under visible light, which accounts for 45% of energy in the solar spectrum [5]. Hence, several research groups now focus on the development of photoelectrodes that can meet such needs [16]. Figure 11.2 represents the band structure of several photoelectrodes that can allow the conversion of water into H_2 and O_2 when it is under illumination [13,22].

When a semiconducting photoelectrode is placed in an electrolyte containing a redox couple, an electronic equilibrium process takes place by electron transfer across the interface (reaction 11.5) [23–25] by establishing a potential difference at the interface. For simplicity, a one-electron process will be considered [24]:

$$O + e^- \rightarrow R \qquad (11.5)$$

For an n-type semiconductor immersed in a redox electrolyte that has a redox Fermi level below the Fermi level in the solid, electrons will be transferred to the oxidized species (O) in the solution until equilibrium is reached. This leads to the formation of a depletion layer (a space charge region (SCR) due to the removal of electrons).

The positive charge at the SCR in the semiconductor is balanced by negative charges in the electrolyte, mainly ions, located at the outer Helmholtz plane of the electrical double layer. This promotes a potential drop at the semiconductor/electrolyte junction at the SCR that results in an electric field that penetrates the surface, and a potential difference is established at the interface [25–27]. At the SCR, the energy band edges in the semiconductor vary continuously because the electric field varies in function of the electrical potential value. It must be pointed out that the electrical potential depends on charge transfer taking place, which causes the band structure of the semiconductor to bend. This phenomenon is known as band bending between the phases (Figure 11.3) [2,21,26]. When equilibrium is reached, the Fermi levels in both phases are the same. Hence, the electrochemical potential of electrons is equivalent to such value [7,28].

In order to describe several properties at the electrode/electrolyte junction (such as the potential, charge distribution and recombination probability) a unique potential value between the bulk and the surface of the semiconductor known as flat-band potential (V_{fb}), can be used. At this potential there is no excess charge on the semiconductor side of the junction, and the band is flat (Figure 11.3)

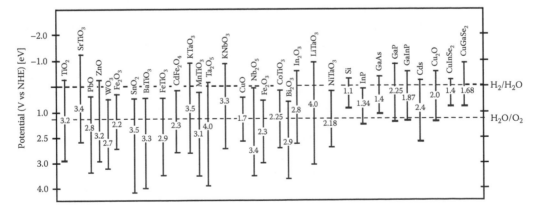

FIGURE 11.2 Band structures of several semiconducting photoelectrodes suitable for PEC devices.

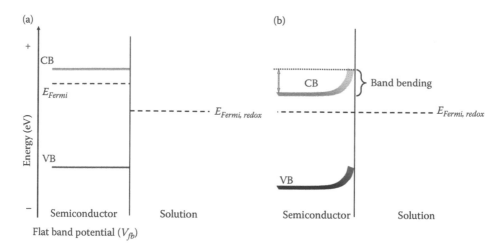

FIGURE 11.3 (a) Flat band potential and (b) band bending of the energy bands.

[21,24,29]. It must be noticed that the electron concentration at the surface is orders of magnitude lower than at the bulk of the semiconductor [24]. This translates into a great sensitiveness of the potential across the Helmholtz layer to ionic surface charges at the semiconductor [24].

In the case where p-type semiconductors are placed in the presence of a redox electrolyte, an analogue situation is appreciated. Holes will be transferred to the reduced species (R) in the solution, which also results in the generation of a depletion layer due to the removal of holes. Band bending and flat band potentials are also observed in these systems.

For a given n-type semiconductor used at a photoelectrochemical process, if the V_{fb} is positive, electrons are attracted toward SCR, hence this region decreases and leads to increases in the recombination charge carriers. Thus, it is desired to have the V_{fb} large and negative (relative to the redox potential of the H$^+$/H$_2$ couple, which depends on pH); to split water molecules without the imposition of a bias, the SCR is broadened and recombination of charge carriers is decreased [21].

If the potential becomes more positive than V_{fb}, electrons leave the electrode (depletion condition), and a positive SCR is left. The positive charge in the SCR in the semiconductor is balanced by a negative charge in the electrolyte, mainly ions, located at the outer Helmholtz plane of the electrical double layer. This promotes a potential drop at the semiconductor/electrolyte junction at the SCR, which generates an electric field. This electric field varies in function of the electrical potential value. The change in the potential across the SCR affects the energy of electrons and holes and is reflected in the band bending [16].

Semiconductor photoelectrodes should efficiently allow water oxidation or reduction reactions in aqueous solutions at a sufficient rate at current densities of (10–15) mA cm^{-2} under 1 Sun illumination showing chemical and physical stability for more than 2000 h [17]. To date, not a single semiconductor has been found that can drive water reduction and oxidation reactions at the same electrode and at the same time exhibit stability in the aqueous electrolytes. Since the water-splitting process consists of two separate half-reactions, it seems adequate to use two semiconductor photoelectrodes: one, for the separate OER and a second for the HER [13].

In a PEC cell where the photoelectrode is an n-type semiconductor (photoanode), photoexcited holes will migrate to the interface to promote OER (Figure 11.4) [13]. On the other hand, if a p-type semiconductor is used as photoelectrode (photocathode), photoexcited electrons will migrate to the interface and allow HER. Following this scheme, only the conduction or valence band position should be considered for HER or OER, respectively [13]. Nevertheless, the current STH conversion efficiency is too low to develop a commercial device for large-scale water splitting.

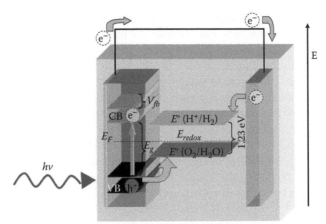

FIGURE 11.4   n-Type semiconductor (photoanode) promoting OER.

11.2.1   The Use of Photoanodes

In order to promote water splitting using photoanodes, the VB edge of the n-type semiconductor should be more positive than the oxygen evolution potential in order for the photoanode to oxidize water. When the photoanode is coupled with a metallic catalytic counter electrode, the majority carrier electrons are transferred to it via an external circuit to produce hydrogen (Figure 11.4). Charge separation is driven by the electric field in the SCR at the surface of the electrode, which must be sufficient to drive oxidation and reduction reactions ($\Delta E_{H_2O}^\circ = 1.23V$) [18,19,30].

In the case of photoanodes for water oxidation, numerous n-type semiconductors have been investigated [13]. The most studied semiconductors to be used as photoanodes for PEC water splitting have been the metal oxides TiO_2, $BiVO_4$, WO_3, Fe_2O_3, ZnO, etc. [5,18] These n-type semiconductors are of interest due to their low VB energy positions with more positive VB potential for water oxidation reaction [22]. However, they do not produce high photocurrent densities under AM 1.5 illumination [15].

Among the studied materials for PEC application, TiO_2 has been a main focus of research due to its low cost, chemical and optical stability in aqueous media in the presence of electrolytes and under light, nontoxicity [18], and suitable band edge position to allow the oxidation/reduction of water [18]. However, the STH conversion efficiency of TiO_2 is low due to the low photocurrent generation [18,31]. Its large band gap ($E_g = 3.2\,eV$) limits its light utilization to the UV region that accounts for only ~4% of the solar energy, resulting in low STH efficiency [21].

$BiVO_4$ with $E_g = 2.4\,eV$ and CB edge (0.46 eV vs NHE) placed below the water reduction energy level can allow photogenerated holes to photooxidize water. However, $BiVO_4$ presents a fast electron/hole recombination, poor charge transport properties, and poor water oxidation kinetics [15].

WO_3 presents good electron transport properties and stability against photocorrosion. However, the band gap of WO_3 (~2.8 eV) is still too large to absorb sufficient visible light, considering that the desired value is ca. 2.0 eV [15]. The CB edge of WO_3 is close to 0.4 eV, which is below the hydrogen redox potential [15,32]. Since CBs are more positive than the potential of water reduction, in some cases it is necessary to apply an external bias to compensate for the electrochemical potential solution deficiency and quickly separate excited charges to avoid photocorrosion of the photoanode [18,19,22].

Metal sulfides are considered efficient photocatalysts, but they suffer photocorrosion [18]. Transition metal oxides are of interest because of their chemical stability. However, in general they present large band gap values ($E_g \geq 3\,eV$), limiting the amount of visible light that can be absorbed. On the other hand, transition metal nitrides have smaller band gaps than transition metal oxides, but they quickly oxidize in aqueous solutions. Alternatively, oxynitrides present smaller band gaps than oxides and better stability than nitrides. For example, the band gap of tantalum oxynitride (TaON) is $E_g \approx 2.5\,\text{eV}$, between those of Ta_2O_5 ($E_g = 3.9\,\text{eV}$) and Ta_3N_5 ($E_g = 2.1\,\text{eV}$). It has a high quantum yield for visible-light water splitting. It also shows appropriate alignment of its valence and conduction band edges relative to the OH^-/O_2 oxidation and H^+/H_2 reduction potentials, allowing the production of H_2 and O_2 from water without applying an external bias [15]. Oxynitride semiconductors are visible-light responsive and could be used for photocatalytic water splitting; however, they have shown low solar energy conversion efficiency [5].

Several strategies have been made to enhance the photocurrent and to avoid charge recombination, such as extending the absorption spectrum by decreasing its dimension to nanoscale and changing its morphology, by doping the semiconductors with metal or nonmetal ions, by hydrogenation or creation of oxygen vacancies in its structure [18], and/or by surface sensitization with small band gap materials [5,19,33]. Equally, the use of cocatalysts to accelerate interfacial oxidation-reduction kinetics and the use of heterojunction structures to achieve spatial separation of electrons and holes [34] has also been tried. All of these will be explained in this chapter.

11.2.1.1 Nanodimensional Structures of Semiconductors

PEC systems are both thermodynamically challenging and kinetically slow, where electron–hole recombination limits the quantum efficiency of photosynthetic systems [34]. For PEC systems, dimensional scale of particle size has been reduce in order to decrease electron–hole recombination losses. Decreasing size to nanoscale reduces the distance electrons and holes must travel before reaching surface-reactive sites. Nanostructured semiconductors are of interest in PEC water splitting, because they present new chemical and physical properties that bond to quantum effects. This implies that by controlling their size and morphology, several characteristics can be tuned or designed, such as band gap energy positions and carrier collection pathways [22,35,36].

Well-oriented one-dimensional (1-D), two-dimensional (2-D) and three-dimensional (3-D) nanostructures have been studied. The 1-D nanostructures like nanowires, nanorods, and nanotubes provide both efficient light harvesting and superior charge transport for PEC water splitting. Meanwhile, 2-D nanostructures, in particular nanosheets of metal oxides and chalcogenides, present efficient charge separation and migration in the water-splitting process. 3-D hierarchical nanostructures of 1-D and 2-D structures present large surface areas for light harvesting without inhibiting charge transfer and separation [37]. Hence, water splitting can take place due to broader light absorption, rapid charge migration, and separation. On the other hand, it has been reported that for zero-dimensional (0-D) systems, hydrogen production is diminished due to the reversal of reactions, where oxygen reacts with excess electrons and hydrogen with excess holes to reform water [38].

Among the mentioned nanostructures, 1-D nanotube arrays have gained much attention because of their intrinsic large surface area and unidirectional flow of charges [5,39,40] due to quantum confinement phenomenon [35]. For example, due to these effects, TiO_2 nanomaterials exhibit a wider band gap than their bulk counterpart. Furthermore, their high surface area facilitates interaction with the surrounding media. 1-D TiO_2 nanostructures have one quantum, unconfined direction for electrical conduction, rather than tunneling transport, which proves beneficial for PEC applications. Also, large internal surface areas without a concomitant decrease in geometric and structural order provide direct channels for charge and mass transport to the back contact [41].

11.2.1.2 Doping

Doping is one strategy to improve the conductivity of semiconductors and reduce recombination of the photogenerated electron–hole pairs [22]. When doping a semiconductor with non-metal ions (N, S, C, B, P, I, F) band gap narrowing is observed. In other words, it exhibits a red shift of the absorption spectrum and possesses a higher photocatalytic activity than the bare semiconductor, especially in the visible part of the solar spectrum. This shift is due to the existence of additional electron levels above the VB (Figure 11.5b) [42,43].

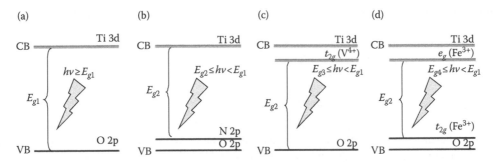

FIGURE 11.5 (a) Undoped TiO_2 material, (b) doping TiO_2 with nonmetal ions, and (c, d) doping TiO_2 with metal ions.

The visible light response of metal-doped semiconductors (metals like Fe, Zn, Zr, and Cr) happens due to the presence of new energy levels in the band gap (Figure 11.5 c,d). In this situation, electrons are excited due to the absorption of photons from the defect state to the CB. Furthermore, doping improves trapping of electrons and avoiding electron-hole recombination during irradiation, which results in enhanced photoactivity in visible light [43]. Doping n-type semiconductors with carbon has also been reported to increase their performance. For example, carbon-doped TiO_2 nanotubes can generate photocurrents 20 times higher than bare TiO_2 [5].

11.2.1.3 Heterojunctions

Another strategy to harvest visible light is by modifying the n-type nanosemiconductor's surface with a second nanomaterial of lower band gap to form a heterojunction. The second nanomaterial serves as a photosensitizer and builder for internal electric field across the interface. The internal potential bias generated between them promotes the excited electrons and holes' separation and transportation across the interface of the heterojunction. By this strategy, recombination of electrons and holes is reduced [5]. The nanostructured morphology and crystallinity of semiconductors with suitable CB and VB positions for water splitting and the interfacial properties in the heterojunction are designed to be properly aligned to generate the desired bias [21].

Three main heterojunction constructions can generally be found (Figure 11.6), where PS corresponds to a semiconductor (p-type or n-type) or another kind of photosensitizer and n-SC corresponds to our n-type semiconductor. Type I heterojunction consists of (two) semiconductors where the CB of component PS is higher than n-SC (Figure 11.6a). The VB of PS is lower than n-SC; therefore, holes and electrons will transfer and accumulate on components. In the second, type II, junction, n-SC has a more negative CB position compared to PS (Figure 11.6b). Holes then shift from the more positive VB of n-SC to PS, allowing efficient charge separation and enhanced photocatalytic

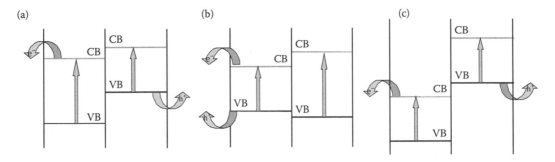

FIGURE 11.6 (a) Type I heterojunction, (b) type II heterojunction and (c) type III heterojunction, where PS corresponds to a p-type or n-type semiconductor or another kind of photosensitizer; n-SC corresponds to the n-type semiconductor of interest.

activity. Type III and Type II are very similar, except that the energy difference in VB and CB positions is greater, which gives a higher driving force for charge transfer (Figure 11.6c) [19].

Heterojunctions formed between two materials, such as semiconductor–semiconductor (S–S), semiconductor–metal (S–M), and semiconductor–carbon (S–C) (carbon nanotubes, graphene, etc.) increase STH conversion efficiency [19].

11.2.1.3.1   Heterojunction n-SC|Semiconductor

It has been observed that p-n type heterojunctions allow the separation of the excited electrons and holes. Furthermore, it assists the charge carrier transfer and separation across the interface because of band gaps overlapping. For example, p-type $CdTe$ and Cu_2O semiconductors form p–n junctions with TiO_2. Both composites have shown a higher efficiency in water splitting than only TiO_2 [5]. Also n-n type heterojuctions have been studied. In general, both the CB and VB of one semiconductor are more negative than those of the counterpart [44]. For example, at the TiO_2/Bi_2WO_6 heterojunction, narrow band gap excited electrons from TiO_2 are quickly transferred to Bi_2WO_6 because its CB is more negative than TiO_2. The VB of Bi_2WO_6 is more positive than the one TiO_2; hence, the excited holes from Bi_2WO_6 can be quickly transferred to TiO_2. As a result, electrons and holes can be separated and transferred quickly for efficient water splitting [5]. n-Si|n-Fe_2O_3, n-Si|TiO_2 [45], and n-Si|ZnO [46] are some other examples that can be found in the literature.

11.2.1.3.2   Heterojunction n-SC|Metal

In addition, plasmonic nanoparticle (NP) coupling is another strategy for utilizing the optical properties of nanometals for better light harvesting [5]. In 1980, Sato and White reported the increased photoconversion of H_2O to H_2 and O_2 by using the catalytic composite system Pt–TiO_2 [47]. Since then, researchers intensified their study on composite metal/semiconductor properties [48].

The surface plasmon resonance (SPR) phenomenon observed in noble metal is believed to help in the generation of photoactive catalysts at the visible light region [48]. SPR is an intrinsic property of metal NPs, in which the oscillation frequency is very sensitive to the metal size and shape as well as the dielectric constant of the surrounding environment [5]. It has been proposed that metal NPs capture light and transfer the excitation energy through the localized surface plasmon to the semiconductor with a large band gap [18].

However, to date it is uncertain whether the electrons are flowing from the semiconductor to the metal due to the Schottky barrier or from the metal to the semiconductor due to the SPR effect [48]. Still, it is known that the employment of noble metal NPs such as Pt, Au, Pd, and Ag avoids the recombination of photo-generated electrons and holes and consequently increases the overall photocatalytic process efficiency [35].

For example, Au NPs-decorated TiO_2 nanowire electrodes exhibited enhanced photoactivity in both the UV and the visible regions (up to 710 nm) and showed a large photocurrent generation [5,35].

11.2.1.4   Hydrogenation or Oxygen Vacancies

In general, semiconductors with wide band gap values present poor electrical conductivity and low energy conversion efficiency, a problem that limits charge separation. This problem has been confronted by the incorporation of hydrogen in the material, which is effective on most semiconductors with a CB minimum that lies below a certain level (i.e., electron affinity > 3.0 ± 0.4 eV) [49]. Hydrogenation has shown an increase in reactivity and surface disorder of the semiconductor lattice, which in turn allows the material to produce a heightened electrical conductivity and conversion [49].

This highly disordered surface layer translates into the generation of a large amount of oxygen vacancies, which are shallow donors that can significantly improve the conductivity of semiconductors [34]. A high concentration of oxygen vacancies results in a high value of donor concentration in the final semiconductor. This value could reflect the response of the semiconductor [33]. It can lead to the improvement of optical absorption in the visible light and near-infrared region, along with change in electrical resistance [50]. Introduction of hydrogen into the metal lattice also promotes the formation of interstitial hydrogen and passivation of dangling bonds defects [49]. The procedure also induces the generation of mixed valance in an oxide structure, which produces shallow donor levels. This way, mid-gap states are yielded.

Hydrogenation, the black colored H–TiO$_2$, shows a higher activity for water splitting. Analytical data allows us to believe that Ti–H and Ti–O–H bonds on the surface of H–TiO$_2$ NPs enhance the separation of photogenerated electrons and holes. In this material, Ti is mostly found as Ti^{4+} but also possesses a titanium lower oxidation state (Ti^{3+}) (Figure 11.7). The presence of Ti^{3+} species and oxygen vacancies provides shallow donor levels below the CB minimum, improving TiO$_2$ photocatalytic activity [43,51,52].

This can also be observed on iron or tungsten oxides, where the mixed valence leads to efficient electron transfer, which in turn is responsible for the strong visible absorption. Combining disorder engineering with elemental doping could allow absorption in the visible range and reduce charge carrier recombination for many wide-band gap materials. For example, H–TiO$_2$ and H–WO$_3$ generates OER even at neutral pH [34], while regular TiO$_2$ and WO$_3$ NPs require highly concentrated acid media to achieve such results.

11.2.1.5 Dye Sensitization

In the early 1980s, Grätzel summarized three suitable light-harvesting units for water splitting using sensitizers and semiconductors or electron relays [53,54]. Grätzel reported the bifunctional redox cocatalyst Pt and RuO$_2$ co-deposited on colloidal Nb-doped TiO$_2$ particles. The system worked under visible light in the presence of Ru(bppy)$_3^{2+}$ (bpy=2,2′-bipyridine), a photosensitive organometallic complex. It was proposed that the sensitizer adsorbed at the semiconductor and injected the electron into the TiO$_2$ CB, affording a quantum yield of ~5% for H$_2$ production under $\lambda \geq 400$ nm light irradiation. Grätzel's work was the starting point for the research of dye sensitizers that could improve PEC development [54].

In general, these systems contain three main components: the photoanode (to which dye molecules are adsorbed), redox electrolyte, and counter electrode. When light strikes the dye, it photoexcites, which results in the injection of an electron into the CB of the semiconductor. The oxidized dye is regenerated by the redox mediator at the electrolyte. The electron in the CB flows through the external circuit to the counter electrode, where the oxidized species of the redox pair in the electrolyte are reduced again, and the circuit is complete [55].

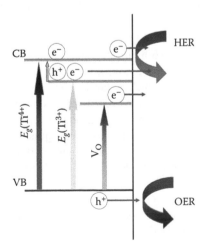

FIGURE 11.7 H–TiO$_2$ material, representing its band structure for water splitting.

The dyes that are commonly used for PEC H_2 production can be classified into metal complexes, organic dyes, and quantum dots. From this classification, metal complexes can be divided into Ru-complexes and other transition metal complexes. Quantum dots (QDs) are semiconducting nanocrystals that show attractive tunable properties due to quantum confinement in all three dimensions [56]. The sensitization of TiO_2 and ZnO with InP, CdS, CdSe, CdTe, PbS, etc., has shown improved performance in PEC water splitting [54,56].

QDs have a diameter less than or equal to the exciton Bohr radius of the corresponding bulk semiconductor. Because of their size effects, they present properties that contribute to generate tunable energy band gaps. They possess a large extinction coefficient due to quantum confinements, large intrinsic dipole moments, and increased number of carriers [56].

Currently, most metal complex dyes and organic dyes can only respond to visible light in the region of $\lambda = 400–600$ nm [54]. This way, dye sensitization has improved PEC reactions by enhancing visible-light absorption and photoconversion. For example, 1,1-binaphthalene-2,2-diol sensitized TiO_2 showed H_2 production under visible light, whereas the bare TiO_2 did not. A recent study found that nanoparticulate anatase TiO_2 sensitized with ruthenium polypyridyl dyes showed visible light induced water splitting. Significantly, the dye molecules serve not only as a sensitizer for visible-light absorption but also as an effective stabilizer and a molecular bridge to connect water oxidation catalyst IrO_2.

Dye sensitization of semiconductors has been studied extensively for H_2 production. Its low cost and easiness of generation makes them very attractive. However, their activity is still low [54].

11.2.2 The Use of Photocathodes

Photocathodes normally consist of p-type semiconductors, where the CB edge of the photocatalysts should be more negative than the hydrogen evolution potential in order to generate hydrogen. In Figure 11.2, several p-type materials with conduction/valence bands appropriate to perform the reactions at the H^+/H_2 potentials can be observed [15]. It must be pointed out that there are fewer reports based on p-type semiconductors as photocathodes for PEC water splitting. However, increasing efforts have been made in the study of Cu_2O, NiO, and chalcogenides in this matter [57].

Cu_2O is a typical p-type semiconductor used as a photoelectrode for PEC water splitting. However, it suffers from self-reduction due to photogenerated electrons [5]. Additionally, some chalcogenides containing Cu(I) ions, such as $CuInS_2$, $CuGaSe_2$, Cu_2ZnSnS_4, and $Cu(Ga, In)(S, Se)_2$, function as p-type photocathodes for PEC water splitting under visible light. However, they suffer from photocorrosion [58].

Spinel oxides such as $CaFe_2O_4$, $NiCo_2O_4$ behave as p-type semiconductors as well and have also been studied for PEC devices [59,60]. p-type semiconductors such as boron-doped Si, Cu_2O, and $GaInP_2$ exhibit small band gaps; thereby, a significant proportion of visible light can be harvested. However, severe challenges such as large overpotential, instability, and a low absorption coefficient have been observed. Thus, recent strategies have been focused on improving the photostability through loading suitable protective layers or coupling with n-type semiconductors to form heterojunctions [5]. The creation of the p-n heterojunction between CdS and Cu(In, Ga) Se_2 allowed the photocurrent to increase, which also increased the charge separation. WO_3/Cu_2O p–n junctions have also been generated to decrease the self-reduction of Cu_2O and enhance the PEC properties.

Incorporation of noble metal NPs to the heterojunction has also been studied. The photocathode $Cu_2O|Al:ZnO–TiO_2$ decorated with Pt NPs showed photocurrents of up to -7.6 mA cm^{-2} at a potential of 0 V vs. RHE at mild pH. In the case of the photocathode, Pt NPs/CdS/$CuGaSe_2$ showed a stable photocurrent (about 4 mA cm^{-2}) under reductive conditions for more than 10 days under visible-light irradiation [5].

QDs have not been reported for the sensitization of p-type semiconductors; however, dyes have been extensively used [54]. Dye sensitization of a p-type semiconductor can be achieved. After dye photoexcitation, the dye injects a hole into the VB of a p-type semiconductor. The injected hole migrates to the counter electrode. The reduced dye molecule is restored to its ground state by delivering an electron to the oxidized species of a redox mediator in the electrolyte. The reduced form of a redox mediator finally passes an electron to an external load to complete a whole circuit. The highest efficiency of p-type PEC has only reached 1.30%, where a mesoporous NiO film was used [54]. Ru-complexes and metal free organic dyes are commonly used because of their variety, low cost, tunable structure, and spectral responsive ability [54].

11.2.2.1 Z-Scheme Structure

At the Z-scheme PEC device, an n-type semiconductor functions as photoanode and a p-type semiconductor is used as photocathode, Figure 11.8. In this system, the mismatching Fermi levels can produce a self-bias that can drive the excited electrons from photoanode to combine with the excited holes from photocathode. The self-bias acts as an extra driving force for carriers' charge transfers and transportation while their performance is still governed by the materials and the competition between chemical reaction and recombination. Water oxidation and reduction take place over the photoanode and photocathode, respectively [5,61].

Z-scheme devices can proceed in the absence of redox pairs due to interparticle electron transfers during the physical contact between the hydrogen and oxygen evolution photocatalyst [58]. In these devices, the maximum photocurrent and the working potential of the photoelectrodes are theoretically determined by the intersection of the steady current–potential curves of the respective photoelectrodes [58]. Wang et al. [62] reported a Z-scheme PEC visible light responsive cell where WO_3 was the photoanode and p-$GaInP_2$ the photocathode. The CB edge of WO_3 (ca. 0.25 V vs. NHE) is more positive than the VB of p-$GaInP_2$ (ca. 0 V vs. NHE), which allows charge transfer from WO_3 to p-$GaInP_2$ via an external circuit [5]. However, the WO_3 presents poor charge separation properties and weak visible-light absorbance and can only split water with high light intensity applied.

A dye molecule or visible-light-absorbing semiconductor is usually employed as the excitation site. The absorption wavelengths are tuned by modifying the dye structure or designing the semiconductor's electronic structure. The electron donor material must meet two principal requirements. First, its energy level must be more negative than the excited state reduction potential of the photosensitizer but more positive than the water oxidation potential (+0.82 V vs NHE, pH 7). Second, the

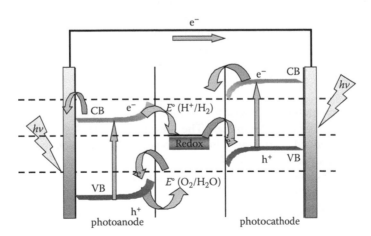

FIGURE 11.8 Z-scheme PEC device with an n-type semiconductor as photoanode and p-type semiconductor as photocathode.

donor must be connected to the photosensitizer to induce a swift electron transfer reaction prior to the decay of the photosensitizer excited state. Similar requirements must also be met for the electron acceptor: its potential energy level must be between the photosensitizer excited state oxidation potential and the water reduction potential (−0.41 V vs NHE, pH 7). Cocatalysts are usually introduced to accelerate water-splitting reactions [34].

Zeng et al. reported a Z-scheme PEC cell with a photoanode constructed by Bi_2S_3-decorated TiO_2 nanotubes and a PV cell Pt-modified Si electrode as photocathode, obtaining a response to visible light. The CB edge of TiO_2 (~ −0.05 V vs. NHE) is more negative than the VB edge of Pt/$Si_{photovoltaic\ cell}$ (~ 0.8 V vs. NHE), a self-bias is formed between the photoelectrodes. However, this system needs the addition of sacrificial agents in order to reduce photocorrosion [5,63]. The photosensitizers available with excited-state reduction potential and the excited-state oxidation potential that meet the requirements are limited. Furthermore, a drawback of these molecules is that only a fraction of the sunlight can be utilized in the water-splitting process [34].

Assuming such processes require 0.3 eV as a reaction driving force, and given that the energy difference between water oxidation and reduction potentials is 1.23 eV, the excitation energy required for the overall process is more than 1.83 eV (<677 nm) [34].

11.2.2.2 Scavengers or Sacrificial Agents

PEC water splitting into H_2 and O_2 has a very low efficiency, not only because of charge carriers recombination, but also due to its rapid reverse reaction. However, the use of scavengers or sacrificial agents can enhance hydrogen production [8,64,65].

These scavengers, which have lower oxidation potentials than water, act like reducing agents, consuming the photogenerated holes accumulated at the photoelectrode's surface [8,65]. The scavenger is oxidized into products that are less reactive toward hydrogen. In addition, as it is not generated in the medium, the reduction reaction of O_2 that competes with the reduction of water is inhibited [8,64,65], allowing photogenerated electrons to become more readily available to reduce H^+ to produce hydrogen [8,65,66].

Alcohols, such as ethylen glycol, methanol, and ethanol, are examples of sacrificial agents that are commonly used in photocatalytic water splitting [64]. In a water–alcohol mixture, methanol is commonly used as a hole-scavenger and usually yields a better efficiency than other compounds. Since the oxidation potential of methanol is 0.02 V (for water $E_{O_2/H_2} = 1.23$ V at pH 0), the reaction products result into less toxic substances [64]. Ethanol is also used as sacrificial agent because it is more sustainable and renewable than methanol. However, ethanol forms acetaldehyde as a reaction intermediate, which can derivate into acetic acid, limiting hydrogen production and favoring methane formation [64]. Every scavenger has a different effect in the production of hydrogen. Nada et al. [67] investigated the effects of ethylendiaminetetraacetic acid (EDTA), methanol, ethanol, CN^-, lactic acid, and formaldehyde in terms of enhancement capability. The ranking found was: EDTA > methanol > ethanol > lactic acid. It should be noted that the decomposition of these hydrocarbons could also contribute to a higher hydrogen yield since hydrogen is one of their decomposed products [8].

Other organic compounds that are widely used for this matter include cysteine, triethanol amine, hydroquinone, and Fe(II)-EDTA complexes. The drawback is that during the PEC reaction, the scavenger needs to be added continuously in order to sustain the reaction, since they will be consumed during photocatalytic reaction [8,64].

Inorganic redox pairs have also been used as sacrificial agents for hydrogen production, such as S^{2-}/SO_3^{2-}, Ce^{4+}/Ce^{3+}, $Fe(CN)_6^{4-}/Fe(CN)_6^{3-}$, and IO_3^-/I^-. For example, when CdS is used as a photocatalyst for water splitting hydrogen production, photocorrosion occurs, as shown in reaction (11.6) [8]:

$$CdS + 2h^+ \rightarrow d^{2+} + S \tag{11.6}$$

Using the redox pair S^{2-}/SO_3^{2-}, S^{2-} reacts with two holes to form S, while SO_3^{2-} reacts with S to generate $S_2O_3^{2-}$. This way S will not deposit at the surface of CdS, and at the same time CdS photocorrosion is prevented [8].

In the case where two photocatalysts are used, the IO_3^-/I^- redox pair has been used. For hydrogen production on the photocathode, I^- can scavenge holes, and, thus, CB electrons are available to reduce protons to hydrogen molecules. For oxygen production on the photocathode, IO_3^- can react with CB electrons to form I^-; thus, VB holes can oxidize water to oxygen. In this system, PEC water splitting produces both hydrogen and oxygen without consumption of the sacrificial reagent. Similarly, Ce^{4+}/Ce^{3+} and Fe^{3+}/Fe^{2+} pairs are effective for water splitting hydrogen production [5,8,68]. Halides, specially Cl^-, are effective hole scavengers. Oxidation of Cl− ions by photogenerated holes at the semiconductor surface achieve halide radical formation. This has also been observed when using Br^- and IO_4^- in an aerated solution [18].

11.3 PHOTOCATALYTIC WATER SPLITTING: THE USE OF COLLOIDS

Colloidal NPs present optical and electronic properties that have caught the attention of researchers. They possess greater surface area to volume ratio, have the ability to carry out all reactions that previously associated with massive semiconductor electrodes [69], and can potentially be scaled into large systems [18]. Furthermore, they are easy to functionalize and synthesize [70]. For these characteristics, they are of interest for colloidal photocatalytic devices for hydrogen and oxygen generation [69].

The development of semiconducting NPs can be optimized by varying the size, shape, or surface ligands of the NPs [71]. The band gap of a visible-light-driven photocatalyst should be narrower than 3.0 eV ($\lambda > 415$ nm). Therefore, suitable band engineering is necessary for tuning the band positions of photocatalyst materials for water splitting under visible light irradiation [14].

In general, photocatalytical systems can be classified into two kinds of systems, depending on their design and architecture. The first type of system consists of a single waterbed colloidal suspension of PV NPs producing a mixture of H_2 and O_2 product gases [14]. It could consist of colloidal n-type semiconductor NPs for the generation of oxygen or the use of colloidal p-type semiconductor NPs for the generation of hydrogen [34]. For example, oxynitride photocatalysts such as GaN–ZnO and $ZnGeN_2$–ZnO solid solutions are active for water splitting as single-particulate photocatalysts [14].

The second is a two-step process and is analogous to the NPs Z-scheme where two photons are employed to drive the overall water-splitting reaction [34]. It consists of a dual waterbed colloidal suspension of PV NPs, with one bed carrying out OER and the other carrying out HER (Figure 11.9) [14]. However, in the two-step process, the kinetic balance is more difficult to control for the whole electron-transfer process without losing the energy through charge recombination reactions [34]. Furthermore, it is difficult to collect oxygen and hydrogen at separate regions, as these are likely to be generated simultaneously. However, separate gas collection could be achieved by employing the membranes [34].

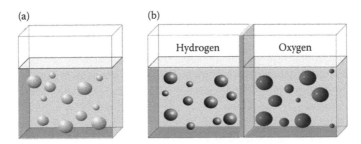

FIGURE 11.9 (a) Single waterbed colloidal suspension of PV NPs and (b) two-step process where two photons are employed to drive the water-splitting reaction.

11.3.1 STRATEGIES TO INCREASE PHOTOCATALYTIC WATER SPLITTING

Conversion efficiency has been estimated to be around 15% in order to cover electrical and mechanical energy costs, such as stirring. However, colloidal NP's aqueous suspensions of promising semiconductors like n-Fe_2O_3, CdS, CdSe, ZnS, ZnSe, and TiO_2 reach an energy efficiency of not higher than 3% [5]. As in the case of electrode PEC devices, strategies must be used in photocatalytic water splitting in order to increase STH efficiency. Basically the same tactics are applied to these systems. In other words, efficiency can be increased by changing their dimension and morphology, by doping the semiconductors, by constructing heterojunctions and oxygen vacancies, and by surface sensitization with small band gap materials.

Morphology of NPs has attracted interest because it allows the exploitation of novel properties in the material. 1-D morphological structures are of common interest. Recently, hierarchical structures like dendrites and flowers have received attention. Wang et al. have reported the use of flower like CdS decorated by histidine [72], which presents almost 13 times more activity than pure CdS.

A way to increase the spectral absorption is to create donor levels in the band gap by doping the material. Metal-oxide semiconductors are intrinsically stable in aqueous media, but they are only capable of absorption in the UV spectrum. Doping with Ce, V, Cu, Sn, Nd, Fe, Cr, and Co has been shown to red shift the absorption spectra of the metal oxide semiconductors. TiO_2 doped with Sb and Cr has shown improvement in photocatalytic activity [18]. Following this strategy, doped oxide photocatalysts that show reasonable activities for H_2 evolution under visible light irradiation are $SrTiO_3$:Cr, Ta, and $SrTiO_3$:Rh [14].

Codoped structures can increase the exciton charge separation and present anion vacancies in the lattice, which translate into visible-light absorption. Such properties have been observed for TiO_2 codoped with Sn^{4+}/Eu^{3+} [73]. Alloying of semiconductor materials can also improve photocatalytic activity of NPs, such as $SrTiO_3$, $CaTiO_3$, and $LaTi_2O_7$. All of these are photo-stable structures [74].

Narrow band gap semiconductor nanostructures like d^{10} semiconducting chalcogenides (e.g., sulfides, selenides, and tellurides) can harvest visible light energy [75]. Chalcogenides are of common study because they possess the ideal edge position for hydrogen production under visible light. However, they suffer from reverse reactions and photocorrosion, showing low conversion efficiencies [18,69,76]. Furthermore, photocorrosion has a strong dependence on particle size in the colloid suspension [18]. On the other hand, the presence of sacrificial agents, such as Na_2S, Na_2SO_3, can avoid these problems.

The use of scavengers cannot be avoided. For example, WO_3, $BiVO_4$, Cr, and Sb codoped TiO_2 (TiO_2:Cr, Sb); $AgNbO_3$, Cr and Ta codoped $SrTiO_3$ ($SrTiO_3$:Cr, Ta); Rh-doped $SrTiO_3$ ($SrTiO_3$:Rh); $SnNb_2O_6$, Cr^- doped $PbMoO_4$ ($PbMoO_4$:Cr); Rh and Sb codoped TiO_2 (TiO_2:Rh, Sb); $Sm_2Ti_2S_2O_5$, Ta_3N_5 and TaON are inactive photocatalysts for overall water splitting into H_2 and O_2 in the absence of scavengers [14]. Hole scavengers lead to an increase in the H_2 generation [77]. p-type semiconductor NPs increase H_2 generation, in the presence of hole scavengers in the suspension media [77]. Ferro-cyanide solutions can be used for this purpose [19].

Sacrificial redox scavengers widely used to avoid photocorrosion are halides, EDTA, $Fe(CN)_6^{4-}$, Fe(II)-EDTA complexes, S^{2-}, SO_3^{2-}, $S_2O_3^{2-}$, cysteine, triethanolamine, and hydroquinone. All are electron donors, which are more energetically favorable for a photo-induced hole to combine with [18].

The effects of hydrogenation on the photocatalytic performance, as mentioned before, are believed to arise from disorder in the structures, resulting in changes to the electronic and optical properties of the photocatalyst [78]. The defects appear to extend from the surface to the bulk phase, resulting in visible light absorption. Control of hydrogenation can allow a material with a CB edge placed above the hydrogen reduction potential, increasing the activity of the photocatalyst [78].

Electron mediators are normally used to accelerate the electron transfer and improve the photocatalytical colloidal performance. An electron mediator plays an important role in electron transfer from an O_2^- evolving photocatalyst to a H_2-evolving photocatalyst. However, they sometimes present color, which has a negative effect in STH efficiency, since they can absorb part of the visible light. A suitable combination between photocatalysts and the electron mediator is required [14]. In general, as electron mediators the most common redox pairs are IO_3^-/I^- and $Fe^{3+/2+}$ [14].

Overall water splitting can be attained by constructing a z-scheme colloidal photocatalysis system composed of H_2^- and O_2^- photocatalysts and a suitable electron mediator [14]. For HER, the systems $Pt/SrTiO_3$: Cr, Ta; $Pt/TaON$; $Pt/CaTaO_2N$; $Pt/BaTaO_2N$; $Pt/SrTiO_3$:Rh; Coumarin/$K_4Nb_6O_{17}$; and $Pt/ZrO_2/TaON$ are used. For OER, the photocatalysts WO_3; Pt/WO_3; $RuO_2/TaON$; $BiVO_4$; and Bi_2MoO_6 are employed [14].

11.3.1.1 Surface Modification

A stable aqueous colloidal suspension of semiconducting NPs is a challenge. It is clear that in order to prevent photocorrosion of NPs, passivation of surface NPs could achieve such a goal; the use of organic ligands was first proposed to passivate the surface [54]. A passivating surface reduces non-radiative recombination, and the semiconductor can be exposed to the local environment. Organic passivating ligands, however, do not display significant long-term stability, can be reactive in the solvent, and due to steric hindrances, can at best passivate only about 80% of all available surface sites. These interfere with the development of promising photocatalysts [18].

Hence, more robust, physically and chemically, passivation strategies were studied. Among these, inorganic core/shell has been a technique mostly employed [18].

Core|shell structures of QDs–quantum wells trap the charge carriers in the "well,", increasing the probability of irradiative recombination [18].

11.3.1.2 Dye Sensitizing Nanoparticle Surface

Sensitizing semiconductors by dyes to improve solar light absorption has been intensely investigated [79]. Light absorbing organic dyes can effectively use the solar spectrum. Dyes with molar extinction coefficients (ε) of $<10^5$ $M^1 \cdot cm^1$ have been examined for light-driven generation of H_2. It was found that those with heavier halogen substituents such as Rose Bengal and Eosin Y can promote H_2 formation. However, all exhibited poor photostability, decomposing within 3–5 h [80].

Among many luminescent d^6 metal complexes, the iridium(III) complexes are of interest. The substituent group in the ligand influence their catalytic performance. Carboxylate groups lead to the formation of electron transfer channels promoting an increase in the electron transfer from the sensitizer to colloidal semiconductor NPs enhancing the catalytic activity [71,80]. The use of phenyl-piridine (ppy) ligands and/or bpy has been studied, along with other sets of charge transfer chromophores that have been used in related systems, like [Pt(terpyridil)(arylacetylide)]+ complexes [80,81]. The use of iridium complexes photosensitizers possess remarkable activity. However, they present a limited operational lifetime, which is a problem for the formation of hydrogen from water [71]. Many approaches have been employed to improve the durability of iridium PSs, such as placing bulky pendant groups on the backbone of the cationic iridium complexes due to their steric protection or using neutral triscyclometalated iridium complexes due to their intrinsic stability to photolysis [71].

Dye-sensitized water-splitting systems are of recent interest (Figure 11.10). The Dye complex injects electrons into the semiconductor. The injected electron is transferred to the counterelectrode to produce hydrogen. Parallel to the process, the hole generated at the dye cation is used to obtain oxygen using the metal oxide NPs attached to the dye [34].

Ru-dyes have also been used to sensitize metal oxide NPs. Spiccia et al. [82] reported the use of TiO_2/Ru-dye in the presence of the Mn-oxo cluster catalyst $[Mn_4O_4L_6]^+$, L = $(MeOPh)_2PO_2^-$. Mn-oxo was suspended in the electrolyte, where it promotes the generation of O_2 after light absorption. Electrons are then transferred to the Ru-dye, which then flows to the TiO_2 semiconductor. Furthermore, Ru dye is also capable of light absorption, so this system is capable of increasing light harvesting. A similar system was proposed by Youngblood [83], where the Ru dye sensitized TiO_2 and anchors hydrated iridium oxide NPs as the oxygen evolving catalyst. Using this system water splitting was achieved.

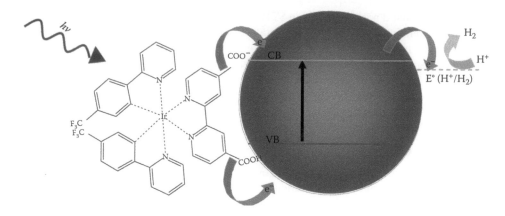

FIGURE 11.10   Dye-sensitized semiconductor NPs water-splitting system.

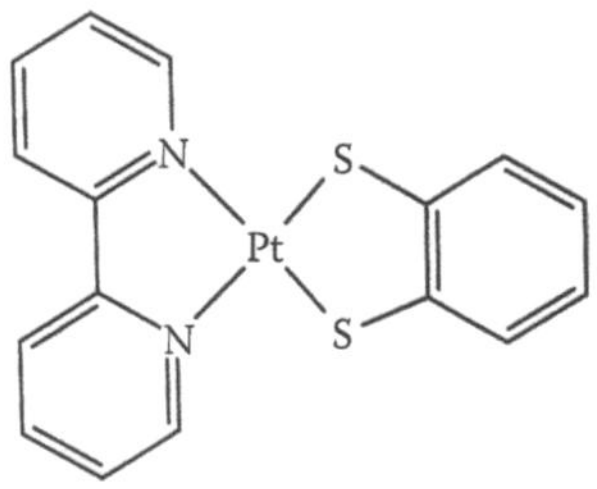

FIGURE 11.11   Schematic representation of PtN_2S_2 complexes.

Platinum diimine dithiolate complexes (PtN_2S_2) (Figure 11.11) constitute another class of charge photosensitizer. The excited state energies of these complexes are significantly lower than those of the Ru(bby), Ir(ppy), and Pt(terpyridyl) complexes. The deprotonated carboxylate groups of di(carboxy)bipyridine (dcbpy) ligands served to link the photosensitizer to the semiconductor NPs. Zheng et al. [80] reported on the system $TiO_2/Pt/PtN_2S_2$ for the production of hydrogen. The system was stable for more than 70 h, using light of $\lambda < 450$ nm. However, activity of the system was low, based in part on the low absorptivity of the PtN_2S_2 and the small amount of the complex bound to TiO_2 [71]. To improve absorptivity (a) the authors linked strongly absorbing organic dye directly to the PtN_2S_2 moiety such as bipyridine linked to a dipyrromethene-BF2 (Bodipy) dye. This allowed the system to promote the production of H_2 from the aqueous protons in conjunction with Pt–TiO_2 as both electron conduit and catalyst.

11.3.1.3   Decoration with Cocatalyst

A cocatalyst plays an important role in providing reaction sites for H_2 formation on the surface of a photocatalyst [14], accelerating interfacial oxidation-reduction kinetics of water splitting. It has been found that there is a relationship between NPs morphology and water-splitting activity under visible light absorption. The semiconductor NPs ($Ga_{1-x}Zn_x$)($N_{1-x}O_x$) system with a cocatalyst provides a water-splitting quantum efficiency of 2.5% at wavelengths of 420–440 nm [80].

The decoration of semiconductor NPs with cocatalysts in colloidal suspension are a system of constant study. In this system, semiconductor NPs absorb light to generate the electron hole-pair, which then is quickly separated. The electron flows to the cocatalyst where it reduces water to generate hydrogen, and the hole is removed by a sacrificial hole scavenger. This charge transfer suppresses recombination, the major rate-limiting step for efficient hydrogen generation.

In the charge carrier's rate separation the key parameters are the energy bands' difference among the electronic states of the semiconductor, the cocatalyst, and the hole scavenger's redox potential. All these properties define the driving force ΔG for the transfer/scavenging reaction. In SNCs these differences between the energy bands can be controlled by designing particle size and morphology, in order to achieve the desired optical and electronic properties via quantum confinement. Recent studies investigated the size-dependence of electron transfer kinetics to cocatalysts [84].

Following this strategy, Li et al. were able to confirm the dependence of cocatalyst size on external quantum efficiencies. The authors used a colloidal dual-band gap cell, in which the two compartments were electrically connected through a membrane sharing the same electrolyte. In this system, external quantum efficiencies increase from 9.4% (2.8 nm particles only) to 14.7% (2.8 and 4.6 nm particles) under visible light illumination (350–480 nm). The improved efficiencies in the dual-band gap cell originate from the smaller nanococatalyst converting photons more efficiently into hydrogen by providing larger driving forces ΔG for the charge transfer reaction [84].

Recently, the loading of wide band gap semiconductor NPs over narrow band gap metal sulfide (ZnS, CuS, FeS, Bi_2S_3, or CdS) photocatalysts has been studied. This arrangement makes it possible to absorb longer wavelengths compared to wide band gap semiconductors of conventional metal oxide systems. In the ZnS/TiO_2 composite system, the evolution of H_2 from methanol/water (1:1) was enhanced dramatically compared to that of pure TiO_2 and ZnS [85].

Effectiveness of the composite is dependent on several parameters, such as the amount of cocatalyst loaded onto the semiconductor's surface. If there are too many cocatalyst NPs on the surface, the active sites of the catalyst can be blocked, while if the cocatalyst content is too low, the desired activity enhancement might not be achieved. Also, it has been reported that photocatalytic activity is highly dependent on the average cocatalyst NPs diameter [85].

Colloidal cocatalyst-decorated semiconductors are commonly stabilized by using ligands with hydrophilic thiol moieties. Ligands such as cysteine, MPA, etc. are commonly used. However, they can be photooxidized and partially removed as soon as the material is illuminated in a matter of minutes to hours depending on illumination intensity, nanoparticle size, type of ligands, pH and ionic strength of the solution, and stirring rate. The continuous removal of ligands promotes aggregation of NPs, which eventually reduces photocatalytic efficiency [70].

Successively supplied thiols or closely packed branched thiol ligands have been used to provide enhanced stability, but their close packing blocks efficient interfacial charge transfer and reduces photocatalytic activity. However, the use of triethanolamine (TEOA) provides colloidal stability, and efficient interfacial charge transfer can be achieved simultaneously [70].

This has been observed on Pt-decorated CdS (Pt/CdS) NPs where Na_2SO_3 was used as a hole scavenger. The presence of TEOA prolonged the H_2 evolution period by more than a factor of 10, without making a significant contribution during H_2 generation [70].

11.4 CONCLUSIONS

Solar water splitting with heterogeneous photocatalysis to generate the clean energy carrier hydrogen is a promising reaction for a sustainable and clean future energy supply [86]. As can be noticed, it has attracted tremendous attention as a potentially significant alternative for sustainable and clean energy in the future. Enormous work is being carried out in order to solve or avoid any problems they present. Recent approaches have utilized nanostructured systems to manipulate quantum confinement, plasmonics, etc., to enhance carrier dynamics. Also, strategies like orbital engineering and intermediate band electronic structure have been studied. As a result of such research, new materials of quantum-confined nanostructures with tuned band gap and band edges for efficient water splitting have been reached. These nanostructures show significant efficiency improvement, since they are capable of absorbing light in the visible region [34].

Presently, renewable energy contributes only about 5% of the commercial hydrogen production, primarily via water electrolysis, while the other 95% of hydrogen is mainly derived from fossil fuels; this is because STH energy conversion efficiency using these techniques is too low from the economic point of view [1]. Commercial solar water splitting devices are still not achievable. Several reasons include short lifetime, photostability, overvoltage losses, and low conversion efficiencies [87].

The highest STH efficiencies, exceeding 10%, have been achieved using very expensive and unstable III–V group semiconductors. Therefore, in order to have a higher contribution from renewable hydrogen production, costs must decrease. Research is focused in finding economically viable, efficient, and stable small band gap semiconducting photocatalysts for PEC water splitting based on earth-abundant elements in order to avoid serious life cycle environmental impacts [8,17].

REFERENCES

1. Ameta, R. et al., Nano-sized photocatalytic materials for solar energy conversion and storage. *Materials Science Forum*, 2016. 855: p. 20.
2. Twidell, J. and T. Weir, *Renewable Energy Resources*. 2015, Taylor & Francis Group, New York.
3. Sivula, K., Solar-to-chemical energy conversion with photoelectrochemical tandem cells. *Chimia (Aarau)*, 2013. 67(3): p. 7.
4. Chow, W.S., Photosynthesis: From natural towards artificial. *Journal of Biological Physics*, 2003. 29: p. 13.
5. Chen, X., Z. Zhang, and L. Chi, Recent advances in visible-light-driven photoelectrochemical water splitting: catalyst nanostructures and reaction systems. *Nano-Micro Letters*, 2016. 8(1): p. 12.
6. Augugliaro, V. et al., Conversion of solar energy to chemical energy by photoassisted processes- II. Influence of the iron content on the activity of doped titanium dioxide catalysts for ammonia photoproduction. *International Journal of Hydrogen Energy*, 1982. 7(11): p. 5.
7. Kamat, P.V. and J. Bisquert, Solar fuels. Photocatalytic hydrogen generation. *Journal of Physical Chemistry C*, 2013. 117(29): p. 3.
8. Ni, M. et al., A review and recent developments in photocatalytic water-splitting using TiO_2 for hydrogen production. *Renewable and Sustainable Energy Reviews*, 2007. 11(3): p. 25.
9. Barelli, L. et al., Hydrogen production through sorption-enhanced steam methane reforming and membrane technology: A review. *Energy* (Oxford), 2008. 33(4): p. 17.
10. Van der Drift, A. and H. Boerrigter, *Synthesis Gas from Biomass*. ECN Biomass, Coal and Environmental Research, Petten, Netherlands, p. 31.
11. Zhou, L. et al., Aluminum nanocrystals as a plasmonic photocatalyst for hydrogen dissociation. *Nano Letters*, 2016. 16(2): p. 6.
12. Zoulias, E. et al., A review on water electrolysis. *TCJST*, 2004. 4(2): p. 31.
13. Chen, Y. et al., Towards efficient solar-to-hydrogen conversion: Fundamentals and recent progress in copper-based chalcogenide photocathodes. *Nanophotonics*, 2016. 5(4): p. 524.
14. Chen, X. et al., Semiconductor-based photocatalytic hydrogen generation. *Chemical Reviews*, 2010. 110: p. 68.
15. Bhatt, M.D. and J.S. Lee, Recent theoretical progress in the development of photoanode materials for solar water splitting photoelectrochemical cells. *Journal of Materials Chemistry A*, 2015. 3: p. 10632.
16. van de Krol, R. and M. Grätzel (eds), *Photoelectrochemical Hydrogen Production*. 1st ed. 2012: Springer, New York, 324 pages.
17. Juodkazytė, J. et al., Solar water splitting: Efficiency discussion. *International Journal of Hydrogen Energy*, 2016. 41(28): p. 8.
18. Best, J.P. and D.E. Dunstan, Nanotechnology for photolytic hydrogen production: Colloidal Anodic Oxidation. *International Journal of Hydrogen Energy*. 34(18): p. 17.
19. Moniz, S.J.A. et al., Visible-light driven heterojunction photocatalysts for water splitting—A critical review. *Energy and Environmental Sciences*, 2015. 8: p. 19.
20. Radecka, M. et al., Importance of the band gap energy and flat band potential for application of modified TiO_2. *Journal of Power Sources*, 2008. 181(1): p. 10.
21. Peerakiatkhajohn, P. et al., Review of recent progress in unassisted photoelectrochemical water splitting: from material modification to configuration design. *Journal of Photonics for Energy Paper*, 2017. 7(1): p. 21.
22. Huang, Q., Z. Ye, and X. Xiao, Recent progress in photocathodes for hydrogen evolution. *Journal of Materials Chemistry A*, 2015. 3(31): p. 14.
23. Cendula, P. et al., Calculation of the energy band diagram of a photoelectrochemical water splitting cell. *Journal of Physical Chemistry C*, 2014. 118(51): p. 9.

24. Peter, L.M., Photoelectrochemistry: From basic principles to photocatalysis, in *Photocatalysis: Fundamentals and Perspectives*, J. Schneider et al., Editors. 2016, Royal Society of Chemistry, London, UK, p. 28.

25. Chazalviel, J.-N., Screening of a static charge distribution beyond the linear regime, in *Coulomb Screening by Mobile Charges: Applications to Materials Science, Chemistry, and Biology*, p. 72, 1st ed. 1999, Publisher Birkhäuser Basel, New York, 355 p.

26. Evans, D.R., Lecture 5. Point defect equilibrium, higher dimensionality defects in microelectronic device fabrication I, P.S. University, 2016. http://web.pdx.edu~davide/Lecture5.pdf.

27. Schmickler, W. and E. Santos, *Interfacial Electrochemistry*, 2nd ed. 2010, Springer, Berlin, Germany, 1 p.

28. Coronado, J.M. et al., Design of *Advanced Photocatalytic Materials for Energy and Environmental Applications*. 2013, Springer, London. 355 p.

29. Kahn, A., Fermi level, work function and vacuum level. *Materials Horizons*, 2016. 3(1): p. 4.

30. A. Smith, W. et al., Interfacial band-edge energetics for solar fuels production. *Energy and Environmental Science*, 2015. 8: p. 12.

31. Chou, J.-C. et al., Photoexcitation of TiO_2 photoanode in water splitting. *Materials Chemistry and Physics*, 2014. 143(3): p. 6.

32. Suzuki, T. et al., Tuning of photocatalytic reduction by conduction band engineering of semiconductor quantum dots with experimental evaluation of the band edge potential. *Chemical Communications (Camb)*, 2016. 52(36): p. 8.

33. Palmas, S. et al., TiO_2 photoanodes for electrically enhanced water splitting. *International Journal of Hydrogen Energy*, 2010. 35: p. 10.

34. Tachibana, Y., L. Vayssieres, and J.R. Durrant, Artificial photosynthesis for solar water-splitting. *Nature Photonics*, 2012. 6: p. 7.

35. Mohamed, A.E.R. and S. Rohani, Modified TiO_2 nanotube arrays (TNTAs): progressive strategies towards visible light responsive photoanode, a review. *Energy and Environmental Science*, 2011. 4: p. 22.

36. Wang, T. et al., Controllable fabrication of nanostructured materials for photoelectrochemical water splitting via atomic layer deposition. *Chemical Society Reviews*, 2014. 4343: p. 16.

37. Lin, J. et al., Anodic nanostructures for solar cell applications, in *Green Nanotechnology—Overview and Further Prospects*, M.L. Larramendy and S. Soloneski, Editors. 2016, InTech Publisher, Rijeka, Croatia.

38. Li, L., P.A. Salvador, and G.S. Rohrer, Photocatalysts with internal electric fields. *Nanoscale*, 2013. 6: p. 19.

39. Qin, G. and A. Watanabe, Surface texturing of TiO_2 film by mist deposition of TiO_2 nanoparticles. *Nano-Micro Letters*, 2013. 5: p. 129.

40. Su, Z. et al., Anodic formation of nanoporous and nanotubular metal oxides. *Journal of Materials Chemistry*, 2012. 22: p. 10.

41. Zhao, A., *Optical Properties, Electronic Structures and High Pressure Study of Nanostructured One Dimensional Titanium Dioxide by Synchrotron Radiation and Spectroscopy, in Chemistry*. 2013, University of Western Ontario: London, ON, p. 103.

42. Piskunov, S. et al., C-, N-, S-, and Fe-doped TiO_2 and $SrTiO_3$ nanotubes for visible-light-driven photocatalytic water splitting: Prediction from first principles. *The Journal of Physical Chemistry C*, 2015. 119(32): p. 11.

43. Zaleska, A., Doped-TiO_2: A review. *Recent Patents on Engineering*, 2008. 2(3): p. 8.

44. Xu, M. et al., Rationally designed n-n heterojunction with highly efficient solar hydrogen evolution. *ChemSusChem*, 2015. 8(7): p. 8.

45. van de Krol, R. and Y. Liang, An n-Si/n-Fe_2O_3 heterojunction tandem photoanode for solar water splitting. *Chimia* (Aarau), 2013. 67(3): p. 4.

46. Ji, J. et al., High density Si/ZnO core/shell nanowire arrays for photoelectrochemical water splitting. *Journal of Materials Science: Materials in Electronics*, 2013. 24(9): p. 7.

47. Sato, S. and J.M. White, Photoassisted water-gas shift reaction over platinized titanium dioxide catalysts. *Journal of the American Chemical Society*, 1980. 102(24): p. 5.

48. Khan, M.R. et al., Schottky barrier and surface plasmonic resonance phenomena towards the photocatalytic reaction: study of their mechanisms to enhance photocatalytic activity. *Catalysis Science and Technology*, 2015. 5: p. 10.

49. Qiu, J., J. Dawood, and S. Zhang, Hydrogenation of nanostructured semiconductors for energy conversion and storage. *Chinese Science Bulletin*, 2014. 59(18): p. 18.

50. Yang, Y. et al., Oxygen deficient TiO_2 photoanode for photoelectrochemical water oxidation. *Solid State Phenomena*, 2016. 253: p. 30.

51. Yan, Y. et al., Slightly hydrogenated TiO_2 with enhanced photocatalytic performance. *Journal of Materials Chemistry A*, 2014. 2: p. 9.

52. Amano, F. et al., Effect of Ti^{3+} ions and conduction band electrons on photocatalytic and photoelectrochemical activity of rutile titania for water oxidation. *The Journal of Physical Chemistry C*, 2016. 120(12): p. 8.

53. Graetzel, M., Artificial photosynthesis: Water cleavage into hydrogen and oxygen by visible light. *Accounts of Chemical Research*, 1981. 14(12): p. 9.

54. Zhang, X., T. Peng, and S. Song, Recent advances in dye-sensitized semiconductor systems for photocatalytic hydrogen production. *Journal of Materials Chemistry A*, 2016. 4: p. 38.

55. Yu, Z., F. Li, and L. Sun, Recent advances in dye-sensitized photoelectrochemical cells for solar hydrogen production based on molecular components. *Energy and Environmental Science*, 2015. 8: p. 16.

56. Sahai, S. et al., Quantum dots sensitization for photoelectrochemical generation of hydrogen: A review. *Renewable and Sustainable Energy Reviews*, 2017. 68: p. 19.

57. Bonomo, M. and D. Dini, Nanostructured p-type semiconductor electrodes and photoelectrochemistry of their reduction processes. *Energies*, 2016. 9(5): p. 32.

58. Hisatomi, T., J. Kubota, and K. Domen, Recent advances in semiconductors for photocatalytic and photoelectrochemical water splitting. *Chemical Society Reviews*, 2014. 43(22): p. 16.

59. Kim, H.G. et al., Fabrication of $CaFe_2O_4/MgFe_2O_4$ bulk heterojunction for enhanced visible light photocatalysis. *Chemical Communications*, 2009. (39): pp. 5889–5891.

60. Wang, Z. et al., Recent developments in p-type oxide semiconductor materials and devices. *Advanced Materials*, 2016. 28(20): p. 62.

61. McShane, C.M. and K.-S. Choi, Junction studies on electrochemically fabricated p–n Cu_2O homojunction solar cells for efficiency enhancement. *Physical Chemistry Chemical Physics*, 2012. 14: p. 7.

62. Wang, H., T. Deutsch, and J.A. Turner, Direct water splitting under visible light with nanostructured hematite and WO_3 photoanodes and a $GaInP_2$ photocathode. *Journal of the Electrochemical Society*, 2008. 155(5): p. 6.

63. Liu, C. et al., Highly efficient photoelectrochemical hydrogen generation using $Zn_xBi_2S_{3+x}$ sensitized plate like WO_3 photoelectrodes. *ACS Applied Materials and Interfaces*, 2015. 7(20): p. 8.

64. Ahmad, H. et al., Hydrogen from photo-catalytic water splitting process: A review. *Renewable and Sustainable Energy Reviews*, 2015. 43: p. 12.

65. Jafari, T. et al., Photocatalytic water splitting—The untamed dream: A review of recent advances. *Molecules*, 2016. 21(7). doi:10.3390/molecules21070900.

66. Chiarello, G.L. and E. Selli, Photocatalytic hydrogen production. *Recent Patents on Engineering*, 2010. 4(3): p. 15.

67. Nada, A. et al., Studies on the photocatalytic hydrogen production using suspended modified TiO_2 photocatalysts. *International Journal of Hydrogen Energy*, 2005. 30(7): p. 5.

68. Kumar, S.G. and L.G. Devi, Review on modified TiO_2 photocatalysis under UV/visible light: Selected results and related mechanisms on interfacial charge carrier transfer dynamics. *The Journal of Physical Chemistry A*, 2011. 115(46): p. 31.

69. Kasem, K.K., Selective photolysis of aqueous colloidal nano-particles of some metal-oxide semiconductors for hydrogen generation. *Oriental Journal of Chemistry*, 2009. 25(1): p. 7.

70. Li, W., J.R. Lee, and F. Jaeckelm, Simultaneous optimization of colloidal stability and interfacial charge transfer efficiency in photocatalytic Pt/CdS nanocrystals. *ACS Applied Materials and Interfaces*. 2016, 8(43): p. 29434.

71. Yuan, Y.-J. et al., Hydrogen photogeneration promoted by efficient electron transfer from iridium sensitizers to colloidal MoS_2 catalysts. *Scientific Reports*, 2014. 4: p.1, doi:10.1038/srep04045.

72. Wang, Q. et al., Highly efficient photocatalytic hydrogen production of flower-like cadmium sulfide decorated by histidine. *Scientific Reports*, 2015. 5: p. 1, doi:10.1038/srep13593.

73. Kubacka, A., M. Fernández-García, and G. Colón, Advanced nanoarchitectures for solar photocatalytic applications. *Chemical Reviews*, 2012. 112(3): pp. 1555–1614.

74. Xu, Y. and M.A.A. Schoonen, The absolute energy positions of conduction and valence bands of selected semiconducting minerals. *American Mineralogist*, 2000. 85: p. 14.

75. Klimm, D., Electronic materials with a wide band gap: Recent developments. *IUCrJ*, 2014. 1(Pt 5): pp. 281–290.

76. Shen, S. et al., Enhanced photocatalytic hydrogen evolution over Cu-doped $ZnIn_2S_4$ under visible light irradiation. *The Journal of Physical Chemistry C*, 2008. 112 (41): p. 8.

77. Simon, T. et al., Redox shuttle mechanism enhances photocatalytic H_2 generation on Ni-decorated CdS nanorods. *Nature Materials*, 2014. 13: p. 6.

78. Xue, J. et al., Nature of conduction band tailing in hydrogenated titanium dioxide for photocatalytic hydrogen evolution. *ChemCatChem*, 2016. 8(12): p. 5.

79. Li, Y. and J.Z. Zhang, Hydrogen generation from photoelectrochemical water splitting based on nano-materials. *Laser and Photonics Reviews*, 2010. 4(4): p. 517.

80. Zheng, B. et al., Light-driven generation of hydrogen: New chromophore dyads for increased activity based on Bodipy dye and Pt(diimine)(dithiolate) complexes. *PNAS*, 2015. 112(30): p. 10.

81. Kapturkiewicz, A. et al., Electrochemiluminescence studies of the cyclometalated iridium(III) complexes with substituted 2-phenylbenzothiazole ligands. *Electrochemistry Communications*, 2005. 6(8): p. 5.

82. Brimblecombe, R. et al., Solar driven water oxidation by a bioinspired manganese molecular catalyst. *Journal of the American Chemical Society*, 2010. 132 (9): p. 3.

83. Youngblood, W.J. et al., Photoassisted overall water splitting in a visible light-absorbing dye-sensitized photoelectrochemical cell. *Journal of the American Chemical Society*, 2009. 131 (3): p. 2.

84. Li, W. et al., Colloidal dual-band gap cell for photocatalytic hydrogen generation. *Nanoscale,* 2015. 7: p. 5.

85. Im, Y. et al., Dynamic hydrogen production from methanol/water photo-splitting using core@shell-structured CuS@TiO$_2$ catalyst wrapped by high concentrated TiO$_2$ particles. *International Journal of Photoenergy*, 2013. 2013: p. 10.

86. Weller, T., J. Sann, and R. Marschall, Pore structure controlling the activity of mesoporous crystalline CsTaWO$_6$ for photocatalytic hydrogen generation. *Advanced Energy Materials*, 2016. 6(16), doi:10.1002/aenm.201600208" 201600208.

87. Döscher, H. et al., Solar-to-hydrogen efficiency: Shining light on photoelectrochemical device performance. *Energy and Environmental Science*, 2016. 9(1): p. 7.

12 Novel Solar Fuels from H_2O and CO_2

Christian Sattler, Anis Houaijia, and Martin Roeb
German Aerospace Center (DLR)

CONTENTS

12.1 INTRODUCTION

Fossil fuels are currently the world's primary energy source. Electricity and fuel production from those resources leads to the emission of greenhouse gases and other pollutants. According to the Environmental Protection Agency, the burning of fossil fuels was responsible for 80.9% of U.S. greenhouse gas emissions in 2014 [1]. This raises the need to develop more sustainable and cleaner energy systems in order to decrease the emissions of greenhouse gases and to minimize the impacts on the global warming. The conversion of solar energy into chemical energy carriers, also well known as solar fuels, represents an interesting pathway to the long-term storage of solar energy and the long-range transport from sunny and desert regions to the industrialized and populated centers of the earth. Solar fuels cover a wide range of possible applications, e.g., electricity and heat generation and can be further processed to other chemicals via Fischer–Tropsch synthesis. Hydrogen is the most representative of solar fuels and can be sustainably generated by using renewable energy sources. As compared to other renewable energy sources, e.g., wind and geothermal, solar energy has the greatest potential to be used in the generation of fuels since it is safe, clean, and unlimited. Figure 12.1 shows the different pathways for solar fuel production, mainly hydrogen and synthesis gas.

Direct thermal processes use concentrated solar energy without intermediary energy conversion systems, which are needed to produce electricity to run water electrolysis. The main pathway for the direct thermal production of solar hydrogen is thermochemical water splitting, in particular through the application of redox materials. In addition to that, CO_2 can be simultaneously split, which leads (if combined with water splitting) to the production of synthesis gas. The high-temperature heat, which is generated by the concentration of solar irradiation, initiates a series of chemical reactions that produce hydrogen and/or synthesis gas in a typical temperature range of 500°C–2000°C depending on the redox materials used [2].

The most interesting representative of the indirect (electrochemical) processes is high-temperature electrolysis, which is more efficient than ambient temperature electrolysis, especially from an economic point of view due to the fact that a part of the total energy demand is supplied as heat, which

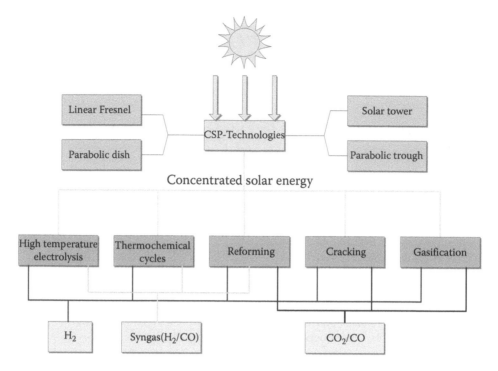

FIGURE 12.1   Solar hydrogen production processes.

is cheaper to generate than electricity [3]. The leading technologies in the field of high-temperature electrolysis are the Solid Oxide Electrolysis (SOE) and the Molten Carbonate Electrolysis (MCE) [4]. The focus of this chapter will be solar high-temperature electrolysis, including the SOE and MCE technologies and solar thermochemical H_2O and CO_2 splitting. Other processes, solar steam methane reforming, solar gasification, and solar cracking of natural gas, will not be a part of the analysis within this chapter. The concentration of solar irradiation can be carried out using many technologies, well known as concentrating solar power (CSP) technologies. Concentrating technologies for scale-up plants are available in four common forms, namely, parabolic trough, linear Fresnel, solar dish, and solar power tower.

12.2   CONCENTRATING SOLAR POWER TECHNOLOGIES

CSP technologies are able to generate electricity by concentrating solar irradiation. The electricity is generated by concerting the high-temperature solar heat that drives a turbine and is finally connected to an electrical power generator. There are several CSP technologies, namely, the solar tower, parabolic trough and linear Fresnel [5]. The solar tower technology is large scale and uses many large, computer-controlled mirrors, known as heliostats, to focus sunlight on a receiver at the top of a tower. The receiver converts the solar radiation into heat. A heat transfer fluid heated in the receiver absorbs the highly concentrated radiation reflected by the heliostats and converts it into thermal energy. This heat is usually transferred in a heat exchanger to a conventional steam cycle to produce electricity. The technology enables operation high temperatures and is characterized by the ability using a storage system to store heat. The process design of the plant depends on the heat transfer fluid, which can be water, air, or molten salt. Table 12.1 shows the state of the art regarding the main specifications of CSP technologies [6–8].

The solar tower technology is the most suitable energy source for coupling to the processes for hydrogen production from H_2O and CO_2 because higher temperature can be achieved than with the

TABLE 12.1

Main Specifications of CSP Technologies

Technology	Heat Transfer Fluid	Temperature (°C)
Solar tower	Air	750–980
	Water	550
	Molten salt	565
Linear Fresnel	Thermal mineral oil	150–400
Parabolic trough	Thermal mineral oil	150–400

other CSP technologies, and this leads to higher overall system efficiency. The solar tower technology will be taken into account in Chapters 12.3 through 12.5., where the analysis on the coupling to the hydrogen generation processes will be carried out.

12.3 SOLAR HIGH TEMPERATURE ELECTROLYSIS

High-temperature steam electrolysis is a very promising pathway for highly efficient large-scale hydrogen production. The main advantage of performing the electrolysis of water at higher temperatures is that the dissociation of steam requires less energy than the electrolysis of liquid water. Moreover, a part of the total required energy for electrolysis is supplied as heat, and this contributes to the decrease of the hydrogen production cost due the fact that heat generation is cheaper than electricity production. Generally, two electrolysis technologies are available regarding the electrolysis of water steam, namely, SOE and MCE. SOE operates in a temperature range of 700°C–1000°C [9], while MCE requires lower operating temperature in the range of 500°C–600°C [10]. Moreover, the MCE process needs a CO_2 source to run the electrochemical reaction of water splitting.

12.3.1 SOLID OXIDE ELECTROLYSIS

The electrolysis of water steam can be performed in a solid oxide electrolysis cell (SOEC). The SOEC consists of a three-layer solid structure (cathode, anode, and electrolyte) and an interconnect plate (Figure 12.2). The electro-chemical reactions, which take place at the anode and the cathode, are given by the next equations:

$$H_2O + 2e^- \rightarrow H_2 + O^{2-} \tag{12.1}$$

$$2O^{2-} \rightarrow O_2 + 4e^- \tag{12.2}$$

Both chemical reactions lead to the overall water splitting reaction:

$$H_2O \rightarrow H_2 + 1/2O_2 \tag{12.3}$$

The electrolyte is a gas-tight ceramic membrane that can conduct ions and is located between two porous electrodes that can conduct electrons: steam/hydrogen electrode (cathode) and the air/oxygen electrode (anode). The steam is fed to the cathode by applying an electrical potential, at which the reduction reaction takes place. The water molecules dissociate to form hydrogen and oxygen ions. The product gas, which flows out from the cathode, contains mainly hydrogen and nonconverted water that can be easily separated by cooling. In order to maintain reducing conditions on the cathode side, the feed steam is mixed with hydrogen. The inlet mixture generally consists of

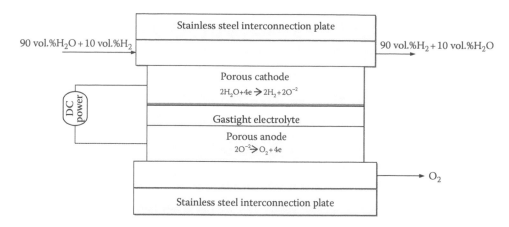

FIGURE 12.2 Layout of SOEC.

90 mol% H_2O and 10 mol% H_2. Sweep gas is also required on the anode side in order to remove the oxygen molecules and to control the cell temperature. Possible sweep gases are oxygen, steam, and air. The advantages and the disadvantages of each sweep gas are summarized in Table 12.2 [11].

SOEC materials have to fulfill many requirements. The materials of the stack have to withstand high operating temperature and ensure high conductivity and catalytic activity. The main materials specifications of the electrodes and electrolytes are listed in Table 12.3 [12].

The solar tower technology, including air, water, and molten salt as heat transfer fluids, is able to deliver the SOE process with the required heat and electricity demand. Storage systems can be implemented for each technology in order to increase the availability of the plant and to compensate for energy fluctuations. Figure 12.3 shows the flowchart of the coupling of the SOE process to the tower technologies.

Electricity generation can be carried out directly or indirectly. Solar irradiation is concentrated by the heliostat field onto a receiver, which is mounted on top of the tower. The air and the molten salt solar towers allow the indirect generation of power. Both heat transfer fluids are heated in the solar receivers by concentrated solar energy. The heat is then transferred to water, which is introduced to the Rankine cycle where electricity is produced. The steam solar tower, also known as a direct steam generation (DSG) power plant, uses the superheated steam in a Rankine cycle that is generated in the solar receiver. The three tower technologies deliver heat and electricity to

TABLE 12.2
Advantages and Disadvantages of Sweep Gases

Sweep Gas	Advantages	Disadvantages
Oxygen	• Pure oxygen generation → oxygen can be considered a valuable by-product	• Decrease of stack efficiency due to the increase of the Nernst potential across cell and safety issues concerning handling of pure hot oxygen
Steam	• Increase of stack efficiency (Nernst potential decreases across the cell) • Separation of water at the anode can be easily performed by condensation	• Additional heat is required to provide water steam at higher temperatures
Air	• Freely available	• Oxygen separation is very expensive

TABLE 12.3
Materials of the SOEC Components

Component	Material	Sintering Temperature (°C)
Steam/hydrogen electrode (cathode)	Nickel Oxide-Yttria stabilized Zirconia (NiO-YSZ)	1400
Electrolyte	Yttria stabilized Zirconia (Zr-YSZ)	1200
Air/oxygen electrode (Anode)	Doped Lanthanum-Manganite(SrLaMnO+YZrO)	1200

the SOE process. The heat provided by the Rankine cycle is used for the feed water preheating of the electrolysis process. Since the SOE is operated at higher temperatures, the feed water needs to be overheated, and this is performed within a heat recovery system, which is directly coupled with the SOE process. Figure 12.4 shows the flowsheet of the SOE process coupled with a DSG power plant.

The electricity generation for the SOE process is performed within a Rankine cycle, which consists of a multistage turbine. Depending on the capacity of the plant, e.g., thermal energy input of the receiver, a turbine with 3–5 multistages is used for electricity generation. Water is preheated in the heat exchangers HX3 and HX4 within the Rankine cycle and evaporation, and overheating is carried out by concentrated solar irradiation in the solar receiver. The superheated steam is introduced to the Rankine cycle, where electricity is generated and fed to the electrolyser. A heat recovery system is integrated into the overall process by using the outlet gases of the electrolyser to overheat water steam up to the operating temperature of the electrolyser. This heat recovery system consists

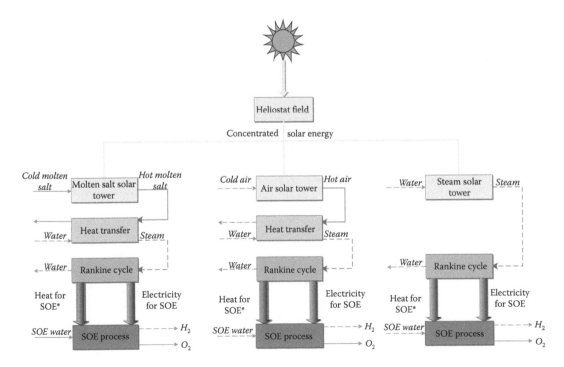

FIGURE 12.3 Flowchart of coupling solar tower technologies to the solid oxide electrolysis (SOE) process. *: Heat for SOE feed water preheating and evaporation. Overheating is performed within a heat recovery system after the SOE process by the outlet gases.

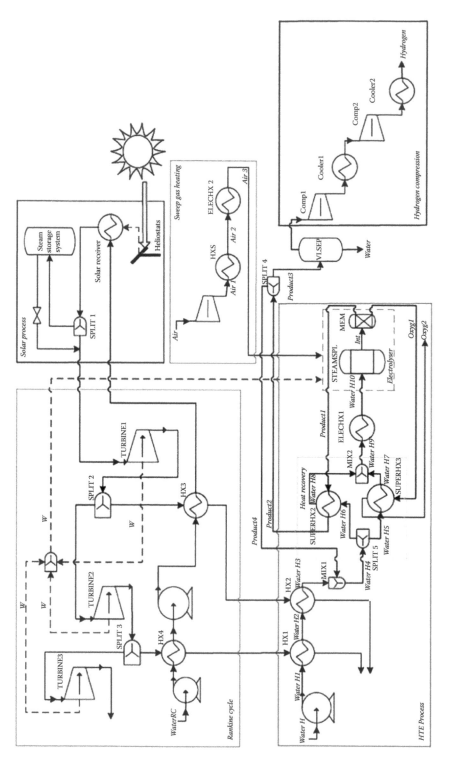

FIGURE 12.4 Flowsheet of the SOE process coupled with DSG power plant.

of two high-temperature heat exchangers (SUPERHX2, SUPERHX3). After water separation by condensation, the hydrogen product is compressed by a two-stage compressor for storage.

Another interesting technology that can be coupled to the SOE process using the solar tower is the molten salt solar tower. This type of CSP tower has the advantage of storing hot molten salt. In the process, the molten salt is heated to a temperature of 565°C. Most of the molten salt (*MS 1.2*) is used to generate steam for the Rankine cycle, which then generates electricity to supply the parasitic loads as well as the electrolysis unit. A smaller part of the molten salt (*MS 2.1*) is used to preheat the air that is used for sweeping the oxygen from the electrolysis unit and afterwards to partially evaporate the feed water. The feed water is pumped to the required pressure level, preheated, and partially evaporated (by approximately one-third) by the first extraction from the turbine (*Ext 1*). Subsequently, molten salt evaporates the remaining water. The steam is mixed with a recycle stream, and they are superheated by the hot sweep air and oxygen flow from the electrolyser. In order to achieve the required inlet temperature, the steam flow is further heated by an electric heater. The water is partially split in the electrolyser to hydrogen and oxygen. The hydrogen leaves the electrolyser with the remaining water on the anode side of the cell. The water/hydrogen mix heats the incoming air and is further cooled in a separator in order for the water to condensate. The hydrogen is then compressed to the required pressure in an intercooled multistage compressor. The oxygen is transported through the membrane to the cathode side and swept with hot air. The oxygen-enriched air leaves the cathode side and is cooled by heating the steam. In the Rankine cycle, there are two turbine extractions that are used for preheating the feed water. Figure 12.5 shows the flowsheet of a SOE process coupled with a molten salt plant.

12.3.2 Molten Carbonate Electrolysis

In order to reduce the temperature gap with current solar thermal fluids, development of electrolysis processes within a temperature range of 500°C–600°C would be an interesting option. Additionally, the operation of the electrolysis process at lower temperatures will decrease the degradation rate of the stack, which is considered the main challenge regarding SOE technology. Another advantage of MCE is that molten salt is used as an electrolyte. This could be considered an ideal medium to lower process temperatures with respect to solid oxide electrolysers since overall ionic conductivity and transport of liquid salts are usually higher than solid-type electrolytes [13,14]. The overall electrolysis process can be described by the next chemical reactions at the cathode and the anode, respectively:

$$H_2O + CO_2 \rightarrow H_2 + CO_3^{2-} \tag{12.4}$$

$$CO_3^{2-} \rightarrow CO_2 + 1/2O_2 + 2e \tag{12.5}$$

The technical feasibility of the MCE was successfully demonstrated on laboratory scale using a molten carbonate fuel cell, which was operated in a reverse electrolysis mode at 600°C [15]. The common electrolyte used within the MCE process is a ternary carbonate eutectic electrolyte based on a salt mixture of Li_2CO_3–Na_2CO_3–K_2CO_3.

Several peculiar aspects of a water MCE process are worthy of note. Firstly, it may be observed that the anode reaction does not produce pure oxygen, but rather a CO_2:O_2 gas mixture, which is ideal for use in oxycombustion processes. In fact, the anodic off-gas is composed of a 2:1 CO_2:O_2 mixture that is comparable in terms of adiabatic temperature to an air stoichiometric combustion [15]. Secondly, since the cathode reaction needs a CO_2 source, a CO_2 closed-loop system with CO_2 capture could be easily realized by an integrated electrolysis-oxycombustion process. Part of the postcombustion CO_2 could be, in fact, reinjected into the cathode, whereas the excess CO_2 could be easily captured or used.

Figure 12.6 shows a flowsheet of a molten salt tower plant coupled with the MCE process.

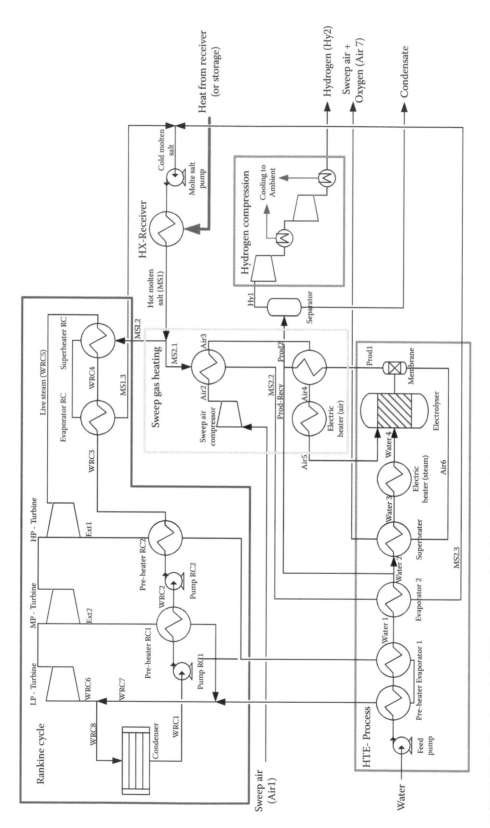

FIGURE 12.5 Flowsheet of the SOE process coupled with a molten salt plant.

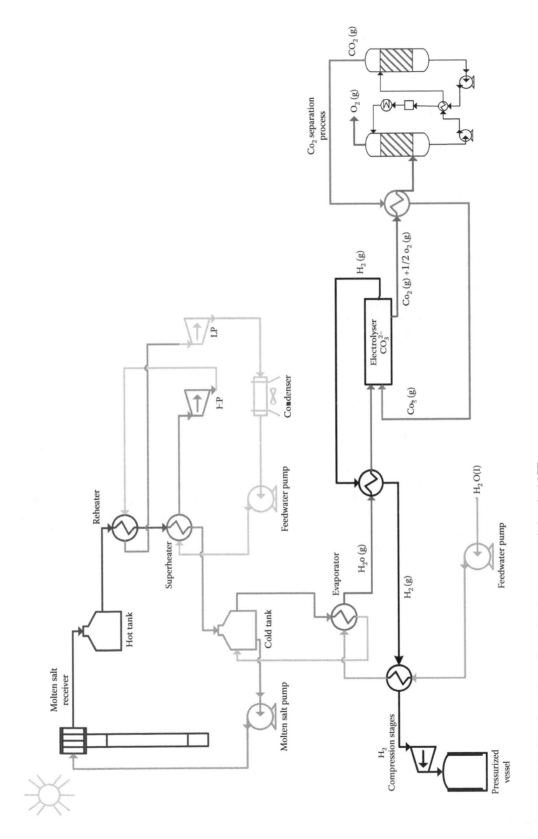

FIGURE 12.6 Flowsheet of a molten salt solar tower coupled to the MCE.

Molten salt is heated up to a temperature of about 565°C in the receiver and absorbs the highly concentrated solar radiations reflected by the heliostats to convert radiation into thermal energy. A part of this thermal energy is used in the boiler to generate steam, which is transported to a steam turbine. The steam expands in the turbine to produce electricity, which is introduced to the electrolyser. A part of this steam is used to evaporate the feed water of the MCE process. A second part of the thermal energy is sent to load the thermal storage system in order to resolve the time-dependent performance of the solar collector field as well as the solar transients and fluctuations and to have a constant feed for the electrolyser. The heat recovery system aims to overheat steam to the operating temperature of the electrolyser as well as the CO_2, which is separated from the oxygen in a suitable CO_2 purification unit through physical and chemical processes.

The high-temperature electrolysis of water is one interesting candidate for solar thermal water splitting due to its higher efficiency compared to room temperature alkaline or PEM electrolysis [16]. But a major effort is necessary to avoid or at least significantly reduce the degradation of the materials caused by the high operating temperature.

12.4 SOLAR THERMOCHEMICAL CYCLES

Thermochemical cycles use high heat sources to split water into its main components hydrogen and oxygen. There are several cycles, which take place in temperatures between 500°C and 2000°C. Figure 12.7 shows a classification of the different thermochemical cycles.

The focus of this section will be redox cycles, which are the most investigated and advanced thermochemical processes.

12.4.1 REDOX CYCLES

Redox cycles are very promising processes for "clean" hydrogen mass production. In the literature, up to 3000 thermochemical cycles are reported, but very few are proven to be relevant for the bulk production of hydrogen. These thermochemical cycles require heat for the water-splitting process. This heat can be provided by concentrated solar energy in a CRS, which consists of mirrors, so-called heliostats, which concentrate the sunlight onto the receiver. The receiver located on the top of the tower is made up of honeycomb-structured ceramics acting as absorbers for solar radiation and as chemical reactors.

With the oil crisis in the late 1970s and early 1980s, the interest in thermochemical water splitting increased a lot, but then research slowed down all over the world (except in Japan because of national concern about foreign energy dependence). However, the integration of such thermochemical cycles with a solar central receiver system is quite new.

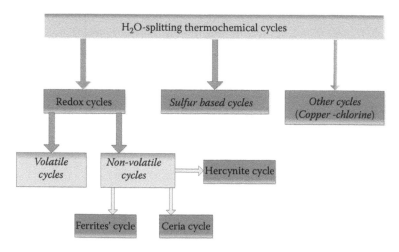

FIGURE 12.7 Classification of the thermochemical cycles.

One process for thermochemical hydrogen production is based on metal oxides (MO). Metal oxide cycles offer a unique alternative to the direct splitting of water, and recent solar thermochemical research has focused on them. The MO cycles are attractive because they involve fewer and less complex chemical steps than lower temperature processes and have the possibility of reaching higher cycle efficiency. The reaction scheme is as follows:

$$MO_{reduced} + H_2O \rightarrow MO_{oxidized} + H_2 \quad \text{(splitting)} \tag{12.6}$$

$$MO_{oxidized} \rightarrow MO_{reduced} + O_2 \quad \text{(regeneration)} \tag{12.7}$$

During the first step of the cycle (water splitting), the reduced and therefore activated material is oxidized by abstracting oxygen from water and producing hydrogen. In the next step (reduction), the material is reduced again, setting some of its lattice oxygen free. Major advantages are that pure hydrogen is produced and hydrogen and oxygen are produced in separate steps (i.e., no separation of hydrogen and oxygen is needed). Moreover, MO cycles involve fewer and less complex chemical steps than lower temperature processes. In the same way, CO$_2$ can be split to produce CO.

Several redox materials consisting of oxide pairs of multivalent metals (e.g., Fe$_3$O$_4$/FeO [17,18] Mn$_3$O$_4$/MnO [19]) or systems of MO/metal (e.g., ZnO/Zn [20]) have been evaluated for such applications. A lot of MO redox systems need too-high temperatures for the solar reduction step ($\Delta G < 0$ for $T > 2500$ K), which is the case for cycles based on MoO$_2$/Mo, SnO$_2$/Sn, TiO$_2$/TiO$_{2-x}$, MgO/Mg or CaO/Ca redox pairs. For some other redox pair cycles, the splitting step (production of H$_2$) is not feasible, as predicted by thermodynamics in the case of Co$_3$O$_4$/CoO and Mn$_3$O$_4$/MnO cycles.

The redox material Mn$_3$O$_4$/MnO can be reduced at 1900 K, but the hydrogen yield in the water splitting step of the process using this redox material is too low to be of any practical interest [21].

A solution is to make the reaction of the solar-produced MnO and NaOH and to consider the Mn$_2$O$_3$/MnO cycle. This cycle was proposed relatively recently:

$$MnO + NaOH \rightarrow NaMnO_2 + 1/2H_2 \tag{12.8}$$

$$NaMnO_2 + 1/2H_2O \rightarrow 1/2Mn_2O_3 + NaOH \tag{12.9}$$

$$1/2Mn_2O_3 \rightarrow MnO + 1/4O_2 \tag{12.10}$$

Studies of two-step cycles at temperatures of about 2300 K were related to CdO/Cd. This cycle is closely related to the zinc oxide cycle, but cadmium's toxicity makes its use in a technical application very unlikely. It produces hydrogen with solar energy at 1723 K.

$$CdO(s) \rightarrow Cd(g) + 1/2O_2 \tag{12.11}$$

$$Cd(l) + H_2O(g) \rightarrow CdO_2(s) + H_2(g) \tag{12.12}$$

Both reactions have been demonstrated in laboratory studies, but until now no feasibility evaluation of the closed cycle operation was made.

Another interesting MO cycle is based on SnO/SnO$_2$, recently demonstrated using solar energy.

$$SnO_2(s) \rightarrow SnO(g) + 1/2O_2 \tag{12.13}$$

$$SnO(s) + H_2O(g) \rightarrow SnO_2(s) + H_2(g) \tag{12.14}$$

The thermochemical hydrogen production process based on MOs was originally proposed by Nakamura in 1977 using Fe_3O_4/FeO as redox material. It consists of two steps:

$$Fe_3O_4 \rightarrow 3FeO + 1/2O_2 \tag{12.15}$$

$$3FeO + H_2O \rightarrow Fe_3O_4 + H_2 \tag{12.16}$$

12.5   FLOWSHEET OF A SOLAR THERMOCHEMICAL WATER SPLITTING PLANT

This section is dedicated to the flowsheet elaboration of a solar thermochemical water-splitting plant based on a redox cycle. The activated redox reagent (usually the reduced state of an MO) is oxidized in a first step by taking oxygen from water and producing hydrogen according to the reaction:

$$3FeO + H_2O(g) \rightarrow Fe_3O_4 + H_2(g) \quad @T = 980°C \tag{12.17}$$

During the second step, the oxidized state of the reagent is reduced to be used again (also known as the regeneration step of the redox material), delivering oxygen according to the next equation:

$$Fe_3O_4 \rightarrow 3FeO + 1/2O_2 \quad @T = 1100°C \tag{12.18}$$

The process consists generally of the following units:

- Water splitting and regeneration
- Feed water evaporation and overheating
- Water separation from the product gas
- Nitrogen separation from the product gas
- Hydrogen compression

Figure 12.8 shows the flowsheet of the overall process.

Deionized water enters the plant as stream WATER1. Then the water is evaporated in an electric evaporator ELECHX. The evaporator can provide slightly overheated steam to prevent condensation between the evaporator and the superheater (SUPER-HX), the component following downstream. There, the steam will be superheated to about 980°C. The superheated steam in WATER4 then enters a three-way valve, which directs the flow to the reactor in water-splitting mode. In that reactor, the MO reacts to the oxidized state, thus converting parts of the steam to hydrogen. The product stream leaving the reactor enters another three-way valve, which directs the stream to PRODUCT1. For safety reasons, a small amount of nitrogen (NITROGEN5) is mixed with the product stream. In SUPER-HX, heat from the product stream is transferred to the feed steam as explained above. As the condition of stream PRODUCT1 varies during the cycle and due to solar conditions, the temperature of PRODUCT2 may still be higher than a reasonable condenser inlet temperature would be. Therefore, an air cooler AC-HX is introduced to release potential waste heat. After that, the product stream PRODUCT3 enters the condenser CONDENS1, where the water content in the gas is reduced. The remaining water in the gas phase (corresponding to the partial pressure of water at the given temperature) and the nitrogen need to be separated from the hydrogen. As the condition of the stream PRODUCT4 is still subject to strong transient behavior, especially concerning molar composition, the gas is fed to a buffer tank, where the composition variations are evened out in the scale of a few hours. For the subsequent gas separation, a pressure swing adsorption system is applied. The system has an inlet pressure of 15 bar, taking into consideration a pressure drop of 3 bar in the PSA. Hence, the product gas is compressed. This compression can be realized in one or two stages. In the flowsheet, two-stage compression is used.

While developing solar high-temperature processes, several challenges regarding materials, components, separation issues, and operational strategies are addressed to make such processes

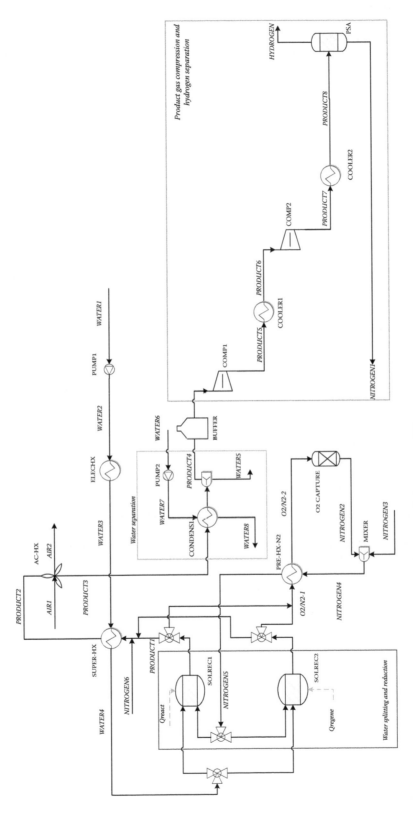

FIGURE 12.8 Flowsheet of a solar thermochemical water-splitting process [2].

practicable and commercially available. Other challenges are the temporal change of load and spatial temperature gradients. Selection, qualification, and certification of suitable construction, catalyst, and redox materials for key components are major objectives of research. Commercial plants with more than 100 MW$_{th}$ are envisaged in the long term. Almost all carbon-free solar thermal hydrogen production processes are in early development stages. First commercial plants for mass production are expected to be available after 2020.

REFERENCES

1. U.S. Environmental Protection Agency, Inventory of U.S. Greenhouse Gas Emissions and Sinks: 1990–2014, 2016.
2. A. Houaijia, C. Sattler, M. Roeb, M. Lange, S. Breuer and J.P. Säck, Analysis and improvement of a high-efficiency solar cavity reactor design for a two-step thermochemical cycle for solar hydrogen production from water, *Solar Energy*, vol. 97, pp. 26–38, 2013.
3. W. Doenitz, R. Schmidberger, E. Steinheil and R. Streicher, Hydrogen production by high temperature electrolysis of water vapour, *International Journal of Hydrogen Energy*, vol. 5, pp. 55–63, 1980.
4. W. Dönitz and E. Erdle, High temperature electrolysis of water vapor-status of development and perspectives for application, *International Journal of Hydrogen Energy*, vol. 10, pp. 291–295, 1985.
5. R. Pitz-Paal, Concentrating solar power. In *Future Energy: Improved Sustainable and Clean Options for Our Plants* (ed. T.M. Letcher), Oxford: Elsevier, pp. 171–192, 2008.
6. M. Romero and A. Steinfeld, Concentrating solar thermal power and thermochemical fuels, *Energy and Environmental Science*, vol. 5, pp. 9234–9245, 2012.
7. R. Dunn, P.J. Hearps and M. Wright, Molten salt power towers: newly commercial concentrating solar energie, *IEEE*, vol. 100, no. 2, pp. 504–515, 2010.
8. H. Price, E. Luepfert. D. Keaney, E. Zarza, G. Cohen, R. Gee and R. Mahoney, Advances in parabolic trough solar power technology, *International Journal of Solar Energy*, vol. 124, pp. 109–125, 2002.
9. A. Brisse, J. Schefold and M. Zahid, High temperature water electrolysis in solid oxide cells, *International Journal of Hydrogen Energy*, vol. 22, pp. 5375–5382, 2008.
10. S. Frangini, C. Felici and P. Tarquini, A novel process for solar hydrogen production based on water electrolysis in alkali molten carbonates, *The Electrochemical Society*, vol. 61, pp. 13–25, 2014.
11. J.O. O'Brien, Thermodynamics and transport phenomena in high temperature steam electrolysis cell, *Journal of Heat Transfer*, vol. 134, pp. 031017–031017, 2012.
12. S. Badwal, F. Ciacchi and D. Milosevic, Scandia-Zircona electrolytes for intermediate temperature solid oxide fuel cell operation, *Solid State Ionics*, vol. 136, pp. 91–99, 2000.
13. S. Licht and H. Wu, STEP Iron, a chemistry of iron formation without CO_2 emission: Molten carbonate solubility and electrochemistry of iron ore impurities, *Journal of Physical Chemistry C*, vol. 115, pp. 25138–25147, 2011.
14. H. Yin, X. Mao, D. Tang, W. Xiao, L. Xing, H. Zhu, D. Wang and D. Sadoway, Capture and electrochemical conversion of CO_2 to valued-added carbon and oxygen by molten salt electrolysis, *Energy & Environmental Science*, vol. 6, pp. 1538–1545, 2013.
15. L. Hi, I. Rexed, G. Lindbergh and C. Lagergren, Electrochemical performance of reversible molten carbonate fuel cells, *International Journal of Hydrogen Energy*, vol. 39, pp. 12323–12329, 2014.
16. C. Graves, S.D. Ebbesen, M. Mogensen and K.S. Lackner, Sustainable hydrocarbon fuels by recycling CO_2 and H_2O with renewable or nuclear energy, *Renewable Sustainable Energy Review*, vol. 15, pp. 1–11, 2011.
17. A. Steinfeld, S. Sanders and R. Palumbo, Design aspects of solar thermochemical engineering: A case study: Two-step water-splitting cycle using the Fe_3O_4/FeO redox system, *Solar Energy*, vol. 65, pp. 43–53, 1999.
18. M. Tsuji, T. Togawa, Y. Wada, T. Sano and Y. Tamaura, Kinetic study of the formation of cation-excess magnetite, *Journal of Chemical Society, Faraday Transactions*, vol. 91, pp. 1533–1538, 1995.
19. M. Sturzenegger and P. Nüesch, Efficiency analysis for manganese-oxide-based thermochemical cycle, *Energy*, vol. 24, pp. 959–970, 1999.
20. A. Steinfeld, Solar hydrogen production via a two-step water-splitting thermochemical cycle based on Zn/ZnO redox reactions, *International Journal of Hydrogen Energy*, vol. 27, pp. 611–619, June 2002.
21. M. Lundberg, Model calculations on some feasible two-step water splitting processes, *International Journal of Hydrogen Energy*, vol. 18, pp. 369–376, 1993.

13 Tannins as Precursors of Supercapacitor Electrodes

Vanessa Fierro and Angela Sánchez-Sánchez
National Scientific Research Centre

Alain Celzard
University of Lorraine

CONTENTS

13.1 INTRODUCTION TO TANNINS

Tannins are complex phenolic substances that can be extracted from many kinds of plants, especially from dicotyledonous ones (Constabel et al., 2014), and distributed in different parts of the plant structure, depending on the species: roots, trunk, barks, seeds, fruits, and leaves at diverse concentrations (Feng et al., 2013). Because of their bioactivity, tannins are essential products for plants to combat pathogens and herbivores. Moreover, tannins are bitter, and most herbivorous predators are disinclined to consume them (Constabel et al., 2014). Tannins bind to and precipitate proteins and various other organic compounds including amino acids and alkaloids, and they have been traditionally used in the tanning of animal skins.

The terms "hydrolysable" and "condensed" tannins are used to distinguish between the two important classes of vegetable tannins, namely, gallic acid-derived and flavan-3, 4-diol-derived tannins, respectively. This chapter will deal only with condensed tannins, which are characterized by aromatic rings bearing hydroxyl groups. Due to such polyphenolic structure, tannins are able to undergo the same kinds of chemical reactions as those known for resorcinol or phenol, two synthetic molecules of petrochemical origin used to prepare commercial resins. There are four types of polyflavonoids composing condensed tannins: (1) prorobinetinidin, (2) prodelphinidin, (3) profisetinidin,

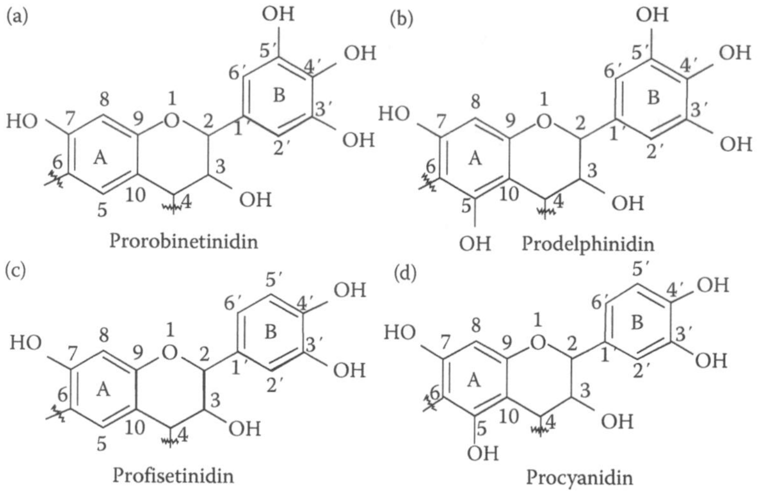

FIGURE 13.1 The four main polyflavonoids present in condensed tannins: (a) prorobinetinidin, (b) prodelphinidin, (c) profisetinidin, and (d) procyanidin. (Reprinted from Celzard, A. et al., *J. Cell. Plast.* 51 (1): 89–102, 2014. With permission from SAGE publications.)

and (4) procyanidin as shown in Figure 13.1. The concentration of these polyflavonoids in tannin depends on the plant species, and the number and the position of –OH functionalities change their reactivity. Thus, for mimosa tannin, resorcinol A-ring and pyrogallol B-ring (constituting a robinetinidin flavonoid unit) are the main patterns at about 90% of the phenolic content of the tannin itself (Pizzi and Mittal, 2003). For quebracho tannin, resorcinol A-ring and catechol B-ring (constituting a profisetinidin flavonoid unit) are the main patterns constituting more than 80% of the phenolic content of the tannin itself (Pizzi and Mittal, 2003). Pine tannins are based on a phloroglucinol A-ring and a catechol B-ring (giving a procyanidin-type tannin) or on a phloroglucinol A-ring and a pyrogallol B-ring (leading to a prodelphinidin-type tannin), see again Figure 13.1. Consequently, the A-rings of these condensed tannins are about 6–7 times more reactive than the A-rings of mimosa and quebracho-type tannins (Pizzi and Mittal, 2003).

Tannin extraction by striping takes place in autoclaves in series, using water at around 70°C and a very small percentage of sodium bisulphite. The latter additive reduces tannin autocondensation and improves tannin solubility (Sealy-Fisher and Pizzi, 1992). After striping, the solution contains 50% of extractives that are concentrated by spray drying. The as-obtained light-brown powder generally contains 80%–82% of polyphenolic flavonoid materials, 4%–6% of water, and 1% of amino and imino acids, the remainder being monomeric and oligomeric carbohydrates, in general broken pieces of hemicelluloses. The polyphenolic molecules have a rather low molecular weight ranging from 500 to 3500 g mol^{-1} (Pizzi, 1992). The older the bark is, the darker the color of tannin. Its color also changes because hydroxyl groups easily oxidize into quinones (Feng et al., 2013).

Table 13.1 gathers some information on three different types of condensed tannins from mimosa (*Acacia mearnsii*), quebracho (*Schinopsis balansae* and *Schinopsis lorentzii*), and maritime pine (*Pinus pinaster*). Maritime pine tannin is more reactive than quebracho and mimosa tannins, the latter presenting similar reactivity. Those condensed tannins are the most abundant and represent more than 90% of the world production (Stokke et al., 2013). Other tannins are those extracted from chestnut (*Castanea sativa*), tara (*Caesalpinia spinosa*), or aleppo oak (*Quercus infectoria*), which represent together 27,300 tons/year. The worldwide production of mimosa tannin is around 80,000 tons/year, and its price, while having significantly increased in the past 10 years, remains quite affordable, from 1800 to 2800 US$/ton depending on the origin and the supplying company.

TABLE 13.1

Summary of the Main Characteristics of Mimosa, Quebracho, and Maritime Pine Tannins

Tannins	Species of Dominant Flavonoid	Reactivity	Annual Production (tons year^{-1})	Price (US$ ton^{-1})
Mimosa	Prorobinetinidin	Low	80000[a]	1800–2800[a]
Quebracho	Profisetinidin	Low	60000[a]	2150[a]
Maritime Pine	Procyanidin	High	300[a]	34500[b]

[a] Indicative figures of tannin's market today provided by Silva Team Company (2017).

[b] Value obtained by DRT Company (2013).

Maritime pine tannin from France is, unlike the former materials, produced with a high purity for high-added value applications and is therefore available in much lower amounts and at a far higher cost, 34,500 US$/ton.

After the end of World War II, tannin commercialization declined with the emergence of various synthetic substances for treating animal skins. However, the demand increased again with the industrial production of condensed tannin-formaldehyde adhesives, which started in the 1970s. These adhesives were successfully applied to wood for producing particleboard, plywood, and glulam (horizontal wood pieces glued together), amongst others, for furniture and construction (Finch, 1984). Apart from the classical application of tannin as leather conditioner (Sreeram and Ramasami, 2003) other minor applications exist and/or were suggested such as wine additives (Harbertson et al., 2012), adsorbents for pollutant removal (Ali et al., 2012; Sánchez-Martín et al., 2010), or antioxidants in health supplements (Aron and Kennedy, 2008; Galati and O'Brien, 2004), amongst others.

13.2 TANNINS AS PRECURSORS OF HIGH ADDED-VALUE MATERIALS

Because of their chemical structure, explaining their reactivity, condensed tannins are versatile chemicals based on which high-quality thermoset phenolic/furanic resins can be obtained (Tondi et al., 2009b; Basso et al., 2011; Li et al., 2012b). Those resins can lead to a number of materials such as those summarized in Figure 13.2, which is far from being exhaustive. Indeed, crosslinking tannins in different conditions allows us to obtain various structures, whether the polymerization is carried out in the presence of a gas phase, leading to foams (Tondi et al., 2009a; Li et al., 2012c; Zhao et al., 2010; Li et al., 2012a; Lacoste et al., 2013; Basso et al., 2013; Szczurek et al., 2014b; Li et al., 2015; Celzard et al., 2014), or in a solvent, leading to gels (Szczurek et al., 2011b; Szczurek et al., 2011a; Amaral-Labat et al., 2012a; Pizzi et al., 2013; Amaral-Labat et al., 2013b; Grishechko et al., 2013; Amaral-Labat et al., 2013a; Rey-Raap et al., 2015, 2016), or in emulsion, leading to a range of materials from cellular monoliths (Szczurek et al., 2013; Szczurek et al., 2014a; Szczurek et al., 2015a; Szczurek et al., 2016; Seredych et al., 2016) to microspheres (Grishechko et al., 2016) and hollow spheres (Li et al., 2016). Tannins are also prone to autocondensation, which is a very interesting polymerization route since no cross-linker is required. This kind of reaction is favored by low pH or can be forced in hot pressurized water. Compared to more usual phenolic molecules, especially resorcinol, tannins have the advantage of being reactive in a much broader range of pH, and playing with this parameter further widens the range of structures that are obtained. Of course, dilution, additives, and temperature are additional conditions that can be changed, leading to new families of materials. Especially, tannin molecules are small enough for infiltrating porous structures and obtaining the corresponding replica after removal of the templates. Tannin-based polymers can also be mesostructured by use of relevant surfactants.

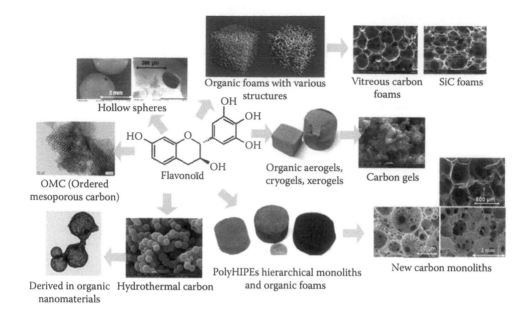

FIGURE 13.2   Examples of materials derived from condensed (flavonoid) tannins.

Interestingly, tannins also present a high carbon yield, around 50 wt%, due to their polyphenolic (hence aromatic) nature. In other words, cross-linked tannin-based structures, which are insoluble and infusible because they are made of thermoset polymers, can be converted into carbon materials having the same structure. Thus, all the materials listed above have their carbonaceous counterparts. Because the weight loss occurring during pyrolysis is associated with shrinkage of the same order of magnitude, about 50 vol%, the bulk density of the materials remains more or less unchanged after conversion into carbon. The latter is glasslike because of the nongraphitizing nature of the phenolic precursor. It is therefore a hard, shiny, and brittle carbon, presenting isotropic properties, and characterized by a good electrical conductivity (although lower than that of graphite) and a good reactivity through which surface functions can be easily created. This carbon is also known to be biocompatible and can be activated in conventional ways (physical or chemical) for opening and developing its porosity and surface area further. In some cases, ceramic materials or oxide nanoparticles can also be obtained by reaction of tannin-based carbons, e.g., with molten silicon (Amaral-Labat et al., 2013c), or by calcination of metal-loaded tannin-based carbons (Braghiroli et al., 2017), respectively.

Considering the above, it is clear that the range of applications of tannin-derived materials, whether in organic, carbon, or ceramic forms, is huge: purification and separation of gases, abatement of pollutants in water, thermal insulation or heat management, heterogeneous catalysis, electromagnetic interference shielding, biomedical applications, and energy conversion and storage. The latter is the purpose of the present chapter and is detailed below.

13.3   TANNINS AS PRECURSORS OF CARBON ELECTRODES FOR SUPERCAPACITORS

13.3.1   FUNDAMENTALS ON SUPERCAPACITORS

Figure 13.3 shows a roadmap for the future energy scenario, which demonstrates the development trend of energy technologies with the goal of replacing energy based on fossil resources by renewable energy (Gao et al., 2017). Therefore, it is essential to explore natural and renewable energy sources that might replace conventional, petrochemical ones, as well as greener and more efficient

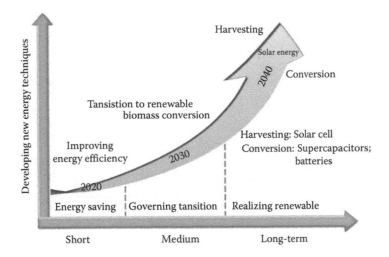

FIGURE 13.3 Indicative roadmap of future energy scenario. (Reprinted from Gao, Z. et al., *Mater. Res. Lett.* 5 (2): 69–88, 2017. With permission from Taylor and Francis.)

energy technologies able to meet the increasing energy demands. Energy conversion and storage indeed play a key role in achieving global energy sustainability.

Supercapacitors and batteries have been proven the most effective electrochemical energy conversion and storage devices for practical applications. Briefly, supercapacitors store charge at the electrode/electrolyte interface via electrical double layers or reversible Faradaic reactions, while batteries directly convert chemical energy into electrical energy by exothermal redox reactions (Conway, 1991).

A supercapacitor consists of two electrodes, an electrolyte, and a separator that electrically isolates the two electrodes (see Figure 13.4). In such a device, each electrode–electrolyte interface represents a capacitor so that the complete cell can be considered two capacitors in series. For a symmetrical capacitor (similar electrodes), the cell capacitance (C_{cell}) is calculated as

$$1/C_{cell} = 1/C + 1/C = 2/C \qquad (13.1)$$

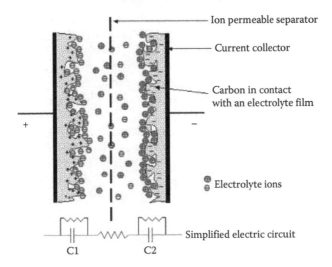

FIGURE 13.4 Representation of an electrochemical double-layer capacitor (in its charged state). (Reprinted from Pandolfo, A.G. and Hollenkamp, A.F., *J. Power Sources* 157 (1): 11–27. With permission from Elsevier.)

where C represents the capacitance of the two identical electrodes. (NB: literature values of specific capacitance often quote the capacitance of a single carbon electrode). From now on, we will refer to C for the electrode capacity and C_{cell} for the cell capacity.

Electrodes are fabricated from materials that have high surface area and high porosity. Charges are stored and separated at the interface between the conductive solid particles and the electrolyte. This interface can be treated as a capacitor with an electrical double-layer (EDL) capacitance, which can be expressed as in Equation 13.2

$$C = A\varepsilon / (4\phi d) \qquad\qquad (13.2)$$

where C is the capacity of each electrode; A is the area of the electrode surface, which for a super-capacitor should be the active surface of the electrode porous layer; ε is the medium (electrolyte) dielectric constant, which will be equal to 1 for vacuum and larger than 1 for all other materials, including gases; and d is the effective thickness of the EDL.

Increasing the capacitance of the electrode has mainly been approached by selecting materials with high surface, thus with highly developed porosity. In this sense, carbon materials have been largely used because of their high surface areas, low fabrication cost, low electrical resistivity, high physicochemical stability, and easy processability. Several studies have demonstrated that micro-pores (<2 nm) are involved in the charge storage process, while mesopores (2–50 nm) favor the charge propagation at high current load. In addition, the presence of mesopores stimulates charge accumulation, effective electrolyte wetting, and fast ion transport, thus increasing power densities and rate capabilities of the supercapacitor. When the carbon structures are ordered, low transport resistance and short ion distance provide an additional power enhancement. Thus, tailoring and controlling the pore size are crucial for designing efficient electrode materials (Aricò et al., 2005).

Besides EDL electrostatic interactions, redox processes with electron transfer at the electrode–electrolyte interface may also contribute to energy enhancement by pseudo-capacitive charge storage processes in an aqueous medium. Nitrogen and oxygen functional groups increase the capacity of the supercapacitors by Faradaic transfer reactions (Nasini et al., 2014; Sánchez-Sánchez et al., 2016) as shown in Figure 13.5.

13.3.2 Electrodes from Tannin-Derived Carbons

In this section, we report a selection of results showing that tannin can be an excellent precursor of carbon electrodes for supercapacitors. Several synthesis strategies are presented below, leading to disordered or ordered mesoporous materials. On the one hand, disordered mesoporous carbons were prepared from tannin-formaldehyde (TF) organic gels or from tannin submitted to hydrothermal treatments. On the other hand, ordered mesoporous carbons were prepared by hard or soft templating. In order to increase the capacity of these materials by pseudo-capacitive reactions, heteroatoms were sometimes introduced by using ammonia solutions as reaction media in hydrothermal treatments.

The carbon electrodes were prepared by spreading a paste, around 3 mg, containing the carbon material, polytetrafluoroethylene binder, and carbon black onto graphite foils. The resultant electrodes were pressed and impregnated with the electrolyte, H_2SO_4. Specific capacities were calculated by cyclic voltammetry (CV) measurements carried out either in 3-electrode or 2-electrode cells. Energy and power densities, as well as cycling stability, were derived from galvanostatic charge-discharge (GCD) measurements performed using H_2SO_4 as the electrolyte. Electrochemical impedance spectroscopy (EIS) measurements were carried out in a symmetric two-electrode cell with the same electrolyte.

The specific gravimetric capacitance of a single electrode (C, F g^{-1}) and of the two identical electrodes (C_{cell}, F g^{-1}) was calculated from the CV curves in a three- or two-electrode cell, respectively, according to Equation 13.3:

Hydroxyl groups: R–OH ↔ R=O+H⁺+e

Carboxylic groups: R–COOH ↔ R–COO⁻+H⁺+e

Carbonyl groups: R–C=O+H⁺+e ↔ R–CH–O

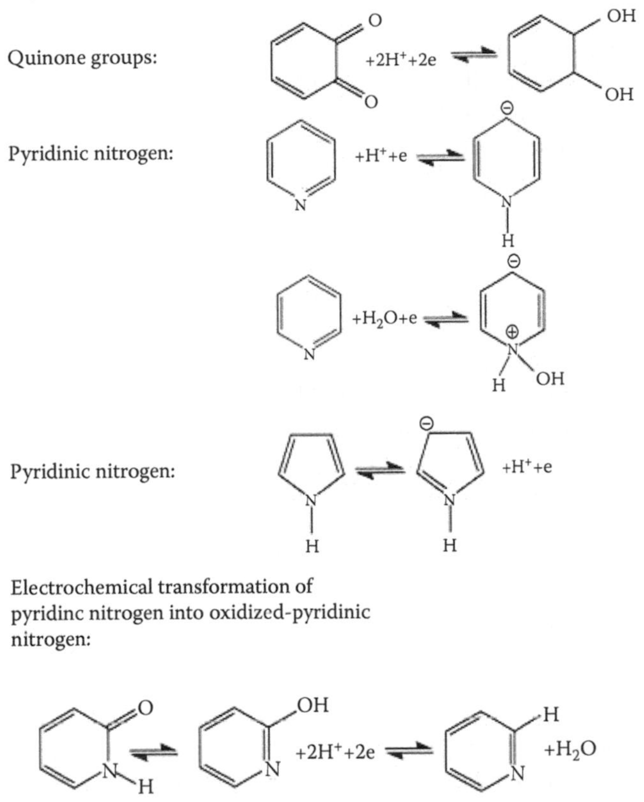

Quinone groups: (structure) +2H⁺+2e ⇌ (structure)

Pyridinic nitrogen: (structure) +H⁺+e ⇌ (structure)

(structure) +H₂O+e ⇌ (structure)

Pyridinic nitrogen: (structure) ⇌ (structure) +H⁺+e

Electrochemical transformation of
pyridinc nitrogen into oxidized-pyridinic
nitrogen:

(structure) ⇌ (structure) +2H⁺+2e ⇌ (structure) +H₂O

FIGURE 13.5 Faradaic reactions responsible for pseudocapacitance. (Reproduced from Sánchez-Sánchez, A. *J. Mater. Chem. A* 4 (16): 6140–48, 2016a. With permission from The Royal Society of Chemistry.)

$$C, C_{cell} = \left(\int I\Delta t \right) / (s\Delta V m) \qquad (13.3)$$

where I (A) is the constant current, s (V s⁻¹) is the scan rate, ΔV (V) is the potential window, and m (g) is the carbon mass of the single electrode or the two electrodes in the three- or two-electrode cells, respectively.

GCD tests were carried out at variable charge-discharge current densities in a three- or two-electrode cell, and the specific capacitances (C and C_{cell}, respectively) were calculated through Equation 13.4:

$$C, C_{cell} = (I \, dt) / (m \, dV) \qquad (13.4)$$

where I (A) represents the discharge current, (dV/dt) (V s⁻¹) is the slope of the discharge curve, and m is the carbon mass of either the single electrode or the two identical electrodes in the three- and two-electrode cells, respectively.

The cycling stability was studied in the same potential range at a given current density, based on the total mass of the two electrodes. From these measurements, gravimetric capacitance (C_{cell}, F g⁻¹), energy density (*E*, Wh kg^{-1}), and power density (*P*, W Kg^{-1}) were calculated by applying Equations 13.4, 13.5, and 13.6, respectively:

$$E = C_{cell} / 8 \times (\Delta V - IR)^2 \tag{13.5}$$

$$P = E / \Delta t \tag{13.6}$$

where ΔV (V) is the potential difference within the discharge time Δt (s), and *IR* (V) is the voltage drop due to the inner resistance at the beginning of the discharge process.

EIS measurements were performed at open-circuit voltage in the frequency range of 100 kHz–1 mHz with a 10 mV alternating current amplitude. The gravimetric capacitance (C_{cell}, F g^{-1}) was calculated according to Equation 13.7:

$$C_{cell} = -2 \, \text{Im}(Z) / (\pi f |Z|^2) m \tag{13.7}$$

where Im(Z) (Ω) is the imaginary part of the impedance, *f* (Hz) is the operating frequency, |Z| (Ω) is the impedance modulus, and *m* (g) is the carbon mass for the two electrodes.

13.3.2.1 Disordered Micro/Mesoporous Materials: Carbon Gels

TF carbon cryogels were prepared from mimosa tannin (Amaral-Labat et al., 2012b). Hydro/alcogels were first obtained by mixing one part in weight of mimosa tannin (T) powder with an aqueous solution containing methanol to which formaldehyde (F) in water (stabilized with methanol) was added. The resultant mixture, having an initial pH of 4.3, was adjusted separately in order to obtain pH values varying unit by unit from 3.3 to 7.3. Water as exchanged by tert-butanol and gels were freeze-dried. TF carbon cryogels were obtained after carbonization at 900°C.

TF carbon cryogels presented extremely low bulk densities, mainly due to a considerable fraction of macropores. Figure 13.6 shows Brunauer–Emmett–Teller (BET) surface area, S_{BET}, total pore volume, $V_{0.99}$, volume of mesopores, V_{meso}, and micropore volume, V_{micro} α, as a function of pH. At low pH, the values of S_{BET} were typical of carbon cryogels, but above pH 6; they were much higher than what is usually encountered with this kind of material. Compared to carbon aerogels prepared from the same TF organic gels (Szczurek et al., 2011b), the S_{BET} obtained with TF cryogels were higher, except at pH 4–5. Figure 13.6a shows a minimum of S_{BET} and pore volumes at such pH, at which the reactivity of phenolic molecules with formaldehyde is the lowest (Finch, 1996; Lin and Ritter, 1997).

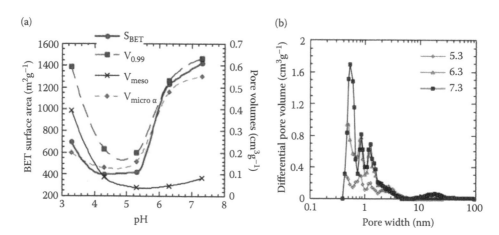

FIGURE 13.6 (a) Surface area and total, mesopore, and micropore volumes of TF-derived carbon cryogels as a function of pH of the hydro/alcogels and (b) Pore size distributions of TF-derived carbon cryogels prepared at pH higher than 5. (Reprinted from Amaral-Labat, G., *Biomass Bioenergy*, 39: 274–82, 2012b. With permission from Elsevier.)

Surprisingly, the position of such minimum exactly coincides with the maximum of both surface area and total pore volume observed in carbon cryogels derived from tannin—resorcinol—formaldehyde (Szczurek et al., 2011a) and in carbon aerogels derived from TF formulations of similar composition (Szczurek et al., 2011b). Figure 13.6b shows the pore size distributions of the samples prepared at pH higher than 5 where it can be easily observed that samples were essentially ultramicroporous (pore size < 0.7 nm). Only samples prepared at pH above 6 possessed high S_{BET} and micropore volumes, but all materials exhibited a very low mesopore volume, which could be a drawback for electrochemical applications.

Once ground and pressed into pellets with a binder, the TF-derived carbon cryogels were tested as electrodes by cyclic voltammetry in a Teflon Swagelok cell working at room temperature, as seen in Figure 13.7 (Braghiroli et al., 2015). The positive and negative electrodes of identical mass were electrically isolated by a glassy fibrous paper separator. The Teflon Swagelok cell was finally filled with 4 mol L^{-1} H$_2$SO$_4$ aqueous solution (Amaral-Labat et al., 2012b). The reference electrode was Ag/AgCl (KCl-saturated). The electrochemical characterizations were performed by cyclic voltammetry, and the voltammograms were obtained within the range 0.05–1.05 V versus reference electrodes at scan rates ranging from 2 to 50 mV s^{-1}.

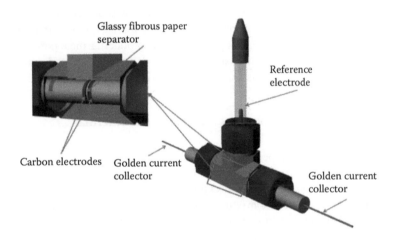

FIGURE 13.7 Two- and three-electrode cell for measuring electrochemical properties of TF-derived carbon cryogels (Reprinted from Braghiroli, F. L., *Ind. Crops Prod.* 70: 332–40, 2015. With permission from Elsevier.)

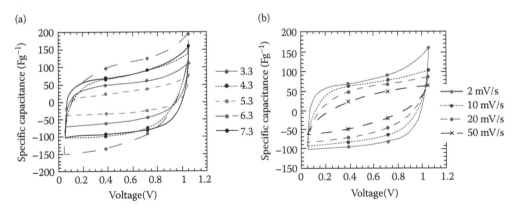

FIGURE 13.8 Cyclic voltammograms: (a) of porous electrodes made from TF carbon cryogels prepared at different values of pH written on the plot, at 2 mV s^{-1} and (b) obtained at different scan rates for porous electrodes made from TF carbon cryogels prepared at pH 7.3. (Reprinted from Amaral-Labat, G, *Biomass Bioenergy*, 39: 274–82, 2012b. With permission from Elsevier.)

Figure 13.8a shows the corresponding cyclic voltammograms. TF carbon cryogels prepared at pH higher than 6 presented high specific capacitances, close to those of more expensive carbon aerogels due to their higher S_{BET}. The square shape of voltammograms at low scan rates suggested a pure electrostatic attraction, and hence ideal capacitive behavior of TF carbon cryogels. Figure 13.8b shows the effect of the scan rate on the specific capacitance: voltammograms progressively lost their square form when increasing scan rate up to 50 mV s⁻¹, indicating transport limitations. Micropores unambiguously controlled the capacitance, with a dominant role of ultramicropores and supermicropores at low and high scan rates, respectively.

Given that micropores participate in the charge storage process, whereas mesopores are required for a fast accessibility of ions, improving the present tannin-based materials is required. A higher mesopore volume, thanks to which a high energy can be stored and delivered at a high rate, can be obtained through different ways:

- Modification of the initial formulation of the gel, through the use of higher solid/solvent fractions, leading to less macroporosity and more mesoporosity. This method requires the investigation of many materials in order to build a phase diagram like the one that has been done for polyurethane-based organic aerogels (Biesmans et al., 1998).
- Supercritical drying in acetone at 250°C. This method produces mesoporosity (Szczurek et al., 2011b) but is more complex and a little more expensive than freeze drying.
- Steam activation, creating new micropores but also producing mesopores by micropore widening. Additionally, oxygenated surface functions will be created, causing pseudocapacitance due to redox reactions and hence improving the total capacitance (Frackowiak and Béguin, 2001).

13.3.2.2 Disordered Micro/Mesoporous Materials: Hydrothermal Carbons from Mimosa Tannin

Mesoporous disordered carbons doped with nitrogen were prepared from mimosa tannin using two different amination methods described elsewhere (Braghiroli et al., 2012). Briefly, evaporated aminated tannin (EAT) was prepared with mimosa tannin mixed with 28% aqueous ammonia in air in room conditions. For that purpose, the mixture was placed in a closed vessel for preventing evaporation and was stirred for 1 h at room temperature. Then, the solution was poured in a Petri dish and left for complete evaporation in a fume hood for 2 days at room temperature. The obtained solid was then mixed with bidistilled water at room temperature. Aminated tannin (AT) was prepared by mixing 2 g of tannin with 16 cm³ of 28% aqueous ammonia in a glass vial until total dissolution. Tannin (T) dissolved in bidistilled water was also used as precursor of a reference material. EAT, AT, or T solutions were placed in a Teflon-lined autoclave of 50 cm³ for hydrothermal carbonization (HTC) in an oven for 24 h. After HTC, the samples were labelled H-EAT, H-AT, and H-T, respectively. The HTC was carried out at either $T_1 = 180°C$ or $T_2 = 210°C$ and, depending on such temperature, the number 1 or 2, respectively, was added to the samples' labels, e.g., H-EAT1 or H-EAT2 for H-EAT materials. These materials were pyrolyzed at 900°C under nitrogen atmosphere, and the resultant carbon materials had the letter C added to their label, i.e., CH-EAT were carbon materials prepared from H-EAT.

Table 13.2 shows the textural parameters, determined by N_2 adsorption at −196°C, together with C, N, H, and O contents determined by elemental analysis. CH-T1 and CH-T2 were essentially microporous materials and had almost the same BET surface areas: 665 and 684 m² g⁻¹, respectively. CH-AT and CH-EAT materials showed a reduction of S_{BET} when the HTC temperature increased. All the materials showed an increase of mesopore volume with the increase of HTC temperature. The mesopore volume was especially important for CH-EAT materials, ranging from 66% to 73%. The carbon content was around 80–82 wt% for the aminated materials, and around 88–92 wt% for reference materials CH-T1 and CH-T2. Nitrogen contents varied from 1.9 to 8.0 wt% for aminated materials and were around 0.7 wt% for reference materials since tannin contains low levels of amino and imino acids. Table 13.2 also gathers the electrochemical performances of these

TABLE 13.2

Main Characteristics of the Six Carbon Materials Tested

	Textural Properties				EA				Electrochemical Performances	
	S_{BET} (m² g⁻¹)	$V_{0.97}$ (cm³ g⁻¹)	V_μ (cm³ g⁻¹)	V_{meso} (cm³ g⁻¹)	C (%)	N (%)	H (%)	O (%)	Specific capacitance (F g⁻¹)	Normalized capacitance (μF cm⁻²)
CH-T1	665	0.26	0.25 (96.2%)	0.01	88.3	0.7	0.9	10.1	246	37
CH-T2	684	0.30	0.26 (86.7%)	0.04	92.1	0.7	0.9	6.3	181	27
CH-AT1	547	0.25	0.21 (84.0%)	0.04	81.7	5.6	1.0	11.7	303	55
CH-AT2	494	0.24	0.19 (79.2%)	0.05	80.0	8.0	1.0	11.0	253	51
CH-EAT1	552	0.62	0.21 (34.0%)	0.41	80.4	2.9	0.8	15.9	322	58
CH-EAT2	442	0.62	0.17 (27.4%)	0.45	81.7	1.9	0.9	15.5	230	52

Source: Reprinted from Braghiroli, F.L. et al., *Ind. Crops Prod.*, 70, 332–40, 2015, with permission from Elsevier.

materials obtained at 2 mV s⁻¹ in a two-and-three electrode cell using 4M H_2SO_4 as electrolyte (Figure 13.7). The specific capacitance varied from 181 to 322 F g⁻¹. The normalized capacitance, obtained by dividing the specific capacitance by S_{BET}, is also reported and the highest value was close to 60 F cm⁻².

Figure 13.9a shows the normalized capacitance at 2 mV s⁻¹ as a function of oxygen content. Although the micropore volume was not constant, the normalized capacitance seemed to increase linearly with the oxygen amount. The effect of nitrogen content was more complex. For example, the sample CH-AT2 presented a normalized capacitance 1.4 times higher than that of CH-T1, whereas its micropore volume was lower and its oxygen content was very similar to that of CH-T1.

The main difference between these two samples was the nitrogen content, which was close to 8 wt% in CH-AT2 against less than 1 wt% in CH-T1. But excessively high nitrogen content did not seem to be favorable either, unlike what was found with oxygen, as an optimum of nitrogen content can be seen in Figure 13.9b. Intermediate nitrogen content ranging from 3 to 6 wt%, combined with an appropriate pore size distribution and a high oxygen amount, can thus enhance the capacitor performances through extra redox reactions of Faradaic pseudo-capacitance.

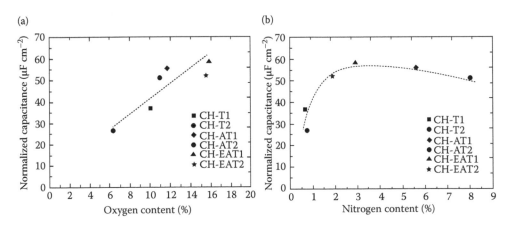

FIGURE 13.9 Normalized capacitance of N-doped tannin-based carbons at 2 mV s⁻¹ as a function of: (a) oxygen content and (b) nitrogen content. (Reprinted from Braghiroli, F. L. *Ind. Crops Prod.* 70: 332–40, 2015. With permission from Elsevier.)

13.3.2.3 Disordered Micro/Mesoporous Materials: Hydrothermal Carbons from Pine Tannins

Nitrogen- and oxygen-doped carbon materials were also successfully prepared by HTC from two types of condensed tannins, extracted from barks of pine tree (*Pinus pinaster*) either with pure water or with an aqueous solution of sodium salts, and subsequent carbonization (Sánchez-Sánchez et al., 2017b). Neither oxidation nor activation post-treatment was applied for functionalizing the materials further or for developing their porosity. The materials were prepared by following three procedures: (1) *HTC/water*: tannin was dissolved in distilled water and subsequently submitted to HTC; (2) *HTC/NH₃*: tannin was dissolved in a 28 wt% ammonia solution and then submitted to HTC; (3) *NH₃ evaporation+HTC*: tannin was dissolved in 28 wt% ammonia solution, then the mixture was stirred in a sealed flask and subsequently left for evaporation in a fume hood until complete ammonia evaporation. The resultant powder was finally dissolved in distilled water and submitted to HTC. The resultant hydrochars were washed with distilled water, dried under vacuum, and carbonized in nitrogen atmosphere at 900°C. The final carbons were designated as *Te-x*, where *e* refers to the extraction method (W=water or Na=sodium salt solution) and *x* to the HTC process: *nothing* for *HTC/water*, *A* for *HTC/NH₃* and *EA* for *NH₃ evaporation+HTC*. Therefore 6 materials were prepared: TW, TW-A, TW-EA, TNa, TNa-A, and TNa-EA.

Figures 13.10a and b show the nitrogen adsorption-desorption isotherms determined at −196°C for TW and TNa materials, respectively. The corresponding pore size distributions are shown in Figures 13.10c and d. Important differences were found, depending on the extraction method and the post-treatment. In general, TW series showed higher surface areas than the TNa series. Amination treatments produced a decrease of the microporosity and an increase of the mesoporosity in the case of TW-EA.

Concerning the heteroatom content, TW and TNa, prepared by *HTC/water*, exhibited the lowest N/C, O/C and H/C atomic ratios, whereas these atomic ratios were the highest for TW-A and TNa-A, synthesized through *HTC/NH₃*. This clearly indicates that HTC/water gave rise to the least

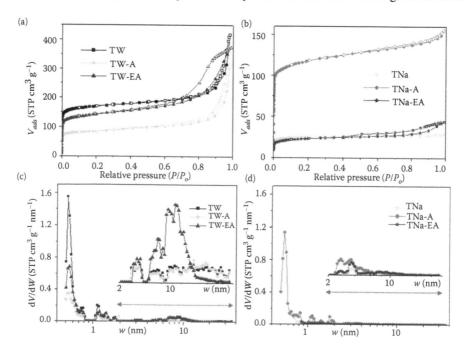

FIGURE 13.10 Nitrogen adsorption isotherms measured at −196°C for the carbons obtained from pine tannins extracted with: (a) water, referred to as W–series, or (b) with sodium salt solution, referred to as Na–series. Pore size distributions of (c) W–series samples and (d) Na–series samples.

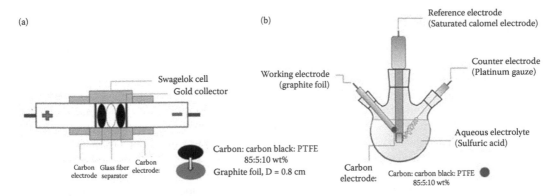

FIGURE 13.11 Assembly of the carbon electrodes in the two-electrode cell (a) and in the three-electrode cell (b) used for electrochemical measurements.

functionalized samples regarding N and O groups, while *HTC/NH₃* led to the most functionalized ones. Among the two series of prepared carbons, those presenting more developed porous textures and high bulk nitrogen and oxygen contents were selected in order to study their electrochemical performances. The selected samples were TW, TW-A, TW-EA, and TNa-A.

Further characterization by Raman spectroscopy allowed showing that the carbons of the W–series (TW, TW-A, and TW-EA) exhibited higher *I(D)/I(G)* ratios than those of the Na–series (TNa-A), so that they possessed larger crystallites and, consequently, slightly less disordered nano-textures than the latter. It should indeed be recalled here that we deal with non-graphitizing carbons, for which the *I(D)/I(G)* ratio is roughly proportional to the square of the crystallite size (Ferrari and Robertson, 2000). Among the W–series materials, TW-A exhibited the most ordered nanotexture, since it displayed the highest *I(D)/I(G)* ratio. A higher-ordered nanotexture is expected to increase supercapacitor performances.

Electrochemical performances of the carbon materials were determined by using the two-electrode (a) and the three-electrode (b) cells displayed in Figure 13.11.

The electrochemical behavior of those carbon materials was investigated by cyclic voltammetry and galvanostatic charge-discharge in a three-electrode cell using 1M H_2SO_4 as electrolyte. The resultant CV and GCD curves within the potential window of 0–0.9 V are shown in Figure 13.12. Figure 13.12a shows the CV curves of the four materials at 1 V s⁻¹. W–series electrodes (TW, TW-A, and

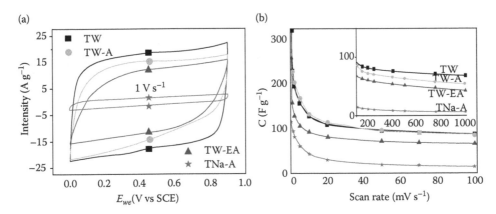

FIGURE 13.12 Electrochemical results obtained by CV with a 3 electrode-system in 1M H_2SO_4 electrolyte: (a) CV curves at 1 V s⁻¹ and (b) Specific capacitance, calculated from the CV curves, versus scan rate. (Reproduced from Sánchez-Sánchez, A. et al., *Green Chem.*, 19, 2653–2665, 2017b. With permission from The Royal Society of Chemistry.)

TW-EA) still retained a nearly rectangular shape at such a high scan rate, which indicates an excellent rate capability of the materials. However, the slightly spindle-like shape of the CV curves for TW-A and TW-EA, as compared with that of TW, denoted the existence of much more kinetic limitations for the electrolyte ions to get into the smaller pores, unlike what was observed for the W–series materials. The performances of TNa-A were worse than those showed by W-series materials, as its CV curve was extremely narrow. High capacitance values of 321 F g^{-1} were achieved by TW at a 0.5 mV s^{-1} scan rate, which exhibited the highest surface area and the lowest density of surface functional groups (Figure 13.12b). At this scan rate, TW-A and TW-EA yielded similar capacitance values of 252 and 253 F g^{-1}, and the lowest capacitance value of 114 F g^{-1} was achieved by TNa-A. The samples of the W–series presented high capacitance retentions at 100 mV s^{-1} of 27.4, 33.9 and 26.4% for TW, TW-A and TW-EA, respectively, while that of the Na–series, TNa-A, exhibited the most pronounced capacitance decrease up to 20 mV s^{-1} and only a capacitance retention of 12.9% at 100 mV s^{-1}.

The energy and power densities and the cycling stability of the supercapacitors were evaluated through GCD in a two-electrode cell using 1M H$_2$SO$_4$ as electrolyte. As shown in the Ragone plot of Figure 13.13a, the W-series carbons exhibited maximal energy densities of 6.7–25.22 Wh kg^{-1} under power outputs of 223.6 W kg^{-1}–19.3 kW kg^{-1}, which clearly exceed those values obtained for other biomass-derived carbons (Chen et al., 2015; Jain et al., 2015; Sánchez-Sánchez et al., 2016b). TNa-A yielded lower energy and power densities than the rest of the materials in the entire current density interval, in agreement with the results discussed in the previous paragraph. Exceeding 3 A g^{-1}, the energy density decreased dramatically, and the power density was limited to a maximum value of ~4.35 kW kg^{-1}.

The cycling stability of the cells was studied at the current density of 0.5 A g^{-1} up to 5000 cycles of charge-discharge (Figure 13.13b). TW exhibited high capacitance retention of 95% up to 2500 cycles of charge-discharge. Above 2500 cycles, the capacitance retention decreased with the number of charge-discharge cycles, probably due to the degradation of electroactive functional groups (Hsieh and Teng, 2002; Pandolfo and Hollenkamp, 2006) and reached 77.2% of the initial capacitance after 5000 cycles.

The frequency response of the carbon electrodes was studied by EIS (Figure 13.14). The TW-EA electrode presented a short 45° Warburg region in the medium frequency range (10^3 – 8 Hz), the highest equivalent distributed resistance of 1.84 Ω, and the highest equivalent series resistance (ESR) of 2.36 Ω, which indicates that this cell possesses the lowest electrical conductivity. TW-EA and TNa-A electrodes exhibited the highest values of electrolyte resistance of 0.37 and 0.38 Ω, respectively, which are likely due to the high percentages of narrow porosity existing in both carbons

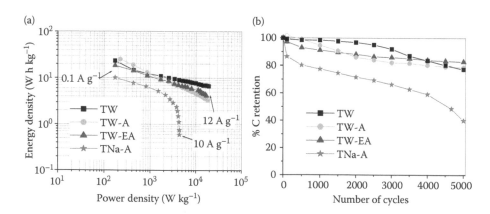

FIGURE 13.13 Electrochemical results obtained by GCD with a two-electrode system in 1M H$_2$SO$_4$ electrolyte: (a) Ragone plot calculated in the current density interval of 0.1–12 A g^{-1} and (b) Cycling stability, determined from GCD tests at 0.5 A g^{-1} until 5000 cycles. (Reproduced from Sánchez-Sánchez, A. et al., *Green Chem.*, 19, 2653–2665, 2017b. With permission from The Royal Society of Chemistry.)

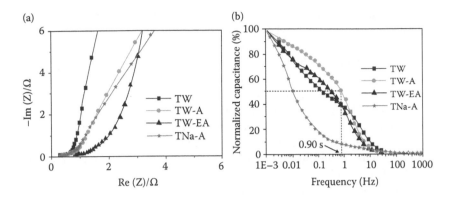

FIGURE 13.14 Results obtained by EIS in a two-electrode system with 1M H_2SO_4 electrolyte: (a) Nyquist plot and (b) Frequency response plot. (Reproduced from Sánchez-Sánchez, A. et al., *Green Chem.*, 19, 2653–2665, 2017b. With permission from The Royal Society of Chemistry.)

together with either low mesoporosity (in TW-EA) or total absence of mesopores (in TNa-A). The maximal specific capacitance values for the two-electrode system, calculated at 1 mHz, were ~ 50, 34, 32, and 22 F g^{-1} for TW, TW-A, TW-EA, and TNa-A, respectively.

The frequency response plots show that ion transport was particularly hindered for the TNa-A electrode, since a sharp drop of the normalized capacitance occurred even at very low frequency (Figure 13.14b). This is explained by the high percentage of narrow porosity and the low mesopore volume of this carbon material; it agrees with the CV and GCD results discussed in the last three previous paragraphs. The relaxation time constant for the TNa-A electrode, calculated as $1/f_{0.5}$, (where $f_{0.5}$ is the frequency at which 50% of the normalized capacitance is achieved) was 65.7 s, which prevents the use of this material in competitive commercial devices. On the contrary, the TW-A electrode exhibited an extraordinarily fast frequency response of 0.9 s. To the best of our knowledge, this value is one of the lowest ever reported for a broad range of doped and nondoped carbon materials (Wang et al., 2013; Wei et al., 2011; Xing et al., 2006) and makes the TW-A carbon an excellent candidate as electrode material in real supercapacitors.

Carbons prepared by HTC in ammonia (TW-A and TNa-A) possessed the highest concentrations of surface functionalities, up to 7.96 mmol g^{-1}, but exhibited pore size distributions that did not favor good electrochemical performances. The porosity of the carbon derived from the tannin extracted with sodium salts and submitted to HTC in water (TNa) was mainly composed of narrow pores and a low fraction of mesopores, which limited the access of the electrolyte ions to the porosity and to the surface functionalities. For this reason, this material yielded the lowest rate capability under high sweep rate regimes, as well as the lowest energy and power densities and cycle stability. Moreover, its capability for ion storage vanished at 10 A g^{-1} current density, and its frequency response was too slow for use in a commercial device. The carbon obtained from the water-extracted tannin submitted to HTC in water (TW) presented a low percentage of narrow porosity, a high mesopore fraction, and a wide pore size distribution (PSD) in the mesopore range. These properties, together with the hydrophilic character of the surface, the high density of surface functionalities, and the high nanotexture order are responsible for its high rate capability (up to 1 V s^{-1}), its high energy density up to ~1500 mA g^{-1}, its low cell resistance, and its extremely fast frequency response, making it a suitable material for practical devices (Sánchez-Sánchez et al., 2017b).

13.3.2.4 Ordered Mesoporous Materials: Hard-Templating Applied to Tannins

In order to create optimized and tailored porous structures, many synthetic routes and carbon precursors have been investigated (Liang et al., 2008). Nanocasting, also called nanotemplating or hard-templating, is a promising route to preparing carbon materials with controlled hierarchical and ordered porous structures by selecting the carbon precursor and the ordered mesoporous silica used as a template. Glucose, sucrose, furfuryl alcohol, or resorcinol, among others, have been commonly used as carbon precursors for preparing ordered mesoporous carbons (OMCs) (Jun et al., 2000; Kruk et al., 2000; Wang et al., 2008; Wu and Liu, 2016). However, nanocasting processes using these molecules typically require multiple infiltration steps for completely filling the pores of the template, as well as long polymerization times and/or the use of toxic reagents.

Sánchez-Sánchez et al. (2017a) demonstrated that different types of plant-derived polyphenols, and especially tannin-related secondary metabolites (phloroglucinol, gallic acid, catechin, and mimosa tannin), may efficiently be used as biosourced precursors for obtaining OMCs through a nanocasting route. Figure 13.15 illustrates the nanocasting route used where three main steps can be distinguished: (1) infiltration, i.e., introduction of the carbon precursors inside the porosity of the silica; (2) carbonization, i.e., thermal treatment in inert atmosphere after which a silica-carbon composite can be obtained; and (3) leaching, i.e., dissolution of the silica template (SBA-15) and recovery of the OMC.

The templating process is properly accomplished as soon as the pore system of the carbon material is the exact inverse replica of the silica template. The obtained OMCs were labelled PhC, GaC, CatC, and TanC, depending on the polyphenol used as precursor in the synthesis: phloroglucinol, gallic acid, catechin, and mimosa tannin, respectively. The template structure was well replicated by means of a single infiltration step, even for the most bulky precursors such as catechin and mimosa tannin, whose size largely exceeds that of the commonly used molecules. Nevertheless, the smallest molecules, phloroglucinol and gallic acid, allowed obtaining ordered carbons at longer ranges and with slightly higher surface areas (see Figure 13.16).

The CV curves, shown in Figure 13.17a, maintained their rectangular shape at sweep rates as high as $100\,mV\,s^{-1}$, which indicates an ideal capacitive behavior allowing fast charge–discharge and ion transport. Slight deviations from the rectangular shape are symptomatic of Faradaic processes of the oxygen functional groups in H_2SO_4 electrolyte. This assumption was supported by the high specific

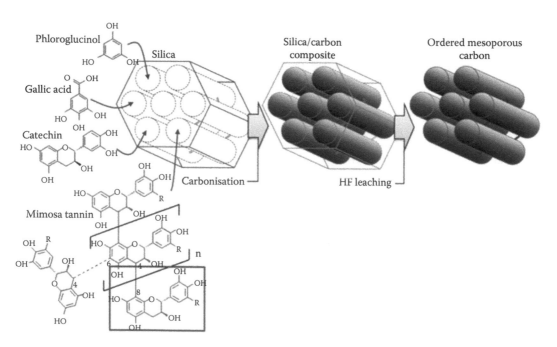

FIGURE 13.15 Schematic procedure for the synthesis of ordered mesoporous carbons (OMCs) through a nanocasting route based on tannin-related carbon sources. (Reprinted from Sánchez-Sánchez, A. et al., *J. Power Sources*, 344: 15–24, 2017a. With permission from Elsevier.)

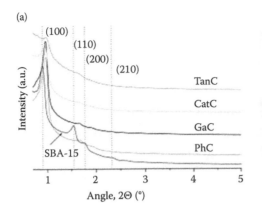
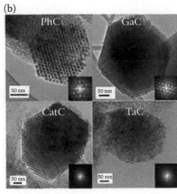

FIGURE 13.16 (a) Small-angle X-ray patterns of parent silica template and derived carbon materials and (b) Transmission Electron Microscopy (TEM) and FFT (inset: Fast Fourier Transform) images of the carbon materials viewed along the cross-sectional direction of the rods. (Reprinted from Sánchez-Sánchez, A. et al., *J. Power Sources*, 344: 15–24, 2017a. With permission from Elsevier.)

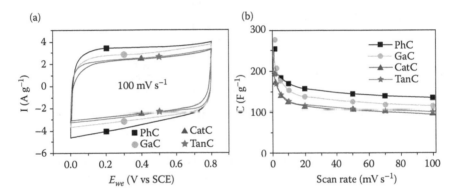

FIGURE 13.17 Results of cyclic voltammetry and galvanostatic charge-discharge tests for OMCs tested in a three-electrode system: (a) CV curves at the sweep rate of 100 mV s⁻¹ in the potential window of 0–0.8 V and (b) Evolution of specific capacitance with scan rate. (Reprinted from Sánchez-Sánchez, A. et al., *J. Power Sources*, 344: 15–24, 2017a. With permission from Elsevier.)

capacitances, up to 277 F g⁻¹ for GaC electrodes at 0.5 mV s⁻¹ sweep rate (Figure 13.17b). These values largely exceed those previously obtained in the same electrolyte for other OMCs prepared by liquid infiltration of sucrose or by chemical vapor deposition of propylene (Jurewicz et al., 2004).

PhC, GaC, and TanC yielded slightly lower energy densities between 24.8 and 31.9 Wh kg⁻¹ under exceptionally high power outputs between 77.3 and 88.5 W kg⁻¹, respectively (Figure 13.18a). After 5000 charge–discharge cycles, GaC led to the highest capacitance retention of 80.6%; PhC and TanC presented capacitance retention values of 79.2% and 78.2%, respectively; and CatC exhibited the lowest value of 71.8% (Figure 13.18b).

Analysis of the EIS results showed that the resistances for ion transport into the electrode bulk increased in the order PhC < TanC < CatC, which is in agreement with the increasing percentage of narrow porosity of the materials (Figure 13.19a). The GaC electrode exhibited the easiest diffusion into the pores and the highest electrical conductivity, presenting an ESR value, 0.48 Ω, which is among the lowest ever reported for OMCs and other carbon materials in aqueous electrolytes, typically higher than 1.0 Ω. This electrode showed one of the fastest frequency responses ever reported, 0.9 s, which correlates with the low volume of narrow porosity, the high electrical conductivity, and the excellent capacitance retention at both high sweep rates and high current densities of this

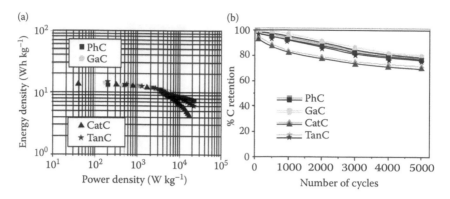

FIGURE 13.18   (a) Ragone plot and (b) cycling stability up to 5000 cycles of charge–discharge. (Reprinted from Sánchez-Sánchez, A. et al., *J. Power Sources*, 344: 15–24, 2017a. With permission from Elsevier.)

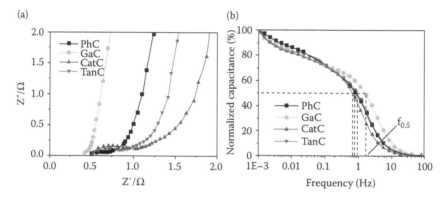

FIGURE 13.19   Results of EIS for all OMCs: (a) Nyquist plots and (b) frequency responses. (Reprinted from Sánchez-Sánchez, A. et al., *J. Power Sources*, 344: 15–24, 2017a. With permission from Elsevier.)

electrode material. The corresponding OMC-based electrodes exhibited better electrochemical performances in terms of specific capacitances, rate capabilities, and energy and power densities than those of a broad variety of OMCs and commercial or biomass-derived activated carbons previously reported (Jain et al., 2015; Long et al., 2016; Wang et al., 2013; Wei et al., 2011; Xing et al., 2006; Zhang et al., 2016).

If these results were excellent in terms of electrochemical performances, the use of a SBA-15 as a template that needs to be removed with a harmful chemical, HF, for recovering the mesostructured carbon is neither economic nor ecologic. A greener alternative to prepare OMCs is soft-templating.

13.3.2.5   Ordered Mesoporous Materials: Soft-Templating Applied to Tannins

Soft-templating is based on the self-assembly of a surfactant in a solution containing the carbon precursor. Surfactant molecules have a hydrophilic head, which spontaneously places itself at the outer part of the micelles for being in direct interaction with the aqueous solution, and a hydrophobic core, in the center of the micelles (Ma et al., 2013). The carbon precursor, usually in the form of monomers or oligomers, interacts with the hydrophilic part of the surfactant and therefore leads to composite micelles with the carbon precursor located preferentially in the hydrophilic coronas of the micelles (Liang and Dai, 2006). After polymerization of the precursor, self-assembly of the composite micelles, and their further organization into the liquid-crystal phase, a last step is pyrolysis in inert atmosphere for: (1) decomposing the template, which is the porogen agent and (2) carbonizing the precursor. The resulting carbon nanostructures display monodisperse mesopore size distribution and pore architecture related to the carbon precursor/surfactant mesophase (see Figure 13.20).

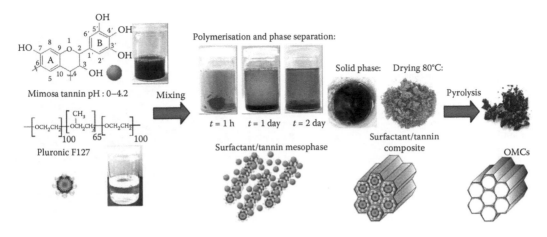

FIGURE 13.20 Scheme of the synthesis of ordered mesoporous carbon materials based on tannin, water, and F127. (Reproduced from Braghiroli, F. L. et al., *Green Chemistry* 18 (11): 3265–71, 2016. With permission from The Royal Society of Chemistry.)

Phenolic molecules containing hydroxyl groups (–OH) have been extensively used as carbon precursors, among them phenol (Huang et al., 2008), resorcinol (Wang et al., 2008) or phloroglucinol (Liang and Dai, 2006), together with formaldehyde as a cross-linker and an amphiphilic triblock copolymer Pluronic F127 (here simply referred to as F127) as a soft-template for the preparation of OMCs.

Braghiroli et al. (2016) showed that interaction between Pluronic F127 and mimosa tannin immediately takes place in water solutions at room temperature and is nearly independent of the initial pH. This synthesis was carried out without any hydrothermal process or the use of any kind of cross-linker (such as formaldehyde or hexamethylenetetramine), organic solvent (e.g., ethanol or THF), or other toxic reagent. Highly porous materials were thus obtained, having surface areas up to 720 m^2 g^{-1}, and total and mesopore volumes equal to 0.64 and 0.36 cm^3 g^{-1}, respectively. Unlike what happened with former preparation methods, the mesoporosity was fully maintained after carbonization at high temperatures, and narrow microporosity was developed. Figure 13.21a shows the total pore volume ($V_{0.97}$) as a function of BET surface area for all tannin-based materials synthesized in the present work, compared with other results from the literature (de Souza et al., 2013; Feng et al., 2014; Hao et al., 2011; Huang et al., 2008; Kubo et al., 2011; Liang and Dai, 2009; Schlienger et al., 2012; Tanaka et al., 2011; Wang et al., 2008). Figure 13.21b shows that these OMCs presented an ordered mesostructure and a good thermal stability after pyrolysis at 900°C. The dashed straight line indicates the trend of the textural properties when the pyrolysis temperature increased. The highest surface areas and pore volumes were obtained after pyrolysis at 900°C when using tannin and F127 as the only precursors.

Unlike previous studies, no formaldehyde or ethanol was used, and the as-obtained OMCs had very good thermal stability, much better than those obtained from tannin dissolved in ethanol and F127, and prepared through the evaporation-induced self-assembly method. Therefore, producing OMCs having both high surface area and high mesoporosity from tannin and F127 after carbonization at 900°C was shown to be feasible. The carbon obtained by direct carbonization of mimosa tannin (MT sample) exhibited a BET area of 392 m^2 g^{-1}, and its porosity was mainly composed of micropores with diameters between 0.45 and 1.25 nm (Figure 13.22a and b). Mesostructuration with Pluronic F127 (MT+F127, pH1 sample) allowed increasing the BET area up to 720 m^2 g^{-1} and generating ordered mesopores with narrow pore size distributions centered on 6.72 nm, according to the BJH-KJS method and on 5.15 nm, according to the NLDFT method applied to slit pores.

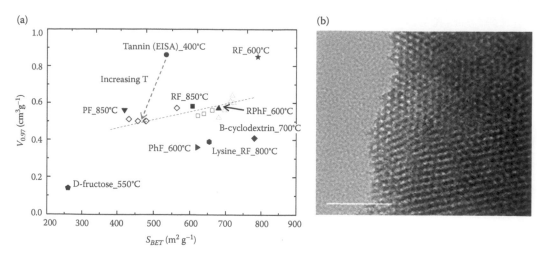

FIGURE 13.21 (a) Total pore volume ($V_{0.97}$) as a function of BET surface area for all mesoporous materials made from mimosa tannin (samples pyrolyzed at ◇ 400°C; □ 700°C and ▲ 900°C), compared to other mesoporous materials prepared from various precursors: ■ Resorcinol/formaldehyde (Wang et al., 2008); ▶ Phenol/formaldehyde (Huang et al., 2008); ▼ Phloroglucinol/formaldehyde (Liang and Dai, 2009); ⬟ D-Fructose (Kubo et al., 2011); ◆ β-cyclodextrin (Feng et al., 2014); ● Mimosa tannin by evaporation-induced self-assembly (Schlienger et al., 2012); ★ Resorcinol/formaldehyde (de Souza et al., 2013); ⬣ Lysine+Resorcinol/formaldehyde (Hao et al., 2011); ▲ Resorcinol+Phloroglucinol/formaldehyde (Tanaka et al., 2011). (b) TEM images of OMC prepared from mimosa tannin and pyrolyzed at 900°C (Braghiroli et al., 2016). (Reproduced from Braghiroli, F. L. et al., *Green Chemistry* 18 (11): 3265–71, 2016. With permission from The Royal Society of Chemistry).

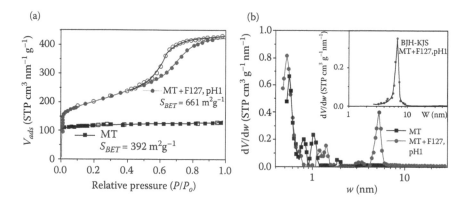

FIGURE 13.22 (a) Nitrogen adsorption isotherms measured at −196°C of carbons obtained from directly carbonized mimosa tannin (MT) and from the OMC obtained from mimosa tannin and Pluronic F127 at pH = 1 (MT+F127, pH1) and (b) Pore size distributions of MT and MT+F127, pH1 obtained by applying the NLDFT method for slit pores to the adsorption branch of the nitrogen isotherms; the PSD obtained by applying the BJH method (KJS correction) to the desorption branch of the nitrogen isotherm of MT+F127; pH1 is also shown in the inset.

The porous textures were responsible for the different electrochemical performances, obtained in a three-electrode cell configuration with 1M H_2SO_4 electrolyte, while MT yielded a maximum specific capacitance of 19 F g^{-1} at 0.5 mV s^{-1}, the ordered mesoporous carbon (MT+F127, pH1) exhibited a maximum specific capacitance of 126 F g^{-1} (Figure 13.23a).

The rate capability was considerably higher for MT+F127, pH1 than for MT, having capacitance retentions at 1 V s^{-1} of 19.6% and 3%, respectively. Moreover, MT+F127, pH1 exhibited remarkable

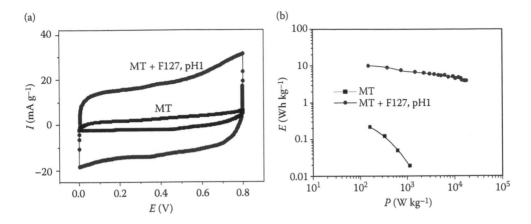

FIGURE 13.23 Results of cyclic voltammetry and galvanostatic charge-discharge tests of MT and MT + F127, pH1 samples: (a) CV curves at the sweep rate of 0.5 mV s^{-1} in the potential window of 0–0.8 V, tested in a three-electrode cell; (b) Ragone plot obtained through galvanostatic charge-discharge tests between 0.1 and 12 A g^{-1} current density in a two-electrode cell.

energy densities between 10 and 4 Wh kg^{-1} under power outputs of 154 W kg^{-1} and 17.5 kW kg^{-1}, respectively; in contrast, the energy and power densities of MT were as low as 0.2–0.02 Wh kg^{-1} and 162 W kg^{-1} – 1.16 kW kg^{-1}, respectively (Figure 13.23b).

13.3.3 Perspectives in Tannin-Based Electrodes

Other tannins, condensed or hydrolysable, can be used to produce microporous-mesoporous materials by using surfactants, which multiply the possibilities of using tannins as precursors of supercapacitor electrodes. Thus, chestnut tannin, which is a hydrolysable tannin, has been used to produce microporous-mesoporous carbons doped with nitrogen (Nelson et al., 2016). The interactions between Pluronic F127 and chestnut tannin (i.e., gallic and ellagic acids) produced hexagonal-type mesostructures such as those produced with mimosa tannin and F127. The porous carbon materials were activated with ammonia to increase the available surface area up to 750 m^2 g^{-1} and to incorporate nitrogen-containing functionalities.

Table 13.3 shows the electrochemical performances of different carbon electrodes, doped or not with heteroatoms, reported in the open literature. Table 13.4 summarizes the results reported in this chapter concerning tannin-derived carbon electrodes. Electrochemical performances obtained with tannin-based electrodes were as good as those obtained with precursors of petrochemical origin such as resorcinol (Tanaka et al., 2015; Wickramaratne et al., 2014). Soft-templating also allowed ordering at the mesoporous scale and hence electrochemical performances as good as those obtained by hard-templating (Jain et al., 2015; Jurewicz et al., 2004; Vix-Guterl et al., 2005). Supercapacitance being related to adsorption, the main strategy to improve tannin-based supercapacitors is increasing the surface area by chemical or physical activation. The highest surface areas reported for tannin-based electrodes were around 1390 m^2 g^{-1} (Amaral-Labat et al., 2012b) but those materials had a disordered mesostructure, and the electrochemical performances were not as good as those obtained by hard-templating, despite the fact that they had surface areas lower than 1100 m^2 g^{-1} (Sánchez-Sánchez et al., 2017a). Mesostructuration of heteroatoms-containing carbon precursors followed by activation to increase the surface area up to 1600–1700 m^2 g^{-1} seems to be a good option to improve electrochemical performances. Higher activation degrees would imply mesostructure destruction and would thus be counterproductive. Additional increasing of the microstructural ordering degree might improve the electrical conductivity, and indeed carbonization at temperatures higher than 900°C (Liang and Dai, 2009) or nickel doping (Szczurek et al., 2015b) have been

TABLE 13.3

Experimental Conditions and Electrochemical Data Reported in Previous Works

Precursor	S_{BET} (m² g⁻¹)	V_{mes} (cm³ g⁻¹)	Heteroatom Concentration	Electrochemical Measurements	Electrochemical Performance: C (F g⁻¹); E (Wh kg⁻¹); P (W kg⁻¹)	Reference
Non-doped OMCs:						
Sucrose/SBA-15	1470	a	a	Two-electrode 1M H_2SO_4	C: 167[b]	Vix-Guterl et al. (2005)
Pitch/ SBA-15	923				C: 87[b]	
Propylene CVD/SBA-15	713				C: 66[b]	
Non-doped OMCs:						
Sucrose/SBA-15	1470	a	a	Two-electrode 1M H_2SO_4	C: 162 (2 mV s⁻¹)	Jurewicz et al. (2004)
Propylene CVD/SBA-15	713				C: 62 (2 mV s⁻¹)	
Non-doped OMCs:						
Sucrose/SBA-15, CO_2 activation	2749	1.13	—	Three-electrode 6M KOH	C: 223 (2 mV s⁻¹)	Xia et al. (2008)
Non-doped OMCs:						
Furfuryl alcohol/SBA-15	1703	a	—	Three-electrode 30 wt% KOH	206.2 (at 5 mV s⁻¹); 4–6; 800–1050	Xing et al. (2006)
Commercial activated carbon (Maxsorb)	3310				333.9 (at 5 mV s⁻¹); 9.4–2.1; 190–410	
N, O, S—OMCs, pyrrole and KIT-6 silica template	693	0.69	10.1 at.% N / 4.4 at.% O / 0.9 at.% S	Three-electrode 2M KOH	C: 320 (1 A g⁻¹)	Jain et al. (2015)
Ethylenediamine, resorcinol, formaldehyde/CO_2 activation	1184	a	3.6 wt% N	Three-electrode 1M H_2SO_4	C: 388 (at 1 A g⁻¹)	Wickramaratne et al. (2014)
Oxygen-doped OMCs:						
Soft-templating of resorcinol, phloroglucinol, formaldehyde and Pluronic F127	430	0.36	0.0749 mmol g⁻¹ (phenols+lactones +carboxylic acids)	Three-electrode 1M H_2SO_4	270 (at 0.1 A g⁻¹); 17.4–25.8; 53.9–5230	Tanaka et al. (2015)
Oxidation with HNO_3						
Potassium gluconate, melamine	950	0.03	5.9 wt% N / 11.6 wt% O	Two-electrode 1M H_2SO_4	186 (at 0.1 A g⁻¹); 10.2; 5.7	Fuertes et al. (2014) and Jain et al. (2015)
Coconut shells/H_2O_2 and $ZnCl_2$ pre-treatments/CO_2 activation	2440	c	a	Three-electrode 1M H_2SO_4	246 (at 0.25 A g⁻¹); 7.6; 4.5	Jain et al. (2015)

a   Data not provided in the publications.

b   Average values between GCD and EIS measurements (at 1 mHz), expressed for a single electrode.

c   Mesopore surface area of 1100 m² g⁻¹.

TABLE 13.4
Experimental Conditions and Electrochemical Data Reported in this Work

Precursor	S_{BET} (m²/g)	V_{mes} (cm³/g)	Heteroatom Concentration	Electrochemical Measurements	Electrochemical Performance: C (F/g); E (Wh/Kg); P (W/Kg)	Reference
Carbon cryogels derived from tannin-resorcinol-formaldehyde	1390	0.08	[a]	Two- and-three electrode cell 4M H_2SO_4	124 F g⁻¹ (three-electrode, 2 mV s⁻¹)	Amaral-Labat et al. (2012b)
Nitrogen-doped mesoporous disordered carbons	442–684	0.01–0.45	1.9–8.0 wt% N	Three- and two-electrode 4M H_2SO_4	181–322 F g⁻¹ (three-electrode, 2 mV s⁻¹)	Braghiroli et al. (2012)
N- and/or O-doped carbons obtained from pine tannins by HTC	336–729	0.36–0.40	3.86–7.96 mmol g⁻¹ (N+O)	Three- or two-electrode 1M H_2SO_4	252–321 F g⁻¹ (0.5 mV s⁻¹, pine tannins extracted with water) 6.7–25.22 Wh kg⁻¹; 223.6 W kg⁻¹ –19.3 kW kg⁻¹	Sánchez-Sánchez et al. (2017b)
OMCs obtained by hard-templating of tannins:						
Gallic acid/SBA-15	1045	0.82	3.3 wt% O	Three- or two-electrode 1M H_2SO_4	277, 254, 237 and 242 F g⁻¹ (GaC, PhC, CatC and TanC; 0.5 mV s⁻¹) 4.2–14.1 ± 0.8 Wh kg⁻¹; 200 W kg⁻¹–22.1 kW kg⁻¹	Sánchez-Sánchez et al. (2017a)
Phloroglucinol/SBA-15	1094	0.92	4.9 wt% O			
Catechin/SBA-15	917	0.65	4.1 wt% O			
Mimosa condensed tannin/SBA-15	1006	1.15	2.8 wt% O			
OMCs obtained by soft templating of tannins:						
Mimosa tannin/Pluronic F127	661	0.34	3.1 at. % O	Three-electrode 1M H_2SO_4	126 F g⁻¹ (0.5 mV s⁻¹)	Unpublished results

[a] Data not provided in the publications.

shown to be efficient for increasing the graphitization level of such carbon materials. Furthermore, nickel would give enhanced capacitance by pseudocapacitance reactions (Yaddanapudi et al., 2016).

13.4   CONCLUSION

Tannins are biosourced phenolic molecules having similar or higher reactivity than those of petrochemical origin, such as resorcinol or phenol, which are commonly used as precursors for supercapacitor electrodes. The electrochemical performances of tannin-based electrodes have been considerably improved by using hydrothermal treatments before carbonization as well as by softtemplating; both strategies allowed obtaining mesoporous materials that are also "greener" than those derived from hard-templating. Moreover, hydrothermal treatments allow the stabilization of heteroatoms in the carbon matrix during carbonization, and these heteroatoms account for Faradaic reactions that increase the electrodes' performances through pseudocapacitance. Soft-templating allows mesostructuration, avoiding any hydrothermal process as well as the use of any kind of cross-linker (such as formaldehyde or hexamethylenetetramine), organic solvent (e.g., ethanol or THF), or any other toxic reagent. Further improvements are expected by increasing the carbon surface area, through physical or chemical activation, and by increasing the graphitization level in order to improve electrode electrical conductivity.

ACKNOWLEDGMENTS

The authors gratefully acknowledge the financial support of the CPER 2007–2013 "Structuration du Pôle de Compétitivité Fibres Grand'Est" (Competitiveness Fibre Cluster), through local (Conseil Général des Vosges), regional (Région Lorraine), national (DRRT and FNADT), and European (FEDER) funds. Part of this work was supported by CHEERS project (FEDER funds). Dr. Angela Sánchez-Sánchez acknowledges the University of Lorraine, the Region Lorraine, and the CNRS for financing her post-Doctoral contract.

REFERENCES

Ali, I., M. Asim, and T. A. Khan. 2012. Low cost adsorbents for the removal of organic pollutants from wastewater. *Journal of Environmental Management* 113: 170–83. doi:10.1016/j.jenvman.2012.08.028.

Amaral-Labat, G., L. I. Grishechko, V. Fierro, B. N. Kuznetsov, A. Pizzi, and A. Celzard. 2013a. Tannin-based xerogels with distinctive porous structures. *Biomass and Bioenergy* 56 (September): 437–45. doi:10.1016/j.biombioe.2013.06.001.

Amaral-Labat, G., L. Grishechko, A. Szczurek, V. Fierro, A. Pizzi, B. Kuznetsov, and A. Celzard. 2012a. Highly mesoporous organic aerogels derived from soy and tannin. *Green Chemistry* 14 (11): 3099–3106. doi:10.1039/C2GC36263E.

Amaral-Labat, G., A. Szczurek, V. Fierro, A. Pizzi, and A. Celzard. 2013b. Systematic studies of tannin–formaldehyde aerogels: Preparation and properties. *Science and Technology of Advanced Materials* 14 (1): 015001. doi:10.1088/1468–6996/14/1/015001.

Amaral-Labat, G., A. Szczurek, V. Fierro, N. Stein, C. Boulanger, A. Pizzi, and A. Celzard. 2012b. Pore structure and electrochemical performances of tannin-based carbon cryogels. *Biomass and Bioenergy*, 39: 274–82. doi:10.1016/j.biombioe.2012.01.019.

Amaral-Labat, G., C. Zollfrank, A. Ortona, S. Pusterla, A. Pizzi, V. Fierro, and A. Celzard. 2013c. Structure and oxidation resistance of micro-cellular Si–SiC foams derived from natural resins. *Ceramics International* 39 (2): 1841–51. doi:10.1016/j.ceramint.2012.08.032.

Aricò, A. S., P. Bruce, B. Scrosati, J.-M. Tarascon, and W. van Schalkwijk. 2005. Nanostructured materials for advanced energy conversion and storage devices. *Nature Materials* 4 (5): 366–77. doi:10.1038/nmat1368.

Aron, P. M. and J. A. Kennedy. 2008. Flavan-3-Ols: Nature, occurrence and biological activity. *Molecular Nutrition and Food Research* 52 (1): 79–104. doi:10.1002/mnfr.200700137.

Basso, M. C., S. Giovando, A. Pizzi, A. Celzard, and V. Fierro. 2013. Tannin/furanic foams without blowing agents and formaldehyde. *Industrial Crops and Products* 49: 17–22. doi:10.1016/j.indcrop.2013.04.043.

Basso, M. C., X. Li, V. Fierro, A. Pizzi, S. Giovando, and A. Celzard. 2011. Green, formaldehyde-free, foams for thermal insulation. *Advanced Materials Letters* 2 (6): 378–82. doi:10.5185/amlett.2011.4254.

Biesmans, G., D. Randall, E. Francais, and M. Perrut. 1998. Polyurethane-based organic aerogels' thermal performance. *Journal of Non-Crystalline Solids* 225: 36–40. doi:10.1016/S0022–3093(98)00103-3.

Braghiroli, F. L., V. Fierro, M. T. Izquierdo, J. Parmentier, A. Pizzi, and A. Celzard. 2012. Nitrogen-doped carbon materials produced from hydrothermally treated tannin. *Carbon* 50 (15): 5411–20. doi:10.1016/j.carbon.2012.07.027.

Braghiroli, F. L., V. Fierro, J. Parmentier, A. Pasc, and A. Celzard. 2016. Easy and eco-friendly synthesis of ordered mesoporous carbons by self-assembly of tannin with a block copolymer. *Green Chemistry* 18 (11): 3265–71. doi:10.1039/C5GC02788H.

Braghiroli, F. L., V. Fierro, A. Szczurek, P. Gadonneix, J. Ghanbaja, J. Parmentier, G. Medjahdi, and A. Celzard. 2017. Hydrothermal treatment of tannin: A route to porous metal oxides and metal/carbon hybrid materials. *Inorganics* 5 (1): 7. doi:10.3390/inorganics5010007.

Braghiroli, F. L., V. Fierro, A. Szczurek, N. Stein, J. Parmentier, and A. Celzard. 2015. Electrochemical performances of hydrothermal tannin-based carbons doped with nitrogen. *Industrial Crops and Products* 70: 332–40. doi:10.1016/j.indcrop.2015.03.046.

Celzard, A., A. Szczurek, P. Jana, V. Fierro, M. C. Basso, S. Bourbigot, M. Stauber, and A. Pizzi. 2014. Latest progresses in the preparation of tannin-based cellular solids. *Journal of Cellular Plastics* 51 (1): 89–102. doi:10.1177/0021955X14538273.

Chen, H., D. Liu, Z. Shen, B. Bao, S. Zhao, and L. Wu. 2015. Functional biomass carbons with hierarchical porous structure for supercapacitor electrode materials. *Electrochimica Acta* 180 (October): 241–51. doi:10.1016/j.electacta.2015.08.133.

Constabel, C. P., K. Yoshida, and V. Walker. 2014. Diverse ecological roles of plant tannins: Plant defense and beyond. In *Recent Advances in Polyphenol Research*, edited by A. Romani, V. Lattanzio, and S. Quideau, 115–42. John Wiley & Sons, Ltd, Chichester. doi:10.1002/9781118329634.ch5.

Conway, B. E. 1991. Transition from 'supercapacitor' to 'battery' behavior in electrochemical energy storage. *Journal of the Electrochemical Society* 138 (6): 1539–48. doi:10.1149/1.2085829.

de Souza, L. K. C., N. P. Wickramaratne, A. S. Ello, M. J. F. Costa, C. E. F. da Costa, and M. Jaroniec. 2013. Enhancement of CO_2 adsorption on phenolic resin-based mesoporous carbons by KOH activation. *Carbon* 65: 334–40. doi:10.1016/j.carbon.2013.08.034.

Feng, S., S. Cheng, Z. Yuan, M. Leitch, and C. Xu. 2013. Valorization of bark for chemicals and materials: A review. *Renewable and Sustainable Energy Reviews* 26 (October): 560–78. doi:10.1016/j.rser.2013.06.024.

Feng, S., W. Li, J. Wang, Y. Song, A. A. Elzatahry, Y. Xia, and D. Zhao. 2014. Hydrothermal synthesis of ordered mesoporous carbons from a biomass-derived precursor for electrochemical capacitors. *Nanoscale* 6 (24): 14657–61. doi:10.1039/C4NR05629A.

Ferrari, A. C. and J. Robertson. 2000. Interpretation of Raman spectra of disordered and amorphous carbon. *Physical Review B* 61 (20): 14095–107. doi:10.1103/PhysRevB.61.14095.

Finch, C. A. 1984. Wood adhesives: Chemistry and technology. Edited by A. Pizzi, Marcel Dekker, New York and Basel, 1983. ISBN 0814715799. *British Polymer Journal* 16 (4): 324–324. doi:10.1002/pi.4980160434.

Finch, C. A. 1996. Advanced wood adhesives technology. A. Pizzi. Marcel Dekker, New York, Basel, 1994. ISBN 0-8247-9266-1. *Polymer International* 39 (1): 78–78. doi:10.1002/pi.1996.210390117.

Frackowiak, E. and F. Béguin. 2001. Carbon materials for the electrochemical storage of energy in capacitors. *Carbon* 39 (6): 937–50. doi:10.1016/S0008–6223(00)00183-4.

Fuertes, A. B., G. A. Ferrero, and M. Sevilla. 2014. One-pot synthesis of microporous carbons highly enriched in nitrogen and their electrochemical performance. *Journal of Materials Chemistry* 2 (35): 14439–48. doi:10.1039/C4TA02959C.

Galati, G. and P. J. O'Brien. 2004. Potential toxicity of flavonoids and other dietary phenolics: Significance for their chemopreventive and anticancer properties. *Free Radical Biology and Medicine* 37 (3): 287–303. doi:10.1016/j.freeradbiomed.2004.04.034.

Gao, Z., Y. Zhang, N. Song, and X. Li. 2017. Biomass-derived renewable carbon materials for electrochemical energy storage. *Materials Research Letters* 5 (2): 69–88. doi:10.1080/21663831.2016.1250834.

Grishechko, L. I., G. Amaral-Labat, V. Fierro, A. Szczurek, B. N. Kuznetsov, and A. Celzard. 2016. Biosourced, highly porous, carbon xerogel microspheres. *RSC Advances* 6 (70): 65698–708. doi:10.1039/C6RA09462G.

Grishechko, L. I., G. Amaral-Labat, A. Szczurek, V. Fierro, B. N. Kuznetsov, A. Pizzi, and A. Celzard. 2013. New tannin–lignin aerogels. *Industrial Crops and Products* 41 (January): 347–55. doi:10.1016/j.indcrop.2012.04.052.

Hao, G.-P., W.-C. Li, S. Wang, G.-H. Wang, L. Qi, and A.-H. Lu. 2011. Lysine-assisted rapid synthesis of crack-free hierarchical carbon monoliths with a hexagonal array of mesopores. *Carbon* 49 (12): 3762–72. doi:10.1016/j.carbon.2011.05.010.

Harbertson, J. F., G. P. Parpinello, H. Heymann, and M. O. Downey. 2012. Impact of exogenous tannin additions on wine chemistry and wine sensory character. *Food Chemistry* 131 (3): 999–1008. doi:10.1016/j.foodchem.2011.09.101.

Hsieh, C.-T. and H. Teng. 2002. Influence of oxygen treatment on electric double-layer capacitance of activated carbon fabrics. *Carbon* 40 (5): 667–74. doi:10.1016/S0008-6223(01)00182-8.

Huang, Y., H. Cai, D. Feng, D. Gu, Y. Deng, B. Tu, H. Wang, P. A. Webley, and D. Zhao. 2008. One-step hydrothermal synthesis of ordered mesostructured carbonaceous monoliths with hierarchical porosities. *Chemical Communications* 23 (June): 2641–43. doi:10.1039/B804716B.

Jain, A., C. Xu, S. Jayaraman, R. Balasubramanian, J. Y. Lee, and M. P. Srinivasan. 2015. Mesoporous activated carbons with enhanced porosity by optimal hydrothermal pre-treatment of biomass for supercapacitor applications. *Microporous and Mesoporous Materials* 218: 55–61. doi:10.1016/j.micromeso.2015.06.041.

Jun, S., S. H. Joo, R. Ryoo, M. Kruk, M. Jaroniec, Z. Liu, T. Ohsuna, and O. Terasaki. 2000. Synthesis of new, nanoporous carbon with hexagonally ordered mesostructure. *Journal of the American Chemical Society* 122 (43): 10712–13. doi:10.1021/ja002261e.

Jurewicz, K., C. Vix-Guterl, E. Frackowiak, S. Saadallah, M. Reda, J. Parmentier, J. Patarin, and F. Béguin. 2004. Capacitance properties of ordered porous carbon materials prepared by a templating procedure. *Journal of Physics and Chemistry of Solids* 65 (2–3): 287–93. doi:10.1016/j.jpcs.2003.10.024.

Kruk, M., M. Jaroniec, R. Ryoo, and S. H. Joo. 2000. Characterization of ordered mesoporous carbons synthesized using MCM-48 silicas as templates. *The Journal of Physical Chemistry B* 104 (33): 7960–68. doi:10.1021/jp000861u.

Kubo, S., R. J. White, N. Yoshizawa, M. Antonietti, and M.-M. Titirici. 2011. Ordered carbohydrate-derived porous carbons. *Chemistry of Materials* 23 (22): 4882–85. doi:10.1021/cm2020077.

Lacoste, C., M. C. Basso, A. Pizzi, M.-P. Laborie, A. Celzard, and V. Fierro. 2013. Pine tannin-based rigid foams: Mechanical and thermal properties. *Industrial Crops and Products* 43: 245–50. doi:10.1016/j.indcrop.2012.07.039.

Li, X., M. C. Basso, F. L. Braghiroli, V. Fierro, A. Pizzi, and A. Celzard. 2012a. Tailoring the structure of cellular vitreous carbon foams. *Carbon* 50 (5): 2026–36. doi:10.1016/j.carbon.2012.01.004.

Li, X., M. C. Basso, V. Fierro, A. Pizzi, and A. Celzard. 2012b. Chemical modification of tannin/furanic rigid foams by isocyanates and polyurethanes. *Maderas. Ciencia Y Tecnología* 14 (3): 257–65. doi:10.4067/S0718-221X2012005000001.

Li, S., A. Pasc, V. Fierro, and A. Celzard. 2016. Hollow carbon spheres, synthesis and applications—A review. *Journal of Materials Chemistry* 4 (33): 12686–713. doi:10.1039/C6TA03802F.

Li, X., A. Pizzi, C. Lacoste, V. Fierro, and A. Celzard. 2012c. Physical properties of tannin/furanic resin foamed with different blowing agents. *BioResources* 8 (1): 743–52. doi:10.15376/biores.8.1.743–752.

Li, X., A. Pizzi, X. Zhou, V. Fierro, and A. Celzard. 2015. Formaldehyde-free prorobitenidin/profi setinidin tannin/furanic foams based on alternative aldehydes: Glyoxal and glutaraldehyde. *Journal of Renewable Materials* 3 (2): 142–50. doi:10.7569/JRM.2014.634117.

Liang, C. and S. Dai. 2006. Synthesis of mesoporous carbon materials via enhanced hydrogen-bonding interaction. *Journal of the American Chemical Society* 128 (16): 5316–17. doi:10.1021/ja060242k.

Liang, C. and S. Dai. 2009. Dual phase separation for synthesis of bimodal meso-/macroporous carbon monoliths. *Chemistry of Materials* 21 (10): 2115–24. doi:10.1021/cm900344h.

Liang, C., Z. Li, and S. Dai. 2008. Mesoporous carbon materials: Synthesis and modification. *Angewandte Chemie International Edition* 47 (20): 3696–3717. doi:10.1002/anie.200702046.

Lin, C. and J. A. Ritter. 1997. Effect of synthesis pH on the structure of carbon xerogels. *Carbon* 35 (9): 1271–78. doi:10.1016/S0008-6223(97)00069-9.

Long, C., J. Zhuang, Y. Xiao, M. Zheng, H. Hu, H. Dong, B. Lei, H. Zhang, and Y. Liu. 2016. Nitrogen-doped porous carbon with an ultrahigh specific surface area for superior performance supercapacitors. *Journal of Power Sources* 310: 145–53. doi:10.1016/j.jpowsour.2016.01.052.

Ma, T.-Y., L. Liu, and Z.-Y. Yuan. 2013. Direct synthesis of ordered mesoporous carbons. *Chemical Society Reviews* 42 (9): 3977–4003. doi:10.1039/C2CS35301F.

Nasini, U. B., V. G. Bairi, S. K. Ramasahayam, S. E. Bourdo, T. Viswanathan, and A. U. Shaikh. 2014. Phosphorous and nitrogen dual heteroatom doped mesoporous carbon synthesized via microwave method for supercapacitor application. *Journal of Power Sources* 250 (March): 257–65. doi:10.1016/j.jpowsour.2013.11.014.

Nelson, K. M., S. M. Mahurin, R. T. Mayes, B. Williamson, C. M. Teague, A. J. Binder, L. Baggetto, G. M. Veith, and S. Dai. 2016. Preparation and CO_2 adsorption properties of soft-templated mesoporous carbons derived from chestnut tannin precursors. *Microporous and Mesoporous Materials* 222 (March): 94–103. doi:10.1016/j.micromeso.2015.09.050.

Pandolfo, A. G. and A. F. Hollenkamp. 2006. Carbon properties and their role in supercapacitors. *Journal of Power Sources* 157 (1): 11–27. doi:10.1016/j.jpowsour.2006.02.065.

Pizzi, A. 1992. Tannin structure and the formulation of tannin-based wood adhesives. In *Plant Polyphenols*, edited by R.W. Hemingway and P.E. Laks, 991–1003. *Basic Life Sciences* 59, Springer, Boston, MA. doi:10.1007/978-1-4615-3476-1_60.

Pizzi, A. and K. L. Mittal. 2003. *Handbook of Adhesive Technology, Revised and Expanded,* Marcel Dekker, New York. https://www.crcpress.com/Handbook-of-Adhesive-Technology-Revised-and-Expanded/Pizzi-Mittal/p/book/9780824709860.

Pizzi, A., H. Pasch, A. Celzard, and A. Szczurek. 2013. Oligomers distribution at the gel point of tannin–formaldehyde thermosetting adhesives for wood panels. *Journal of Adhesion Science and Technology* 27 (18–19): 2094–2102. doi:10.1080/01694243.2012.697669.

Rey-Raap, N., A. Szczurek, V. Fierro, A. Celzard, J. A. Menéndez, and A. Arenillas. 2016. Advances in tailoring the porosity of tannin-based carbon xerogels. *Industrial Crops and Products* 82 (April): 100–106. doi:10.1016/j.indcrop.2015.12.001.

Rey-Raap, N., A. Szczurek, V. Fierro, J. A. Menéndez, A. Arenillas, and A. Celzard. 2015. Towards a feasible and scalable production of bio-xerogels. *Journal of Colloid and Interface Science* 456 (October): 138–44. doi:10.1016/j.jcis.2015.06.024.

Sánchez-Martín, J., M. González-Velasco, J. Beltrán-Heredia, J. Gragera-Carvajal, and J. Salguero-Fernández. 2010. Novel tannin-based adsorbent in removing cationic dye (methylene blue) from aqueous solution. Kinetics and equilibrium studies. *Journal of Hazardous Materials* 174 (1–3): 9–16. doi:10.1016/j.jhazmat.2009.09.008.

Sánchez-Sánchez, A., V. Fierro, M. T. Izquierdo, and A. Celzard. 2016a. Functionalized, hierarchical and ordered mesoporous carbons for high-performance supercapacitors. *Journal of Materials Chemistry A* 4 (16): 6140–48. doi:10.1039/C6TA00738D.

Sánchez-Sánchez, A., M. T. Izquierdo, J. Ghanbaja, G. Medjahdi, S. Mathieu, A. Celzard, and V. Fierro. 2017a. Excellent electrochemical performances of nanocast ordered mesoporous carbons based on tannin-related polyphenols as supercapacitor electrodes. *Journal of Power Sources* 344 (March): 15–24. doi:10.1016/j.jpowsour.2017.01.099.

Sánchez-Sánchez, A., M. T. Izquierdo, S. Mathieu, J. González-Álvarez, A. Celzard, and V. Fierro. 2017b. Outstanding electrochemical performances of highly N- and O-doped carbons derived from pine tannin. *Green Chemistry* 19: 2653–665. doi: 10.1039/C7GC00491E.

Sánchez-Sánchez, A., A. Martinez de Yuso, F. L. Braghiroli, M. T. Izquierdo, E. D. Alvarez, E. Pérez-Cappe, Y. Mosqueda, V. Fierro, and A. Celzard. 2016b. Sugarcane molasses as a pseudocapacitive material for supercapacitors. *RSC Advances* 6 (91): 88826–36. doi:10.1039/C6RA16314A.

Schlienger, S., A.-L. Graff, A. Celzard, and J. Parmentier. 2012. Direct synthesis of ordered mesoporous polymer and carbon materials by a biosourced precursor. *Green Chemistry* 14 (2): 313–16. doi:10.1039/C2GC16160E.

Sealy-Fisher, V. J., and A. Pizzi. 1992. Increased pine tannins extraction and wood adhesives development by phlobaphenes minimization. *Holz Als Roh- Und Werkstoff* 50 (5): 212–20. doi:10.1007/BF02663290.

Seredych, M., A. Szczurek, V. Fierro, A. Celzard, and T. J. Bandosz. 2016. Electrochemical reduction of oxygen on hydrophobic ultramicroporous PolyHIPE carbon. *ACS Catalysis* 6 (8): 5618–28. doi:10.1021/acscatal.6b01497.

Sreeram, K.J and T Ramasami. 2003. Sustaining tanning process through conservation, recovery and better utilization of chromium. *Resources, Conservation and Recycling* 38 (3): 185–212. doi:10.1016/S0921-3449(02)00151-9.

Stokke, D. D., Q. Wu, and G. Han. 2013. Adhesives used to bond wood and lignocellulosic composites. In D. D. Stokke, Q. Wu, G. Han, and C. V. Stevens (Series Editor) *Introduction to Wood and Natural Fiber Composites*, 169–207. John Wiley & Sons Ltd, Chichester, UK. doi:10.1002/9780470711804.ch6.

Szczurek, A., G. Amaral-Labat, V. Fierro, A. Pizzi, and A. Celzard. 2011a. The use of tannin to prepare carbon gels. Part II. Carbon cryogels. *Carbon* 49 (8): 2785–94. doi:10.1016/j.carbon.2011.03.005.

Szczurek, A., G. Amaral-Labat, V. Fierro, A. Pizzi, E. Masson, and A. Celzard. 2011b. The use of tannin to prepare carbon gels. Part I: Carbon aerogels. *Carbon* 49 (8): 2773–84. doi:10.1016/j.carbon.2011.03.007.

Szczurek, A., V. Fierro, A. Pizzi, and A. Celzard. 2013. Mayonnaise, whipped cream and meringue, a new carbon cuisine. *Carbon* 58: 245–48. doi:10.1016/j.carbon.2013.02.056.

Szczurek, A., V. Fierro, A. Pizzi, and A. Celzard. 2014a. Emulsion-templated porous carbon monoliths derived from tannins. *Carbon* 74: 352–62. doi:10.1016/j.carbon.2014.03.047.

Szczurek, A., V. Fierro, A. Pizzi, M. Stauber, and A. Celzard. 2014b. A new method for preparing tannin-based foams. *Industrial Crops and Products* 54 (March): 40–53. doi:10.1016/j.indcrop.2014.01.012.

Szczurek, A., V. Fierro, M. Thébault, A. Pizzi, and A. Celzard. 2016. Structure and properties of poly(furfuryl alcohol)-tannin polyHIPEs. *European Polymer Journal* 78: 195–212. doi:10.1016/j.eurpolymj.2016.03.037.

Szczurek, A., A. Martinez de Yuso, V. Fierro, A. Pizzi, and A. Celzard. 2015a. Tannin-based monoliths from emulsion-templating. *Materials and Design* 79: 115–26. doi:10.1016/j.matdes.2015.04.020.

Szczurek, A., A. Ortona, L. Ferrari, E. Rezaei, G. Medjahdi, V. Fierro, D. Bychanok, P. Kuzhir, and A. Celzard. 2015b. Carbon periodic cellular architectures. *Carbon* 88: 70–85. doi:10.1016/j.carbon.2015.02.069.

Tanaka, S., H. Fujimoto, J. F. M. Denayer, M. Miyamoto, Y. Oumi, and Y. Miyake. 2015. Surface modification of soft-templated ordered mesoporous carbon for electrochemical supercapacitors. *Microporous and Mesoporous Materials* 217 (November): 141–49. doi:10.1016/j.micromeso.2015.06.017.

Tanaka, S., N. Nakatani, A. Doi, and Y. Miyake. 2011. Preparation of ordered mesoporous carbon membranes by a soft-templating method. *Carbon* 49 (10): 3184–89. doi:10.1016/j.carbon.2011.03.042.

Tondi, G., V. Fierro, A. Pizzi, and A. Celzard. 2009a. Tannin-based carbon foams. *Carbon* 47 (6): 1480–92. doi:10.1016/j.carbon.2009.01.041.

Tondi, G., W. Zhao, A. Pizzi, G. Du, V. Fierro, and A. Celzard. 2009b. Tannin-based rigid foams: A survey of chemical and physical properties. *Bioresource Technology* 100 (21): 5162–69. doi:10.1016/j.biortech.2009.05.055.

Vix-Guterl, C., E. Frackowiak, K. Jurewicz, M. Friebe, J. Parmentier, and F. Béguin. 2005. Electrochemical energy storage in ordered porous carbon materials. *Carbon* 43 (6): 1293–1302. doi:10.1016/j.carbon.2004.12.028.

Wang, H., Z. Li, J. K. Tak, C. M. B. Holt, X. Tan, Z. Xu, B. S. Amirkhiz et al. 2013. Supercapacitors based on carbons with tuned porosity derived from paper pulp mill sludge biowaste. *Carbon* 57: 317–28. doi:10.1016/j.carbon.2013.01.079.

Wang, X., C. Liang, and S. Dai. 2008. Facile synthesis of ordered mesoporous carbons with high thermal stability by self-assembly of resorcinol–formaldehyde and block copolymers under highly acidic conditions. *Langmuir* 24 (14): 7500–7505. doi:10.1021/la800529v.

Wei, L., M. Sevilla, A. B. Fuertes, R. Mokaya, and G. Yushin. 2011. Hydrothermal carbonization of abundant renewable natural organic chemicals for high-performance supercapacitor electrodes. *Advanced Energy Materials* 1 (3): 356–61. doi:10.1002/aenm.201100019.

Wickramaratne, N. P., J. Xu, M. Wang, L. Zhu, L. Dai, and M. Jaroniec. 2014. Nitrogen enriched porous carbon spheres: Attractive materials for supercapacitor electrodes and CO_2 adsorption. *Chemistry of Materials* 26 (9): 2820–28. doi:10.1021/cm5001895.

Wu, K. and Q. Liu. 2016. Nitrogen-doped mesoporous carbons for high performance supercapacitors. *Applied Surface Science* 379: 132–39. doi:10.1016/j.apsusc.2016.04.064.

Xia, K., Q. Gao, J. Jiang, and J. Hu. 2008. Hierarchical porous carbons with controlled micropores and mesopores for supercapacitor electrode materials. *Carbon* 46 (13): 1718–26. doi:10.1016/j.carbon.2008.07.018.

Xing, W., S. Z. Qiao, R. G. Ding, F. Li, G. Q. Lu, Z. F. Yan, and H. M. Cheng. 2006. Superior electric double layer capacitors using ordered mesoporous carbons. *Carbon* 44 (2): 216–24. doi:10.1016/j.carbon.2005.07.029.

Yaddanapudi, H. S., K. Tian, S. Teng, and A. Tiwari. 2016. Facile preparation of nickel/carbonized wood nanocomposite for environmentally friendly supercapacitor electrodes. *Scientific Reports* 6 (September): 33659. doi:10.1038/srep33659.

Zhang, H., L. Zhang, J. Chen, H. Su, F. Liu, and W. Yang. 2016. One-step synthesis of hierarchically porous carbons for high-performance electric double layer supercapacitors. *Journal of Power Sources* 315: 120–26. doi:10.1016/j.jpowsour.2016.03.005.

Zhao, W., A. Pizzi, V. Fierro, G. Du, and A. Celzard. 2010. Effect of composition and processing parameters on the characteristics of tannin-based rigid foams. Part I: Cell structure. *Materials Chemistry and Physics* 122 (1): 175–82. doi:10.1016/j.matchemphys.2010.02.062.

14 Recent Contributions in the Development of Fuel Cell Technologies

José Luis Reyes-Rodríguez, Heriberto Cruz-Martínez,
Miriam Marisol Tellez-Cruz, Adrián Velázquez-Osorio,
and Omar Solorza-Feria
National Polytechnic Institute

CONTENTS

14.1 INTRODUCTION

Fossil fuels are the dominant energy sources of our modern society and will undoubtedly be depleted in the near future. Global energy production is almost entirely associated with the use of highly contaminant fossil fuels, responsible for greenhouse emissions that contribute to climate change. In order to diminish the impact of human activities related to conventional energy production, we must transition to more environmentally friendly sources such as sustainable renewable energy (IPCC, 2012; EIA, 2016). However, the development of renewable energy technologies such as solar, wind, and geothermal is not without challenges; much work is needed to solve problems related to production, storage, and use. In combination, renewable systems could transform primary renewable energy into chemical energy like solar fuels or hydrogen as proposed by artificial photosynthesis (Listorti et al., 2009; Vlcek, 2012; Seh et al., 2017) to be later used for power generation in fuel cells, which have great potential for direct conversion of chemical energy into electricity. Such compound systems would effectively address some of the challenges that have hindered the transition of fossil to renewable energy and aid in environmental protection (Dunn et al., 2011; Shao, 2015).

Hydrogen, a clean energy source, with the highest specific energy density of all fuels, is the best alternative for replacing fossil fuels when used in fuel cells. Fuel cells are electrochemical devices that efficiently convert diverse fuels into electricity and are key elements in the construction of a reliable and sustainable clean energy economy. This technology has drawn significant attention due to its simplicity, viability, and fast start-up and shut-off times and because its versatility allows for virtually unlimited applications in transportation, material-handling equipment, and backup power storage (Bayindir et al., 2011; Pasupathi et al., 2016; Wang et al., 2016; Wilberforce et al., 2016).

The chapter is organized as follows: the first section covers the importance of a hydrogen economy. The second section discusses general thermodynamics of fuel cells. The third section shows the operating principle, components, and perspectives of different types of available fuel cells. Finally, the last section offers a summary of the advances and prospective of this technology.

14.2 HYDROGEN ECONOMY

Until now, global energy needs have revolved around a "fossil fuel economy"; gasoline, coal, natural gas, etc., have been the primary energy sources, determining the types of transportation, consumption, and development possible. This fossil fuel economy has led to global instability as diminishing resource availability has pushed energy prices to ever-higher levels, resulting in political and military strife. The consequences of the extraction and use of fossil fuels on the environment cannot be understated; climate change, oil spills, and social unrest are merely some examples of the long-term damage caused by the fossil fuel economy.

The scientific breakthroughs of the twenty-first century now allow the possibility of evolving from a fossil fuel economy to a "hydrogen economy." The term hydrogen economy refers to a global energy system reliant on the use of hydrogen as the main energy source and offers the possibility of a sustainable global economy. The use of hydrogen as an energy carrier offers many benefits, including its great abundance, simple production, and storage or transportation in metal hydrides for conversion into electricity at relatively high efficiencies without the harmful emissions associated with fossil fuels (Sherif et al., 2014).

Hydrogen is the basis of a truly sustainable energy system, but is not available in its natural state and must be manufactured using hydrogen-rich compounds as raw material (Morgan, 2006; Jamesh, 2016; Lamy, 2016). Water decomposition enables large-scale hydrogen production; methods developed for this type of production are thermolysis, photolysis from biomass, thermochemical processes, and the most commonly used method: electrolysis (Sørensen, 2005). Water electrolysis has efficiencies between 72% and 82% and can provide storage options for intermittently produced electricity from other renewable technologies such as wind and solar.

Water electrolysis is conducted by splitting water molecules into hydrogen and oxygen through the application of an external electrical current. There are three main approaches for electrochemical water splitting: (1) alkaline electrolysis, which uses a liquid electrolyte at temperatures below 100°C; (2) electrolysis by solid polymer membrane electrolyzers, which use an acidic solid ionomer electrolyte at temperatures below 100°C; (3) electrolysis by steam electrolyzers, which use a solid oxide electrolyte at temperatures around 700°C–1000°C (Anantharaj et al., 2016).

The decomposition reaction of water into its basic components (oxygen and hydrogen) is an endothermic process that requires electricity and heat as described by Equation 14.1:

$$H_2O + 237 \text{ kJ mol}^{-1} + 48.6 \text{ kJ mol}^{-1} \rightarrow H_2 + \tfrac{1}{2} O_2 + 2e^- \tag{14.1}$$

deduced from the enthalpy of water electrolysis $\Delta H = \Delta G + \Delta Q$, where ΔG is the change in Gibbs free energy, and ΔQ is the heat energy exchanged under standard conditions (Smolinka et al., 2015). A reversible potential $E_r = 1.23$ V and a thermoneutral potential $E_t = 1.48$ V for adiabatic conditions are derived by considering either the change in Gibbs free energy or the reaction enthalpy, respectively. Hence, the thermoneutral value is the minimum voltage required for electrolysis to take place; some materials like molybdenum chalcogenides are promising electrocatalysts for electrochemical water decomposition.

Small volume and low weight materials capable of storing large amounts of hydrogen are required for the development of hydrogen-powered vehicles and other mobile applications. Hydrogen storage in solids is the subject of worldwide research; challenges in this area include high dehydrogenation temperature, reversibility, and thermal management (Aricó et al., 2005; Zhang et al., 2013). The use of nanostructured materials is advantageous, as it offers significant surface-to-volume ratios and can affect the dehydrogenation temperature and manipulate the reaction mechanisms. Optimally designed and fabricated nanocomposites are desired for their thermal conductivity, which mediates the absorption or release of heat during the hydrogenation or dehydrogenation process.

14.3 THERMODYNAMIC ASPECTS OF THE FUEL CELLS

Fuel cell technology requires an energy balance analysis that should take into account energy conversion processes, electrochemical reactions, heat loss, etc. (Pilatowsky et al., 2011). The reaction in a fuel cell corresponds to a chemical process that is divided into two electrochemical half-cell reactions. The overall reaction in most fuel cells is $H_2 + \frac{1}{2}O_2 \rightarrow H_2O$.

From the thermodynamic point of view, the maximum electric work obtained from this reaction corresponds to the Gibbs free energy (Srinivasan et al., 2006; Vielstich, 2010; Pilatowsky et al., 2011). This is a spontaneous reaction and is thermodynamically favored because the free energy of the products is less than that of the reactants (Srinivasan et al., 2006). The standard free energy change of the fuel cell reaction is represented by the Equation 14.2:

$$\Delta G = -nFE_r \tag{14.2}$$

where n is the number of electrons involved, E_r is the reversible potential, and F is Faraday's constant.

The enthalpy change (ΔH) of a fuel cell reaction represents the entire heat released by the reaction at constant pressure. The cell potential based on ΔH is defined in terms of the thermoneutral potential (E_t):

$$\Delta H = -nFE_t \tag{14.3}$$

where E_t has a value of 1.48 V for the electrochemical reaction (Pilatowsky et al., 2011). The operating efficiencies of fuel cells are lower than the expected theoretical values due to activation, ohmic, and mass transport overpotentials. All available energy from the reaction can be converted into electrical energy in an ideal process. The theoretical efficiency of the fuel cell can be calculated as

$$\varepsilon_r = (\Delta G / \Delta H) * 100 \tag{14.4}$$

where ΔH is the total energy available in the process, and ΔG is the theoretical energy that can be converted into electrical energy (Srinivasan et al., 2006). Table 14.1 summarizes thermodynamic parameters and theoretical efficiencies for a hydrogen–oxygen fuel cell operating under standard conditions.

TABLE 14.1
Thermodynamic Data for Fuel Cell Reactions and Theoretical Efficiency

Parameter	Value
Gibbs free energy	−237 kJ/mol
Enthalpy	−286 kJ/mol
Reversible potential	1.23 V
Theoretical efficiency	83%

14.4 TYPES OF FUEL CELLS

There are at least six types of fuel cells suitable for different power demands, namely: polymer electrolyte membrane fuel cells (PEMFCs), direct alcohol fuel cells (DAFCs), alkaline fuel cells (AFCs), phosphoric acid fuel cells (PAFCs), molten carbonate fuel cells (MCFCs), and solid oxide fuel cells (SOFCs). The characteristics of each of these cells are summarized in Table 14.2.

14.4.1 POLYMER ELECTROLYTE MEMBRANE FUEL CELLS

PEMFCs are electrochemical devices that generate electricity from oxidation–reduction reactions at the solid-electrolyte interfaces. The heart of this fuel cell is the membrane electrode assembly (MEA), constituted by two catalytic materials that act as anodic and cathodic electrodes, respectively. Both electrodes are separated by a proton exchange polymer membrane (PEM) that works as a solid electrolytic barrier between the electrodes to transfer protons (H^+) from the anode to the cathode (U.S. Department of Energy, Office of Energy Efficiency and Renewable Energy, Fuel Cells website).

In PEMFCs, hydrogen is fed to the anode and carries out an oxidation process on the catalytic surface of the electrode, releasing electrons that form an electric current, which is channeled by an external circuit for a specific application. Protons generated at the anode diffuse to the cathode through the PEM via a proton exchange mechanism where they combine with pure oxygen, either fed to the system or taken from the air, enabling a reduction process on the catalytic surface to form water and heat as waste products (see Figure 14.1).

PEMFCs experience less electrode corrosion thanks to the use of a solid electrolyte. These cells operate efficiently at temperatures around 20°C–100°C, making them ideal for applications in transport systems (with efficiencies around 60%), stationary power systems (with efficiencies around 35%), and portable low- and medium-power devices (1–250 kW) (Sundmacher, 2010). The use of costly nanoparticulated noble metals, necessary to reduce their high operation overpotential, is the main disadvantage of this type of cell (Hamnett, 2010).

Polymer membranes with structures that allow only protonic transfers, with high mechanical strength, low swelling, and low gas and fuel permeability are essential for the cell (Houchins et al., 2012). Nafion (perfluorosulfonated polymeric ionomer), is the most commonly used membrane in

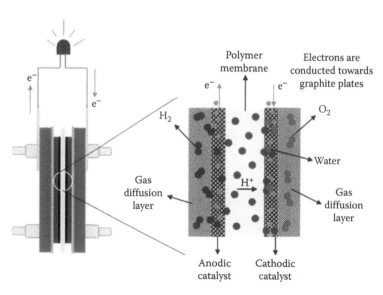

FIGURE 14.1 Representative scheme of a PEMFC.

TABLE 14.2

Types of Fuel Cells and Some of Their Main Characteristics

Fuel Cell Type	Common Electrolyte/ Charge Carrier	Cathode Reaction Anode	Operating Temperature	Typical Stack Size	Efficiency	Applications	Advantages	Challenges
Polymer electrolyte membrane (PEMFC)	Perfluoro sulfonic acid/H^+	$H_2 \rightarrow 2H^+ + 2e^-$ $1/2O_2 + 2H^+ + 2e^- \rightarrow H_2O$	50°C–100°C 122°F–212°F	1–250 kW	60% transportation 35% stationary	• Backup power • Portable power • Distributed generation • Transportation • Speciality vehicles	• Solid electrolyte reduces corrosion and electrolyte management problems • Low temperature • Quick start-up	• Expensive catalysts • Sensitive to fuel impurities
Direct alcohol (DAFC)[a]		$CH_3OH + 6OH^- \rightarrow CO_2 + 5H_2O + 6e^-$ $2/3O_2 + H_2O + 6e^- \rightarrow 6OH^-$						
Alkaline (AMFC)	Aqueous KOH/OH^-	$H_2 + 2OH^- \rightarrow 2H_2O + 2e^-$ $1/2O_2 + H_2O + 2e^- \rightarrow 2OH^-$	90°C–100°C 19°F–212°F	10–100 kW	60%	• Military • Space	• Cathode reaction faster in alkaline electrolyte, leads to high performance • Low cost components	• Sensitive to CO_2 in fuel and air • Electrolyte management
Phosphoric acid (PAFC)	H_3PO_4 soaked in a matrix, Acid Polymer/H^+	$H_2 \rightarrow 2H^+ + 2e^-$ $1/2O_2 + 2H^+ + 2e^- \rightarrow H_2O$	150°C–200°C 302°F–392°F	400 kW 100 kW module	40%	• Distributed generation	• High temperatures • Increased tolerance to fuel impurities	• Pt catalyst • Long start up time • Sulfur sensitivity

(Continued)

TABLE 14.2 (*Continued*)
Types of Fuel Cells and Some of Their Main Characteristics

Fuel Cell Type	Common Electrolyte/ Charge Carrier	Cathode Reaction Anode	Operating Temperature	Typical Stack Size	Efficiency	Applications	Advantages	Challenges
Molten carbonate (MCFC)	Solution of lithium, sodium and/or potassium carbonates soaked in a matrix/CO_3^{2-}	$H_2 + CO_3^{2-} \rightarrow H_2O + CO_2 + 2e^-$ $1/2O_2 + CO_2 + 2e^- \rightarrow CO_3^{2-}$	600°C–700°C 1112°F–1292°F	300 kW–3 MW 300 kW module	50%–60%	• Electric utility • Distributed generation	• High efficiency • Fuel flexibility • Can use a variety of catalysts • Solid electrolyte	• High temperature corrosion and breakdown of cell components • Long start up time • Low power density
Solid oxide (SOFC)	Yttria-Stabilized Zirconia $(Zr_{0.92}Y_{0.08}O_2)/O^{2-}$	$H_2 + O^{2-} \rightarrow H_2O + 2e^-$ $1/2O_2 + 2e^- \rightarrow O^{2-}$	600°C–1000°C 1112°F–1832°F	1 kW–2 MW	50%–60%	• Auxiliary power • Electric utility • Distributed generation	• High efficiency • Fuel flexibility • Can use a variety of catalysts • Solid electrolyte	• High temperature corrosion and breakdown of cell components • Long start up time and limits shutdowns

[a] Direct Alcohol Fuel Cells (DAFC) are a subset of PEM and can operate using Methanol or Ethanol as fuel.

Note: Information from Fuel Cell Technology Office—U.S. Department of Energy.

these cells and is commercially available in the following thickness presentations: N-112 (51 μm), N-115 (127 μm), and N-117 (183 μm). Even though Nafion membranes are the preferred polymer choice, their low operating temperature (<120°C) and high cost make them unattractive for large-scale use. This has sparked research into new low-cost anionic and cationic polymer materials (Bose et al., 2011; Varcoe et al., 2014) for membrane design capable of maintaining high performance levels at intermediated temperatures (100°C–150°C). The use of nanocomposite membranes (Mollá and Compañ, 2015) based on Polybenzimidazole (Shen et al., 2011), Sulfonated Polyether-ether-ketone (SPEEK), or SPEEK nanofiber mats embedded in a polyvinyl alcohol (PVA)/polyvinyl butyral (PVB) copolymer mix (Reyes-Rodríguez et al., 2016) have been reported as stable materials for intermediated temperature operation.

Nanometric catalysts are used to accelerate the reaction kinetics conducted on the electrodes of the fuel cell. A PEMFC's performance is limited by the oxygen reduction reaction (ORR), most commonly promoted through the use of platinum-based catalysts. The scarcity of this noble metal and its high cost have motivated research into new catalytic materials capable of partially or totally substituting the use of pure platinum by alloying it to nonnoble metals; examples are Pt-decorated Ni-Co, Ni-Pd, Fe_2O_3, and Y-OH nanocatalysts (Flores-Rojas et al., 2016; Tellez-Cruz et al., 2016a,b; Tinoco-Muñoz et al., 2016), M@Pt and (M=Ni, Co) core-shell nanocatalysts (Godínez-Salomón et al., 2012; Reyes-Rodríguez et al., 2013) among other materials with a promising ORR activity. Recent advances in catalyst design have focused on the control of particle size distribution to increase active surface sites and in the control of the catalyst's morphology to obtain nanoforms with specific crystallographic orientations that favor ORR activity (Gong et al., 2012; Chen et al., 2014). An example is the hot-injection method, where the presence of organic molecules such as Oleylamine (Oam) and/or Oleic acid (Oac) can be selectively adsorbed on faces of primitive particles during the nucleation stage, thereby allowing preferential growth in specific orientations. Figure 14.2a shows a scanning electron micrograph of polyhedral Ni-Pt nanoparticles obtained through the hot-injection method using Oam and Oac as shape promotors. Figure 14.2b shows a High-Angle Annular Dark-Field (HAADF) micrograph where the bright contrast observed on 20–30 nm nanoparticles reveals higher Pt-atom concentration on the edges and corners. This result was confirmed by a mapping test (Figure 14.2c–e); this Pt arrangement improves the ORR catalytic activity. Materials with adequate catalytic activity and stability are used to prepare MEAs for performance evaluation in a single fuel cell.

PEMFC technology can be found in the transportation sector, where companies such as Toyota have launched the first generation of hybrid vehicles that operate on hydrogen PEMFC and Lithium-battery systems (Figure 14.3). Production and marketing costs are expected to decrease in coming years, making this technology more accessible.

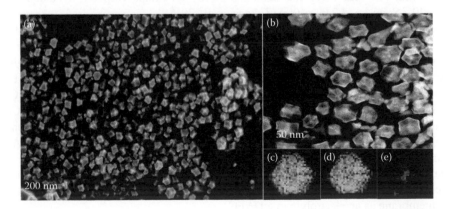

FIGURE 14.2 (a) SEM micrograph for polyhedral Ni-Pt nanoparticles, (b) HAADF micrograph, (c–e) Mapping for Ni-Pt, Ni and Pt atoms.

Power control unit   High pressure
(on top of motor)   H$_2$ tanks

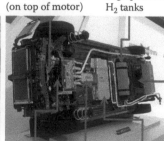

Drive battery
(on top of tank)

Motor   FC boost   Fuel cell
          converter   (FC) stack

FIGURE 14.3 (Left) Toyota FCV vehicle working with hybrid PEM fuel cells and lithium batteries. (Retrieved from https://commons.wikimedia.org/w/index.php?title=File:ITM_Power_Hydrogen_Station_ and_Toyota_Mirai.jpg&oldid=226214718 (Creative Commons Attribution-Share Alike 4.0 International.) (Right) Components of the Toyota FCV (Retrieved from https://commons.wikimedia.org/w/index. php?title=File:Toyota_FCV_Concept_bare_chassis.jpg&oldid=143818306 (Public Domain)).

14.4.2 Direct Alcohol Fuel Cells

DAFCs are power systems that use various types of alcohols as energy carriers to produce electricity. As a liquid fuel, alcohol provides a safe and easy way to store and transport energy. Methanol, being the simplest alcohol, provides the fastest oxidation kinetics, resulting in the highest power density of all alcohols; for this reason, it is often the preferred fuel choice in DAFCs (Corti and Gonzalez, 2014). Although the development of DAFC technology began in 1922, its application is still currently limited to prototypes applied to automotive and portable consumer electronics (Müller et al., 2003).

In the case of alkaline direct methanol fuel cells (DMFCs), the anodic reaction produces CO_2 as a product (see Table 14.2), which reacts with the OH^- of the electrolyte, resulting in the poisoning of the medium to prevent this problem; research has been conducted in the development of anionic membranes inert to CO_2. Some examples include polysiloxane membranes containing quaternary ammonium groups (Antolini and Gonzalez, 2010; Corti, 2014), membranes synthesized by incorporating layered double hydroxides into polysulfone chloromethylate, which exhibits an ionic conductivity of 2.36×10^{-2} S cm^{-1} at 60°C (Liu et al., 2017), membranes based on SPEEK/WC/Zr-SPP, and membranes of PVA/3(trimethylammonium) propyl-functionalized silica, which have low fuel permeability (2.2×10^{-7} cm^2 s^{-1} for methanol and 1×10^{-7} for ethanol respectively) and high conductivity (2.0×10^{-2} S cm^{-1} and 2.09×10^{-2}) (Zakaria et al., 2016).

Two challenges that must be addressed in DAFC technology are the improvement of catalysts used for the methanol oxidation reaction (MOR) and ORR, and the alcohol crossover problem, which causes a reduction in the catalytic activity for ORR. To improve the MOR, highly active alloys, with greater activity than pure platinum, are being researched, including Pt-M bimetallic alloys (M=Ru, Sn, Pd, Au, CeO$_2$, NiO, and V$_2$O$_5$), Pt-Ru-M ternary alloys (M=W$_2$C, Mo, Ir, Ni, Co, Rh, Os, and V), platinum-free catalysts such as Ru, Ni, Ni-M (M=Ti, Ru and Cu), Pd-M (M=Au and Ni), and perovskite oxides. Some Pt-based systems that have received recent interest for the improvement of the ORR in the presence of methanol are Pt-M (M=Pd, Ta, Co, Au, V, and Pb), Pd, and Pd-M (M=Au and Sn) (Corti, 2014; Sheikh et al., 2014).

In contrast to alkaline fuel cells, direct methanol PEMFCs receive more attention due to their high efficiency and extensively developed components. Figure 14.4 shows the use of a proton exchange membrane as an electrolyte in a DAFC.

PEMFCs also present a methanol crossover problem as their alkaline counterparts. To solve this problem, membranes with reduced methanol permittivity are being developed; examples of those

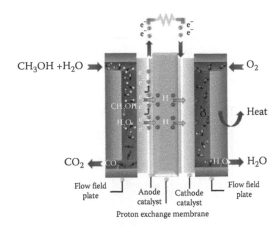

FIGURE 14.4 Representative scheme of a DAFC.

are based on poly (vinyl alcohol) and poly (diallyldimethylammonium chloride) with phosphotungstic acid molecules anchored onto carbon nanotubes (Li et al., 2017a), membranes structured with a bifunctional polymeric nanosieve (polyamide macromolecular proton conductor with single bond COOH end-capped) in composite Nafion (Cai et al., 2016), and membranes with amino-containing sulphonated poly (ether sulphone) with tetra-n-butylammonium bromide and 1-ethyl-3-methylimidazolium tetrafluoroborate (Li et al., 2017b).

In order to improve the MOR in acid media, Pt-based alloys with high electrochemical activity and stability, including Pt-M (M=Ni, Ru, Os, Sn, W, Mo, etc.) (Gonzalez, 2014) and Pt-M-Ni (M=Ru, Sn) (Antolini, 2017) have been developed. Some catalysts with high catalytic activity for the ORR and with improved tolerance to methanol include PtN, Pd-M (M=Pt, Co) (Vecchio et al., 2017), Pd-Co-Pt, Pt@C, and RuSe (Ticianelli and Lima, 2014).

14.4.3 ALKALINE FUEL CELLS

AFCs have shown high efficiencies and power densities, leading to an established application in the aerospace industry as seen by their use in the Apollo missions and the Space Shuttle program (Gülzow, 2004, 2012; O'Hayre et al., 2016). AFCs employ a solution of potassium hydroxide (KOH) in different concentrations as electrolyte (Gülzow, 2004; Cifrain and Kordesch, 2003; O'Hayre et al., 2016). Depending on the KOH concentration in the electrolyte, the AFC can operate at temperatures between 60°C and 250°C (O'Hayre et al., 2016). Figure 14.5 shows the operating principle of AFCs; their anodic and cathodic reactions are described in Table 14.2.

In an AFC cell, the OH⁻ is conducted through alkaline electrolyte from the cathode to the anode, in contrast to acid fuel cells where H⁺ is transported from the anode to the cathode. AFCs have two main advantages: (1) enhanced electro-kinetics in the oxygen reduction reaction, which allow the use of smaller amounts of noble metal electrocatalysts or nonnoble metal catalysts; (2) AFCs operate efficiently at room temperature (Gülzow, 2004, 2012; Cifrain and Kordesch, 2003; O'Hayre et al., 2016). In spite of the great strengths of the AFCs, some problems of this technology are electrode carbonation when using air instead of pure oxygen and by the decline of OH⁻ concentration in the electrolyte over time (Gülzow, 2004, 2012; O'Hayre et al., 2016). Within the last decade, solid-polymer alkaline electrolyte membranes (AEMs) have been developed to address these concerns by replacing the alkaline liquid electrolyte as a way to mitigate CO_2 instability associated with the AFC operation (Slade et al., 2013; O'Hayre et al., 2016). AFCs where the liquid electrolyte is replaced by an AEM are called alkaline membrane fuel cells (AMFCs). Today the greatest challenge for this technology is the creation of new polymeric membrane materials that combine high OH⁻ conductivity and long-term stability when exposed to highly basic and aggressive conditions.

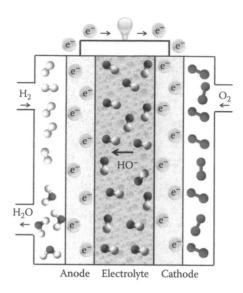

FIGURE 14.5   Schematic diagram of an AFC.

A variety of polymers such as polyethers, polysulfones, polyketones, and polystyrenes have been functionalized with various cationic groups such as quaternary ammoniums, imidazoliums, phosphoniums, and guanidinium to improve the anionic conductivity (Varcoe et al., 2014; Dang et al., 2015). The wide development of AEMs allows for AMFCs to become a promising technology to solve the cost barriers of fuel cell commercialization.

14.4.4   PHOSPHORIC ACID FUEL CELLS

PAFCs were the first type of cells to be commercialized in 1991 and are usually used in stationary power plants and buses. These cells have been escalated to provide up to 11 MW of electric power (Kasahara et al., 2000). During operation, gases are continuously fed to the cell, hydrogen is oxidized at the anode, and oxygen is reduced at the cathode (see Figure 14.6). PAFCs' electrochemical reactions are shown in Table 14.2.

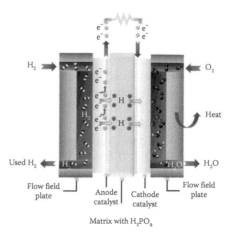

FIGURE 14.6   Representative scheme of a PAFC.

In contrast to alkaline fuel cells, the phosphoric acid electrolytic medium has high temperature stability and does not react with carbon dioxide, making PAFCs tolerant to virtually any concentration of CO_2 (Li, 2006) but not to carbon monoxide. CO poisons the cathode's catalyst, usually prepared from platinum-based alloys such as Pt-M (M = V, Co, Fe, Mn, Mo, Cu, and Cr), Pt-Cr-Co, or Pt-Pd (Stonehart, 1992; Watanabe et al., 1994; Kirubakaran et al., 2009). The performance of PAFCs, around 40% (O'Hayre et al., 2016), is limited by deficient ORR kinetics. However, at temperatures around 170°C–210°C, greater CO tolerance and lower internal resistance (attributed to the acid itself) can be achieved.

At temperatures below 100°C, liquid water is produced, mixing itself with the phosphoric acid, causing a decrease in proton conductivity and instability by cell degradation. Additives like hydrophilic inorganic compounds are used to ameliorate this problem (Lan et al., 2010).

Studies have reported that the addition of small amounts of fluorinated organic or silicone compounds to the phosphoric acid electrolyte enhances oxygen solubility and diffusivity, without the detrimental absorption of the electrolyte in the electrode surface, thereby improving the performance of a conventionally operated PAFC (Bjerrum et al., 1994). Challenges in PAFCs are cost-effectiveness, performance improvements, the need for more efficient electrode arrangements, and better acidic solutions enriched with optimum additives.

14.4.5 MOLTEN CARBONATE FUEL CELLS

MCFCs and SOFCs belong to the group of high temperature fuel cells. Currently, MCFCs have been developed to operate with waste streams from natural gas or coal power plants and have industrial, military, and electrical distribution applications (U.S. Department of Energy, 2017). Their efficiency is around to 50%–60% (Table 14.2) but increases up to 65% when coupled with turbines and may increase even further when using the heat generated in combined cycles to reach efficiencies around 85% (Hamnett, 2010). MCFCs employ a solid electrolyte based on a eutectic mixture of alkali carbonates generally composed of 68% Li_2CO_3/32% K_2CO_3 retained in an inert matrix of porous $LiAlO_2$ ceramic. This type of cell operates at high temperatures (600°C–700°C), where the carbonates form conducting molten carbonate ions (CO_3^{-2}), which are the main charge carriers. At high temperatures, ORR kinetics is sufficiently accelerated following an Arrhenius-type behavior and carbon monoxide (CO) is easily desorbed, preventing the poisoning of the catalysts. MCFC cells may operate with an anode of 90 wt.% Ni/10 wt.% Cr alloy and a lithiated nickel oxide cathode of $Li_xNi_{1-x}O$ (0.022 ≤ x ≤ 0.04) to catalyze the reactions of the electrodes; therefore, the use of noble metals is unnecessary (Hamnett, 2010). Their operation is as follows: at the cathode, a reduction process occurs when oxygen reacts with the catalyst and with carbon dioxide (CO_2) flowing or recirculated from the anode. In this process, carbonate ions are formed and diffused through the solid electrolyte from the cathode to the anode; at the anode, an oxidation process occurs when hydrogen reacts with carbonate ions to form water vapor and carbon dioxide (CO_2) by the reaction shown in Table 14.2. The carbon dioxide (CO_2) formed in the anode chamber is generally recycled to the cathode chamber for consumption after removal of water vapors. Figure 14.7 shows an MCFC schematic.

An important advantage of MCFCs is the flexibility of the fuels used for the anode. In contrast to AFCs, PAFCs, and PEMFCs, MCFCs do not require an external reformer to convert fuels like natural gas and biogas into high purity hydrogen. At high operating temperatures, methane and other light hydrocarbons from these fuels are converted into hydrogen within the MCFC through a process called "internal reforming," which translates into a significant cost reduction (Hamnett, 2010).

The disadvantages of MCFCs lie in long start-up times, where ORR kinetics are considerably dependent on the electrolyte used, since the reduction reaction starts with the formation of peroxides in lithium-rich mixtures or superoxides in potassium-rich carbonates mixtures. Therefore, the use of the eutectic mixture of the electrolyte is preferred. Moreover, the operation's high temperature conditions favor an aggressive chemical environment that promotes the corrosion and dissolution

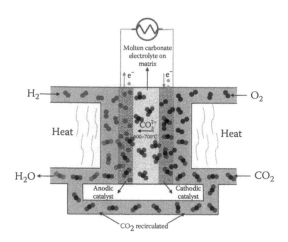

FIGURE 14.7   Representative scheme of a MCFC.

of NiO at the cathode, where Ni (II) species can migrate to the anode and be deposited in metallic form. The metal deposition can cause short circuits by creating electronic channels in the electrolyte (Hamnett, 2010). The ohmic drop in the electrolyte and the loss of its stability are other challenges to overcome. Currently, commercial MCFCs have an average useful life of 40,000 h (5 years) (U.S. Department of Energy, 2017).

14.4.6   SOLID OXIDE FUEL CELLS

SOFCs are useful in large, high-power applications such as industrial stations and large-scale power stations (Stambouli and Traversa, 2002). The basic components of the SOFC are: (1) the cathode, where oxygen is reduced to oxygen ions, O^{2-}, which are transported through the solid electrolyte to the anode, and (2) the anode, where the ions react with the fuel (generally hydrogen or carbon monoxide) to produce water or CO_2 respectively, as well as electricity and heat (Yamamoto, 2000; Ormerod, 2003). The most common SOFC electrolyte material is yttria-stabilized zirconia (YSZ), which is an oxygen ion (oxygen vacancy) conductor (Singhal, 2000; Fergus, 2006; O'Hayre et al., 2016). The most common material for the anode is a nickel–YSZ cermet (cermet is a ceramic and metal mixture). Nickel provides high conductivity and catalytic activity; YSZ adds ion conductivity, thermal expansion compatibility, mechanical stability, and maintains high porosity and high surface area in the anodic structure (O'Hayre et al., 2016). The cathode is usually a mixed ceramic material with ion and electrical conduction properties; typical cathode materials with high oxidation resistance and high catalytic activity include strontium-doped lanthanum manganite, lanthanum–strontium ferrite, lanthanum–strontium cobaltite, and lanthanum–strontium cobaltite ferrite (Jacobson, 2010; O'Hayre et al., 2016). Figure 14.8 illustrates the operating principle of SOFCs; their corresponding reactions are described in Table 14.2.

SOFCs operate at temperatures between 600°C and 1000°C. The high operating temperature presents challenges and advantages. Advantages are fuel flexibility, the use of nonprecious metal catalysts, high-quality waste heat delivered for cogeneration applications, the use of solid electrolytes, and relatively high power density. Disadvantages include the need for materials resistant to high-temperatures and relatively expensive components (O'Hayre et al., 2016).

Special attention has been given to SOFC designs in the intermediate temperature range (400°C–700°C) to remove most of the disadvantages associated with high-temperature operation while maintaining the most significant benefits. Such SOFCs could employ much cheaper sealing technologies and robust, inexpensive metal (rather than ceramic) stack components. At the same time, these SOFCs could provide reasonably high efficiency and fuel flexibility (O'Hayre et al., 2016).

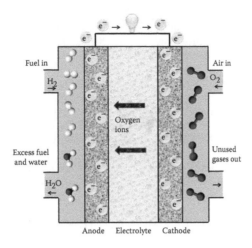

FIGURE 14.8   Representative scheme of a SOFC.

14.5   CONCLUSIONS

This chapter presented research progress on various fuel cell technologies linked to renewable energies toward a hydrogen economy. A description of efforts placed on the development of new materials to improve the performance and fuel efficiency in diverse types of fuel cell applications was presented. A complete understanding of the ORR and alcohol oxidation reaction has not yet been reached; more theoretical and experimental efforts are necessary. Current and future research efforts focus on improving methodologies for electrocatalyst syntheses and membrane preparation, to open up new technological possibilities and fully exploit their tremendous potential for wide varieties of applications, most notably, in stationary power plants, portable devices, and mass transportation. Fuel cells represent the core foundation of a sustainable, renewable, and efficient hydrogen economy.

ACKNOWLEDGMENTS

This study was financially supported by the Mexican Council of Science and Technology, CONACYT (Grant No. 245920). JLRR, HCM, and MMTC thank CONACYT for their doctoral fellowship.

REFERENCES

Anantharaj, S., S. R. Ede, K. Sakthikumar, K. Karthick, S. Mishra, and S. Kundu. 2016. Recent trends and perspectives in electrochemical water splitting with an emphasis on sulfide, selenide, and phosphide catalysts of Fe, Co, and Ni: A review. *ACS Catalysis* 6(12): 8069–8097.

Antolini, E. 2017. Pt-Ni and Pt-M-Ni (M=Ru, Sn) anode catalysts for low-temperature acidic direct alcohol fuel cells: A review. *Energies* 10(1): 42.

Antolini, E. and E. R. Gonzalez. 2010. Alkaline direct alcohol fuel cells. *Journal of Power Sources* 195(11): 3431–3450.

Aricò, A. S., P. Bruce, B. Scrosati, J. M. Tarascon, and V. W. Schalkwijk. 2005. Nanostructured materials for advanced energy conversion and storage devices. *Nature materials* 4(5): 366–377.

Bayindir, K. Ç., M. A. Gözüküçük, A. Teke. 2011. A comprehensive overview of hybrid electric vehicle: Powertrain configurations, powertrain control techniques and electronic control units. *Energy Conversion and Management* 52: 1305–1313.

Bjerrum, N. J., X. Gang, H. A. Hjuler, C. Olsen, R. W. Berg. 1994. Phosphoric acid fuel cell. US 5344722 A. filed Jun 25.

Bose, S., T. Kuila, T. X. H. Nguyen, N. H. Kim, K.-T. Lau, and J. H. Lee. 2011. Polymer membranes for high temperature proton exchange membrane fuel cell: Recent advances and challenges. *Progress in Polymer Science* 36(6): 813–843. doi:10.1016/j.progpolymsci.2011.01.003.

Dunn, B., H. Kamath, J.-M. Tarascon. 2011. Electrical energy storage for the grid: A battery of choices. *Science,* 334: 928–935.

Cai, W., K. Fan, J. Li, L. Ma, G. Xu, S. Xu, L. Ma, and H. Cheng. 2016. A bi-functional polymeric nano-sieve Nafion composite membrane: Improved performance for direct methanol fuel cell applications. *International Journal of Hydrogen Energy* 41(38): 17102–17111.

Chen, C., Y. Kang, Z. Huo, Z. Zhu, W. Huang, H. L. Xin, J. D. Snyder et al. 2014. Highly crystalline multimetallic nanoframes with three-dimensional electrocatalytic surfaces. *Science* (New York, NY) 343(6177): 1339–1343.

Cifrain, M, and K. Kordesch. 2003. Hydrogen/oxygen (Air) fuel cells with alkaline electrolytes. In *Handbook of Fuel Cells – Fundamentals, Technology and Applications.* edited by Vielstich, W., H. A. Gasteiger, A. Lamm, and H. Yokokawa, vol. 1, part 4, Chichester: John Wiley & Sons, pp. 267–280.

Corti, H. R. 2014. Membranes for direct alcohol fuel cells. In *Direct Alcohol Fuel Cells: Materials Performance, Durability and Applications*, edited by Corti, H. R., and E. R. Gonzalez, 121–230, New York: Springer.

Corti, H. R. and E. R. Gonzalez. 2014. Introduction to direct alcohol fuel cells. In *Direct Alcohol Fuel Cells: Materials Performance, Durability and Applications*, edited by Corti, H. R., and E. R. Gonzalez, 1–32. New York: Springer.

Dang, H.-S., E. A. Weiber, and Patric Jannasch. 2015. Poly(phenylene oxide) functionalized with quaternary ammonium groups via flexible alkyl spacers for high-performance anion exchange membranes. *Journal of Materials Chemistry A* 3: 5280–5284.

Energy Information Administration .2016. International Energy Outlook 2016 (IEO2016), Intergovernmental Panel on Climate Change. 2012. Special Report on Renewable Energy Sources and Climate Change Mitigation

Fergus, J. W. 2006. Electrolytes for solid oxide fuel cells. *Journal of Power Sources* 162: 30–40.

Flores-Rojas, E., H. Cruz-Martínez, M. M. Tellez-Cruz, J. F. Pérez-Robles, M. A. Leyva-Ramírez, P. Calaminici, and O. Solorza-Feria. 2016. Electrocatalysis of oxygen reduction on CoNi-decorated-Pt nanoparticles: A theoretical and experimental study. *International Journal of Hydrogen Energy* 41(48): 23301–23311.

Godínez-Salomón, F., M. Hallen-López, and O. Solorza-Feria. 2012. Enhanced electroactivity for the oxygen reduction on Ni@Pt core-shell nanocatalysts. *International Journal of Hydrogen Energy* 37(19): 14902–14910.

Gong, J., G. Li, and Z. Tang. 2012. Self-assembly of noble metal nanocrystals: Fabrication, optical property, and application. *Nano Today* 7(6): 564–585.

Gonzalez, E. R. 2014. Catalysts for methanol oxidation. In *Direct Alcohol Fuel Cells: Materials Performance, Durability and Applications*, edited by Corti, H. R. and E. R. Gonzalez, 33–62. New York: Springer.

Gülzow E. 2004. Alkaline fuel cells. *Fuel cells* 4: 251–255.

Gülzow, E. 2012. Alkaline fuel cells. In *Fuel Cell Science and Engineering.* edited by Stolten, D., and B. Emonts, vol. 1, part 1, Weinheim: Wiley-VCH Verlag GmbH & Co. KGaA, pp. 97–126.

Hamnett, A. 2010. Introduction to fuel-cells types. In *Handbook of Fuel Cells – Fundamentals, Technology and Applications.* edited by Vielstich, W., H. A. Gasteiger, A. Lamm, and H. Yokokawa, vol. 1, New York: John Wiley & Sons, pp. 36–43.

Houchins, C., G. J. Kleen, J. S. Spendelow, J. Kopasz, D. Peterson, N. L. Garland, D. L. Ho et al. 2012. U.S. DOE progress towards developing low-cost, high performance, durable polymer electrolyte membranes for fuel cell applications. *Membranes* 2(4): 855–878.

IPCC, 2012. Managing the Risks of Extreme Events and Disasters to Advance Climate Change Adaptation. A Special Report of Working Groups I and II of the Intergovernmental Panel on Climate Change [Field, C.B., V. Barros, T.F. Stocker, D. Qin, D.J. Dokken, K.L. Ebi, M.D. Mastrandrea, K.J. Mach, G.-K. Plattner, S.K. Allen, M. Tignor, and P.M. Midgley (eds.)]. Cambridge, UK and New York: Cambridge University Press, 582 p.

Jacobson, A. J. 2010. Materials for solid oxide fuel cells. *Chemistry of Materials* 22: 660–674.

Jamesh, M. I. 2016. Recent progress on earth abundant hydrogen evolution reaction and oxygen evolution reaction bifunctional electrocatalyst for overall water splitting in alkaline media. *Journal of Power Sources* 333(1): 213–236.

Kasahara, K., M. Morioka, H. Yoshida, and H. Shingai. 2000. PAFC operating performance verified by Japanese gas utilities. *Journal of Power Sources* 86(1): 298–301.

Kirubakaran, A., S. Jain, and R. K. Nema. 2009. A review on fuel cell technologies and power electronic interface. *Renewable & Sustainable Energy Reviews* 13(9): 2430–2440.

Lamy, C. 2016. From hydrogen production by water electrolysis to its utilization in a PEM fuel cell or in a SO fuel cell: Some considerations on the energy efficiencies. *International Journal of Hydrogen Energy* 41(34): 15415–15425.

Lan, R., X. Xu, S. Tao, and J. T. S. Irvine. 2010. A fuel cell operating between room temperature and 250°C based on a new phosphoric acid based composite electrolyte. *Journal of Power Sources* 195(20): 6983–6987.

Li, Xianguo. 2006. *Principles of Fuel Cells*. New York: Taylor & Francis.

Li, Y., H. Wang, Q. Wu, X. Xu, S. Lu, and Y. Xiang. 2017a. A poly (vinyl alcohol)-based composite membrane with immobilized phosphotungstic acid molecules for direct methanol fuel cells. *Electrochimica Acta* 224(1): 369–377.

Li, Y., M. Hoorfar, K. Shen, J. Fang, X. Yue, and Z. Jiang. 2017b. Development of a crosslinked pore-filling membrane with an extremely low swelling ratio and methanol crossover for direct methanol fuel cells. *Electrochimica Acta* 232: 226–235.

Listorti, A., J. Durrant, and J. Barber. 2009. Artificial photosynthesis: Solar to fuel. *Nature Materials* 8: 929–930.

Liu, W., N. Liang, P. Peng, R. Qu, D. Chen, and H. Zhang. 2017. Anion-exchange membranes derived from quaternized polysulfone and exfoliated layered double hydroxide for fuel cells. *Journal of Solid State Chemistry* 246: 324–328.

Mollá, Sergio and Vicente Compañ. 2015. Nanocomposite SPEEK-based membranes for direct methanol fuel cells at intermediate temperatures. *Journal of Membrane Science* 492: 123–136.

Morgan, T. 2006. The hydrogen economy: A non-technical review. Paris: UNEP/DTIE, ENERGY BRANCH.

Müller, J., G. Frank, K. Colbow, and D. Wilkinson. 2003. Transport/kinetic limitations and efficiency losses. In *Handbook of Fuel Cells: Fundamentals Technology and Applications*, edited by Vielstich, W., A. Lamm, and H. A. Gasteiger. Volume 4: Fuel Cell Technology and Appliactions, 845–855. West Sussex: John Wiley & Sons.

O'Hayre R., S. Cha, F. B. Prinz, W. Colella. 2016. *Fuel Cell Fundamentals*. New Jersey: John Wiley & Sons.

Ormerod, R. M. 2003. Solid oxide fuel cells. *Chemical Society Reviews* 32: 17–28.

Pasupathi, S., H. Su, H. Liang, B. C. Pollet. 2016. Advanced technologies for proton-exchange-membrane fuel cells. In *Electrochemical Energy Advanced Materials and Technologies*, edited by P. K. Shen, Ch. -Y. Wang, S. P. Jiang, S. S. J. Zhang, Boca Raton: CRC Press, pp. 405–420.

Pilatowsky, I., R. J. Romero, C. A. Isaza, S. A. Gamboa, P. J. Sebastian, and W. Rivera. 2011. Thermodynamics of fuel cells. In *Cogeneration Fuel Cell-Sorption Air Conditioning Systems*. London: Springer-Verlag.

Reyes-Rodríguez, J. L., F. Godínez-Salomón, M. A. Leyva, and O. Solorza-Feria. 2013. RRDE study on Co@Pt/C core–shell nanocatalysts for the oxygen reduction reaction. *International Journal of Hydrogen Energy* 38(28): 12634–12639.

Reyes-Rodríguez, J. L., O. Solorza-Feria, A. García-Bernabé, E. Giménez, O. Sahuquillo, and V. Compañ. 2016. Conductivity of composite membrane-based poly(ether-Ether-Ketone) sulfonated (SPEEK) nano fiber ats of varying thickness. *RSC Advances* 6(62): 56986–56999.

Seh, Z. W., J. Kibsgaard, C. F. Dickens, Ib Chorkendorff, J. K. Nørskov, and Th. F. Jaramillo. 2017. Combining theory and experiment in electrocatalysis: Insights into materials design. *Science,* 355: 1–12.

Shao, M. 2015. Electrocatalysis in fuel cells, *Catalysts* 5: 2115–2121.

Sheikh, A. M., K. E. Abd-Alftah, C. F. Malfatti. 2014. On reviewing the catalyst materials for direct alcohol fuel cells (DAFCs). *Journal of Multidisciplinary Engineering Science and Technology* 1(3): 1–9.

Shen, C.-H., L.-C. Jheng, S. L.-C. Hsu, and J. T.-W. Wang. 2011. Phosphoric acid-doped cross-linked porous polybenzimidazole membranes for proton exchange membrane fuel cells. *Journal of Materials Chemistry* 21(39): 15660.

Sherif, S. A., F. Barbir, and T. N. Veziroglu. 2014. Hydrogen economy. In *Handbook of Hydrogen Energy*, edited by Sherif S. A., D. Y. Goswami, E. K. Stefanakos, and A. Steinfeld, 1–16. Boca Raton, FL: CRC Press/Taylor & Francis.

Singhal, S. C. 2000. Advances in solid oxide fuel cell technology. *Solid State Ionics* 135: 305–313.

Slade, R. C. T., J. P. Kizewski, S. D. Poynton, R. Zeng, and J. R. Varcoe. 2013. Alkaline membrane fuel cells. In *Fuel Cells: Selected Entries from the Encyclopedia of Sustainability Science and Technology*. edited by Kreuer K.-D., 372–386. New York: Springer.

Smolinka T., E. T. Ojong, and T. Lickert. 2015. Fundamentals of PEM water electrolysis. In *PEM Electrolysis for Hydrogen Production: Principles and Applications*, edited by Bessarabov D., H. Wang, H. Li, and N. Zhao, 11–33, Boca Raton, FL: CRC Press/Taylor & Francis.

Sørensen, B. 2005. *Hydrogen and Fuel Cells: Emerging Technologies and Applications*. New York: Elsevier Academic Press.

Srinivasan, S., L. Krishnan, and C. Marozzi. 2006. Fuel cell principles. In *Fuel Cells,* edited by Srinivasan, S. 189–233. New York: Springer.

Stambouli, A. B. and E. Traversa. 2002. Solid oxide fuel cells (SOFCs): A review of an environmentally clean and efficient source of energy. *Renewable and Sustainable Energy Reviews* 6: 433–455.

Stonehart, P. 1992. Development of alloy electrocatalysts for phosphoric acid fuel cells (PAFC). *Journal of Applied Electrochemistry* 22(11): 995–1001.

Sundmacher, K. 2010. Fuel cell engineering: Toward the design of efficient electrochemical power plants. *Industrial & Engineering Chemistry Research* 49(21): 10159–10182.

Tellez-Cruz, M. M., M. A. Padilla-Islas, M. Pérez-González, and O. Solorza-Feria. 2016a. Comparative study of different carbon-supported Fe_2O_3-Pt catalysts for oxygen reduction reaction. *Environmental Science and Pollution Research.*

Tellez-Cruz, M. M., M. A. Padilla-Islas, J. F. Godínez-Salomón, L. Lartundo-Rojas, and O. Solorza-Feria. 2016b. Y-OH-decorated-Pt/C electrocatalyst for oxygen reduction reaction. *International Journal of Hydrogen Energy* 41(48): 23318–23328.

Ticianelli, E. A., and F. H. B. Lima. Nanostructured electrocatalysts for methanol and ethanol-tolerant cathodes. In *Direct Alcohol Fuel Cells: Materials Performance, Durability and Applications*, edited by Corti, H. R., and E. R. Gonzalez, 99–119. New York: Springer.

Tinoco-Muñoz, C. V., J. L. Reyes-Rodríguez, D. Bahena-Uribe, M. A. Leyva, J. G. Cabañas-Moreno, and O. Solorza-Feria. 2016. Preparation, characterization and electrochemical evaluation of Ni–Pd and Ni–Pd–Pt nanoparticles for the oxygen reduction reaction. *International Journal of Hydrogen Energy* 41(48): 23272–23280.

U.S. Department of Energy, Office of Energy Efficiency & Renewable Energy, Fuel Cells website. https://energy.gov/eere/fuelcells/fuel-cells (accessed February 24, 2017).

Varcoe, J. R., P. Atanassov, D. R. Dekel, A. M. Herring, M. A. Hickner, P. A. Kohl, A. R. Kucernak et al. 2014. Anion-exchange membranes in electrochemical energy systems. *Energy & Environmental Science* 7: 3135–3191.

Vecchio, C. L., D. Sebastián, C. Alegre, A. S Aricò, and V. Baglio. 2017. Carbon-supported Pd and Pd-Co cathode catalysts for direct methanol fuel cells (DMFCs) operating with high methanol concentration. *Journal of Electroanalytical Chemistry* In Press. doi.org/10.1016/j.jelechem.2017.02.042.

Vielstich, W. 2010. Ideal and effective efficiencies of cell reactions and comparison to carnot cycles. In *Handbook of Fuel Cells – Fundamentals, Technology and Applications*, edited by Vielstich, W., H. A. Gasteiger, A. Lamm, and H. Yokokawa. Chichester: John Wiley & Sons, p. 27.

Vlcek, A. 2012. Solar fuels. *Coordination Chemistry Reviews* 256: 2397–2398.

Wang, Y., D. Y. C. Leung, J. Xuan, and H. Wang. 2016. A review on unitized regenerative fuel cell technologies, part-A: Unitized regenerative proton exchange membrane fuel cells. *Renewable & Sustainable Energy Reviews*, 65: 961–977.

Watanabe, M., K. Tsurumi, T. Mizukami, T. Nakamura, and P. Stonehart. 1994. Activity and stability of ordered and disordered Co-Pt alloys for phosphoric acid fuel cells. *Journal of the Electrochemical Society* 141(10): 2659–2668.

Wilberforce, T., A. Alaswad, A. Palumbo, M. Dassisti, and A. G. Olabi. 2016. Advances in stationary and portable fuel cell applications. *International Journal of Hydrogen Energy* 41: 16509–16522.

Yamamoto, O. 2000. Solid oxide fuel cells: fundamental aspects and prospects. *Electrochimica Acta* 45: 2423–2435.

Zakaria, Z., S. K. Kamarudin, and S. N. Timmiati. 2016. Membranes for direct ethanol fuel cells: An overview. *Applied Energy* 163(1): 334–342.

Zhang, Q., E. Uchaker, S. L Candelaria, and G. Cao. 2013. Nanomaterials for energy conversion and storage. *Chemical Society Reviews* 42(1): 3127–3171.

15 Towards More Sustainable Aeronautics via the Use of Biofuels

Aris Iturbe
Aeronautic University of Queretaro

W. Vicente and J.E.V. Guzmán
National Autonomous University of Mexico

CONTENTS

15.1 INTRODUCTION

Global air traffic has increased considerably over the last few decades. For instance, the demand of jet fuel, which in 2006 represented 6.5% of the total oil refined worldwide, increased to 9.5% the following year (Nygren et al., 2009). This raises serious concerns from an environmental point of view, because nearly 2% of the total pollutants released to the atmosphere are produced by airplanes (Felice et al., 2013).

Biofuels have recently entered the scene as a promising eco-friendly alternative to conventional jet fuels. As a matter of fact, the European Commission started supporting projects to develop green turbines for diverse applications (Janssen, 2004). Successful trials concerning the generation of power with 30 kW turbines have demonstrated the potential of green fuels in these kinds of applications (Khairallah, 2004). Furthermore, multiple airlines are already conducting test flights along commercial routes to evaluate the performance of biofuel blends under regular operational conditions. Table 15.1 contains an excerpt of these tests and includes the mixture percentages pertaining to each case (Felice et al., 2013).

Among other advantages, biofuels do not contain sulfur or fatty acid methyl esters. More importantly, their use produces significant reductions of the CO_2 released to the atmosphere. This fact was amply demonstrated during the test flights indicated in Table 15.1; it is worth noting that the Lufthansa flights alone saved 1480 tons of CO_2. Laboratory experiments also suggest that NO_x emissions are reduced by nearly 30% in the high thermal input part of the operational range (Krishna, 2007). In addition, biofuels have relatively high thermal conversion efficiencies (Habib et al., 2010; Tan and Liou, 2011).

By way of contrast, some major disadvantages include the biofuel's ability to degrade certain materials (e.g., rubber) and their propensity to crystallize at low temperatures. Further challenges

TABLE 15.1

Commercial Flights Using Mixtures of Biodiesel with Conventional Jet Fuels

Airline	Mixture (%)	Biosource	No. of Flights
Lufthansa	50–50	Jatropha	1187
Lufthansa	50–50	Jatropha	1 (transatlantic)
KLM Royal Dutch A.	50–50	UCO	200
Horizon Air	50–50	UCO	64
Alaska Airlines	50–50	UCO	264
Finnair	50–50	UCO	1
Qantas Airlines	50–50	UCO	1
Air France	50–50	UCO	1
United Airlines	40–60	Algae	1
Iberia Airlines	25–75	Camelina	1
Interjet	27–73	Jatropha	1
Aeromexico	50–50	UCO, Camelina	1

remain to be resolved, both from economic and technical points of view. First of all, concerns about the cost effectiveness of a viable commercial solution, as well as the sustainability of the production must be addressed. For example, even in oil-producing countries like Mexico, this is a pressing matter, because the price of jet fuels tends to increase around 70% in periods of 5 years, while the market value of biofuels is nearly 10 times higher than that of conventional fuels (López-Meyer in a private communication).

Several technical problems persist with the engine's internal design and operational envelopes. In this respect, a few obvious examples include the required modifications of the combustion chamber, the fuel injection nozzles, the exhaust diffusers, and the turbine blades, which would prevent engine stall and postcombustion ash-generation problems. At the root of these issues lay the physical and chemical properties of the available biofuels. As opposed to conventional fossil fuels, these substances have higher viscosities and different chemical compositions.

15.2 BIODIESEL: A PROMISING ALTERNATIVE FOR AERONAUTIC APPLICATIONS

The nature of a biofuel makes it an attractive choice to power engines. Broadly speaking, these substances may be obtained from almost any biomass source: among many others one may consider, for example, naturally produced organic matter, waste from diverse farming activities (including animal and vegetation), industrial organic by-products (such as those related with sugar processing), etc. (Macera et al., 2006). As well as being nontoxic and biodegradable, biodiesels have a most convenient lubrication property.

Third generation biofuels may be obtained from easily farmed algae, which can even be genetically manipulated to improve its properties for combustion purposes (Macera et al., 2006). The sustainability studies indicate that production ratios in the interval 4.6–18.4 L/m^2 are possible (Garibay-Hernandez, 2009). Thus, a relatively small surface of 52,000 m^2 could produce over 95 million barrels of fuel per day, at just a fraction of the cost of conventional hydrocarbon fuels.

More important from an environmental perspective is the fact that the combustion of a biofuel does not produce sulfur oxides (SO_x), while its yield of carbon oxides (CO_x) to the atmosphere is much lower (Kamps, 2013). NO_x emissions are consistently lower as the study by Krishna (2007) reveals. Figure 15.1 illustrates the favorable comparison between two low-concentration blends and the regular fossil fuel.

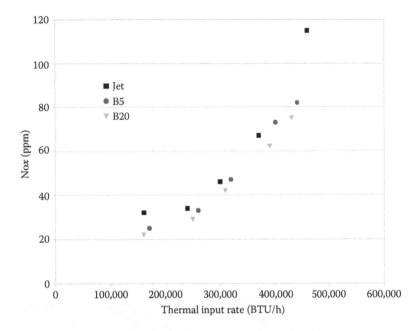

FIGURE 15.1 A typical spray of injected fuel into a combustion chamber. (From Krishna, Report on atomization tests for project titled biodiesel blends in microturbine. Technical Report, Brookhaven National Laboratory, Upton, NY, 2007.)

On the other hand, density provides an indirect measurement of the content of energy per unit volume. Thus, for example, a kerosene with density 0.81 kg/L may provide 2.8E+7 J/L, while jet fuels with densities 0.762 kg/L would provide as much as 3.2E+7 J/L. Since biofuels typically have calorific powers of approximately 4.4E+7 J/L, it is clear that they are quite competitive from an energetic point of view.

Despite these advantages, the stringent requirements imposed by the operational envelope impose serious limitations for full-scale implementation. Table 15.2 summarizes the acceptable values for the properties of a fuel intended for propulsion turbines.

Although biofuels comply with most of these requirements, previous studies show that they may have sodium and potassium concentrations as high as 5 ppm, which is above the allowable limits set by the norm ASTM D6751 (2008). Higher ash levels tend to be the norm, and this translates into shorter life spans of certain mechanical components (e.g., bearings and sensors).

TABLE 15.2
Property Requirements for Jet Fuels

Property	Acceptable Level
Humidity	1.0%
Sediments	1.0%
Viscosity	20 cS
Dew point	20°C
Carbon residues	1.0%
Hydrogen	11%

15.3 CHALLENGES IN ENGINE DESIGN AND OPERATION WITH BIOFUELS

Among the various design challenges encountered during the development of a jet engine at the UNAQ, the following two have required special attention: (1) the hydrodynamic features affecting the fuel injection system and (2) the geometry of the combustion chamber (Figure 15.2). Nonetheless, it is important to acknowledge that other elements of the engine can be modified, albeit at much higher costs. For example, modifying the geometry of the compressor or the turbine stages would involve reengineering the entire blade geometry. Thus, engine redesign and optimization entail not only technical aspects, but also issues of economical relevance.

For concreteness, the present discussion focuses on the fuel-injection process. Thus, we are concerned with the characteristics of the fuel spray generated at the injection nozzle. It has been experimentally observed that the physical properties of the biofuel, including the density ρ, the viscosity μ, and the surface tension σ, modify these characteristics in an essential way. In turn, a proper combustion process cannot be realized and, as a result, the global efficiency of the engine degrades. Figure 15.3 illustrates the simulated evolution of a typical fuel spray.

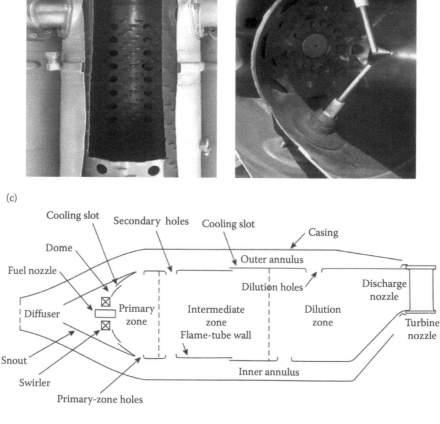

FIGURE 15.2 Snapshots of the combustion chamber (a) and close-up view of the injection port (b). The two rods observed in the photographs are the spark generators. Diagram (c) illustrates the entire configuration. The flow takes place from left to right.

(a)

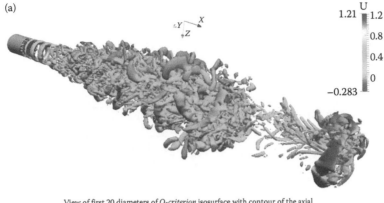

View of first 20 diameters of *Q-criterion* isosurface with contour of the axial
velocity at early times (*Q* = 30).

(b)

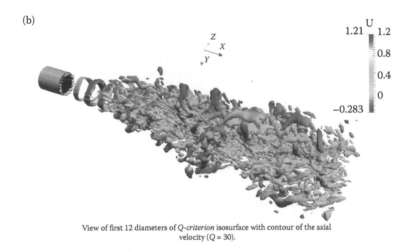

View of first 12 diameters of *Q-criterion* isosurface with contour of the axial
velocity (*Q* = 30).

FIGURE 15.3   Simulated structure of a typical spray. Q-criterion surfaces are shown. Image (a) illustrates
the resulting structure 20 diameters downstream from the nozzle, during the fuel injection start-up process.
Image (b) provides a closer look of the steady state 12 diameters downstream from the nozzle.

As previously mentioned, the operating envelope imposes minimum threshold values on these
properties. For instance, the lowest admissible temperature of the injection system is determined by
the kinematic viscosity of the fuel, whose maximum value should not exceed 1.2E-5 m^2/s.

Proper atomization of the fuel in the combustion chamber begins with the growth of unstable
modes induced on the liquid jet as it emerges from the injection nozzle (Figure 15.4). However, a
combination of other mechanisms also interferes on the configuration of the resulting spray.

In spite of the fact that some authors have concluded that viscosity does not necessarily represent a problem (e.g., Ali et al., 1995), it is generally observed that higher viscosities impede proper
atomization. In order to understand the effect of viscosity on the formation of the spray, we briefly
outline the theoretical considerations by Plateau, Rayleigh, and Chandrasekhar (the details may be
consulted in Drazin and Reid, 2004 and Chandrasekhar, 1961). First, it is assumed that the liquid
jet is discharged inside the combustion chamber to an ambient fluid with pressure p_∞, the inertia of
which may be neglected. Second, the surface tension effects modify the pressure inside the jet, such
that $p_i = p_\infty + \gamma / R$, ($\gamma$ is the surface tension coefficient, and R is the radius of the jet). Thus, the lin-
earized set of equations to be solved for small velocity and pressure perturbations, $\delta \mathbf{u} = \mathbf{u}$ and δp is

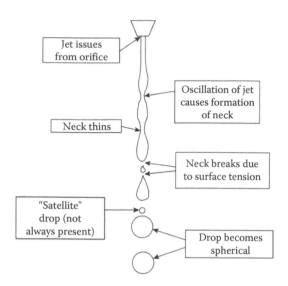

FIGURE 15.4   Characteristics of the jet at the outlet of the injection port. The jet becomes unstable under the action of the relevant forces and breaks up into nearly spherical droplets. Further fragmentation of the initial droplets is accomplished through subsequent turbulent interactions.

$$\nabla \cdot \mathbf{u} = 0 \tag{15.1}$$

$$\frac{\partial \mathbf{u}}{\partial t} = \frac{-1}{\rho} \nabla \delta p + v \nabla^2 \mathbf{u} \tag{15.2}$$

In the perturbed state, the surface of the jet is assumed to undergo symmetrical deformations about the axis, appearing in the form of normal modes

$$\xi = R + \epsilon_0 \, e^{st+i(kz+m\phi)}. \tag{15.3}$$

The stress at the interphase must then express the relevant effects in the following manner

$$p + \delta p - 2v\rho \frac{\partial u_\xi}{\partial \xi} = \gamma \nabla \cdot \mathbf{n} \tag{15.4}$$

where $p = p_i - p_\infty$. By developing this equation in terms of the shape modes (Equation 15.3), one may obtain

$$\frac{\delta p}{\rho} = -\frac{\gamma}{R^2 \rho} \left(i - m^2 - \lambda^2 \right) \epsilon_0 \, e^{st+i(kz+m\phi)} \tag{15.5}$$

which is used in the final part of the solution. On the other hand, the divergence of Equation 15.2, together with Equation 15.1, leads to

$$\frac{d^2 \delta p}{d\xi^2} + \frac{1}{\xi} \frac{d\delta p}{d\xi} - \left(k^2 + \frac{m^2}{\xi^2} \right) \delta p = 0. \tag{15.6}$$

This is a modified Bessel equation of order m, whose solution is

$$\delta p = p \, \epsilon_0 \, I_m \left(k\xi \right) e^{st+i(kz+m\phi)} \tag{15.7}$$

It is noted that the solution $K_m(k\xi)$ is unbounded as $\xi \to 0$ and so it is discarded. Equations 15.5 and 15.7 can be combined in order to solve Equation 15.2 for the velocity. Chandrasekhar (1961) has shown that the velocity may be expressed as a poloidal vector of the form

$$\mathbf{u} = \left(-ik\xi U \mathbf{e}_\xi + \frac{1}{\xi} \frac{\partial}{\partial \xi} \left(\xi^2 U \right) \mathbf{e}_z \right) e^{st+i(kz+m\phi)}. \tag{15.8}$$

Substitution into Equation 15.2 then yields

$$\frac{\partial u}{\partial t} \to sU e^{st+i(kz+m\phi)} \tag{15.9}$$

$$-\nabla \delta p \to \frac{-i}{\xi} \epsilon_0 \, \delta p I_1 \left(k\xi \right) e^{st+i(kz+m\phi)} \tag{15.10}$$

$$\nu \nabla^2 u \to \left(\frac{\partial^2 U}{\partial \xi^2} + \frac{3}{\xi} \frac{\partial U}{\partial \xi} - k^2 U \right) e^{st+i(kz+m\phi)} \tag{15.11}$$

Considering only axial modes (the flow is stable for all nonaxial modes), then $m = 0$ and the general solution for the velocity is given by

$$U = i \left(A \frac{I_1(\zeta\xi)}{\xi} - \frac{\epsilon_0 \, \delta p}{s} \frac{I_1(k\xi)}{\xi} \right) \tag{15.12}$$

where the wave number in the Bessel function $I_1(\zeta\xi)$ is $\varsigma^2 = k^2 + s/\nu$. The constant of integration A is to be determined from the boundary conditions. To this end, it is required that the radial velocity matches the time variation of the shape of the interface. Additionally, the tangential and normal stresses vanish at the interface. Fulfillment of these requirements leads to

$$s^2 = \frac{\delta p}{R} \frac{k^2 - \zeta^2}{k^2 + \zeta^2} \lambda I_1(\lambda) \tag{15.13}$$

and

$$p + \delta p - 2\nu\rho \frac{\partial U}{\partial \xi} = \gamma \left(\frac{1}{R_1} + \frac{1}{R_2} \right). \tag{15.14}$$

It is also found that

$$\left(\frac{\gamma}{R^2 \rho} (1 - \lambda^2) + \delta p I_0(\lambda) \right) \epsilon_0 \, e^{st+i(kz+m\phi)} = 2\nu \frac{\partial U}{\partial \xi} \tag{15.15}$$

Putting together these results, and algebraically manipulating the corresponding expressions, one may obtain the characteristic equation

$$2k^2 R^2 (k^2 R^2 + \zeta^2 R^2) \frac{I_1'(kR)}{I_0(kR)} \left(1 - \frac{2k\zeta R^2}{k^2 R^2 + \zeta^2 R^2} \frac{I_1(kR)}{I_1(\zeta R)} \frac{I_1'(\zeta R)}{I_1'(kR)} \right)$$

$$-(k^4 R^4 - \zeta^4 R^4) = \frac{\gamma R k}{\rho \nu^2} \frac{I_1(kR)}{I_0(kR)} (1 - k^2 R^2) \tag{15.16}$$

This general result accounts for situations whereby the interplay between viscous and capillary forces is present. The following limiting cases are of special interest in our discussion:

a. Viscous forces dominate over capillary forces: Equation 15.16 approximately reduces to

$$s = \frac{\gamma}{6\rho v R}(1 - k^2 R^2) \tag{15.17}$$

Hence in limit $v \to \infty$, there appears no maximum mode of instability: the function decreases monotonously for all axial wave numbers. In other words, if the viscosity is high enough, the jet does not break up into small blobs of liquid, and the formation of a spray inside the chamber is not possible.

b. Capillary forces dominate over viscous forces: by a similar procedure as the one described above, it is possible to obtain the appropriate expression for the limit $v \to 0$. It reads

$$s_0^2 = \frac{\gamma k}{R^2 \rho} \frac{I_1(kR)}{I_0(kR)}(1 - k^2 R^2). \tag{15.18}$$

In contrast with Equation 15.17, this relation exhibits a mode of maximum instability at $kR = 0.697$. To this mode corresponds the wave length

$$\lambda = \frac{2\pi R}{k} = 2(4.51)R = 4.51d \tag{15.19}$$

which indicates that the jet will rupture at distances comparable with its diameter ($d=2R$). Accordingly, the break up times are of order 10^{3s}. The initial droplet size may be inferred from this result. Since incompressibility of the liquid implies volume conservation, it is reasonable to assume

$$4.51\frac{\pi}{4}d^3 \sim \frac{\pi}{6}D^3 \tag{15.20}$$

so that the droplet size is of order

$$D \sim 1.89d. \tag{15.21}$$

All other intermediate cases represented by Equation 15.16 have modes with wavelengths that become larger as the actual value of v increases. A straightforward conclusion follows: if the viscous effects become strong enough, the fragmentation of the jet will be delayed, or at least may be partially impeded. In other words, rather than producing a fine mist, the fuel injection system would spit large blobs of liquid into the chamber. In turn, a proper combustion process is inhibited. This is the reason why modifications of the injection system are needed to accommodate biofuels with higher than normal viscosities.

It must be noted that the preceding discussion does not take into account the effects of turbulence. It is well known that secondary fragmentation of large droplets occurs under the action of the turbulent atmosphere inside the combustion chamber. Therefore, the nature of the turbulent field ultimately determines the size distribution, as well as the general geometry of the spray. Theoretical developments may be conducted along the previously outlined scheme of analysis, or by considering the production of turbulent kinetic energy $k \sim u'^2 / (2g)$ and its corresponding dissipation rate $\epsilon = -dk / dt$. In any case, the variation of the surface energy due to the deformations of the free surface must again take into account the effects related to surface tension and to the oscillatory motions induced by the turbulence. Consideration of these processes is beyond the scope of the present review.

Some technologies are attaining sufficient maturity to help overcome the injection problems in an efficient manner. It is worth noting those producing an ultrasonic type of fuel injection. These are based on piezoelectric devices that operate in the 40 kHz band. The droplet size is found to be inversely proportional to the frequency of oscillation, but fine tuning should be required for specific values of the viscosity. Usually such devices have power consumptions of around 2 W. However, once again this figure may vary according to the viscosity and may even go as high as 8 W if a wide spray is needed. One major drawback related with these systems is the operating temperature; their practical limit is only 425 K. For these reasons, a proper set of benchmark experiments, aimed at quantitatively determining the effect of ultrasonic injection, is required. Finally, another technique explored by Agarwal et al. (2011) deserves consideration. The steam turbine injection gas technique could be further developed to enhance the engine's performance. According to these authors' results, the power output increases to some extent with the technique, although the problem of suitable operation with higher ambient temperatures still poses some problems.

15.4 CONCLUSIONS

An overview concerning the use of biofuels in jet propulsion applications was reviewed. While the potential benefits of biofuels are rather obvious and quite promising, technical and economical obstacles have to be addressed.

The very successful test flights conducted by major air companies hoping to reduce their CO_2 footprints is a clear indication of the true potential and viability of biofuels. It must be emphasized that these flights saved considerable amounts of CO_2, with blends that contained at the most 50 wt% of biofuel.

Moving toward a full fossil fuel replacement with a 100% blend is the future goal. This requires, at least, substantial modifications to some of the main components of jet engines. In this chapter, two of these problems were pointed out, namely: the characteristics of the fuel spray produced by the injection system and the geometry of the combustion chamber. The discussion was subsequently focused on the issues related with the injection of biofuels with higher viscosities. Viscosity is likely responsible for retarding the breaking up of the jet, such that the resulting spray contains undesirably large droplets. An inadequate combustion process is thereby produced. This situation might be overcome with an ultrasonic assisted injection system that enhances jet fragmentation. Nevertheless, further experimentation is required to assess its operational limits within the entire flying envelope.

ACKNOWLEDGMENTS

The images of the simulated fuel sprays were generated by Dr. M. Salinas-Vázquez of the IINGEN-UNAM as part of the ongoing research.

REFERENCES

Agarwal S., Kachhwaha S. S., and Mishra R. S. Performance improvement of a simple gas turbine cycle through integration of inlet air evaporative cooling and steam injection. *Journal of Scientific and Industrial Research*, 70(7): 544, 2011.

Air Group Network of Alaska Airlines and Lufthansa Group. Alaska Air Group Issues Sustainability Report. http://www.aviationpros.com/press_release/10731665/aeromexico-asa-and-boeing-will-operate-a-flight-withbiofuel-during-rio20.

Alaska Airlines. Alaska Airlines Flies on Gevo's Renewable Alcohol to Jet Fuel. http://splash.alaskasworld.com/Newsroom/ASNews/ASstories/AS_20160607_130516.asp, 2016.

Ali Y., Hanna M. A., and Leviticus L. I. *Emissions and Power Characteristics of Diesel Engines on Methyl Soyate and Diesel Fuel Blends*. Lincoln, NE: Elsevier Science Limited, 1995.

ASTM International. http://www.intechopen.com/books/advances-in-gas-turbine technology/biofuel-and-gas-turbine-engines ASTM. Specification for Biodiesel (B100) ASTM D6751-08. Technical Report, ASTM International, 2008.

Chandrasekhar S. *Hydrodynamic and Hydromagnetic Stability*. New York, Dover Publications Inc., 1961.

Drazin P. G. and Reid W. H. *Hydrodynamics Stability*. London: Cambridge Mathematical Lybrary, Cambridge University Press, 2nd Edition, 2004.

Felice J., Kate K., Chris M., Michal R., Robbie B., and Maija S. J., https://www.lufthansagroup.com/en/press/news-releases/singleview/archive/2011/july/15/article/1980.html, 2013.

Garibay-Hernandez A and Biotechnology Institute in the Autonomous University of Mexico. Biodiesel's Microalgae. *BioTecnologia*, 13(3): 38, http://scholar.google.com/scholar?hl=en&btnG=Search&q=intitle:Biodiesel+a+Partir+de+Microalgas#0, 2009.

Habib Z., Parthasarathy R., and Subramanyam G. Performance and emission characteristics of biofuel in a small-scale gas turbine engine. *Applied Energy*, 87(5): 1701–1709. doi:10.1016/j.apenergy.2009.10.024. http://dx.doi.org/10.1016/j.apenergy, 2010.

Janssen R. Biofuel-burning Microturbine—Opportunities of Biofuel-burning Microturbines in the European Decentralizesed-generation Market (Bioturbine). Technical Report, 2004.

Kamps T. Mass model of microgasturbine single spool turbojet engine. *Journal of Kones Powertrain and Transport*, 20(1): 6, 2013.

Khairallah P. Bioturbine Project Workshop, 2004.

Krishna C. R. Report on atomization tests for project titled biodiesel blends in microturbine. Technical Report, Brookhaven National Laboratory, Upton, NY, 2007.

Macera Cerruti O. R., Aguillón Martínez J. E., and Arvizu Fernández J. L. La Bioenergía en México, Un Catalizador del Desarrollo Sustentable, Red Mexicana de Bioenergía, 2006.

Nygren E., Aleklett K., and Höök, M., Aviation fuel and future oil production scenarios. *Energy Policy*, 37: 4003–4010, 2009.

Tan E. I. H. and Liou W. W. *Microgas turbine engine characteristics using biofuel. The Hilltop Review*, 5(1) 46–47, 2011.

16 Solar Cooking for All

Antonio Lecuona-Neumann, José I. Nogueira,
Pedro A. Rodríguez, and Mathieu Legrand
Carlos III University

CONTENTS

16.1 INTRODUCTION

Cooking is a basic need for humankind. Raw food can be insecure and difficult to digest. Only some vegetables and fruits are appropriate to be eaten raw (Sun, 2012) and those only under controlled conditions to avoid illnesses. Boiled water for drinking is essential to avoid deadly infections in many areas of the world, mainly in developing countries and where nontreated water is used.

16.1.1 ENERGY SOURCES FOR COOKING

Cooking is practiced everywhere by humankind, and for that a heat source above 100°C is needed. In developed countries, modern forms of energy are used in order to make cooking fast, clean, and controllable. Mainly gaseous fuels are used, either natural or petroleum gas, in addition to electricity using Joule effect, microwave ovens, or induction ranges. The energy consumption and cost for cooking in an average family is generally minute in comparison to heating, air conditioning, and other electricity consumption by home appliances, except for low-income users (Tiffany and Morawicki, 2013). In the poorest situations in developing countries, only firewood can be used, gathered directly from nature or bought from local markets. According to the International Energy Agency (IEA), about 3 billion people worldwide depend on burning solids for cooking (IEA, 2010). More than 95% of these people are in sub-Saharan African or in developing Asia, and around 80% are in rural areas (IEA, n.d.). A first transition is shifting from firewood to charcoal as it is lighter, burns with fewer fumes, and gives a higher and more consistent heating effect; for this, it is much preferred in urban and peri-urban environments, although the start of cooking is quite slow. Conversion of wood into charcoal wastes at least 2/3 and up to 9/10 of the heating content in wood, needing on average 5–7 kg of dry wood for producing 1 kg of charcoal. Although the net CO_2 released to the atmosphere by charcoal combustion is null (it has been previously absorbed by the plant from which it proceeds); its manufacture is very polluting and increases user cost. Using mineral coal carries similar problems to firewood and worsens some aspects: the formation of even more toxic fumes, the addition of net CO_2 to the atmosphere and the need for long distance transportation. Transition to liquid fuels such as kerosene and paraffin oil reduces ash and indoor fumes substantially but still not enough. In vast territories, its supply is irregular. Also, its cost is high, around 5 to 15 times that of firewood, although the efficiency of cook stoves is around twice that of those burning firewood or charcoal. Accidents with such fuels are still too great a possibility, such as children ingesting fuel, spillovers, and fires. Bioalcohol and other liquid biofuels, such as vegetable oil, have not reached widespread use for cooking. Canned liquefied petroleum gasses (LPGs) such as propane and butane are the next step forward in the path toward safety, cooking control, and air cleanliness. They are costly and scarce in many undeveloped areas in spite of governments' tendency to promote and subsidize LPGs for helping low-income families. Access to electricity is even worse in those areas; it probably will not improve in the foreseeable future in many small and remote communities because of cost. Use of electricity is very clean and controllable, but its production from burning fossil fuels signifies emissions to the atmosphere elsewhere. For both modern fuels and electricity, irregular supply and high cost are the main causes of avoiding them in favor of firewood or charcoal. A detailed study on choices for domestic fuels within this framework is Bisu et al. (2016). In some regions, the cost of firewood can be higher than the cost of the food itself. Worldwide biomass amounts to around 10% of world primary energy consumption (firewood, agricultural residues, dung, and waste). The World Health Organization estimated that the use of biomass fuels and coal for cooking and heating accounts for 10%–15% of global energy use (World Health Organization, 2006). It is the fourth energy source in the world behind oil, coal, and natural gas and can be around 80% of undeveloped countries' primary energy consumption, although data are difficult to gather because of the noncommercial nature of the directly collected biomass. Use of firewood for household cooking is around 50% worldwide, according to diverse sources.

Obtaining biogas from soft organic residues in anaerobic digesters is one possibility of getting a clean fuel for distributing to small communities in a renewable way (Abadi et al., 2017). In some places, feedstock biomass supply competes with fertilizers, but in other areas its use as fuel is used to manage biological residues, although in an elaborate way (Chen et al., 2017). This resource seems feasible only in rural environments. In peri-urban areas, its implementation is difficult if not impossible because of space requirements and environmental issues, such as odors.

16.1.2 HEALTH AND ENVIRONMENTAL ISSUES ASSOCIATED WITH COOKING

One main problem with combustion in poor dwellings is indoor air pollution from short-term toxic gases, such as carbon monoxide (CO) and unburned hydrocarbons (noncondensing HCs). This can be a problem not only in low-income dwellings (Logue and Singer, 2014). Long-term health effects are caused by a low concentration of organic gases and flying particles (ash, tar, and cinder) as well as smoke. These health effects include stroke, ischemic heart disease, chronic obstructive pulmonary disease (COPD), and lung cancer (Smith, 1994; Subramanian, 2014; Nasir et al., 2015, among others). An effective chimney is frequently nonexistent, and only small ventilating apertures are present. From 2 to 4 million premature deaths worldwide per year are attributed to this cause (Foell et al., 2011; World Health Organization, 2017) as well as decreasing GDP (e.g., Naciones Unidas/OCDE, 2016).

Indoor domestic air pollution is exacerbated by the low efficiency of the cook stoves used, mainly open fire frequently on the floor (three-stone cooker). The efficiency is typically not higher than 15% of the lower heating value of wood (Sutar et al., 2015). This translates into an intense formation of hot fumes that eventually are used for indoor space heating and food preservation through smoking. Moreover, it translates also into a high firewood consumption and its corresponding pollution. This makes children and women breathe polluted air for a prolonged time. Also, both demographic groups spend many hours gathering wood, which hinders the initial stages of economic development; increases the risks of accidents, kidnapping, wild animal attacks, and articular lesions; and keeps children out of school. Cultural and gender issues coincide to aggravate this problem (see Ong, 2015). *Deforestation* is another consequence of nonrenewable firewood collection and even more if charcoal is obtained (Tucker, 1999), but note that there are others (Allen and Barnes, 1985).

The first step for helping people reduce both indoor air pollution and firewood consumption is doubling the efficiency of the cooker by simply enclosing the flame in a reradiating insulated container that promotes complete combustion and redirects hot combustion fumes toward the cooking pot(s) (Still and Kness, n.d.). Many models of *improved cook stoves* are available, made of clay pottery, can sheets, welded steel, or a combination, constructed in place or in remote factories (ClimateTechWiki, n.d.). Large campaigns of dissemination have been endeavored for improved cook stoves, reporting good results. This implies advances in fighting *energy poverty* and reducing deforestation, but firewood is still necessary (IEA (OECD-IEA), 2010; IEA, n.d.). Air pollution and greenhouse gases emissions are also reduced but not eliminated, raising doubts about the efficacy of these initiatives in the long term. Lacey et al. (2017) indicate the benefits of complete abatement of domestic solid fuel burning: "Abatement in China, India, and Bangladesh contributes to the largest reduction of premature deaths from ambient air pollution, preventing 198,000 (102,000–204,000) of the 260,000 (137,000–268,000) global annual avoided deaths in 2050." In sub-Saharan Africa, about 730 million people cook with solid fuels, according to the IEA (2014). The shift to charcoal in urban areas in combination with population growth could increase the pressure on declining forests for firewood consumption.

Cooking creates nonnegligible atmospheric greenhouse gas emissions. Xu et al. (2015) review different measures to reduce the carbon footprint of cooking.

16.1.3 COOKING WITH SOLAR ENERGY

The use of solar energy for cooking will eliminate air pollution, the use of scarce firewood as well as its cost and burden (Solar Cookers International, 2016). Many energy-poor areas in the world are very sunny, e.g., SOLARGIS (2010–2016) and PVGIS (Institute for Energy and Transport, n.d.). In spite of the much-studied advantages of solar cooking, it is still underused (MacClancy, 2014).

The practice of solar energy is an old story that dates back to ancient times. Its use for cooking is not frequent, although several millions of solar cookers have been deployed in the last decades. Many people do not realize that it is actually possible. Billions of people still do not have access to modern forms of energy so that the potential benefits are huge, including global climate change prevention. Today, solar cookers are low cost and can be found in many sizes. Some facilities serve several thousand solar meals a day, using thermal technologies, thus an appropriate technical level has been reached. In India and in other countries, food processors use solar cookers in their operations, e.g., Solar Food Processing Network (2017) and International Solar Energy Society (2009). Elsewhere in the world, some restaurants use only the sun to cook meals, e.g., Restaurante Solar Delicias del Sol Villaseca et al. (2016), as a touristic attraction. Although the outdoor space requirement for solar cookers is high, it is still possible in crowded peri-urban areas (Toonen, 2009), especially with innovative photovoltaic cookers (more on that in Section 16.5.2).

Direct exposure of food to solar irradiance, with a maximum of around 1 kW/m^2 at midday, is not strong enough to elevate temperatures for cooking; only drying happens. Special devices have to be employed to reach cooking temperatures under the sun, using a highly absorptive surface (generally a black surface) in addition to either or both

- *Heat insulation* by protecting the cooking utensil with a greenhouse-effect cover, transparent to solar rays but opaque for infrared spontaneous self-radiation and at the same time keeping the hot air and steam from escaping. In non-transparent sides of the cover, conventional heat insulation needs to be practiced using wool, glass fibers, foams, dry grass, leaves, etc.
- *Optical solar concentration* by means of mirrors, and to a lesser extent, lenses; to increase the irradiance impacting in the utensil containing food

The combination of these techniques forms a contemporary *solar cooker*. Some examples are shown in Figures 16.3 through 16.5. Striking features are the possibility of construction with simple tools and elements, even scrap materials, and low maintenance; only ordinary cleaning is needed. They do not consume any fuel and do not produce any residues. They can use traditional metal pots and only require exterior paint to absorb the sun's rays.

Solar cooking also offers a health-improving effect for an enormous population, avoids environmental damage, contributes to food safety, and improves health issues for children and women. For displaced people (e.g., refugees), solar cookers are preferred to fueled cookers for their null fire risk lack of required supplies. Some solar cookers are very light and foldable, thus portable.

Also in developed countries, forest wildfires consume natural spaces every year, sometimes caused by barbecuing. Solar cookers not only eliminate this fire risk in parks and forest rest areas, but they also provide an educational experience about sustainability.

16.1.4 COOKING GROSS REQUIREMENTS AND SOLAR COOKING POSSIBILITIES

Food is mainly water; the simplification of assuming it is only water allows a conservative energy calculation. Let us assume a single meal for a small family:

Bringing 2 L of water from ambient temperature $T_a = 25°C$ to boiling, $T_{bo} \sim 100°C$ requires a supply of heat amounting to $Q = m_w c_w \Delta T = 2\ kg \times 4.2 \dfrac{kJ}{kg\ K} \times 75\ K = 630\ kJ$. Assuming that 37% of the input is lost to ambient (energy efficiency $\eta = \dfrac{630\ kJ}{(630+370)\ kJ} = 0.63$, only attainable with advanced electric and solar cookers) $\approx 1\ MJ \approx 0.28\ kW\ h$ of heat input is required. Less efficient cooking may represent twice this energy consumption, and three stone cookers four times. On the other hand, reducing the water content, during the cooking time (Figure 16.2) requires boiling approximately half of this water (~1 kg) for a representative recipe. This means consuming $Q = m_{bo} L_{bo,w} = 1\ kg \times 2.26\ MJ/kg = 2.3\ MJ$. Applying larger heat losses during boiling, because of a higher average temperature than during heating, makes about 3 MJ = 0.83 kW h required for just boiling, indicating the high energy consumption for reducing the water content in the food for traditional recipes such as rice cooking.

Solar cooking shares with other solar technologies its temporal variability, difficult to forecast in the mid or even short term. Moreover, dinner and breakfast cannot be cooked on-line for the lack of sun at those times. This makes both *heat retention* on the food itself and *thermal energy storage* (TES) in an additional body (Lecuona et al., 2003) very important to delay cooking. Nowadays these techniques are not much developed, except for *hay baskets* (Solar Cookers International Network a, n.d.).

Solar cooking has to be performed mostly outdoors, which can be problematic in some communities (e.g., urban areas) and inconvenient in severe weather. The dramatic price reduction of photovoltaic (PV) panels (UnderstandSolar.com, 2016) opens the possibility for indoor solar cooking with renewable electricity at a reasonable cost, appropriate for both rural and peri-urban areas. Some financial help is required as its amortization by not consuming any fuel can take several years. At the same time, PV opens possibilities for home off-grid electrification in isolated communities, for such services as telecommunication, lamp charging, cold production for food and drugs preservation, etc. Some studies have analyzed the strategy of starting modern energy supply for illumination and mobile charging to 1.2 billion people worldwide without grid access and 1 billion more with unreliable access. This can be achieved by deploying solar home systems of pico-power size PV (1–10 W peak of electricity, W_p) (Salas, 2017) and even expanding them to supply radios, TVs, and fans, with 100 W peak systems. In some studies, like Phadke (2015), the basic needs of cooking, water sterilization, sanitary hot water preparation, and food preservation are not considered in these kinds of campaigns, surprisingly. Sterilization of surgical instruments using solar energy in remote locations has also received some attention (González-Mora et al., 2016).

This chapter offers (1) the fundamentals of solar cookers, (2) the main types, (3) the testing of their performances, and (4) the new ways of offering attractive devices not only for developing countries but also for the developed ones. The new possibilities of TES and PV are also analyzed.

16.2   FUNDAMENTALS

Cooking the inside of the solids is more difficult than cooking liquids. The center of the pieces requires reaching at least temperatures T above 70°C for some time. Otherwise, pathogens can be alive, and cooking does not happen. Humid food preservation needs higher temperatures. When cooking with combustion or with electricity the power is so high (0.5–2 kW) that this is seldom a problem and boiling water ($T_{bo} \approx 100°C$ at sea level) is the current practice to guarantee sterilization and increase fluid agitation by bubbling. Solar cooking devices are not so powerful, so this is not the case; for individual or family use, the typical heating power is around 0.5 kW maximum with a 0.5–1.5 m² aperture area for collecting the sun's rays. Thus, (a) high-temperature cooking such as roasting, deep frying or surface browning is not so easy and cannot be reached except with special solar cookers of high sun concentration or electricity, (b) boiling water is not really necessary, representing both heat and water losses, (c) splitting food into small pieces allows the heat to reach the inside faster, (d) adding a liquid to dry food enhances heat transfer from the containing pot.

Solar *direct cookers* are those that expose the cooking utensil to the sun's rays (Yettou et al., 2014). Large solar cookers are of the *indirect cooker* type, whereas a heat transfer fluid is heated in a solar collector and transports heat indoors, where it is transferred to food (Figure 16.7).

Low-temperature cooking at $T < 100°C$ is possible and convenient to minimize heat losses to ambient, while tenderness and flavor can be enhanced (Wikipedia a, n.d.). At higher temperatures, cooking is faster and operates in a different way because some chemical reactions appear so that temperature is a basic parameter for some recipes. Reducing water or oil in a classical recipe and increasing cooking time is current practice in solar cooking, but it is not always done. This makes people change their cooking habits; what is not always easy, especially in a poor environment, is taking into account the time recovered from firewood collection. In order to understand the many possibilities that solar cooking offers, a simple mathematical model seems fruitful.

16.2.1   THERMAL MODEL

For the basic equation governing the net transient heat transfer to a single body $\dot{Q}_{tr}$ & [W] causing its temperature T change along time t, we consider the utensil (pot), food, and added water (sauce). We assume a lumped parameter model of a single body at a homogeneous temperature T above ambient T_a [K]. Cooking chemical reactions take negligible energy from heating and boiling energy.

A cover with a transparent section reduces the heat loss of the body by greenhouse-effect and favors both capturing sun rays and retaining sun heat (see Figures 16.4 and 16.5). Eventually, an insulated opaque section retains heat (see Figure 16.3). If a cover is used just for keeping heat indoors after sun exposure it is an opaque heat insulating box that can be different from the one used for sun exposure (e.g., hay baskets). Following Figure 16.1:

$$\dot{Q}_{tr} = C_{b+c}\frac{dT}{dt} = F'\left[\overbrace{\underbrace{A_a G_T \eta_o}_{\text{Solar heat}} - \underbrace{UA(T - T_a)}_{\text{Heat loss: } \dot{Q}_{l,dry}}}^{\dot{Q}_{tr,dry}} - \overbrace{\dot{m}_{bo}L_{bo}}^{\substack{\text{Enthalpy loss} \\ \text{by evaporation} \\ \text{or boiling}}}\right] \qquad (16.1)$$

In Equation 16.1:

- C_{b+c} is the heat capacity of the body (food f, water w, and utensil u) and cover c, respectively; $C_{b+c} = \overbrace{C_f + C_w + C_u}^{\text{Utensil}} + C_c$, (Incropera et al., 2007). As the cover usually takes a lower temperature than the body, its thermal capacity $C_C\left[\dfrac{J}{K}\right]$ is a fraction $\delta_c < 1$ of its real one C_{cover}.

- F' is a nondimensional correction coefficient that considers that the external surface, where heat losses occur to ambient, can be at a different temperature T_{ex} than the inner one T. On the body exposed to sun $T_{ex} > T$ and in the shadow $T_{ex} < T$. In a well-posed mathematical model of the body $F' \approx 1$ (Duffie and Beckman, 1991). $F' \to 1$ when the internal heat transfer to A is much more active than the external one interacting with T_a.

- A_a is the area devised for capturing the sun's rays, aperture area $\left[\text{m}^2\right]$.

- G_T is the tilted solar irradiance $\left[\dfrac{W}{m^2}\right]$, normal for the aperture surface, named also in-plane or with null incidence angle (Duffie and Beckman, 1991). For good efficiency all the sun concentrating optics need to be oriented towards the sun and preferably to track it in

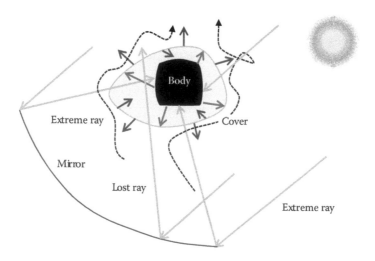

FIGURE 16.1 Schematic representation of a solar cooker with a fully transparent cover. Light gray arrows represent the sun's rays; four possibilities are shown. Extreme rays enclose aperture area A_a. Rays leaving the body represent infrared spontaneous thermal radiation stopped (absorbed or reflected) by the cover. Dark gray rays for spontaneous lower temperature cover thermal radiation. Dash lines represent convection air currents heat losses. Currents inside the cover or inside the body are not indicated.

its trayectory in the sky, at a higher or lower degree. If tracking the sun, it coincides with normal to sun rays irradiance $G_T = G_n$. If concentrating, only the beam irradiance has to be considered $G_T = DNI$ (Direct Normal Irradiance) as diffuse irradiance component cannot be optically concentrated.

- η_o is the optical efficiency of the sun collection device from sun impact down to the body. $\eta_o < 1$ because of mirror reflectivity, the transmittance of cover windows, lost rays, and utensil absorbance. A practical value for individual and family size solar cookers can be $\eta_o \approx 0.5$, although it can be much different depending on design, materials, and sun tracking.

- $A\left[\text{m}^2\right]$ is the area for heat losses, by conduction *cond*, convection *conv*, and radiation *rad*. Normally it is considered the external utensil area.

- $U\left[\dfrac{\text{W}}{\text{m}^2\,\text{K}}\right]$ is the global heat transfer coefficient through A, (Incropera et al., 2007); the barrier of the cover has to be considered. Generically $U = U_{cond} + U_{conv} + U_{rad}$. U_{conv} increases with $T - T_a$ for natural (free) convection, but forced convection must be added, this one increasing with wind velocity v_w. U_{rad} increases with body temperature and can be much reduced by a single or double layer cover made of glass. Generally $U_{conv,air} \approx U_{rad,amb} \approx 5 - 20$ W/K m^2 when cooking at $T_{ext} = 100°$C and there is no wind. A representative value is that of a sphere into air, as the shape of a compact body does not have much influence either to convection or to radiation heat transfer. For elements in a series (i.e., several layers in a single path) $U = \left(\sum_i U_i^{-1}\right)^{-1}$, so that the smallest U controls the overall value. This is why a layer of thermal insulation

- $(U_{in} \sim 0.1 - 1$ W/K m^2) as a cover is usually devised to dominate the value of U and in this way significantly reduces heat loss.

- $\dot{m}_{bo}\left[\dfrac{\text{kg}}{\text{s}}\right]$ is the mass flow rate of evaporated liquid from the foodstuff, typically water w, whose phase change enthalpy is very high $L_{bo,w} = 2.26$ MJ/kg. At $T < T_{bo}$, $\dot{m}_{bo}$ is small, and it is proportional to the liquid/air interphase area, grows rapidly (about exponentially) with T and liquid/gas agitation, and is smaller when liquid contains dissolved or suspended nonvolatile substances, e.g., salt, sugar, or oil. When $T = T_{bo}$ boiling heat balances the net heat applied so that $\dot{m}_{bo}L_{bo,w} = \dot{Q}_{tr}$ and temperature stabilizes. The resulting concentration of less volatile components increases boiling temperature.

The following sections add some explanations and implications:

16.2.2 Some Comments on the Model

- The cooking process is sketched in Figure 16.2. When $T = T_a$ solar heating power $\dot{Q}_{tr}$ is maximum, as losses are null. As temperature increases, losses increase, reducing the net heating power. Actually, if all terms in Equation 16.1 are constant except T and t, the solution for T has an exponential like evolution toward the steady-state temperature.

  Integrating Equation 16.1 with no evaporation, and all variables are considered constants except T and t between (t_1, T_a) and (t_2, T_2), Figure 16.2, one gets

$$\frac{T_{st,dry} - T_2}{T_{st,dry} - T_1} = \exp\left(-\frac{t_2 - t_1}{t_{coo}^*}\right); t_{coo,v}^* \doteq \frac{C_{b+c}}{\left(\overline{F'U}\right)_{hea} A} \qquad (16.2)$$

Here $\left(\overline{F'U}\right)_{hea}$ stands for the average value between t_1 and t_2.

$T_{st,dry}$ is the dry stagnation temperature, formulated in Equation 16.3. $t^*_{coo,v}$ is defined as a virtual characteristic cooling time at the same operating conditions as heating but with $G = 0$, actually $t^*_{coo,v} \approx t^*_{coo}$ (Equation 16.4). $\left(\overline{F'U}\right) >$, can be different during cooling in the shadow, Equation 16.4, because of a different temperature distribution on the body and cover.

The nonconstant rate heating process needs averaging for both power and efficiency during heating for getting a single figure for the user of the heating process, Equation 16.5.

- Maximum temperature is reached when $\dfrac{dT}{dt} = 0$. If water boiling is happening $\dot{m}_{bo} = F'\big|_{T=T_{bo}}\left[A_aG_T\eta_o - U\big|_{T=T_{bo}} A(T_{bo} - T_a)\right]/L_{bo,w}$. For no boiling, neither evaporation (requiring a pot cover or no significant amount of water in the food) the steady-state *dry stagnation temperature* is obtained:

$$T_{st,dry} = T_a + \frac{A_aG_T\eta_o}{A\left(F'U\right)_{T=T_{st}}} \tag{16.3}$$

Here, the possibility of piecewise integration along heating is retained, so that $F'U$ can evolve during the process.

- $t > t_{stopG}$ corresponds to $G_T = 0$ (dusk, heavy cloud, or indoors); cooling rate $\dfrac{dT}{dt} < 0$ is maximum in magnitude at the higher temperature and slows down as ambient temperature approaches. Integrating Equation 16.1 with all the parameters constant except t and T between (t_I, T_I) and (t_{II}, T_{II}), Figure 16.2, one gets

$$\frac{T_{II} - T_a}{T_I - T_a} = \exp\left(-\frac{t_{II} - t_I}{t^*_{coo}}\right); \quad t^*_{coo} \doteq \frac{C_{b+c}}{A\left(\overline{F'U}\right)_{coo}} \tag{16.4}$$

$\left(\overline{F'U}\right)_{coo}$ is the average value between T_I and T_{II}. An average value can be considered along the cooling or piecewise values considered as U depends on temperature.

- The solar cooker efficiency η has to consider that the heating process is transient. The steady-state period in Figure 16.2 is seldom taken into account as (a) boiling is not universal in solar cooking and (b) boiling power can be deduced from the heating period so that a test with steady-state boiling is not very useful, and (c) the capacity of boiling water depends on details of the utensil, more than on the cooker itself, e.g., lid is open or closed.

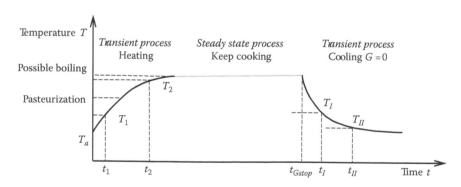

FIGURE 16.2   Representative temperature-time evolution starting and ending at ambient temperature.

- On the grounds of Equation 16.1 optical concentration makes area A for heat loss become much smaller than aperture area A_a. We define the *geometrical solar concentration*: $\mathbf{C} = \dfrac{A_a}{A} > 1$ resulting in a higher dry stagnation temperature, Equation 16.3.
- Thermal insulation reduces U to something slightly larger than the value of the insulation $U = \dfrac{k_{in}}{l}$ where the *heat conductivity* of the insulating materials is $k_{in} = (0.04 - 0.05)$ W/K m and l is the thickness of the insulation layer. Just $l = 5$ cm gives $U_{in} = 1$ W/K m$^2 \ll U_{conv+rad,air}$.
- The average water heating power $\bar{Q}_{tr}$ and efficiency $\bar{\eta}$ during heating can be calculated between (a) a temperature slightly higher than T_a to avoid water cooling, $T_{1,hea}$ and (b) a temperature slightly lower than T_{bo} to limit mass loss m_{bo}, $T_{2,hea}$, indicated in Figure 16.2:

$$\bar{Q}_{tr} = C_w \frac{T_{2,hea} - T_{1,hea}}{t_{2,hea} - t_{2,hea}}; \quad \bar{\eta}_{hea} = \frac{\bar{Q}_{tr} + \overbrace{m_{bo}L_{bo}}^{\text{optional}}}{G_T A_a}; \quad \overbrace{\eta_{bo}}^{\substack{\text{Steady-state} \\ \text{at } T = T_{bo}}} = \frac{m_{bo}L_{bo}}{G_T A_a t_{bo}}; \tag{16.5}$$

In this equation, the little-used boiling efficiency η_{bo} during a time t_{bo} has been introduced.

Generally, a constant G_T is required to obtain consistent results.

Direct thermal cookers generally do not attain $\bar{\eta}_{hea} > 50\%$ except those that are very elaborate. Both box ovens and dish concentrator (e.g., Figure 16.3) solar cookers with no covers yield typical values around 25%–35%.

- Considering the global tilted irradiance G_T (Equation 16.5) the normal one G_n, or the one on the horizontal plane G_h is something that needs to be clarified when performing characterization of solar cookers as there are nontracking and sun-tracking solar cookers, while others

FIGURE 16.3 High concentration solar cookers with dish mirrors. Left is the SK-14 type. It uses a deep paraboloid made with stripes of aluminum and is shown loaded with a black enameled pot of high sides to capture the sun's rays in the morning and afternoon (http://www.atlascuisinesolaire.com/photos-cuisine-solaire.php, 2014). Right: a community solar cooker based on a flexible shallow paraboloidal mirror rotating around a polar axis to automatically track the sun with constant speed, type Scheffler. It is made with flat mirror facets. Located in Museu de la Ciència i la Tècnica de Catalunya, Terrassa, Barcelona, Spain (http://atlascuisinesolaire.free.fr, 2010).

use a horizontal aperture area, such as the box oven. Moreover, the geographical information systems (GISs) typically give integral or average values of G_h and in some cases G_n.

- When cooling in the shadow with no evaporation, Equation 16.1 reduces to

$$C_{b+c}\frac{dT}{dt} = -A\left(F'U\right)_{coo}\left(T-T_a\right) \qquad (16.6)$$

This equation indicates that by piecewise deriving the $T-t$ evolution one can obtain $\left(F'U\right)_{coo} \approx \left(F'U\right)_{hea}$ as a function of T, requiring to specify A. C_b can be experimentally evaluated heating with a constant known power, such as an electrical resistance under adiabatic conditions. There are also several experiments for quantifying η_o (Lecuona-Neumann, 2017; Lecuona et al., 2013).

Pressure pots maintain a higher than atmospheric pressure inside the pot during boiling, resulting in an increased T_{bo}, reducing the time for accomplishing cooking, at the expense of slightly higher heat loss. A pressure pot/cooker is necessary and is usually more expensive than an atmospheric pressure (open) pot.

16.2.3 STANDARDS AND RECOMMENDATIONS FOR SOLAR COOKER EFFICIENCY AND TESTING

The complexities already considered and those of sun optics recommend joining all of the effects in a few nonbiased figures of merit for easy performance quantification that comes out of simple and low-cost testing. Pure water can be used instead of food for homogeneous testing and cleanliness, or sometimes oil to avoid evaporation and boiling. The problem with oil is knowing its specific heat for different temperatures. Two standards and a set of recommendations have been developed for solar cookers, needing only good weather and monitoring: temperature with at least one sensor, monitoring wind with an anemometer (avoidable if the day is calm), and sun radiance with a low-cost semiconductor solar meter (alternatively on a sunny day, data from a nearby meteorological station will serve). In addition, clock time and its translation into solar time are necessary for knowing when it is noon. Following are available standards:

1. American Society of Agricultural Engineers (2003) and American Society of Agricultural and Biological Engineers (2013). This standard puts emphasis on a standard average water heating power (at near maximum irradiance) measured during 10 min below 90°C and centered on $T-T_a = 50$°C, correcting for a global (horizontal) irradiance $G_h = 700$ W/m^2.
2. Bureau of Indian Standards (BIS) (1992) and BIS (2000). This standard was developed mainly for *box cookers*, based on Mullick et al. (1987) and described in Garg (1987). It has been applied in Mullick et al. (1991), among others. It gives two figures of merit: (a) F_1 quantifies losses in front of optical efficiency $F_1 = \dfrac{\eta_o}{(F'U)_{st,dry}} = C\dfrac{\left(T_{st,dry}-T_a\right)}{G}$, according to Equation 16.3, in a test of an empty cooker (or containing some oil just to allow a reliable temperature measurement) attaining the stagnation temperature around midday, and (b) F_2 is an approximation to η_o obtained in a transient heating test, both under approximately constant G_h.
3. CERCS (1994), Grupp et al. (1994), BMZ aktuell 060 (1994), and GTZ (1999). This unofficial standard recommends more exhaustive thermal testing and considers certain other aspects of cookers such as safety factors, ease of access to cooking pot, durability, and ergonomics. It is actually not a standard, but a collection of recommendations.

All of these standards share some philosophy with the water boiling test for firewood cookers (U.S. Environmental Protection Agency) that accounts for the mass of fuel (energy required) needed to bring to boil a specific amount of water (energy delivered), considering useless heating the utensil. The time for boiling the nominal load under standardized irradiance has not caught attention in the solar cooking standards, but has in some papers (e.g., Yettou et al., 2014). A review of standards and a unified proposal for a new world standard can be found in Kundapur and Sudhir (2009).

16.3 TYPES OF DIRECT SOLAR COOKERS

A comprehensive review of solar cookers types is Yettou et al. (2014). Thermal cookers are those that directly convert solar energy into heat, which is the current practice.

16.3.1 HIGH CONCENTRATION SOLAR COOKERS

The type that maximizes solar concentration $\mathbf{C} = A_a / A$ typically uses a moveable *concentrating dish* of reflecting surfaces for concentrating rays on the surface of an ordinary pot (see Figure 16.3). This happens on its bottom near midday but laterally in the morning and the afternoon, owing to the low altitude angle of the sun (e.g., Sonune and Philip, 2003). The high concentration allows avoiding a cover and keeping the pot open for stirring or adding ingredients, (Solar Cookers International Network b, n.d.; alSol, n.d.). Some test results can be found in Manchado-Megía (2010), Chandak et al. (2011), and Chaudhary et al. (2013).

Linear types of concentrating solar cookers have been developed with large diameter evacuated tube solar collectors entirely made of glass, e.g., GOSUN (n.d.) and Solar Cookers International Network d (n.d.). The external tube forms a transparent cover, resulting in a high insulation of the internals as the intermediate space between the external and the internal receiver tube is in a vacuum. Selective coating is applied to the external surface of an internal coaxial tube to simultaneously maximize absorption and minimize emission. This tube is joined with the external one in one of the sides. Between them, there is an evacuated space at high vacuum; as it is well sealed, it does not accumulate dirt or humidity. These tubes can be found in the market as they were initially developed for evacuated tube solar collectors. The food is inserted inside the inner coaxial tube, necessarily in small pieces. This provides them a very high thermal gain not needing high concentration; thus, these solar cookers are a sort of intermediate type; see Section 16.3.3.

High concentration solar cookers frequently use a reflecting dish. With them T_{st} can be up to around 200°C at maximum G, although up to 400°C have been reported, yielding a steady-state power of about 300–800 W with aperture areas of around 1.1–1.5 m². There are two basic types: those that spread the irradiance on the surface of the pot (deep dish) and those that concentrate irradiance in a small spot, thus approaching a shallow paraboloidal mirror shape. With family size types, boiling is possible and even frying lightly, such as eggs and fish. Baking is perfectly possible. Tracking the sun is necessary every 20 min, approximately.

16.3.2 HIGH INSULATION SOLAR COOKERS

These are the simplest, easily made solar cookers; their construction can be very affordable and made with local and simple materials. Actually, they are called *box solar ovens* as the food is heated uniformly, and the utensil is in a closed box. Its construction is shown in Figure 16.4: a well-insulated box has a removable top window where the sun enters. Frequently, heat loss is reduced by a double ordinary glass layer. The box can be made of wood, cardboard, cork, or any suitable material available. Insulation is enough if a several centimeters-thick wall is filled with dry hollow nonmetallic material impeding air motion. Several black pots can be accommodated inside. With basic designs, T_{st} barely can reach 100°C, especially when loaded, as its aperture area is small. The heating power is low, around 20 W for the basic types, so that cooking time is long (El-Sebaii and Ibrahim, 2005); one needs to add all ingredients and wait. To increase heating power, external boost flat mirrors are used (see Figure 16.4). With those, empty stagnation temperatures up to 200°C have

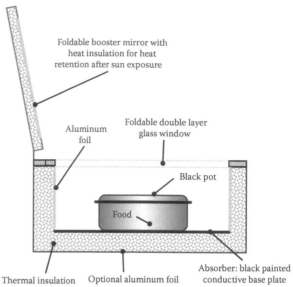

Foldable booster mirror with
heat insulation for heat
retention after sun exposure

Aluminum
foil

Foldable double layer
glass window

Black pot

Food

Thermal insulation   Optional aluminum foil   Absorber: black painted
conductive base plate

FIGURE 16.4   High insulation solar cookers. Left: commercial box solar oven made of injected plastic (Xuaxo, 2007) with collapsible booster mirror that can limit the cooking time, type Suncook. Right: a scheme of a generic solar box oven containing a single pot.

been reported, but power is still low. Some results of testing can be found in Sharma et al. (2000), Buddhi et al. (2003), Kumar (2005), El-Sebaii and Ibrahim (2005), Lahkar (2010), Saxena et al. (2011), and Mahavar et al. (2012).

Unattended cooking is possible if no external boost mirrors are used. A simple reorientation after some hours is enough when just a single mirror oriented toward the equator is used (see Figure 16.4b).

16.3.3   INTERMEDIATE SOLAR COOKERS

The representative type is the popular *panel cooker* (Solar Cookers International Network c, n.d.). This combines moderate solar concentration and insulation. Folded flat cardboard, plastic, or metallic mirrors concentrate the sun in a nonperfect way on a greenhouse, constructed with a bare plastic bag or scrap washing machine glass windows, forming a fairly air-tight enclosure. Inside it, a black pot contains the food. There are models for high latitudes and low latitudes. Net heating power with water is moderate, around 65 W at $T = T_a$, down to null value for $T - T_a = 75°C$. These data were measured in Madrid, Spain, for the Hotpot cooker, Figure 16.5 (Lecuona-Neumann, 2017).

FIGURE 16.5   Panel cookers. Left: Hotpot commercial intermediate solar cooker (Arveson, June 27, 2011). Right: Cookit type with scrap washing machine windows as cover and thermocouples for testing, photo by Manuel Manchado, Universidad Carlos III de Madrid www.uc3m.es.

16.4 HEAT STORAGE FOR SOLAR COOKING

Although heat storage for cooking is possible either with direct or indirect solar cookers, it is introduced in this section and extended to indirect solar cookers in Section 16.5. Cooking time extension is performed currently by wrapping the pot containing the already finished or partly cooked food with a much insulating cover, indoors. Also, this pot warping with insulating material keeps the food warm and ready to serve for some hours. Cooking continues down to 70°C, performing what is called fireless or thermal cooking (Wikipedia c, n.d.). In locations where training on the topic has been performed, ordinary baskets filled with hollow materials are useful and have become popular. It is enough to fill the baskets with used clothes, hay or dry leaves (Wikipedia d, n.d.). They do not insulate very well, as they allow steam and hot air transpiration, but on the other hand, this avoids mold formation. In addition, low-temperature cooking is a fashionable tendency in developed countries (Wikipedia b, n.d.).

There are two possibilities for extending or delaying cooking after sunset or allowing cooking during cloudy periods; sensible and latent TES are described below.

16.4.1 SENSIBLE HEAT THERMAL ENERGY STORAGE (TES)

Storing sensible heat (elevating temperature) in the food itself is not enough to keep it warm for hours; neither is it enough to keep from ruining cooking during a cloudy period. This limitation can be overcome by heating a mass of material at an elevated temperature that later is put in contact with the food. If this mass has stored enough heat at enough of an elevated temperature, it can even cook, called delayed cooking.

Cooking delay needs an intensive TES. This means charging a mass of high heat capacity material (c) with the sun, elevating its temperature above that required for cooking, such as in Schwarzer and Vieira da Silva (2003). This means a slower heating during the morning under the sun until cooking temperature has been reached and cooking in parallel to continue charging. Introducing the TES in an insulating cover maintains its cooking capacity for several hours and even a day. If the thermal battery cools down to a temperature lower than 70°C, sanitary hot water can still be produced.

Integrating Equation 16.6 for $t > t_{Gstop}$, Figure 16.2, considering the utensil and cover in contact with the TES and neglecting evaporation, the temperature results in a similar expression to Equation 16.4:

$$\frac{T_{II} - T_a}{T_I - T_a} = \exp\left[-\frac{A\overline{(F'U)}_{coo}}{C_{b+c+TES}} (t_{II} - t_I) \right] \tag{16.7}$$

This makes $t^{*}_{coo,TES} = \dfrac{m_{TES} c_{TES} + C_{b+c}}{A\overline{(F'U)}_{coo}}$ the time when initial over-temperature $T_I - T_a$ reduces by a factor $e^{-1} = 0.368$. It is the quotient of the increased thermal inertia by $m_{TES} c_{TES}$ over loss rate. If desiring $t^{*}_{coo} \approx 10 - 20$ h, a large mass m_{TES} is required of a material with large enough specific heat capacity c_{TES}.

Putting the food in contact with the TES performs cooking. Storage in sand, used engine oil, rocks, and cast iron has been attempted, but the mechanism involved is sensible heating s, so that to store a significant amount, high temperatures are needed, resulting in low efficiency of the solar cooker η_{hea} when charging with the sun, but also during storage. If bringing to boil a mass of water m_w at 25°C with a mass of TES m_{TES}, a temperature above 100°C ΔT_{TES} is required. It can be calculated with an adiabatic heat balance between masses, neglecting the heat capacity of the utensil and cover: $\Delta T_{TES,s} = \dfrac{m_w c_w}{m_{TES} c_{TES}} (100 - 25)$°C. As $\dfrac{c_w}{c_{TES}} \approx 1 - 10$ (exception made for liquid ammonia as

TES), either large m_{TES} or/and Δ_{TES} are needed. Storing heat in water or any other volatile liquid is inconvenient because of vapor pressure. Storing heat in hot oil accelerates its degradation and gum formation and produces odor.

16.4.2 LATENT HEAT TES

A better choice than sensible heat TES is latent heat TES with *phase change materials, PCMs* (Zalba et al., 2003; Sharma et al., 2009) among others. Some implementations are reported in Sharma et al. (2000) and Lecuona-Neumann (2017). Melting the PCM stores energy, which is given back when solidifying. If melting occurs above 100°C, it allows for boiling water when heat is restituted during solidification. Acetanilide has been attempted as PCM, (Chaudhary et al., 2013). The sugar alcohol erythritol *ER* is a good choice, with phase change at $T_{PC} = 118°C$ involving $L_{PCM,ER} = 340$ kJ/kg, figures similar to ice. It is edible and massively produced as a noncaloric sugar at an indicative retail price of 5 US\$/kg. Performing again an adiabatic heat balance within two identical TES masses, a reduction of over-temperature required for sensible heat storage of

$$\Delta T_{TES,l} = \frac{L}{c_{TES}}$$ is obtained. For erythritol it is around 120°C–250°C as $c_{TES,ER} = 1.38$ kJ/K kg when

solid and 2.76 kJ/K kg when liquid, (Lecuona-Neumann, 2017), resulting in a great advantage over sensible TES. The mass of PCM required to adiabatically heat up to a boiling state a mass of water

m_w only by phase change is $m_{TES} = \dfrac{m_w c_w}{L_{TES}} 75°C \approx m_w$ for erythritol. Some numerical studies can

be found in Lecuona et al. (2003) and in Tarwidi et al. (2016). Experimental results are shown in Lecuona-Neumann (2017). A discussion on figures of merit for TES solar cookers can be found in Lecuona et al. (2013).

With current PCMs, heat conductivity has to be enhanced using submerged metal fins, sponges, or particles, if substantial heating power is required both when charging and discharging, as the heat conductivity of PCMs is moderate, e.g., $k_{ER} = 0.733 / 0.326$ W/K m as solid/liquid. Issues still under review are PCM subcooling and thermal degradation e.g., Shukla et al. (2008). One topic that has not been studied much is the separation of the PCM from the containing walls when contracting as a result of solidification (Figure 16.6).

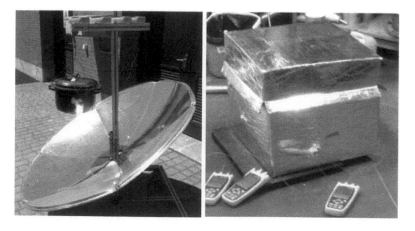

FIGURE 16.6 Left: TES utensil with PCM between outer and inner coaxial pots, being charged on a high concentration solar cooker of the paraboloidal dish type, model Icogen, in a testing campaign. Right: insulating cover for keeping heat for indoor use of the charged TES utensil, made of a cardboard box, plastic film, and polyurethane foam. (Photos by Alberto Barbero at Universidad Carlos III de Madrid, www.uc3m.es.)

16.5 INDIRECT SOLAR COOKERS

In indirect solar cookers, heat is delivered remotely from the solar collection. Heat transport is needed between the solar collector and the actual cooker.

16.5.1 INDIRECT SOLAR THERMAL COOKERS

For building a large power and for cooking indoors, a split layout is convenient. A solar collector of the medium temperature type (collecting heat at >100°C) heats fluid in a closed circuit that delivers the cooking heat remotely. Pumps are needed, or natural convection (*thermosiphon* effect or *heat pipes*) can be relied upon. Additional thermal losses in ducts and tanks must be considered and eventually reduced by adequate insulation. This design allows for a storage tank of the thermal fluid for TES and even with PCM storage (e.g., Kumaresan et al., 2016). Well stirred or stratified liquid storage strategies may be considered. The size of the tank should be optimized to reduce investment and/or enhance efficiency.

As already mentioned in Section 16.3.1, there is a great opportunity for obtaining high efficiency using evacuated glass tubes of the Sydney type from commercial solar collectors. A compound parabolic shaped mirror trough, similar to the one in Figure 16.7, concentrates solar rays on a tilted external glass tube. Oil is heated in an inner coaxial tube, acting as a sun receiver, ascending just by buoyancy forces toward a large bain-marie pot directly connected to the end of the tube, where the cooking pot is partially submerged in the hot free-flowing oil. Heavy insulation around the large pot allows to maximize the net heating power and provide a large t_{coo}^*. The oil that has delivered heat to the cooking pot spontaneously dives into the lower part of the inner tube, as a consequence of its higher density, to regain heat. On example is the blazing tube (2015).

16.5.2 PHOTOVOLTAIC SOLAR COOKERS

PV solar cookers have been attempted elsewhere (e.g., Joshu and Jani, 2013), using large solar panels and batteries as an indirect solar cooker. In this section, ongoing research for the development of a cost-effective PV solar cooker and its principles are described so that engineers and entrepreneurs can bring the ideas to a mature and reliable product for application in developing countries.

The basic design is a standalone electric system. A PV solar panel is fixed on the roof, eventually with some periodic adjustment of its orientation to better gather sun rays. Cables transport DC electricity to the kitchen (Figure 16.8). Batteries, or preferably TES, serve as storage for delaying cooking. Hydrogen-based electric generation, storage, and ulterior burning have been studied (Topriska et al., 2016) but their application for developing countries seems far in the future because of their cost.

Today, solar panels are massively produced with a retail price per watt peak (@$G_n = 1\,kW/m^2$ and $T_a = 25°C$) in the order of 0.7–1 US$/W$_p$ down to 0.5 US$/W$_p$ for large quantities. The largest silicon panels with $A_a \approx 2\,m^2$ yield around 300 W$_p$, at 12 or 24 V nominal. This power is equivalent

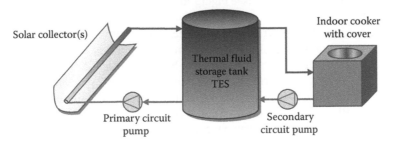

FIGURE 16.7 Layout of an indirect solar thermal cooker of the split type with sensible heat TES. Parabolic trough solar collector.

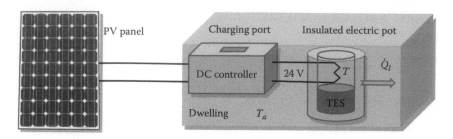

FIGURE 16.8 Layout of a stand-alone PV solar cooker with no batteries, using an electronic controller including a charging port for illumination, mobiles, and small appliances based on small independent batteries.

to the net power of a thermal solar cooker of family size with the advantage of high thermal insulation because apertures do not have to be devised for the sun around the pot. This delineates the possibility of clean and sustainable cooking indoors and provides surplus electricity for other uses (25 W_p are typically considered enough for illumination and mobile charging). Today, the main problem is the batteries. Deep cycle lead/acid ones will equal the 300 Wp panel price for 2.5 kW h nominal charging capacity. Only around 1/10 to no more than 1/2 of this capacity can be used for storage to avoid operational life reduction. Lead/acid batteries have a short duration, ~ 10^3 charging cycles, and have polluting residues if abandoned. The power mentioned for a single panel is enough to cook a meal and adequate for a year of average daily production of 1.5–1.7 kW h of a fixed stand-alone panel in northern Africa, according to PVGIS (2001–2008) (Institute for Energy and Transport, n.d.). This level of production will enter the dwelling into the official classification of "access to electricity" as more than 500 kW h per year will be possible. Two devices are added to the PV panel for a feasible solar cooker: (1) an electrical pot with enough heat insulation that allows cooking at this power level with an internal resistance, (2) a PV panel controller without batteries, which is not current practice.

1–1.2 kW maximum power AC electrical pots including a microprocessor are available at 30–100 US$ retail price in the international market with 5–6 L capacity. They incorporate washable inner liners, insulating lids, and double walls; they are thus heat insulated and even capable of acting as moderate pressure cookers, for reduced cooking time and steam loss. Eventually, heat insulation can be improved with a fiberglass mat. An independent electrical resistance for 24 V/300 W needs to be added in the heating base plate, thus eliminating the necessity of a DC/AC inverter for PV electricity and corresponding energy losses ~10%. With this modification, the pot would be dual electric feeding, retaining the capacity of using the grid AC electricity when available. This electricity can be also solar if in a later time a solar power plant feeds the community (e.g., Price Waterhouse Coopers, 2016; Condé, 2017).

For an efficient DC supply, a PV panel controller is needed to match its $I - V$ characteristic to the heating fixed resistance R in order to capture the maximum power under different irradiances, either throughout the day or in cloudy periods (see Figure 16.9). The simplest technique is an input feedback electronic circuit for constant voltage operation of the panel, using duty cycle intensity modulation, eventually with correction for sun irradiance and panel temperature. A better technique is based on a microcontroller that iteratively tracks the maximum power point in a continuous way (maximum power point tracking) (Salas et al., 2006). In Lecuona-Neumann et al. (2016), there is a description of a PV cooker with no battery or inverter, using a commercial electric pot. The power of such device ($A_a \approx 2 \text{ m}^2$) would be enough to also serve as a charging port for LED illumination devices, mobile phones, and even small appliances in low-income dwellings (e.g., Phadke, 2015). For houses in developed countries, the programmable electric pot could be a modern appliance for sustainable kitchen practices. PCM TES can be an optional add-on block inside the pot for delayed cooking.

With the 300 W applied, bringing 2 L of water to boil and melting 2 kg of erythritol takes about 40 min each. On a sunny day, it will take around 2 h to boil water when starting at sunrise.

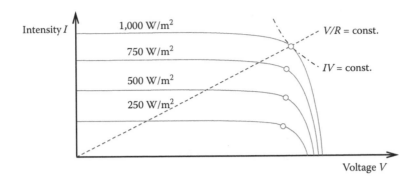

FIGURE 16.9 *I–V* curve of a solar panel at different normal irradiances and constant nominal temperature, showing the operating points of maximum power MPP with a hollow circle ∘. Dash line is for *I–V* relationship of a fixed load resistance *R* matched at peak irradiance. The dash-dot line is for constant power.

16.6 COMMENTS

16.6.1 COMMENT ON HEAT LOSS TO AMBIENT

In Hosny and Abou-Ziyan (1998), a correlation for the overall heat transfer coefficient for a bare cylindrical pot of 20 cm diameter and 9 cm height is given as a function of wind velocity v_w between 2 and 5.6 m s^{-1}, disregarding the effect of water temperature, which was below boiling, thus including evaporation losses, convection, and radiation (lid was not black):

$$\overline{F'U}_{coo} = \left[16.29 + 9.16\left(\frac{v_w}{\text{m s}^{-1}}\right)^{0.88}\right] \text{W/m}^2 \text{ K} \qquad (16.8)$$

This allows calculating the characteristic cooling time t_{coo}^*, Equation 16.4, of this pot at the maximum wind speed considered for standard testing, $v_w = 3$ m s^{-1} and assuming $F'_{coo} = 1$. Let´s consider a content of 1 L of water and neglect the heat capacity of the pot:

$$e^{-1} = \frac{T - T_a}{T_{1,coo} - T_a} = \exp\left[-\frac{AU_{coo}}{C_b}t_{coo}^*\right]$$

$$\rightarrow t_{coo}^* = \frac{m_w c_w}{AU_{coo}} = \frac{1 \text{ kg} \times 4.18 \text{ kJ/kg K}}{\underbrace{\left(\frac{\pi}{2} \times 0.2^2 + \pi \times 0.2 \times 0.09\right)}_{0.119} \text{m}^2 \underbrace{\left[16.29 + 9.16 \times 3^{0.88}\right]}_{40.4} \text{W/m}^2 \text{ K}} = 14.5 \text{ min}$$

With null wind, this time increases to 36 min. With a full cover of 5 cm thick with $k_{in} = 0.04$ W/K m, it increases to about 12 h.

16.6.2 COMMENT ON EFFICIENCY FACTOR F′

In Equation 16.1, the efficiency factor F' has been introduced so that the temperature internal to a pot, generally food, can be used in the equation despite the fact that the external surface is the one losing heat to ambient with radiation and convection heat transfer coefficient $U_{ex} = U_{rad+con}$, no cover for simplicity neither conduction losses. Considering this surface at the same time receiving the sun's rays, we have:

$$\dot{Q}_{tr} = A_a G_T \eta_o - (AU)_{ex} (T_{ex} - T_a) \tag{16.9}$$

The internal heat transfer equation from the pot's external wall to its content with heat transfer coefficient U_{in} can be formulated as

$$\left. \begin{array}{l} + \text{ for heating, } T_{ex} > T_a \\ - \text{ for cooling, } T_{ex} < T_a \end{array} \right\} \pm \overset{>0}{\dot{Q}_{tr}} = A_{ex} U_{in} (T_{ex} - T) \tag{16.10}$$

We neglect the temperature change across the metallic wall of the pot and any area variation.

Using Equation 16.10 to eliminate T_{ex} in Equation 16.9 as the relevant temperature is the one inside the pot and the controlling heat transfer is the external one, we get

$$\dot{Q}_{tr} = F' \left[A_a G_T \eta_o - (AU)_{ex} (T - T_a) \right]; F' = \left(1 \pm \frac{U_{ex}}{U_{in}} \right)^{-1} \tag{16.11}$$

A typical value for a bare pot in still atmosphere is $U_{ex} \approx 10$ W/K m^2 while for a liquid content $U_{in} \approx 100 - 1.000$ W/K m^2 so that $F' \approx 1$. If there is a cover external to A_{ex}, U_{ex} is even lower. If one selects the external surface coinciding with the external one of the cover, then U_{ex} will not change but U_{in} will diminish greatly, making F' differ substantially from unity. The same can happen if in the inside there is air: then $U_{in} \approx U_{ex}$.

REFERENCES

Abadi, N., Gebrehiwot, K., and Techane, K. (2017). Links between biogas technology adoption and health status of households in rural Tigray, Northern Ethiopia. *Energy Policy, 101*, 284–292. doi:http://dx.doi.org/10.1016/j.enpol.2016.11.015.

Allen, J. C. and Barnes, D. F. (1985). The causes of deforestation in developing countries. *Annals of the Association of American Geographers, 75*(2), 163–184. doi:http://dx.doi.org/10.1111/j.1467-8306.1985.tb00079.x.

alSol. (n.d.). *Cocina Solar alSol 1.4*. Retrieved February 1, 2017, from http://alsol.es/productos/cocina-alsol-1-4/.

American Society of Agricultural and Biological Engineers. (2013). *ASAE Standard S-580.1 Testing and Reporting Solar Cooker Performance*. Standard, ASABE, St. Joseph, MI.

American Society of Agricultural Engineers. (2003). *Standard ASAE S580 JAN 03 testing and reporting solar cooker performance*. ASAE, St. Joseph, MI.

Arveson, P. (June 27, 2011). *Wikipedia*. Retrieved June 20, 2016, from Solar cooker: http://en.wikipedia.org/wiki/File:HotPot_d.PNG.

Bisu, D., Kuhe, A., and Iortyer, A. (2016). Urban household cooking energy choice: An example of Bauchi metropolis, Nigeria. *Energy, Sustainability and Society, 6*(15), n.a. doi: 10.1186/s13705-016-0080-1.

Blazing Tube. (2015, n.a. n.a.). *Blazingtubesolar.com*. Retrieved February 14, 2017, from BlazingTubeSolarAppliance: http://blazingtubesolar.com/index.html.

BMZ aktuell 060. (1994). *ECSCR. Second international solar cooker test - Summary of results, ECSCR*. Bonn: Bundesministerium fur wirtschaftliche Zusammenarbeit und Entwicklung.

Buddhi, D., Sharma, S., and Sharma, A. (2003). Thermal performance evaluation of a latent heat storage unit for late evening cooking in a solar cooker having three reflectors. *Energy Conversion & Management, 44*, 809–817.

Bureau of Indian Standards (BIS). (1992). *Indian Standard-Solar Cooker, IS 13429*. New Delhi: BIS, Manak Bhawan.

Bureau of Indian Standards (BIS). (2000). *Indian Standards IS 13429: Solar Cooker e Box Type, First Revision*. New Delhi: BIS, Manak Bhawan.

CERCS. (1994). *Comite europeen pour la recherche sur la cuisson solair*. Systemes Solaires 104, 33–52. Paris.

Chandak, A., Somani, S. K., and Suryaji, P. M. (2011). Comparative analysis of SK-14 and PRINCE-15 solar concentrators. *Proceedings of the World Congress on Engineering 2011 Vol III.* Londres.

Chaudhary, A., Kumar, A., and Yadav, A. (2013). Experimental investigation of a solar cooker based on parabolic dish collector with phase change thermal storage unit in Indian climatic conditions. *Journal of Renewable and Sustainable Energy, 5*(12), n.a. doi:http://dx.doi.org/10.1063/1.4794962.

Chen, Y., Hu, W., Chen, P., and Ruan, R. (2017). Household biogas CDM project development in rural China. *Renewable and Sustainable Energy Reviews, 67,* 184–191. doi:http://dx.doi.org/10.1016/j.rser.2016.09.052.

ClimateTechWiki. (n.d.). *Improved Cook Stoves.* (W. van der Gaast, Editor) Retrieved February 1, 2017, from http://www.climatetechwiki.org/technology/imcookstoves.

Condé, A. (2017). *We can bring electricity to all of Africa in just 10 years.* (World Economic Forum) Retrieved February 1, 2017, from World Economic Forum Annual Meeting 2017: https://www.weforum.org/agenda/2017/01/we-can-bring-electricity-to-all-of-africa-in-just-ten-years?utm_content=buffer4595a&utm_medium=social&utm_source=twitter.com&utm_campaign=buffer.

Duffie, J. A. and Beckman, W. A. (1991). *Solar Engineering of Thermal Processes.* Hoboken, NJ: John Wiley & Sons.

El-Sebaii, A. A. and Ibrahim, A. (2005). Experimental testing of a box-type solar cooker using the standard procedure of cooking power. *Renewable Energy, 30,* 1861–1871.

Foell, W., Pachauri, S., Spreng, D., and Zerriffi, H. (2011). Household cooking fuels and technologies in developing economies. *Energy Policy, 39,* 7487–7496. doi:http://dx.doi.org/10.1016/j.enpol.2011.08.016.

Garg, H. P. (1987). Solar cookers. In H. P. Garg (ed.), *Advances in Solar Energy Technology. Volume 3 Heating, Agricultural and Photovoltaic Applications of Solar Energy* (pp. 1–61). Amsterdam, The Netherlands: Springer.

González-Mora, E., Rincón-Mejía, E., and Lawrence, D. (2016). Using a new solar sterilizer for surgical instruments as a solar oven for cooking. *CYTEF 2016 – VIII Congreso Ibérico | VI Congreso Iberoamericano de las Ciencias y Técnicas del Frío* (p. n.a.). Coimbra: SECYTEF http://secytef.upct.es/. Retrieved from http://www.adai.pt/event/event/home/index.php?target=home&defLang=4&event=2.

GOSUN. (n.d.). *Gosun Sport.* Retrieved February 1, 2017, from Gosun: https://www.gosunstove.com/products/gosun-sport.

Grupp, M., Merkle, T., and Owen-Jones, M. (1994). *Second International Solar Cooker Test.* Lodeve, France: European Committee for Solar Cooking Research & Synopsis.

GTZ. (1999). *Grupp, M. et al., Solarkocher in Entwicklungsländern (Solar cookers in developing countries).* Eschborn: http://www.giz.de.

Hosny, Z. and Abou-Ziyan. (1998). Experimental investigation of tracking paraboloid and box solar cookers under Egyptian environment. *Applied Thermal Engineering, 18,* 1375–1394.

http://atlascuisinesolaire.free.fr, A. B. (2010, May 1). *Wikipedia.* Retrieved February 3, 2017, from Solar cooker: https://en.wikipedia.org/wiki/File:Parabole_de_cuisson_solaire_Scheffler_coccion_solar_cooking.jpg.

http://www.atlascuisinesolaire.com/photos-cuisine-solaire.php. (2014, January 1). *Wikipedia.* Retrieved February 03, 2017, from Solar cooker: http://en.wikipedia.org/wiki/File:ALSOL.jpg.

IEA, International Energy Agency. (2010). *World Energy Outlook 2010.* Paris, France: IEA.

IEA, International Energy Agency. (n.d.). *Role of sustainable energy in ending poverty.* Retrieved February 2, 2017, from http://www.iea.org/topics/energypoverty/.

Incropera, F. P. et al. (2007). *Fundamentals of Heat and Mass Transfer.* Hoboken, NJ: John Wiley & Sons.

Institute for Energy and Transport. (n.d.). *PVGIS.* (European Commission. Center for Joint Research) Retrieved January 31, 2017, from http://re.jrc.ec.europa.eu/pvgis/.

International Energy Agency (OECD-IEA). (2010). *Energy Poverty How to Make Energy Access Universal?* Paris: International Energy Agency.

International Energy Agency. (2014). *Africa Energy Outlook.* (Robert Priddle, Ed.) Paris, France: International Energy Agency (OECD-IEA). Retrieved from http://www.iea.org/.

International Solar Energy Society. (2009). *Solar Food Processing.* International Solar Energy Society (ISES). Freiburg: International Solar Energy Society. Retrieved January 31, 2017, from www.ises.org.

Joshu, S. and Jani, A. (2013). Certain Analysis of a Solar Cooker with Dual Axis Sun Tracker. *2013 Nirma University International Conference on Engineering (NUiCONE).* IEEE.

Kumar, S. (2005). Estimation of design parameters for thermal performance evaluation of box-type solar cooker. *Renewable Energy, 30,* 1117–1126.

Kumaresan, G., Vigneswaran, V., Esakkimuthu, S., and Velraj, R. (2016). Performance assessment of a solar domestic cooking unit integrated with thermal energy storage system. *Journal of Energy Storage, 6,* 70–79. doi:http://dx.doi.org/10.1016/j.est.2016.03.002.

Kundapur, A. and Sudhir.C.V. (2009). Proposal for a new world standard for testing solar cookers. *Journal of Engineering Science and Technology, 4*(3), 272–281.

Lacey, F., Henze, D., Lee, C., van Donkelaar, A., and Martin, R. (2017). Transient climate and ambient health impacts due to national solid fuel cookstove emissions. *PNAS*, 114(6), 1269–1274. doi:10.1073/pnas.1612430114.

Lahkar, P. J. (2010). A review of the thermal performance parameters of box type solar cookers and identification of their correlations. *Renewable and Sustainable Energy Reviews, 14*(6), 1615–1621.

Lecuona, A., Nogueira, J. I., Ventas, R., Rodríguez-Hidalgo, M. C., and Legrand, M. (2003). Solar cooker of the portable parabolic type incorporating heat storage based on PCM. *Applied Energy, 111*, 1136–1146. doi:10.1016/j.apenergy.2013.01.083.

Lecuona, A., Nogueira, J. I., Vereda, C., and Ventas, R. (2013). Solar cooking figures of merit. Extension to heat storage. In A. Méndez-Vilas, & http://www.formatex.info/energymaterialsbook/ (Eds.), *Materials and Processes for Energy: Communicating Current Research and Technological Developments* (pp. 134–141). Badajoz, Spain: Formatex Research Center.

Lecuona-Neumann, A. (2017). *Cocinas solares. Fundamentos y aplicaciones. Herramientas de lucha contra la pobreza energética*. Barcelona, Spain: Marcombo.

Lecuona-Neumann, A., Sanchez-Bodas, M., and Ventas-Garzón, R. (2016). Propuesta de cocina solar fotovoltaica. Análisis simplificado. In CYTEF (Ed.), *VIII Iberian Congres/VI Iberian Congress on Refrigeration Sciences and Technology* (p. 597). Coimbra: CYTEF. Retrieved September 17, 2016, https://www.adai.pt/event/event/home/index.php?target=home&defLang=4&event=2, from file:///C:/Users/ANTONIO/AppData/Local/Temp/Rar$EXa0.107/CYTEF2016%20-%20Proceedings/CYTEF2016_proceedings/CYTEF2016_proceedings/others/4.Chapter08.html.

Logue, J., and Singer, B. (2014). Energy impacts of effective range hood use for all U.S. residential cooking. *HVAC&R Research, 20*(2), 264–275. doi:10.1080/10789669.2013.869104.

MacClancy, J. (2014). Solar Cooking ¿Why is it not yet global? *Food, Culture & Society, 17*(2), 301–318.

Mahavar, S., Sengar, N., Rajawat, P., Verma, M., and Dashora, P. (2012). Design development and performance studies of a novel Single Family Solar Cooker. *Renewable Energy, 47*, 67–76.

Manchado-Megía, M. (2010). *Caracterización de una cocina solar parabólica*. Universidad Carlos III de Madrid, INgeniería Térmica y de Fluidos. Leganés, Madrid, España: Universidad Carlos III de Madrid. Retrieved Agosto 2015.

Mullick, S. C., Kandpal, T. C., and Kumar, S. (1991). Thermal test procedure for a paraboloid concentrator solar cooker. *Solar Energy, 46*(3), 139–144.

Mullick, S. C., Kandpal, T. C., and Saxena, A. K. (1987). Thermal test procedure for box type solar cooker. *Solar Energy, 39*(4), 353–360.

Naciones Unidas/OCDE. (2016). *Evaluaciones del desempeño ambiental, Perú*. (CEPAL, Ed.) Santiago, Chile: Naciones Unidas/OCDE. Retrieved from http://repositorio.cepal.org/bitstream/handle/11362/40171/S1600313_es.pdf?sequence=1.

Nasir, Z. A., Murtaza, F., and Colbeck, I. (2015). Role of poverty in fuel choice and exposure to indoor air pollution in Pakistan. *Journal of Integrative Environmental Sciences, 12*(2), 107–117. doi:10.1080/1943815X.2015.1005105.

Ong, C. (2015). Choice of energy paths: its implications for rural energy poverty in less developed countries. *Society & Natural Resources, 28*(7), 733–748. doi:10.1080/08941920.2015.1020583.

Phadke, A. (2015). *Powering a home with just 25 watts of solar PV: super-efficient appliances can enable expanded off-grid energy service using small solar power systems*. Berkeley, CA: Ernest Orlando Lawrence Berkeley National Laboratory.

Price Waterhouse Coopers. (2016). *Electricity beyond the grid. Accelerating access to sustainable power for all*. Price Waterhouse Coopers. Retrieved from www.pwc.com/utilities.

PVGIS. (2001–2008). *Photovoltaic Geographical Information System*. (E. C. European Commission, Editor). Retrieved 2012, from http://re.jrc.ec.europa.eu/pvgis/apps3/pvest.php.

Restaurante Solar Delicias del Sol Villaseca, Vicuña, Chile. (2016). *Restaurante Solar Delicias del Sol Villaseca*. Retrieved from https://www.facebook.com/pages/Restaurant-Solar-Delicias-Del-Sol-Villaseca/254951814584007.

Salas, V. (2017). Stand-alone photovoltaic systems. In N. Pearsal (Ed.), *The Performance of Photovoltaic (PV) Systems* (pp. 251–296). Elsevier. doi:10.1016/B978-1-78242-336-2.00009-4.

Salas, V., Olías, E., Barrado, A., and Lázaro, A. (2006). Review of the maximum power point tracking algorithms for stand-alone photovoltaic systems. *Solar Energy Materials and Solar Cells, 90*(11), 1555–1578. doi:10.1016/j.solmat.2005.10.023.

Saxena, A., Varun, Pandey, S. P., and Srivastav, G. (2011). A thermodynamic review on solar box type cookers. *Renewable and Sustainable Energy Reviews, 15*, 3301–3318.

Schwarzer, K. and Vieira da Silva, M. (2003). Solar cooking system with or without heat storage for families and institutions. *Solar Energy, 75*(1), 35–41.

Sharma, A., Tyagi, V., Chen, C., and Buddhi, D. (2009). Review on thermal energy storage with phase change materials and applications. *Renewable and Sustainable Energy Reviews, 13*, 318–345.

Sharma, S. D., Buddhi, D., Sawhney, L. R., and Sharma, A. (2000). Design, development and performance evaluation of a latent heat storage unit for evening cooking in a solar cooker. *Energy Conversion & Management*(41), 1497–1508.

Shukla, A., Buddhi, D., and Sawhney, R. (2008). Thermal cycling test of few selected inorganic and organic phase change materials. *Renewable Energy, 33*, 2606–2614.

Smith, K. R. (1994). Health, energy, and greenhouse-gas impacts of biomass combustion in household stoves. *Energy for Sustainable Development, 1*(4), 23–29.

Solar Cookers International. (2016, September). *Solar Cookers International Network*. (S. C. International, Editor). Retrieved September 17, 2016, from SCInet: http://solarcooking.wikia.com/wiki/Scheffler_Community_Kitchen.

Solar Cookers International Network a. (n.d.). *Heat-retention cooking*. Retrieved February 2, 2017, from http://solarcooking.wikia.com/wiki/Heat-retention_cooking.

Solar Cookers International Network b. (n.d.). *Parabolic solar cooker designs*. Retrieved February 1, 2017, from http://solarcooking.wikia.com/wiki/Parabolic_solar_cooker.

Solar Cookers International Network c. (n.d.). *Solar panel cooker designs*. Retrieved February 2, 2017, from http://solarcooking.wikia.com/wiki/Solar_panel_cooker_designs.

Solar Cookers International Network d. (n.d.). *Solar trough cooker designs*. Retrieved February 1, 2017, from http://solarcooking.wikia.com/wiki/Solar_trough_cooker_designs.

Solar Food Processing Network. (2017). *Solar Food Processing Network*. Retrieved January 31, 2017, from http://www.solarfood.org/.

SOLARGIS. (2010–2016). *SOLARGIS*. (SOLARGIS, Editor). Retrieved January 31, 2017, from Solargis s.r.o.: http://solargis.info/imaps/.

Sonune, A. V. and Philip, S. K. (2003). Development of a domestic concentrating cooker. *Renewable Energy, 28*, 1225–1234.

Still, D. and Kness, J. (n.d.). *Capturing HEAT: Five Earth-Friendly Cooking Technologies and How to Build Them (2nd ed.)* (available http://aprovecho.org/publications-3/). Cottage Grove, OR: Aprovecho Research Center.

Subramanian, M. (2014, Mayo 28). Global health: Deadly dinners. *Nature, 509*, 548–551.

Sun, D. W. (2012). *Thermal Food Processing 2nd ed.* Boca Raton, FL: CRC Press.

Sutar, K. B., Ravi, M. R., and Ray, A. (2015). Biomass cookstoves: A review of technical aspects. *Renewable and Sustainable Energy Reviews, 41*, 1128–1166.

Tarwidi, D., Murdiansyah, D., and Ginanja, N. (2016). Performance evaluation of various phase change materials for thermal energy storage of a solar cooker via numerical simulation. *International Journal of Renewable Energy Development, 5*(3), 199–210. doi:10.14710/ijred.5.3.199-210.

Tiffany, J. H. and Morawicki, R. (2013). Energy consumption during cooking in the residential sector of developed nations: A review. *Food Policy, 40*, 54–63.

Toonen, H. M. (2009). Adapting to an innovation: Solar cooking in the urban households of Ouagadougou (Burkina Faso). *Physics and Chemistry of the Earth, Parts A/B/C, 34*(1–2), 65–71.

Topriska, E., Kolokotroni, M., Dehouche, Z., Novieto, D., and Wilson, E. (2016). The potential to generate solar hydrogen for cooking applications: Case studies of Ghana, Jamaica and Indonesia. *Renewable Energy*, 495–509. doi:10.1016/j.renene.2016.04.060.

Tucker, M. (1999). Can solar cooking save the forests? *Ecological Economics, 31*(1), 77–89.

UnderstandSolar.com. (2016, n.a. n.a.). *Cost of solar panels over time*. Retrieved February 09, 2017, from https://understandsolar.com/cost-of-solar/#cost.

US Environmental Protection Agency. (n.d.). *The water boiling test (WBT) version 4.2.2, cookstove emissions and efficiency in a controlled laboratory setting*. US EPA, Partnership for Clean Indoor Air (PCIA), Global Alliance for Clean Cookstoves (GACC), Aprovecho Research Center (ARC). Retrieved from http://www.aprovecho.org (last accessed on September 22, 2017).

Wikipedia a. (n.d.). *Low temperature cooking*. Retrieved February 2, 2017, from https://en.wikipedia.org/wiki/Low-temperature_cooking.

Wikipedia b. (n.d.). *Slow food*. Retrieved February 2, 2017, from https://en.wikipedia.org/wiki/Slow_Food.

Wikipedia c. (n.d.). *Thermal cooking*. Retrieved February 2, 2017, from https://en.wikipedia.org/wiki/Thermal_cooking.

Wikipedia d. (n.d.). *Wonderbag*. Retrieved February 2, 2017, from https://en.wikipedia.org/wiki/Wonderbag.

World Health Organization. (2006). *Fuel for life: household energy and health*. Geneva: World Health Organization. Retrieved from https://www.cabdirect.org/cabdirect/abstract/20063133325

World Health Organization. (2017). *Household air pollution and health*. Retrieved from Fact sheet No 292: http://www.who.int/mediacentre/factsheets/fs292/en/.

Xu, Z., Sun, D., Zhang, Z., and Zhu, Z. (2015). Research developments in methods to reduce carbon footprint of cooking operations: A review. *Trends in Food Science & Technology, 44*(1), 49–57. doi:http://dx.doi.org/10.1016/j.tifs.2015.03.004.

Xuaxo. (2007). *Wikipedia*. (Wikipedia) Retrieved from Solar Cooker: https://commons.wikimedia.org/wiki/File:Solar_oven_Portugal_2007.jpg.

Yettou, F., Azoui, B., Malek, A., Gama, A., and Panwar, N. (2014). Solar cooker realizations in actual use: An overview. *Renewable and Sustainable Energy Reviews*, *37*, 288–306. doi:10.1016/j.rser.2014.05.018.

Zalba, B., Marín, J. M., Cabeza, L. F., and Mehling, H. (2003). Review on thermal energy storage with phase change: materials, heat transfer analysis and applications. *Applied Thermal Engineering, 23*, 251–283.

17 Sustainable Wind Energy Systems

Eduardo Rincón-Mejía
Autonomous University of Mexico City

Ana Rincón-Rubio
Mexico State University

CONTENTS

"The answer is blowing in the wind,"

Bob Dylan, Literature Nobel Prize Laureate 2016

17.1 INTRODUCTION

How can we get a sustainable energy system? Perhaps the answer is not just in the wind, but wind power will always be an important part of all large-scale sustainable energy systems. This chapter is dedicated to providing insight into the potential contribution of wind power. After a brief historical review of the use of wind energy for millennia and the technical aspects of wind harnessing, an estimate is made of which technologies may be the most appropriate, given the environmental problems they could cause and how to avoid or mitigate them. Wind is inexhaustible, abundant, clean, and potentially quite environmentally friendly, and there is an abundance of windy regions: shores, mountains, and the offshore immense extensions, where absence of terrain obstacles allows strong wind currents to be utilized. Wind is also suitable to be caught in relatively small extensions (in contrast to solar plants, which require around one hectare per electric MW installed; in the same extension a multi MW turbine can be erected easily). The environmental issues that the deployment of wind power could bring can be avoided or mitigated, and in schemes of societal organization like "community power" it has resulted in the must sustainable renewable power technology: low cost, no GHG or toxic emissions, safe and secure operation, suitable for appropriate technology, no societal issues, inexhaustible and reliable for the centuries to come. As an example, since 2012, the cost of energy from offshore wind installations in Britain has fallen by a third, and in 2017 wind accounts for more than 40% of new electric capacity in the U.S., representing an annual investment of almost US$15bn, and growing. Now commercial 10 MW wind turbines for offshore application have been announced, and others even bigger are already in the design bureaus of wind engineers. With these big machines, more than 500 GW of wind capacity have been installed worldwide, and this capacity is expected to be doubled by 2024. However, small wind machines, affordable for every farmer and homeowner in the countryside and in cities all around the world, are perhaps the most sustainable option for autonomous, decentralized, resilient, and reliable component of hybrid future global 100% renewable energy systems (Rincón and Duran, 2006).

17.2 HISTORICAL REVIEW OF WIND APPLICATIONS

Wind was the first source of nonanimal, nonhuman power to provide mechanical energy to mankind. It is well known that many ancient civilizations took advantage of wind to propel their boats, with which they made long voyages, populated remote islands in Polynesia and traveled from Scandinavia to North America many centuries before the voyages of Christopher Columbus, Magellan, and other celebrated sailors. Also, winds have always been extensively used with small sailed boats in local-extension mobility of persons and merchandise through rivers, lakes, and lagoons, in addition to a broad cabotage along coasts. There is archaeological evidence of ancient engravings and still colored paintings of Egyptian boats dating back more than three millennia. Also, very well preserved Viking ships more than 1100 years old have been unearthed; these today are exhibited in the Vikingskipshuset Museum in Oslo, Norway.

In addition to propelling boats, wind has been used since innumerable times to ventilate and refresh buildings in torrid areas, smelt metals (Juleff, 1996), grind grains, pump water, and from the end of the nineteenth century, generate electricity, probably its most recognized application. There is even the hypothesis that the famous hanging gardens of Babylon were irrigated 38 centuries ago by wind pumps in the time of the great Hammurabi (Golding, 1976). Despite the recent warlike conflicts that have destroyed much of Afghanistan's and Iran's historic heritages, there are Persian ruins of what were vertical-axis windmills, whose successful design allowed the operation of these types of air machines over at least fifteen centuries, in what was a truly sustainable application. As opposed to contemporary wind turbines, old mills had to be attended to continuously by an operator.

In the European Middle Ages, during the Crusades, European fighters took back to their countries the technology of windmills, which by that time was already widespread throughout all the Middle East, reaching Greece, France, Spain, the Netherlands, England, and farther. Recall the struggles of Don Quixote against windmills in La Mancha, narrated in the masterpiece by Miguel de Cervantes, written more than 400 years ago. In the Netherlands and Mediterranean countries, the technological evolution of windmills continued (according to the constructal law of design evolution treated in Chapter 4 of this book). The Dutch took to these machines, having at one time 10,000 in their tiny country. They used that advanced technology (the first wind farms indeed) to pump water to prevent the ocean from flooding low-lying land, and even to gain new lands from the North Sea. That was undoubtedly another sustainable application of wind machines.

Robert R. Righter, a wind energy historian, wrote that windmills were the "most complex power devices of medieval times," preceding the coming Industrial Revolution. In his book (1996), he relates the history of the efforts to capture the power of wind for electricity, from the first European windmills to California's wind farms of the late twentieth century. Much before these wind farms, between the twelfth and nineteenth centuries, wind power was the main source of mechanical energy, with some not negligible waterwheel contribution. Fernand Braudel, the recognized French historian, wrote: "In the 11th, 12th and 13th centuries, the Occident experienced its first mechanical revolution. We refer to 'Revolution' as the total of all changes that were caused by the increased number of wind—and water—mills. Although the output of these 'primary drivers' was very limited (between 2 and 5 hp for a watermill, 5 to max 10 hp for a windmill), considering the world's very poor supply, this constituted a dramatic increase in power and was decisive for the first grow phase of Europe" (1979). From the commercial introduction of the steam engines of Watt and Boulton in 1774, initiating a second great technologic (and industrial) revolution, these engines began very slowly to displace wind and water mills. However, more than 120 years after the beginning of the massive commercialization of steam—and even internal combustion—machines, most European power was provided by wind and water currents, plus human and animal power. When the massive use of abundant and low-cost fossil fuels for internal combustion engines became unstoppable, this polluting new technology displaced (for less than a century) the old wind machines.

According to the historian Walter Prescott Webb, three inventions made it possible for the WASP pioneers of the nineteenth century to colonize the American prairie: the revolver, barbed wire, and the wind pump. A common saying was: "Women who can't fire a gun or climb a wind pump have no future here" (Webb, 1931). The prairie was described as a land where "the wind pumps the water and the cows chop the wood" (since there was no wood, dried cow dung was used instead of firewood). It is estimated that about six million wind pumps operated in this area from the late 1800s to the Depression of the 1930s. Even today, old-fashioned multibladed wind pumps are manufactured in Mexico and exported to many places in the world. As today, they were considered compatible with nature and part of the local landscape. For the complete colonization of Australia, wind pumps were indispensable, of course.

At the end of the nineteenth century, electricity came into use and windmills gradually became wind turbines as their rotors were connected to electric generators. The first full automatically operated wind turbine was designed and built in 1888 by Charles Brush, a wealthy Ohio inventor and industrialist. He installed in his 2.5 hectare backyard a giant, multibladed, farm-type wind machine, 17 m in diameter. This wind turbine fed a group of storage batteries with power transmitted through low-voltage DC wires—with high losses—to feed 100 light bulbs, two arc lights, and three electric motors (Gipe, 2004). Due to the losses, electricity had to be generated close to the site of consumption. By 1900, Brush had abandoned his machine and taken advantage of the electricity supplied by Cleveland's central power plant. He had developed a machine that in America at the time had no market (Johnson, 2014). The hidden costs of the coal-powered utility of Cleveland have been paid by all who have breathed the polluted air it has released since then.

By 1897 in Denmark, Professor Poul LaCour installed in his school one of the first wind tunnels in the world in order to investigate the rotor's aerodynamics. As a product of his research, he could design and build a four-bladed rotor connected to a 25 kW electric generator, which became the first successful wind turbine to be traded. Hundreds of copies of his model were sold and installed in Europe to generate electric energy. In 1918, at the end of WW1, Denmark had 120 LaCour-type wind turbines covering 3% of its electricity consumption. However, diesel engines and coal-fired steam turbines took over the production of electricity until the early 1940s, when during the Second World War shortage of fuels put wind power back on the agenda, and it flourished again. F. L. Smidth, normally a leading supplier of machinery to the international cement industry, diversified into wind turbine design and manufacturing. He made the first turbines with propeller-like aerodynamic blades (van Solingen, 2015). After WW2, the interest in wind power diminished again, but the development of advanced wind turbines was still pursued by enthusiastic visionaries in several countries such as Germany, Russia, the U.S., France, the UK, and Denmark. "In Denmark, this work was undertaken by Johannes Juul, who was an employee in the utility company SEAS and a former student of LaCour. In the mid1950s, Juul introduced what was later called the Danish concept by constructing the famous Gedser turbine, which had an upwind, three-bladed, stall-regulated rotor connected to an AC asynchronous generator running with almost constant speed" (Hansen, 2008). Just after the oil crisis, the U.S. government granted contracts to its big aerospace enterprises to develop MW turbines that for scale factor reasons were supposed to be the most reliable and cost effective. However, all those companies failed as the energy produced was too expensive and their wind turbines had aerodynamic, fatigue, and other issues not well addressed, in spite of the great development and success they had had at the end of the 1960s with helicopters, airplanes, and space craft: supersonic and hypersonic planes, ultra-maneuverable helicopters, manned artifacts that reached the Moon, etc. Meanwhile in Europe, a few small wind companies that developed Gedser-type turbines in the order of a few hundred watts, which survived the mechanical fatigue and economic issues, became what are today large successful enterprises. Since then, the wind turbine industry has gradually become an extremely important business in Europe, U.S., and Asia, with an annual turnover of many billions of U.S. dollars per year and growing at gigantic paces.

17.3 THE ATMOSPHERIC OCEAN

The Earth's surface is covered by what can be seen as a continuous "atmospheric air ocean." It has the following sublayers: the troposphere starts at the Earth's surface and extends 8–14.5 km high. Almost all winds blow and weather occurs in this very thin layer. Also, most of the air is contained there. The stratosphere starts just above, and it extends to 50 km high. It accommodates the ozone layer, which absorbs and scatters the dangerous solar UV radiation. Then, the mesosphere follows above and extends to 85 km high. The smaller meteorites are burned in this layer, avoiding their falling over us and our homes. Beyond this layer, the thermosphere extends to 600 km high. Auroras occur here, and satellites travel through it.

The ionosphere is a plasmatic air layer that spans from about 48 km to almost 1000 km above the earth's surface, overlapping into the mesosphere and thermosphere. It is able to reflect radio waves, making radio communications possible. This region grows and shrinks dynamically according to solar weather conditions, and it is further divided into the three subregions named D, E, and F, depending on what wavelength solar radiation is absorbed. The ionosphere is the critical link in the chain of Sun–Earth interactions, upon which life and Earth's photonic supply depend. The upper limit of our atmosphere is the exosphere. It extends from the top of the thermosphere up to about 10,000 km. Beyond the exosphere, solar wind prevails, as an extension of the Sun atmosphere, so far away but right above our heads, we can feel and hear the terrestrial wind when blowing, and of course take advantage of the enormous energy it carries in practically every place on Earth, with appropriate technologies, sustainably as our ancient parents did.

Wind flows due to gravitational forces and pressure gradients generated by a number of factors. The most important are uneven solar heating at global and local scales (that create flotation forces on expanded hot air), the large-scale Coriolis effect due to the Earth's rotation around its polar axis (which deflects laterally the trajectory of air flows), and local geographical conditions (shores, mountains, plains, and other orographic factors that result in local flows that overlap with the global ones). As the equatorial zone receives much more solar energy than the high latitude regions, heated air expands and tends to float. This way, hot air with lower density at the equator rises up to the high atmosphere and moves toward the poles, and cold air with higher density flows from the poles toward the Equator along the Earth's surface. Due to Earth's rotation and the consequent induced Coriolis force, in addition to some stability phenomena in this large-scale movement, two major global circulations are originated: the Hadley cells circulation (three in each hemisphere, North and South) shown in Figure 17.1 and the Rossby circulation, as the meandering jet streams formed in upper borders between Hadley cells shown in Figure 17.2.

All planets or satellites endowed with an atmosphere and heated by a star, like the Earth–Sun system, have complex wind patrons on their surfaces, with lots of potentially affordable energy. If plans to install a scientific station on Mars were serious, as an example, Martian wind energy would be a vast and clean power source.

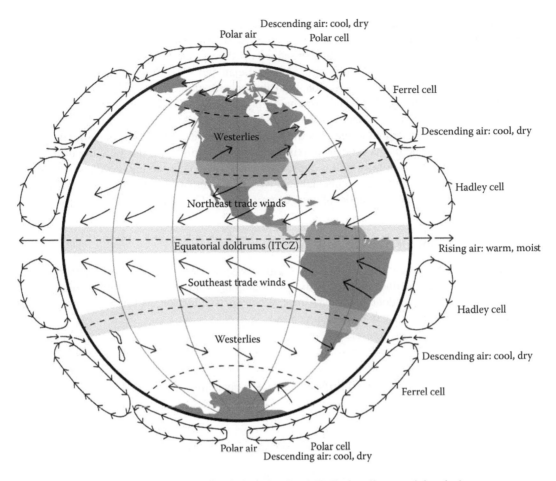

FIGURE 17.1   Idealized atmospheric circulation, showing 3 Hadley's cells on each hemisphere.

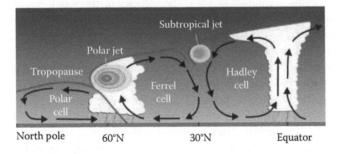

FIGURE 17.2   The three Hadley cells and two jet streams in the Northern hemisphere. (From NOAA, http://www.srh.noaa.govjetstreamglobalimagesjetstream3.jpg.)

17.4   THERMODYNAMIC LIMITS TO THE ATMOSPHERIC THERMAL MACHINE

The Earth's atmosphere is indeed a gigantic thermal machine that converts radiant Sun energy into mechanical energy: the kinetic energy contained in the wind. As discussed above, wind originates when the highly heated and expanded air at low latitude regions rises to the upper atmosphere. When it moves to less-heated zones, it cools, its density increases, and then tends to descend back to the Earth's surface. In this up and down movement, the Coriolis force deviates laterally the trajectory of each air particle, giving rise to the trade winds and westerlies. The treatment of winds formation in terms of thermodynamic cycles was first made by Prof. Jeff Gordon and Yair Zarmi (1989). Alexis De Vos, G. Flatter, and P. van der Wel further refined the theory (De Vos and Flatter, 1991; De Vos and van der Wel, 1993). More recently, Prof. Adrian Bejan and A.H. Reis published a thermodynamic optimization of the global circulation, both atmospheric and marine (Bejan and Reis, 2005). It is to be highlighted that the Gordon–Zarmi model provides a fundamental limit for the efficiency conversion of solar energy into wind, which is valid for any celestial body with atmosphere, heated for a star in any solar system, and that this limit is 7.67%. Due to thermal inertia, combined with a possible rotation of the planet around its axis (as Earth does), night-day temperature differences are to be smoothed out, diminishing this conversion energy efficiency to lower values. According to J. Peixoto and A. Oort, their experimental data for Earth is about 1% (1992). This was the value utilized to estimate the gross potential of wind energy given in Chapter 2.

17.5   THE POWER OF WIND

Wind carries kinetic energy in the air as it flows. This energy can be converted into electrical energy by power converting machines or directly used for pumping water, grinding grains, sailing ships, or in other applications, as done in ancient times. When air is in motion, its kinetic energy per unit mass with a local velocity V can be determined as

$$e_k = \frac{1}{2} V^2 \tag{17.1}$$

The flow mass through a transversal surface of area A is given by:

$$\dot{m} = \rho A V \tag{17.2}$$

So, the flow of kinetic energy crossing through a transversal surface of area A (the power of wind) is given by

$$P_w = \dot{m} e_k = \frac{1}{2}\rho A V^3 \tag{17.3}$$

where ρ is the density (mass/volume) and V is the velocity of the wind. So, the power extractable from the wind is linearly proportional to air density ρ and the transversal swept area A (which is at the same time proportional to the square of the diameter of the turbine, when a turbine is used to catch the wind) and to the cubic power of the velocity V. Because of this last cubic relation, a small variation in wind speed can result in a large change in wind power. As an example, taking advantage of the foregoing fact, the "hill effect" shown in Figure 17.3 consists of an increase in wind velocity due to the smoothness of the hill; two same-sized turbines with same height towers, one on the top of the hill and the other in the lower region of it, provide a power difference in the order of $2^3=8$, if wind velocity is doubled by the hill effect, as shown in the figure.

So, wind speed is the most critical characteristic in wind power generation; it varies amply in time and space. As a matter of fact, in real wind turbine operation, wind velocity can be extremely variable along the length of each blade. Because wind speed is a stochastic parameter, measured wind speed data are usually dealt with using statistical methods. Wind power density is a comprehensive index to evaluate the wind resource at a particular site. It is the available wind power in airflow through a perpendicular cross-sectional unit area in a unit time period. The classes of wind power density at two standard wind measurement heights are listed in Table 17.1, according to an old American Wind Energy Association (AWEA) classification.

Today, power density is assessed at 100m height and higher, as some wind turbine towers have surpassed 180m; it is expected the growth will continue. However many small, sustainable autonomous application wind turbines are installed at less than 50m, and the data of Table 17.1 are still valid.

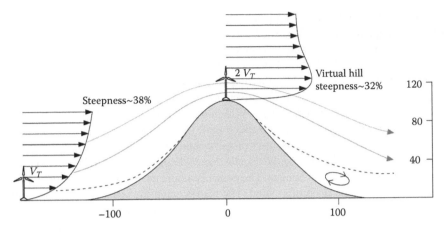

FIGURE 17.3   Taking advantage of the hill effect.

TABLE 17.1
Classes of Wind Power Densities

Wind Power Class	10 m Height		50 m Height	
	Wind Power Density (W/m²)	Mean Wind Speed (m/s)	Wind Power Density (W/m²)	Mean Wind Speed (m/s)
1	<100	<4.4	<100	<5.6
2	100–150	4.4–5.1	200–300	5.6–6.4
3	150–200	5.1–5.6	300–400	6.4–7.0
4	200–250	5.6–6.0	400–500	7.0–7.5
5	250–300	6.0–6.4	500–600	7.5–8.0
6	300–350	6.4–7.0	600–800	8.0–8.8
7	>400	>7.0	>800	>8.8

Wind varies with the local orography and height above the Earth's surface, weather, geographical location, time of day, and season. The understanding of the wind characteristics always helps to optimize the wind turbine design, develop wind-measuring techniques, and select the turbine distribution in wind farms.

17.6   WIND MACHINES TO CATCH ITS KINETIC ENERGY

The most practical devices to capture the kinetic energy of the wind and convert it in rotational movement have always been turbines. They can rotate around a horizontal or vertical axis. A wind turbine transforms part of the kinetic energy carried by the wind into mechanical energy in a shaft. This energy is mechanically transmitted through gearboxes or belts to an alternator where it is finally converted into electrical energy.

17.6.1.   THE POWER COEFFICIENT

The portion of the wind power Pw a turbine can capture and transmit through its axis in a given instant is called the turbine "power coefficient," commonly denoted by C_p. So:

$$C_P = \frac{P_s}{\frac{1}{2} \rho A V_\infty^3} \qquad (17.4)$$

where P_s is the power captured by the machine and V_∞ is the velocity of the wind-free stream. In the denominator appears the wind power P_w as given in Equation 17.3. Of course, the power coefficient has to have a theoretical maximum value; this is known as the "Betz limit" and has a value of $16/27 \sim 0.593$, as discussed in Section 17.6.3. Carefully designed modern wind turbines can operate close to this limit, with C_P over 0.5 for optimal design conditions. Wind flows across the turbine's blades, which rotate around its axis at a relatively low speed, depending on the design and for some turbines, of the instantaneous wind velocity. Within the nacelle, usually a gearbox transfers the low speed rotation of the turbine axis to a high-speed shaft that drives an alternator. Alternating current (AC) electricity is generated and flows, after proper conditioning, through a power line to the power grid and then to users. Sensors outside the nacelle detect changes in wind velocity and direction, and computer controls react to these changes. The controls adjust the blade pitch to the wind speed and face the rotor in front of the oncoming wind. The blades catch the wind and keep the turbine spinning. When the wind speed reaches three or four meters per second, electricity begins to flow. At lower wind velocities the turbine doesn't turn, but as wind velocity increases, the generated power increases—not linearly but exponentially—until reaching optimal design conditions. For stronger winds, the power coefficient must diminish quickly in order to maintain almost uniform energy generation and to protect the whole wind installation.

17.6.2 THE CAPACITY FACTOR

The "capacity factor" of a power system is the ratio of the energy generated in a given time period to the energy the system could provide if it operated at its nominal capacity during the whole period. Geothermal plants usually have the greatest capacity factors, around 87%, among all energy technologies due to the continuous thermal supply from the geothermal wells, and the low maintenance these plants require. In contrast, PV installations usually have capacity factors around 25% because half of each day is indeed night, and generation depends on the very variable sun irradiation along the sunny hours. For wind power installations, because different regions have different wind potentials, average wind farm capacity factors also vary. According to the American Wind Energy Association, these factors range from 22% in New England to more than 32% in Texas. In Tehuantepec, Mexico, capacity factors over 50% are standard for the dozens of wind farms installed there.

17.6.3 THE LANCHESTER–BETZ LIMIT

The theoretical maximum value for the power coefficient of a turbine in ideal conditions (steady-state, incompressible, unidirectional and uniform flow at each transversal flow section before and after crossing the turbine, etc.) was derived by Frederic William Lanchester (1915) and Albert Betz (1926). They revealed that no wind turbomachines could convert more than $16/27$ ($\sim 59.26\%$) of the kinetic energy of wind into mechanical energy. As shown in Figure 17.4, a free stream of wind with a hypothetical uniform velocity V_T approaches a horizontal-axis wind turbine, which will extract part of its kinetic energy by slowing the velocity with no thermodynamic irreversibility. As wind approaches the turbine, its velocity adiabatically diminishes, increasing its static pressure while conserving the enthalpy. When passing the turbine, pressure falls steeply due to the energy extraction from the wind, while velocity smoothly continues decreasing. After passing the turbine, wind must continue slowing in order to recover its static pressure (the ambient pressure), at the expense of the kinetic energy. Under these ideal conditions, the maximum power extracted occurs when wind velocity diminishes by one-third of this free current value, and the wake far away from the turbine diminishes to one-third of that free stream velocity, as shown in Figure 17.4. This limit has an exact value of $4^2/3^3 = 16/27 \sim 0.592593$. For realistic conditions, the turbine rotation causes a counter-rotation of the wake, with important energetic costs. Also, at the tip of each blade, an

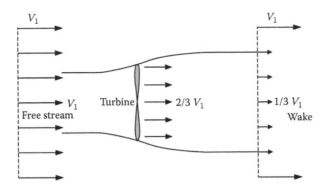

FIGURE 17.4   The flow through and ideal wind turbine at the Lanchester–Betz limit conditions.

inevitable energy-consuming vortex originated by the pressure differences between the intrados and extrados of the blades is detached. Finally, in the boundary layer near the surfaces of the blades, a nonnegligible part of the mechanical energy of the wind is irreversibly dissipated by viscous friction. The Lanchester–Betz limit is in a certain manner analogous to the thermodynamic Carnot limit imposed by thermal machines; in principle, no realistic wind machine can surpass this limit, at least under the conditions stated in its formulation. However, Section 17.7 considers some situations that lead to power coefficients greater than the Lanchester–Betz one.

17.6.4   THE TIP SPEED RATIO

The tip speed ratio λ is an extremely important factor in wind turbine design; it is defined as the ratio of the tangential speed at the blade tip to the free current wind speed, i.e.,

$$\lambda = \frac{\omega R}{V_\infty} \tag{17.5}$$

where ω is the angular velocity of the turbine, R is its radius, and V_∞ is free stream velocity.

17.6.5   POWER COEFFICIENTS OF EMBLEMATIC HORIZONTAL AND VERTICAL AXIS WIND TURBINES

In order to get a very efficient turbine, the drawbacks mentioned in Section 17.6.3 must be addressed. The rotational speed of turbines is characterized by the nondimensional parameter called the tip speed ratio λ (the relation in the tangential velocity of the tip of each turbine blade to the free stream velocity). Low values of the tip speed ratio mean that behind the turbine there is a swirling wake with too much nonutilizable rotational kinetic energy. To reduce rotational losses, the tip speed ratio must be increased, resulting in a power coefficient also increasing. This increase has a limit because viscous dissipation also increases with λ, but in a nonlinear manner. The number of blades is an important parameter; as this number increases, C_P also does, but the cost of each additional blade does not offset the moderate increase in efficiency. As a rule of thumb, horizontal-axis wind turbines have bigger C_P values than those that of vertical-axis turbines, because the first ones work on lift forces usually much bigger than the drag ones, on which most vertical-axis wind turbines rely. Figure 17.5 shows the power coefficients of some emblematic wind turbines. Note the high values for the modern three-bladed horizontal-axis wind turbines, when they work at tip-speed ratios around 6.5. However, high values of λ, from the environmental point of view, are potentially more risky for birds, bats, and insects, so a sustainable design should take all of these factors into account in order to get efficient but environmental friendly wind turbines.

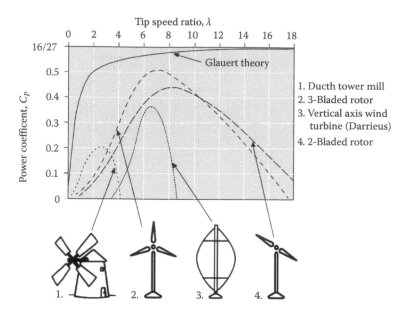

FIGURE 17.5   Power coefficients of emblematic wind turbines as functions of λ.

For a horizontal-axis wind turbine, with n blades with the same profile all along its length optimized according to the Glauert theory (1935) the following empirical equation can be used to estimate the maximum power coefficient (Wilson et al., 1976):

$$C_{P\,max} = 0.593\left[\frac{\lambda n^{0.67}}{1.48+\left(n^{0.67}-0.04\right)\lambda+0.0025\lambda^2}-\frac{1.92\lambda^2 n}{\left(1+2\lambda n\right)\left(L/D\right)}\right] \quad (17.6)$$

where L/D is the ratio of the lift coefficient C_L to the drag coefficient C_D at the design angle of attack α_{opt}. For a fine profile, L/D should be about 100 at an optimum attack angle $\alpha_{opt} \sim 5°$.

17.6.6   The Power Curves

The power delivered by a turbine as a function of the wind velocity is known as its "power curve." This is a very important graph because it shows how much power can be expected from a turbine, given the wind conditions at the place where it is erected. These curves pertain to given air densities (normally at sea level); they are obtained experimentally through numerous measurements. Data are adjusted to a continuous power curve. A wind turbine starts to deliver usable power at a low wind speed defined as the "cut-in speed," about 3.5 m/s. The power output increases nonlinearly with the increase of the wind speed until it reaches a value, defined as the "rated power output." Correspondingly, the wind speed at this point is defined as the "rated speed." Once the rated speed is attained, an additional increase in the wind speed will not increase the power output due to the activation of the power controls. The aerodynamic power coefficient has to diminish quickly as wind exceeds the rated velocity in order to sustain a stationary output power. When the wind speed becomes large enough to potentially damage the wind turbine, the turbine shuts down to avoid damage. This wind speed is defined as the "cut-out speed." Thus, the cut-in and cut-out speeds define the operating limits of a wind turbine.

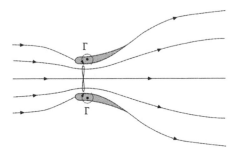

FIGURE 17.6   A shrouded turbine, as proposed by Hansen and others.

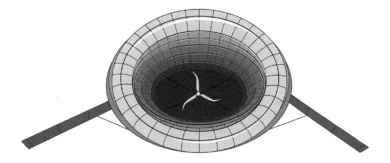

FIGURE 17.7   The MIT buoyant airborne shrouded turbine.

17.7   SURPASSING THE LANCHESTER–BETZ LIMIT?

It has been suggested that it is possible to extract more power from the wind by directing it through a diffuser that could be incorporated to the system (Igra, 1981). This way, the turbine is shrouded by the diffuser as shown in Figure 17.6. With this construction, according to Hansen and others (Hansen et al., 2000; Sharpe, 2004), it is possible to exceed the Lanchester–Betz limit: if the cross-section of the diffuser is shaped with an airfoil profile, a circulation around the diffuser would be created, and a lift force would be generated by the flow through the diffuser. The effect of this lift force is to create a ring vortex, which by the Biot–Savart law will induce a velocity that increases the mass flow through the rotor. The increase in the power coefficient for a shrouded turbine is proportional to the ratio between the mass flow through the turbine in the diffuser and the same turbine without the diffuser. Several CFD analyses and experiment work have been done (Gilbert and Foreman, 1983) with such promising results that researchers at MIT have proposed the idea of a buoyant airborne turbine, a kind of giant floating generator inflated like a helium dirigible, shown in Figure 17.7. It would take advantage, besides the increased power coefficient due to the floating diffuser, of the much-bigger wind velocities present at high altitudes. Several prototypes that can reach altitudes of up to 600 m have been demonstrated.

17.8   NEW WIND MACHINE PROPOSALS. ARE THEY VIABLE?

It took several decades to evolve the most successful design: three-bladed, horizontal axis, with bigger and bigger diameters, for onshore and offshore installations. However, there are often published articles about new and innovative methods for harvesting wind energy. Some with nonnegligible financial support are briefly mentioned the following subsections.

17.8.1 THE WIND DRONE

Figure 17.8 shows the wind drone that the Swiss startup TwingTec Company is developing. According to its promoters, it is able to take off vertically and generate power as it is swept aloft. The drone flies without a human operator, and the entire system is designed to fit into a standard shipping container with ease. It can be quickly shipped to remote locations or areas hit by disasters for swift deployment. The TwingTec Company has as investors Swiss bank Zürcher Kantonalbank and German automation company Festo. A prototype has already been tested.

17.8.2 THE GOOGLE MAKANI KITE

Google subsidiary Makani is testing power-generating kites endowed with eight wind turbines each—as shown in Figure 17.9—that can soar up to 300 m in order to take advantage of stronger winds prevailing at that altitude. The 600 kW kites travel in big circular trajectories, with wing-mounted turbines generating electricity that can be passed down the tether cable to the grid.

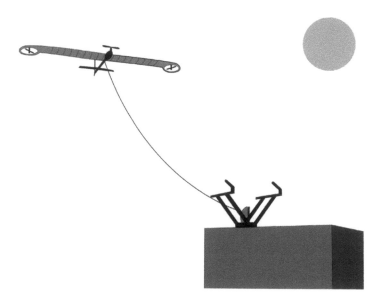

FIGURE 17.8   The wind drone concept.

FIGURE 17.9   The Google Makani kite concept.

In a note published in the electronic magazine *New Atlas*, Mike Barnard wrote of these proposals: "These devices are likely to remain a niche for the simple reason that putting lots of them up high in the atmosphere would require 1–4 km long, effectively invisible cables which stretch over a broad and shifting downwind range. This would require a significant area to be declared a no-fly zone for most forms of aviation, although passenger jets could still fly overhead. If the system failed, and the device fell from the sky, it would drape those kilometers of cables over everything downwind, including roads and buildings, requiring that a large area downwind be fairly free of any human structures. And for the solid flying wing, a very heavy object with rotating propellers would fall out of the sky somewhere between 1 and 10 km downwind in the event of a failure. That's why, after a period of assuming they these could make major onshore contribution, most of these devices are now aimed at servicing remote locations or offshore sites. It doesn't help that the lighter-than-air variants require increasingly rare helium, which is also required for other, arguably much more valuable, uses including as a coolant in medical imaging machines. There are significant scalability issues with such turbines, and, given the increasing height of HAWTs, they are dealing with diminishing returns in any event" (2013). Perhaps he is right in his observations.

17.8.3 EXTRACTING ENERGY FROM WIND WITHOUT TURBINES?

Some attempts have been recently undertaken to generate electric energy with no turbines. It is difficult to prognosticate their success. The movement of the wings of a flying hummingbird has been mimicked to develop a wind machine, shown in Figure 17.10, that according to Tunisian startup Tyler Wind is able to capture much of the power of the wind without the problems that conventional turbines have, such as danger to birds. Its flapping wings occupy less physical space than large turbine blades and move at slower speeds, which should mean fewer bird collisions. The wings are expected to generate power on both the up and down strokes and generate less vortex wake than traditional blades. This could allow for denser, more efficient wind farms, if the innovative design is deemed worthy. Tyler Wind is funded by private investors from Pakistan and Algeria but has not yet demonstrated any large-scale prototypes.

Another proposal consists of simple vibrating towers whose movement can be used to generate electricity. It works by the well-known vortex-shading phenomenon studied by Von-Karman and many others a century ago. The vortex shading from high chimneys used to be a problem some decades ago. Now Spanish startup Vortex is using the same effect to generate electricity from towers with no blades at all. Vortex claims that electricity from its oscillating carbon-fiber towers might one day be 40% less expensive than that from conventional turbines because it requires far less material. Vortex also says the towers are virtually silent and pose very little risk to birds (Figure 17.11).

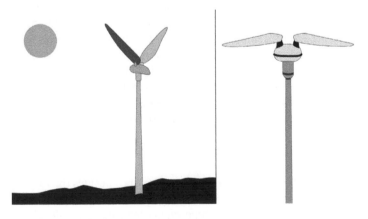

FIGURE 17.10   The Tunisian Hummingbird wind converter.

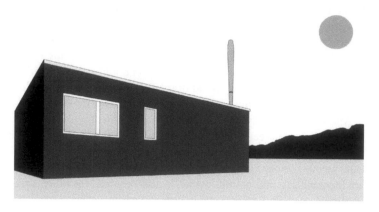

FIGURE 17.11   The vibrating tower concept.

17.9 ENVIRONMENTAL CONCERNS

All around the world there is a growing number of new wind farms, with larger wind turbines. Some of these wind parks are placed in biodiverse regions, such as near tropical forests or shores where some birds migrate. Therefore, environmental issues must be addressed if wind energy systems are to be sustainable. Wind turbines do kill some birds, bats, and insects, and there are problems involved with the necessary destruction of remnant native vegetation and possible erosion problems connected with the building of roads and hardstands on which the turbines are placed. Modern wind turbines might sweep areas of tens of thousands of m² whose blade tips could have tangential speeds higher than 250 km/h. At such high speeds, animals flying through the area swept by the blades may be killed by them. Even though there are official reports that the number of birds killed by cars each years is three orders of magnitude greater than the number of birds the wind turbines kill, these deaths can be avoided with some changes in the design of turbines: diminishing the tip speed ratio (with a light reduction of the turbine efficiency); softening the material of their leading edges in order to reduce the possible traumatisms; stopping the turbines when a flock of birds is detected; "warning" the birds with sound or other means, etc. Other concerns relate to noise that originates from lots of vortex when blades don't work efficiently. Improved designs have diminished the noise levels (and have boosted the efficiencies of the turbines) in such a way that noise is no longer a big issue. Before installing a new wind farm, a series of environmental assessments has to be completed including bird migration routes as well as how to minimize bat and insect collisions and visual landscape contamination. To reduce the visual effect, wind turbines usually use colors such as green for the lower part of the tower and light blue and off-white for the upper part of the tower and the blades. For some people, a slender and elegant tower crowned with an impressive majestic wind turbine is a touristic amenity. Once a wind farm is commissioned, rigorous monitoring should be undertaken to better understand the ongoing relationship of birds, bats, insects, and the wind farm to diminish its environmental impact. When the life of a 25-year-old turbine ends, the turbine must be recyclable. New blades could be made of biopolymers, which are easily recycled and even biodegradable in a few decades. The metal components of the wind turbines are easily recoverable, as well as the materials of all its subsystems, from the foundation to the control electronics.

17.10 EPILOGUE

In terms of sustainability, the most relevant wind energy issues besides the technical, environment, and economical are societal and political ones. In underdeveloped countries, when a wind farm is proposed, usually communities are split into supporters (financed by the developer companies) and opponents (land defenders); very often, bad feelings arise. As an example, in Tehuantepec, Mexico, one of the more windy regions in the world, the poor local peasants have been deprived of their communal terrains by powerful transnational companies who install huge wind farms for private electricity supply, of course with the complicity of government authorities. This has led to bloody clashes and widespread rejection of wind farms by the deprived peasants, who not only did not benefit from the wind parks, but were visited with more misery and death. When a wind farm is planned, local communities should be ensured of a direct benefit of hosting the project; they should be allowed to invest and receive generous dividends from the park, or even own it. In Denmark and Germany, for example, many local communities have total control over wind projects, as well as financial stakes in their success. Locals can invest directly in projects, guaranteeing themselves a share of income. They have given power to the people!

Finally, every energetic installation has an impact. The only option for avoiding impact would be the renunciation of all energy consumption. Fossil fuels and nuclear power have always had a worse environmental impact than wind installations. We must choose whether to seek and work for a sustainable scheme or a devastating one.

REFERENCES

Barnard, M. 2013. Dodgy wind? Why "innovative" turbines are often anything but. New Atlas. http://newatlas.com/dodgy-wind-turbines/27876/.

Bejan, A., and A. H. Reis. 2005. Thermodynamic optimization of global circulation and climate. *International Journal of Energy Resources* 29:303–316.

Betz, A. 1926. *Windenergie und ihre Ausnutzung durch Windmühlen (Wind Energy and Its Utilization through Windmills)*. Götingen: Vandehoek & Ruprech.

Braudel, F. 1979. *La Méditerranée et le monde méditerranéen à l'époque de Philippe II*, Vol. 1. Paris: A. Colin.

De Vos, A. and G. Flater. 1991. The maximum efficiency of the conversion of solar energy into wind energy. *American Journal of Physics* 59:751–754.

De Vos, A. and P. van der Wel. 1993. The efficiency of the conversion of a solar energy into wind by means of Hadley cells. *Theoretical and Applied Climatology* 46:193–202.

Gilbert, B. L., and K. M. Foreman. 1983. Experiments with a diffuser-augmented model wind turbine. *Journal of Energy Resources Technology* 105:46–53.

Gipe, P. 2004. *Wind Power: Renewable Energy for Home, Farm, and Business*. White River Junction: Chelsea Green Publishing Company.

Glauert, H. 1935. Airplane propellers. In *Aerodynamic Theory*, ed. W. F. Durand, 169–360. Berlin: Julius Springer.

Golding, E. W. 1976. *The Generation of Electricity by Windpower*. London: E. & F.N. Spon Lt.

Gordon, J., and Y. Zarmi. 1989. Wind energy as a solar-driven heat engine: A thermodynamic approach. *American Journal of Physics* 57:995–998.

Hansen, M. O. L. 2008. *Aerodynamics of Wind Turbines*, 2nd ed. London: Earthscan.

Hansen, M. O. L., N. N. Sorensen, and R. G. J. Flay. 2000. Effect of placing a diffuser around a wind turbine. *Wind Energy* 3:207–213.

Igra, O. 1981. Research and development of shrouded wind-turbines. *Energy Conversion and Management* 21:13–48.

Johnson, R. 2014. *Chasing the Wind. Inside the Alternative Energy Battle*. Knoxville: The University of Tennessee Press.

Juleff, G. 1996. An ancient wind-power smelting technology in Sri Lanka. *Nature* 379 (6560):60–63.

Lanchester, F. W. 1915. A contribution to the theory of propulsion and the screw propeller. *Transactions of Institution of Naval Architects* XXX:330.

Peixoto, J., and A. Oort. 1992. *Physics of Climate*. New York: American Institute of Physics, 365–400.

Righter, R. W. 1996. *Wind Energy in America: A History*. Norman: University of Oklahoma Press.

Rincón, E., and M. D. Duran. 2006. Small wind machines for the well-being of humankind. In *Proceedings of ANES/ASME Solar Joint 2006 XXX Mexican National Solar Energy Week Conference*. Veracruz: ASME Press-ANES.

Sharpe, D. J. 2004. A general momentum theory applied to an energy-extracting actuator disc. *Wind Energy* 7:177–188.

Van Solingen, E. 2015. *Control Design for Two-Bladed Wind Turbines*. Bernice: Technische Universiteit Delft.

Webb, W. P. 1931. *The Great Pleins*. Lincoln and London: 1981 printing of the University of Nebraska Press.

Wilson, R. E., P. B. S. Lissaman, and S. N. Walker. 1976. *Aerodynamic Performance of Wind Turbines*, ERDA/NSF/04014-7611, Washington, DC: U.S. Department of Energy.

18 Tidal Current Technologies
Brief Overview and In-Depth Coverage of the State of the Art

Vanesa Magar
Centre of Scientific Research
and Higher Education of Ensenada (CICESE)

CONTENTS

18.1 INTRODUCTION

The use of ocean energy technologies goes back thousands of years, to the times when humans invented sailing boats to tap into wind power for ocean transportation, or river wheels to tap into current stream power for wheat grinding. In contrast, tapping the power of the ocean for electricity generation was pursued much later in human history, with evidence of first-wave energy converter patent registrations starting in the middle of the nineteenth century (Leishman and Scobie, 1976). Indeed, the ocean being a highly corrosive and extreme environment where all ocean infrastructure is at the mercy of storm waves and high winds and where operation and maintenance is therefore more costly than it is inland, makes the pace of development of ocean energy technologies slower compared to land-based renewable energy infrastructure (Pelc and Fujita, 2002). By 2014, ocean technologies had a global cumulative installed capacity of 0.53 GW only, in contrast with more than 10 GW of global cumulative installed capacity for offshore wind. Of these 0.53 GW, 0.494 GW correspond to two large tidal hydroelectric dams: la Rance in France, with 240 MW installed capacity, has been in operation since 1966; the Sihwa lake tidal power station in Korea, with 254 MW installed capacity, has been in operation since 2011. The remaining global cumulative installed capacity of 36 MW is distributed into smaller projects exploiting other types of ocean energies: tidal coastal lagoons, wave energy farms, tidal turbine farms, and ocean temperature and ocean salinity energy conversion systems.

Ocean energy technologies can extract energy from five different types of phenomena: the rise and fall of the tides, as in la Rance and Sihwa tidal hydroelectric dams; ocean and tidal currents; surface gravity waves (and for some devices, infragravity waves); temperature gradients; and salinity gradients. The global offshore potential for each of these ocean energy technologies can be estimated using available models and databases, providing either a purely theoretical resource assessment based on the energy available per unit area, a technical resource assessment based on the power that can be extracted by a single device or an array of devices in a given area, or a practical resource assessment based on the technical resources available, together with some additional constraints (O'Rourke et al., 2010). Practical resource assessments may take into account, for example, power extraction restrictions such as exploitability thresholds or ocean ice coverage, distance to shore, seabed characteristics, water depth, conflicting maritime activity, or proximity to transmission lines, amongst others (Mørk et al., 2010; Cavazzi and Dutton, 2016; Magar et al., 2016). Each ocean energy technology has different theoretical and technical resource availability that is worth discussing in more detail. We will start with the case of wave energy and move on to tidal energy, then comment on current temperature and salinity gradient technologies, and finish with other offshore energy resources such as wind or geothermal energy.

Mørk et al. (2010), using the WORLDWAVES wave database by Barstow et al. (2003), evaluated the available theoretical and technical wave energy resources to be 3702 GW (theoretical), 3475 GW (excluding areas with less than 5 kW/m), and 2985 GW (excluding ice-covered areas and areas with less than 5 kW/m). The WORLDWAVES database was generated using the operational ECMWF WAM model data (Komen et al., 1996) on a 0.5° spherical (lat/lon) grid and sampled every six hours over the 10-year period between 1997 and 2006. The coastline and bathymetry database used in the model were the GMT Global, Self-consistent, Hierarchical, High-resolution Shoreline Database at a working scale of 1:250,000 (Wessel and Smith, 1996) and the Digital Bathymetric Database—Variable Resolution (DBDB-V) developed by the Naval Oceanographic Office (NAVOCEANO, 2002), respectively. The model was validated against long-term in-situ buoy wave monitoring networks and altimeter data (Barstow et al., 2003; Mørk et al., 2010). Although a strong emphasis is put on resource characterization, many other factors need to be taken into account to make wave energy device deployments economically viable. Thus, Wave Energy is lagging some years behind Tidal Energy, although some investors are starting to become interested in these technologies as well (World Energy Council, 2013).

While estimates of global potential vary, it has been estimated that tidal stream energy could theoretically supply more than 150 TWh per annum (Legrand, 2009), well in excess of all domestic electricity consumption in the UK. This represents a potential total global market size of up to 90 GW of installed capacity. It is worth noting that a number of SMEs who started with device development and testing in the early 2000s, such as Marine Current Turbines (originally based in Bristol on the West Coast of Britain), OpenHydro (based in Greenore on the East Coast of Ireland), or Hammerfest Strom (now called Andritz Hydro Hammerfest, based in Norway), have been bought by large companies such as Siemens AG, DCNS SA, and Andritz Hydro, respectively (World Energy Council, 2013). This clearly indicates that tidal stream technologies are reaching high technology readiness levels (TRL), based on NASA's TRL definitions (Mai, 2015).

In tropical regions, renewable energy may be generated with salinity and temperature gradients. Ocean Thermal Energy Conversion (OTEC) uses the temperature difference between the warm surface waters and the cool deep water of the ocean to turn a liquid such as alcohol or ammonia into vapor to run an engine and produce electricity. OTEC plants can be land-based, offshore fixed to the seafloor, or offshore floating able to be moved to areas with higher temperature differences. The energy produced may be directly delivered to the grid or stored in different ways. Ocean salinity gradient power, on the other hand, is generated from differences in salt concentration between two fluids, such as river flowing into the sea (Kempener and Neumann, 2014). There are methods for converting salinity gradients into energy, using membranes to separate a concentrated salt solution (like sea water) from freshwater: pressure retarded osmosis (PRO) or reversed electro dialysis (RED). PRO works by increasing the pressure in the salty water chamber. A turbine is spun as the pressure is compensated and electricity is generated. In contrast, RED consists of a stack of alternating cathode and anode exchanging selective membranes through which ions are transported. The salinity gradient difference between the two chambers of the device is the driving force that results in an electric potential, which is then converted to electricity.

In this chapter we will discuss first, in Section 18.2, the fundamentals of tidal stream technologies, followed in Section 18.3 by numerical modeling at regional and local scales. Section 18.4 on final comments focuses on the impact of boundary conditions, specifically the bathymetric data used and the mesh size, on tidal energy resource assessments. The chapter principally provides in-depth coverage of renewable energy by way of tidal current energy. Other technologies will be highlighted as appropriate.

18.2   FUNDAMENTALS

After a brief historical introduction, resource assessment protocols and technological advances are described. The latter two should be considered simultaneously to identify optimal tidal energy development sites.

18.2.1   INTRODUCTION

Evidence suggests that the French, the Britons, and the Spanish have used tidal power mills to grind wheat since at least the eighth century; archeological findings of tidal mills show they were used as far back as 787 BCE in England. Tidal storage ponds were filled with seawater from the incoming tide through sluice gates. At high tide, the sluice gates were closed and the water was redirected through a water wheel. There are only two working tide mills in the world. They are both in the UK, one in Woodbridge, Suffolk (geographical coordinates: 52°5′24.60″N, 1°19′15.32″E), and the other in Eling Hill, Totton (geographical coordinates: 50°54′39.54″N, 1°28′55.74″W), close to Southampton (http://www.elingexperience.co.uk/elingmill.html; last accessed: 28/10/2016) (Figure 18.1).

Seawater is denser than air (usually around 833 times denser), giving ocean and tidal currents an extremely high energy density. Therefore, for an equivalent power rating and installed capacity, tidal energy converters have a smaller rotor size and require less seabed area than offshore wind energy converters. Moreover, tidal turbines can be installed closer together within each array; as most devices are submerged, visual pollution is minimal compared to other offshore energy methods. Unlike tidal barrage schemes, tidal energy devices are modular, so the infrastructure required for their installation is minimal, and the impact on the natural environment is also minimal (Bahaj, 2011). Tidal energy is generated by tides, which are in turn generated by astronomical forces among our planet, its moon, and the Sun in its solar system. Tides are also affected to a much smaller degree by other celestial bodies; hence, those forces are generally neglected. Since the astronomical forces are 100% predictable, tidal energy is also 100% predictable, which is an enormous advantage of tidal energy over other renewable energy forms; wind, wave, solar, and land-based hydroelectricity are subject to unpredictable climatic fluctuations that occur at many different temporal and spatial scales such as storms, droughts, clouds, or luminosity changes. This can generate challenges for grid management and balancing of the transmission system.

As with all energy resources estimation, measurement, analysis, and reporting of tidal stream resource assessments need to follow standard procedures in order to provide bankable information (i.e., reliable, reproducible, and useful information for other researchers, decision-makers, and funders) at sites that could be suitable for the installation of tidal energy conversion systems (TECs).

FIGURE 18.1   Eling Tidal Mill. (a) Low tide. (b) High tide. (c) and (d) Close-ups of buildings.

18.2.2   Resource Assessment Protocols

In this section, we summarize the resource assessment protocols developed by Black and Veatch for the European Marine Energy Centre (Legrand, 2009). Except when otherwise stated, definitions, terms, symbols, units, and abbreviations used in this chapter follow those in that work. It is assumed that the technology is mature and commercially available for the selected site.

18.2.2.1   Objective and Nature of Resource Assessment

Although it may be obvious that the objective of a resource assessment is, ultimately, to install commercially viable devices connected to the electrical grid, it is always important to state at the beginning of any resource assessment the objective and the nature of the resource assessment.

18.2.2.2   Resource Assessment Stages

Table 18.1 shows the three different stages (stages 1, 2a, 2b, and 3) a resource assessment study undergoes. The first stage is the regional site screening. A site screening is considered as regional if it includes many potential development sites. The site screening is an initial characterization of the potential sites and leads to the identification of the optimal one. A prefeasibility assessment then follows in which the resource at the optimal site is characterized in more detail, and major constraints for development are identified. If these constraints turn out to be very numerous and they limit the technical, environmental, or economic viability, then another site needs to be selected, until the least constrained site is found. The next part of the resource assessment protocol consists of a full feasibility assessment whereby a full economic model is applied to the selected site, and all constraints have been identified and assessed. If the site is promising, then applications for development permits are filed.

Resource assessments need to consider from an early stage the technology that will be deployed at the site. Legrand (2009) mentions the rotor diameter, d, and the top and bottom clearance, c_t and c_b respectively, as the three device-related dimensions of importance in site suitability assessments. c_t is the distance between the lowest astronomical tide and the highest point of the capture area, while c_b is the distance between the lowest point of the capture area and the seabed. The device should be chosen as early as Stage 2a of the resource assessment in order to take these characteristics into account. However, if d, c_t or c_b is unknown, then a generic TEC device may be used.

Changing bathymetry and coastal topography can have an impact on the tidal stream resource and its suitability. Therefore, it is important to follow bathymetric survey standards when performing bathymetric surveys and if possible, analyze long-term or seasonal seabed changes. This

TABLE 18.1

Resource Assessment Stages

Stage	Category and Aim	Area	Constraints	Permitting
Stage 1	Regional site screening	Region or country	Limited constraints identified	No
Stage 2a	Prefeasibility assessment	Whole estuary, channel, etc.	Major constraints identified	No
Stage 2b	Full feasibility assessment	Localized area in estuary, channel, etc.	All constraints identified and assessed	Applied for
Stage 3	Design development	Localized area in estuary, channel, etc.	All constraints fully assessed	Obtained

requires a history of bathymetric monitoring surveys, which in most coastal areas will be difficult to obtain, unless the area has been of commercial or military interest. This was the case, for example, in previous work in Great Yarmouth, for instance in Scroby Sands, one of the pilot offshore wind farms built during the UK's Round 1 offshore wind development phase (Power Technology, 2016). In this region the port of Great Yarmouth is also located, which due to its commercial importance has been surveyed by the UK Hydrographic Office since 1848 and has been surveyed regularly, every 3–10 years, between 1894 and 2006 (Reeve et al., 2008). From these surveys, it has been established that the Scroby Sands sandbanks have always been present, but they have changed significantly in shape and position, typically at interannual timescales. However, significant variations in the location, height, and width of individual banks occur in the area, with the "outer" banks, less sheltered from storm waves and surges, suffering more variations than the "inner" banks (Reeve et al., 2001). Considering that offshore windfarm lifespans are of around 25 years, such seabed variations can have an important impact on the stability of the turbines, especially when the turbines are in shallower zones with noncohesive sediments, where waves can have a significant effect on sediment transport. In fact, wave-induced liquefaction of sediments and scour at the base of the turbines have been observed in Scroby Sands, where the upper levels of the seabed are composed of medium-sized sand (Whitehouse et al., 2011), endangering turbine stability.

It is clear that seabed composition and seabed slope are two additional significant factors in offshore renewable resource assessments, specifically for devices with gravity base or monopile foundations. Also, the level of resolution of the surveys needs to be aligned with the resource assessment stage. At stage 1, resolutions of 1–2 km are sufficient, whereas at stage 2a and 2b both water depth and sediment composition should be known at scales of 100–200 m. By stage 3, the survey resolution needs to be of the order of 5 m, so that the seabed slope and the sediment composition can be closely analyzed. The surveys should be carried out according to International Hydrographic Office (IHO) standards (Soediono, 1989), with calibrated (Mann, 1998), multibeam sonar systems preferred over single beam or side scan sonar. At stage 3 a remotely operated vehicle should be used to uncover any uncertainties in the sonar measurements, accompanied with subbottom profiling in areas with large quantities of suspended sediments (Legrand, 2009). The surveys' metadata should report the date of the survey, the method of measurement, and the precision. The bathymetry should be processed following the approach described in Gardner et al. (1999). Finally, in regions where significant suspended sediments or layers of liquefied mud have been observed, subbottom profiling may be required. In addition to the in-situ measurements described above, a thorough regional numerical modeling effort based on a model with shallow-water approximations (see Section 18.3.2.2), both in two and three dimensions, should be undertaken. A three-dimensional model is usually necessary to identify the tidal current device array configurations that maximize power extraction and

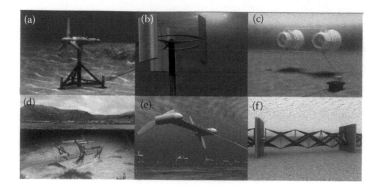

FIGURE 18.2 Illustrations of the six different broad types of tidal in-stream turbines. (a) HATT (Alstom, 2013). (b) VATT (http://www.esru.strath.ac.uk/). (c) Venturi-effect HATT. (d) Oscillating hydrofoil. (e) Tidal kyte. (f) Tidal fence.

minimize negative impacts on the environment. A review on the regional hydrodynamic modeling methodology and rationale behind resource assessment can be found in Blunden and Bahaj (2007).

18.2.3 TECHNOLOGICAL ADVANCES

Tidal stream technologies have seen great advances in the last two decades, with a number of small and medium companies (SMEs) such as Marine Current Turbines making their first prototypes in the early 2000s and being bought by large enterprises such as Siemens by the second half of the 2010s. Marine Current Turbines come with different turbine types and turbine geometries. EMEC (2016) has catalogued tidal energy converters into six different types: the most mature technology is the horizontal axis tidal turbine (HATT), where the rotors rotate around a horizontal axis and generate power; the vertical axis tidal turbines (VATT), where the turbine rotates around a vertical axis; the Venturi-effect HATT, where the turbine tips are enclosed by a funnel or shrouds that streamline and concentrate the flow toward the turbine; the oscillating hydrofoil, where the tidal stream current flowing either side of a hydrofoil wing results in lift and power generation through a hydraulic pump inside the oscillating arm; the "tidal kite," where a device tethered to the seafloor "flies" in the tidal stream to increase the speed of the water flowing through the turbine; and other devices, with designs that are very unique or too specific to fall within any of the more established technological categories.

Besides the different turbine designs, tidal stream converters also can have different sea mounting designs, depending on the locations where they will be deployed. Devices may be bottom mounted or floating devices. The methods of fixing seabed mounted, pile mounted, or gravity base tidal devices are similar to any other seabed mounted, pile mounted, or gravity base structure, respectively. Similarly, floating tidal energy converters could have flexible moorings, rigid moorings, or be positioned on a floating structure. Hydrofoil-type devices also may stay in position through hydrofoil-induced downforces, provided that the downforce exceeds the overturning moment. Figure 18.2a through f shows examples for each of the different tidal turbine types mentioned, developed by a number of different companies.

18.3 NUMERICAL MODELING

One of the most important research issues on numerical modeling is the physics-dynamics coupling for all science and engineering problems, not the least in renewable energy. The physical parameterizations in dynamical models can be improved through a better understanding of the physics at model scales, through laboratory experimentation or computational fluid dynamics simulations that resolve the full equations. One of the most fundamental aspects in fluid dynamics is similarity equations that permit the generalization of dynamics throughout 1:200 up to 1:1 prototype scales. This is achieved through nondimensional analysis.

18.3.1 Nondimensional Analysis and Similarity Equations

In order to compute the performance of different types of tidal turbines at different scales, it is important to identify the similarity laws that result from the nondimensional analysis that allows the comparison of the eight classical parameters in turbomachinery: the head H or pressure rise p, the flow rate Q, the angular speed ω, the power P, the density ρ, the diameter of the device D, the viscosity μ, and the speed of sound a (sometimes replaced by elasticity e). The Buckingham-PI theorem then leads to the identification of five similarity laws defining the Reynolds number, the speed coefficient, the head coefficient, the flow coefficient, and the power coefficient,

$$\frac{\rho \omega D^2}{\mu}, \frac{\omega D}{a}, \frac{gH}{\omega^2 D^2}, \frac{Q}{\omega D^3}, \frac{P}{\rho\, \omega^3 D^5},$$

respectively.

The power coefficient, usually denoted as C_p in the literature, is the turbine's efficiency, as it is the ratio of the actual electrical power produced divided by the tidal power into the turbine. The tidal power into the turbine is generally written as

$$P_{in} = \frac{1}{2}\rho A_d u^3,$$

where A_d is the blade's swept area. The flow and power coefficients of turbines are very useful for performance analyses of single turbines.

18.3.2 CFD and Shallow-Water Models for Tidal Energy Applications

The numerical modeling for marine renewable energies, and in particular for tidal energy research and development, could be divided into two important categories that depend on the goals of the modeling exercise. The fundamental tools used are the same as those for all computational fluid mechanics applications, with an emphasis on fluid-structure-environment interactions (Hu et al., 2016; Rivier et al., 2015, 2016), turbine design and representation in regional models, turbulence-induced wake modeling, or bottom-drag parameterizations, to name but a few (Roc et al., 2013; Kramer and Piggott, 2016; Laws and Epps, 2016; Shives and Crawford, 2016). In order to maximize energy extraction and minimize environmental impacts, a central research topic in the Geophysical Fluid Dynamics and Environmental Modeling Laboratory at the Physical Oceanography Department of CICESE, regional models with porous-disk type representations of the turbines are necessary to analyze the hydro-environmental impacts of tidal arrays installed in areas with promising tidal energy resources. First the arrays need to be designed so upstream turbines' wakes disturb as little as possible the flow reaching the downstream turbines; then, bottom drag needs to be minimized in order to avoid substantial seabed disturbances. However, if such disturbances cannot be avoided, and if such regions have a social, economic, or ecological importance, then mitigation measures will need to be implemented, or a different tidal array emplacement of configuration will need to be considered.

18.3.2.1 Actuator Disk Approximation in Computational Fluid Dynamics (CFD) Models

The actuator (or porous) disk is widely used in wind energy; that is also applicable to horizontal-axis tidal turbines (Burton et al., 2001). The concept is illustrated in Figure 18.3 and relies on assuming that the flow is incompressible and that no mixing occurs between the flow that comes into contact with the turbine and the flow that has no interaction with it. The symbol ∞ in Figure 18.3 refers

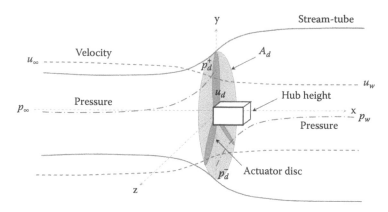

FIGURE 18.3 Actuator disk sketch.

to conditions far upstream, d refers to conditions at the disk, and w refers to conditions in the far wake. This defines a "stream-tube" of water that goes across the turbine for energy generation. The presence of the turbine causes a deceleration of the flow in the stream tube even before the flow reaches the turbine, and so the stream tube expands due to the incompressibility assumption. In turn, the static pressure of the flow approaching the turbine has to rise, given that no energy has been extracted yet from the flow and the decrease in kinetic energy associated with the flow's deceleration needs to be compensated. As the flow passes the turbine, there is by design a step decrease in the static pressure associated with the pressure energy extracted with the rotor, so the air flows downstream with reduced speed and reduced static pressure; this region is defined as the wake. Further downstream, the static pressure of the wake, p_w, needs to increase back to the atmospheric pressure. The actuator disk concept helps one understand the energy extraction process without the need for a specific turbine design, and the stream tube concept helps express the mass, momentum, and energy conservation equations in an approximate manner.

The mass of air that passes through a given cross-section of the stream tube in a unit length of time is $\rho A u$, where ρ is the air density, A is the cross-sectional area of the stream tube, and u is the flow speed. So due to mass flux rate conservation,

$$\rho A_\infty u_\infty = \rho A_d u_d = \rho A_w u_w,$$

it is usual to consider that the disk induced a reduction in flow speed so that the net streamwise flow speed at the disk is

$$u_d = u_\infty(1-a),$$

where a is the axial flow induction factor, or inflow factor.

On the other hand, the rate of change of momentum over the disk is the overall change in flow speed times the mass flux rate, and the rate of change of momentum is caused entirely by the change of pressure across the disk. So from the above,

$$F = (p_d^+ - p_d^-)A_d = \left(u_\infty - u_w\right)\rho A_d u_d = \left(u_\infty - u_w\right)u_\infty(1-a)\rho A_d,$$

where F is the force concentrated in the actuator disk. The pressure difference is obtained via Bernoulli's equation applied separately to the upstream and downstream sections of the stream tube. Bernoulli's equation states that, under steady conditions, the total energy in the flow, comprising kinetic energy, static pressure energy, and gravitational potential energy, remains constant provided no work is done on or by the fluid. Thus, upstream we have:

$$\frac{1}{2}\rho_\infty u_\infty^2 + \rho_\infty g h_\infty + p_\infty = \frac{1}{2}\rho_d u_d^2 + \rho_d g h_d + p_d^+,$$

which for incompressible and horizontal flow reduces to

$$\frac{1}{2}u_\infty^2 + p_\infty = \frac{1}{2}u_d^2 + p_d^+.$$

Downstream, a similar equation may be obtained, and since in the far wake $p_w = p_\infty$,

$$\frac{1}{2}u_w^2 + p_\infty = \frac{1}{2}u_d^2 + p_d^-,$$

then the pressure difference across the actuator disk is

$$p_d^+ - p_d^- = \frac{1}{2}\left(u_\infty^2 - u_w^2\right).$$

Going back to the momentum conservation equation, we see it can be reduced to

$$u_w = u_\infty(1 - 2a).$$

Hence, half of the axial speed loss occurs upstream of the disk, and half downstream.

The power $P = Fu_d$ extracted by the turbine naturally becomes

$$2\rho a A_d u_\infty^3 (1 - a)^2.$$

And the power coefficient, $c_p = P / P_{in}$, then, can be expressed as

$$c_p = 4a(1 - a)^2.$$

From this expression one can easily deduce that the Betz limit (1919), the limit when the power coefficient is maximum, is achieved for an inflow factor of $a = 1/3$, corresponding to

$$c_p = \frac{16}{27} = 0.593.$$

From the momentum theory presented above an expression of the thrust coefficient (Myers and Bahaj, 2010),

$$c_T = \frac{T}{\frac{1}{2}\rho u_\infty^2 A} = 4a(1 - a),$$

may also be derived. Given the differences between fluid densities of water and air, thrust values per unit area are approximately 50 times greater for tidal current turbines than for wind turbines at typical operating speeds (Myers and Bahaj, 2010). It is worth mentioning that if the inflow factor is equal to or larger than one-half, a problem arises, with the wake velocity vanishing or even turning negative. Under such conditions, the momentum theory presented here no longer applies and an empirical modification has to be made; this modification is taken into account, for example, in rotor blade theories, including a rotor blade design that provides an accurate description of both torque and thrust (Burton et al., 2001). Despite this, some authors have successfully applied the actuator disk assumption in 3D high-resolution CFD models and have captured, for example, the wake decay process accurately (Sun et al., 2008; Harrison et al., 2010), but at an extremely high computational cost. In fact, computational costs of CFD models are so high that they only can be implemented for realistic flows with large spatial coverage and complex bathymetry and forcing if the timescales are short, e.g., of the order of days. Therefore, for realistic applications with large spatial coverage, regional models with adequate tidal current turbine parameterizations are more appropriate than CFD models (Roc et al., 2013). Finally, it is worth noting that, although the actuator disk theory has been deduced using the assumptions made for wind turbines (Burton et al., 2001), the same assumptions may be used for a turbine in a tidal stream when the tideway is wide and deep compared with the turbine rotor. Thus, corrections need to be implemented when the bathymetry or the free surface constrains the flow and causes flow acceleration, as required to ensure mass conservation (Blunden and Bahaj, 2007).

18.3.2.2 Shallow-Water Models

Shallow-water conservation equations are an approximation of the Navier–Stokes equation, which assumes the fluid motion in the horizontal dimensions dominates over the vertical ones. This approximation is typical in atmospheric and oceanographic sciences. The depth-averaged mass conservation equation under the shallow-water assumptions is the following:

$$\frac{\partial \zeta}{\partial t} + \frac{\partial\left[(\zeta+d)U\right]}{\partial x} + \frac{\partial\left[(\zeta+d)V\right]}{\partial y} = S,$$

where ζ is the water level measured from the reference level corresponding to mean sea level (MSL), d is the water depth from MSL, defined as positive downwards, $\vec{U} = (U,V)$ is the depth-averaged velocity, and S is a mass source/sink term. The coordinates (x,y) correspond to the two coordinates of the horizontal plane. They may be Cartesian and expressed in meters, as in local UTM coordinate systems, or spherical (lon,lat) coordinates. The vertical frame of reference may be a terrain-following σ-level coordinate system (where the number of layers is the same everywhere regardless of the water depth) or a z-level coordinate system (which is strictly horizontal; the bottom is usually represented as a staircase). The difference between these two coordinate systems is illustrated in Figure 18.4, showing as well the differing vertical coordinate conventions. Similarly, the momentum conservation equations in depth-averaged form,

$$\frac{\partial U}{\partial t} + U\frac{\partial U}{\partial x} + V\frac{\partial U}{\partial y} + \frac{\omega}{\zeta+d}\frac{\partial U}{\partial \sigma} - fV = -\frac{P_x}{\rho_0} + F_x + M_x + \frac{1}{(\zeta+d)^2}\frac{\partial}{\partial \sigma}\left(\nu_V \frac{\partial U}{\partial \sigma}\right)$$

$$\frac{\partial V}{\partial t} + U\frac{\partial V}{\partial x} + V\frac{\partial V}{\partial y} + \frac{\omega}{\zeta+d}\frac{\partial V}{\partial \sigma} - fU = -\frac{P_y}{\rho_0} + F_y + M_y + \frac{1}{(\zeta+d)^2}\frac{\partial}{\partial \sigma}\left(\nu_V \frac{\partial V}{\partial \sigma}\right),$$

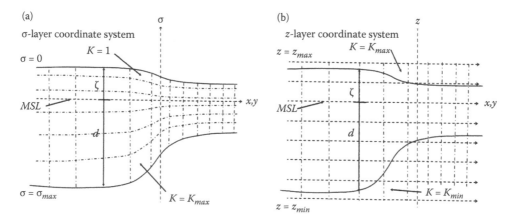

FIGURE 18.4 Vertical coordinate conventions: σ-level (left) and z-level (right).

express the balance between the inertia term $\partial \vec{U} / \partial t$, the advection terms $\vec{U}.\nabla\vec{U}$, the sigma-correction terms $\left[\omega /(\zeta+d)\right] \partial \vec{U} / \partial \sigma$, and the Coriolis terms $-f(V,U)$, and the pressure gradient terms $-\vec{P} / \rho_0$, the horizontal Reynolds stresses $\vec{F}$, the external forcing terms $\vec{M}$, and the vertical diffusion terms. It is worth noting that the pressure gradient terms in hydrostatic form depend only on the water elevation, being expressed as $-g\nabla\zeta$ in their simplest form. However, some compressibility effects should be taken into account, and so pressure gradients are in balance with an additional term, that depends on vertically averaged density gradients

$$\frac{\vec{P}}{\rho_0} = g\nabla\zeta + g \frac{\zeta+d}{\rho_0} \int_\sigma^0 \nabla\rho + \nabla\sigma' \frac{\partial\rho}{\partial\sigma'} d\sigma'$$

where the density ρ is a multiphase density that considers the water density ρ_w, as well as the sediment concentration $c^{(i)}$ and the density $\rho_s^{(i)}$ of each fraction i of the sediments in suspension,

$$\rho = \rho_w + \sum_{i=1}^{LSED} c^{(i)}\left(\rho_s^{(i)} - \rho_w\right).$$

From the velocity solution computed through the momentum conservation equation one may obtain the velocity magnitude and use it for the computation and analysis of the tidal power density into the turbine, P_{in}. Finally, some constituent transport equations need to be implemented if it is of interest to find, amongst others, the turbulence kinetic energy, k; the turbulence dissipation rate, ε; salinity or temperature fields; sediment fraction concentrations; or wave or roller energy fields (for 2D surf beat applications). Let C denote a generic constituent, then the constituent transport equation in the (x,y,σ) coordinate system, with the shallow-water approximation, may be expressed as

$$\frac{\partial\left[(\zeta+d)C\right]}{\partial t} + \frac{\partial\left[(\zeta+d)UC\right]}{\partial x} + \frac{\partial\left[(\zeta+d)VC\right]}{\partial y} + \frac{\partial(\omega C)}{\partial\sigma}$$

$$= (\zeta+d)\left[\frac{\partial}{\partial x}\left(D_H \frac{\partial C}{\partial x}\right) + \frac{\partial}{\partial y}\left(D_H \frac{\partial C}{\partial y}\right)\right] + \frac{1}{(\zeta+d)} \frac{\partial}{\partial\sigma}\left(D_V \frac{\partial C}{\partial\sigma}\right) + (\zeta+d)S$$

The vertical diffusivity D_V is of the form

$$D_V = \nu_V / \sigma_c$$

where σ_c is the Prandtl–Schmidt number. In sediment transport application, σ_c is a constant usually taken as 1, implying that sediment diffusion has the same efficiency as turbulent diffusion. However, a wide range of values may be found in the literature (van Rijn, 1984; Cellino, 1998; Ogston and Sternberg, 2002; Malarkey et al., 2015). Both the vertical diffusivity and the water viscosity are determined from the turbulence closure model and include density effects.

For resource assessments, the 2D formulation of the shallow-water equations may be sufficient, provided that the bathymetrys and the grid are known and defined, respectively, with a horizontal resolution that is sufficient for identifying the areas with highest theoretical tidal power density. However, for environmental impact resource assessments, a 3D model is recommended, so that the vertical velocity profile is described with sufficient detail. The erosion and sedimentation processes need to be captured well enough to assess the sediment transport processes near the seabed, and so the σ-levels near the bed need to be thinner than the levels higher up in the water column. However, if wind-driven flows are relevant in the study, the σ-levels near the sea surface also need to be thinner. In both cases, it is recommended that the thickness of the level closest to the boundary is at most 2% of the total water depth (Deltares, 2014).

18.4   FINAL COMMENTS

Although tidal speed monitoring is ideal and essential from stage 3 onwards at resolutions of the order of 5 m, it is practically impossible to survey a country's coastline at those resolutions; even in numerical models, such resolutions are only used in micrositing studies. On the other hand, the resolutions of 5 to 12 km used in regional and global models are too large to identify the locations of the most promising tidal current exploitation sites with enough confidence (O'Rourke et al., 2010). So, a numerical experiment was designed to assess the largest grid resolution that could be used. In the experiment, a sensitivity analysis was implemented to assess as well the effect of the bathymetric data on the resource assessment results. The indicator used for the analysis was the tidal power density average over a tidal cycle.

Figure 18.5a shows the site used for illustration: the Infiernillo Channel, a narrow strait between mainland Mexico and Mexico's largest island, Tiburón Island. The geographical position of the Infiernillo Channel is (29°7'556.44"N, 112°13'9.03"W); it is about 41 km long between the points marked "Southern End" and "Northern End," with geographical positions (28°58'16.16"N, 112°11'1.60"W) and (29°19'52.45"N, 112°16'59.61"W), respectively. It is 8 km wide at its widest, in the middle of the Channel and at the Northern entrance, closed by the sand spit of Cabo Tapioca. It is located in the Gulf of California, on the Pacific side of Mexico. First, we analyzed the sensitivity of the model predictions to the bathymetric data used. In order to reduce the dependency of the results on the mesh site, a very fine curvilinear mesh with cells of around 100 m × 200 m was designed. The model was forced at the open boundary with 11 tidal components from the Topex-Poseidon TPXO 8.0 global inverse tidal solution (Egbert et al., 1994; Egbert and Erofeeva, 2002). After a couple of days of spin-up time, the model reached convergence, and then once-monthly averaged tidal speeds were determined from the velocities computed by the model. The sensitivity of the tidal speeds to bathymetric data conditions is shown in the other subplots of Figure 18.5. This sensitivity was tested with the General Bathymetric Chart of the Oceans released in 2014, the GEBCO14 bathymetry (GEBCO, 2014), see the Figure 18.5b; with the GEBCO08 bathymetry, the 2008 release of GEBCO (GEBCO, 2008), see Figure 18.5c; and the Lancín85 bathymetry reproduced by Lancín (1985) and digitized in-house, see Figure 18.5c. Both GEBCO14 and GEBCO08 have a resolution of 30 arc-seconds; they are the most authoritative, publicly available bathymetric data sets for the world's oceans. GEBCO14 and GEBCO08 come from the same sources, with GEBCO14 offering some improvements in relation to GEBCO08 by including some new ship surveys and a seamless connection between surface and land (Carabajal et al., 2011). The bathymetric part of the data comes from a database of ship track soundings with interpolation between soundings guided by satellite-derived gravity data. It must be noted that GEBCO products are essentially deep-sea products and do not include, in general, detailed bathymetries for shallow shelf seas. Moreover, for the most part, bathymetric mapping is an interpretation based on random track lines of data from differing quality and coverage. In fact, a close inspection of the GEBCO bathymetries showed some inaccuracies. In the Infiernillo Channel, the strait is known to be shallow (Lancín, 1985), with average mean water depths of or under 12.8 m, but the GEBCO bathymetries show depths of 100 m in some areas.

Figure 18.5b and c place the locations with largest theoretical tidal energy potential in the same areas, namely South of "Punta Santa Rosa" and "Punta San Miguel" at the Southern End of the Infiernillo Channel; Figure 18.5d places these locations to the South West of Laguna Sargento, or West of "Punta Perla," the sand spit on Tiburon Island, on the Northern End of the Infiernillo Channel. Without a better representation of the bathymetry, it is not possible to decide which of these two possible locations is better suited for tidal energy exploitation. However, the location and magnitude of the monthly averaged tidal speed maximum using Lancín85's bathymetry both agree qualitatively well with those found in a previous study by Merifield et al. (1970). However, it is unclear whether these authors did measurements at the location at Southern End of the Channel identified in Figure 18.5b and c as well, for comparison.

(a)   (b)

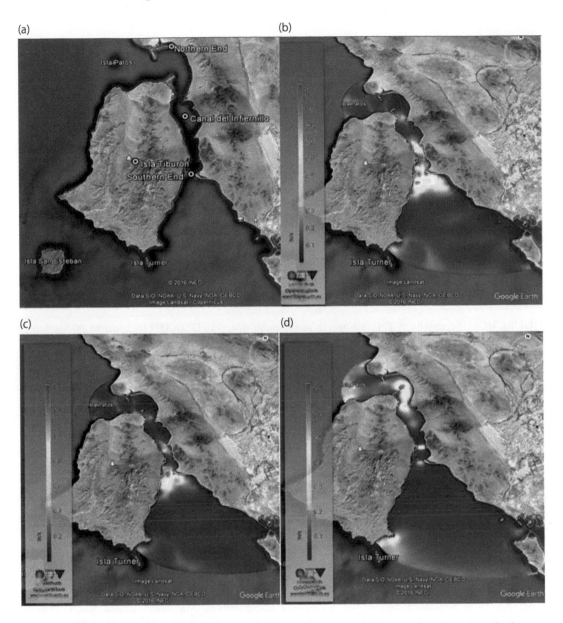

(c)   (d)

FIGURE 18.5 Site chosen and sensitivity of monthly averaged tidal speeds to bottom boundary (bathymetric) conditions, with the same curvilinear mesh. (a) Site: Infiernillo Channel. Predicted monthly averaged tidal speeds with (b) GEBCO14, (c) GEBCO08, and (d) Lancín85 bathymetries.

Now that the sensitivity to bathymetric conditions has been demonstrated, we show the sensitivity of the predictions to the numerical grid mesh size. For illustration purposes, in this part of the analysis the bathymetry conditions used are fixed to GEBCO14, as only one parameter can be varied at a time to assess how it affects the indicator estimates. This time the indicator chosen is the tidal power density (TPD). Figure 18.6a and b show the TPD in kW/m^2, with a mesh of 3 km × 3 km and 1 km × 1 km, respectively. A comparison with Figure 18.5b clearly shows that a resolution of 1 km × 1 km, as shown in Figure 18.6b, is sufficient to find the locations where the indicators take maximum values, as the Southern End is identified as the optimal site in both case, whereas a resolution of 3 km × 3 km, as shown in Figure 18.6a, clearly is insufficient for tidal energy resource assessments.

(a)   (b)

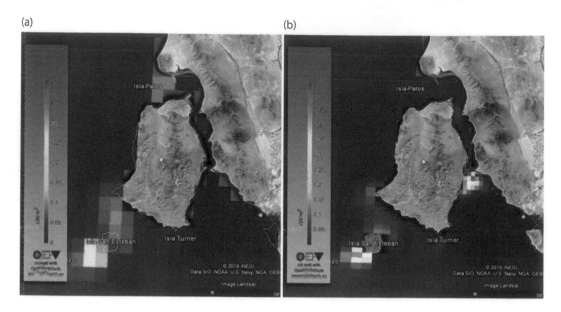

FIGURE 18.6   Sensitivity of TPD (kW/m^2) to model mesh size, with GEBCO14 bathymetry. (a) Mesh of 3 km × 3 km. (b) Mesh of 1 km × 1 km.

ACKNOWLEDGMENTS

We thank Mario Nieto-Oropeza for creating the figures shown in Figure 18.5b through d, and. Figures 18.6a and b. Thanks to Dr. Scott Brown for revising an earlier version of the manuscript.

REFERENCES

Alstom (2013) Alstom's Tidal Turbine Reaches 1MW in Offshore Conditions. Available at: http://www. alstom.com/press-centre/2013/7/alstoms-tidal-turbine-reaches-1mw-in-offshore-conditions/ (Accessed December 23, 2016).

Bahaj, A. S. (2011) Development of marine current turbines for electricity production, *IEEE Power and Energy Society General Meeting*, pp. 0–3. doi:10.1109/PES.2011.6039067.

Barstow, S., M-rk, G., L-nseth, L., Schjølberg, P., Machado, U., Athanassoulis, G., Belibassakis, K., Gerostathis, T. and Spaan, G. (2003) WORLDWAVES: High quality coastal and offshore data within minutes for any global site, in *Proceedings of the 16th Australasian Coastal & Ocean Engineering Conference in Auckland, New Zealand*, September 2003. Available at: http://www.oceanor.no/Services/Worldwaves/.

Betz, A. (1919) Schraubenpropeller mit geringstem Energieverlust. Mit einem Zusatz von l. Prandtl, *Nachrichten von der Gesellschaft der Wissenschaften zu Göttingen, Mathematisch-Physikalische Klasse*, 1919, pp. 193–217.

Blunden, L. S. and Bahaj, A. S. (2007) Tidal energy resource assessment for tidal stream generators, *Sage Journals*, 221, pp. 137–146. doi:10.1243/09576509JPE332.

Burton, T., Sharpe, D., Jenkins, N. and Bossanyi, E. (2001) *Wind Energy Handbook*, John Wiley & Sons, Chichester, UK.

Carabajal, C. C., Harding, D. J., Boy, J.-P., Danielson, J. J., Gesch, D. B. and Suchdeo, V. P. (2011) Evaluation of the Global Multi-Resolution Terrain Elevation Data 2010 (GMTED2010) using ICESat geodetic control, *Proceedings of SPIE*, 8286, pp. 1–13. doi:10.1117/12.912776.

Cavazzi, S. and Dutton, A. G. (2016) An Offshore Wind Energy Geographic Information System (OWE-GIS) for assessment of the UK's offshore wind energy potential, *Renewable Energy*, 87(P1), pp. 212–228. Available at: http://www.econpapers.repec.org/RePEc:eee:renene:v:87:y:2016:i:p1:p:212–228.

Cellino, M. (1998) *Experimental Study of Suspension Flow in Open Channels*. École Polytechnique Fédérale de Lausanne, Switzerland.

Deltares (2014) *Delft3D-FLOW, User Manual*. Delft, The Netherlands: Deltares. Available at: http://www. deltaressystems.nl.

Egbert, G. D., Bennett, A. F. and Foreman, M. G. G. (1994) TOPEX/POSEIDON tides estimated using a global inverse model, *Journal of Geophysical Research*, 99852(15), pp. 821–824. doi:10.1029/94JC01894.

Egbert, G. D. and Erofeeva, S. Y. (2002) Efficient inverse modeling of barotropic ocean tides, *Journal of Atmospheric and Oceanic Technology*, 19(2), pp. 183–204. doi:10.1175/1520–0426(2002)019<0183:EIMOBO>2.0.CO;2.

EMEC (2016) EMEC: European Marine Energy Centre—Tidal Energy Devices. Available at: http://www.emec.org.uk/tidal_devices.asp.

Gardner, J. V., Dartnell, P., Mayer, L. A. and Hughes Clarke, J. E. (1999) *Bathymetry and Selected Perspective Views of Lake Tahoe, California and Nevada*. United States Geological Survey Water-Resources Investigations Report 99-4043, USA.

GEBCO (2008) GEBCO_08 Grid—version 20100927. Available at: http://www.gebco.net/data_and_products/gridded_bathymetry_data/version_20100927/#.WGH2O4luSSE.mendeley (Accessed December 27, 2016).

GEBCO (2014) GEBCO_2014 Grid—version 201411037. Available at: http://www.gebco.net/data_and_products/gridded_bathymetry_data/version_20141103/#.WGH3xEpnd1s.mendeley (Accessed December 27, 2016).

Harrison, M. E., Batten, W. M. J., Myers, L. E. and Bahaj, A. S. (2010) Comparison between CFD simulations and experiments for predicting the far wake of horizontal axis tidal turbines, *IET Renewable Power Generation*, 4(6). doi:10.1049/iet-rpg.2009.0193.

Hu, Z. Z., Greaves, D. and Raby, A. (2016) Numerical wave tank study of extreme waves and wave-structure interaction using OpenFoam® *Ocean Engineering*, 126, pp. 329–342. doi:10.1016/j.oceaneng.2016.09.017.

Kempener, R. and Neumann, F. (2014) *Salinity Gradient Energy: Technology Brief, International Renewable Energy Agency (IRENA)*. doi:10.1109/OCEANS.1979.1151215.

Komen, G. J., Cavaleri, L., Donelan, M., Hasselmann, K., Hasselmann, S. and Janssen, P. A. E. M. (1996) *Dynamics and Modelling of Ocean Waves*. Cambridge University Press, Cambridge, UK.

Kramer, S. C. and Piggott, M. D. (2016) A correction to the enhanced bottom drag parameterisation of tidal turbines, *Renewable Energy*, 92, pp. 385–396. doi:10.1016/j.rcnene.2016.02.022.

Lancín, M. (1985) Geomorfología y génesis de las flechas litorales del Canal de Infiernillo, Estado de Sonora, *Revista del Instituto de Geología, UNAM*. Universidad Nacional Autónoma de México, Instituto de Geología, 6(1), pp. 52–72.

Laws, N. D. and Epps, B. P. (2016) Hydrokinetic energy conversion: technology, research, and outlook, *Renewable and Sustainable Energy Reviews*. 57, pp. 1245–1259. doi:10.1016/j.rser.2015.12.189.

Legrand, C. (2009) *Assessment of Tidal Energy Resource*. Marine Renewable Energy Guides Series. London: The European Marine Energy Centre Ltd (EMEC).

Leishman, J. M. and Scobie, G. (1976) The Development of Wave Power. A Techno-economic Study. Available at: http://www.homepages.ed.ac.uk/v1cwaveg/0-Archive/EWPP archive/1976 Leishman and Scobie NEL.pdf.

Magar, V., González-García, L. and Gross, M. S. (2016) Evaluación técnico-económica del potencial de desarrollo de parques eólicos en mar: el caso del Golfo de California, *BIOtecnia*.

Mai, T. (2015) Technology readiness level. Available at: https://www.nasa.gov/directorates/heo/scan/engineering/technology/txt_accordion1.html (Accessed November 28, 2016).

Malarkey, J., Magar, V. and Davies, A. G. (2015) Mixing efficiency of sediment and momentum above rippled beds under oscillatory flows, *Continental Shelf Research*, 108. doi:10.1016/j.csr.2015.08.004.

Mann, R. (1998) *Field Calibration Procedures for Multibeam Sonar Systems*. Report TEC-0103. Alexandria, VA. US Army Coprs of Engineers Topographic Engineer Center.

Merifield, P. M., Marzolf, J. E. and Lamar, D. L. (1970) Marine Sand Waves in El Infiernillo Channel, Gulf of California: Final Report. Earth Science Research Corporation. Available at: https://books.google.com.mx/books?id=5LcPAQAAIAAJ.

Mørk, G., Barstow, S., Kabuth, A. and Pontes, M. T. (2010) Assessing the global wave energy potential, in *Proceedings of OMAE2010 29th International Conference on Ocean, Offshore Mechanics and Arctic Engineering, June 6–11, 2010*. Shanghai, China.

Myers, L. E. and Bahaj, A. S. (2010) Experimental analysis of the flow field around horizontal axis tidal turbines by use of scale mesh disk rotor simulators, *Ocean Engineering*. 37(2–3), pp. 218–227. doi:10.1016/j. ceaneng.2009.11.004.

NAVOCEANO (2002) Digital Bathymetric Data Base Variable Resolution (DBDB-V) from the U.S. Naval Oceanographic Office (NAVOCEANO). Available at: http://gcmd.nasa.gov/records/GCMD_DBDBV. html.

Ogston, A. S. and Sternberg, R. W. (2002) Effect of wave breaking on sediment eddy diffusivity, suspended-sediment and longshore sediment flux profiles in the surf zone, *Continental Shelf Research*, 22(4), pp. 633–655. doi:10.1016/S0278-4343(01)00033-4.

O'Rourke, F., Boyle, F. and Reynolds, A. (2010) Tidal current energy resource assessment in Ireland: current status and future update, *Renewable and Sustainable Energy Reviews*, 14(9), pp. 3206–3212. doi:10.1016/j.rser.2010.07.039.

Pelc, R. and Fujita, R. M. (2002) Renewable energy from the ocean, *Marine Policy*, 26(6), pp. 471–479. doi:10.1016/S0308-597X(02)00045-3.

Power Technology (2016) *Scroby Sands Offshore Wind Farm, United Kingdom*. Available at: http://www.power-technology.com/projects/scrobysands/ (Accessed November 11, 2016).

Reeve, D. E., Horrillo-Caraballo, J. M. and Magar, V. (2008) Statistical analysis and forecasts of long-term sandbank evolution at Great Yarmouth, UK, *Estuarine, Coastal and Shelf Science*, 79(3), pp. 387–399. doi:10.1016/j.ecss.2008.04.016.

Reeve, D. E., Li, B. and Thurston, N. (2001) Eigenfunction analysis of decadal fluctuations in sandbank morphology at Gt Yarmouth, *Journal of Coastal Research*, 17(2), pp. 371–382.

van Rijn, L. C. (1984) Sediment transport, Part {II}: suspended load transport, *Journal of Hydraulic Engineering*, 110(11), pp. 1613–1641.

Rivier, A., Bennis, A. C., Pinon, G., Gross, M. and Magar, V. (2015) Regional numerical modelling of offshore monopile wind turbine impacts on hydrodynamics and sediment transport, in *Renewable Energies Offshore - 1st International Conference on Renewable Energies Offshore, RENEW 2014*.

Rivier, A., Bennis, A. C., Pinon, G., Magar, V. and Gross, M. (2016) Parameterization of wind turbine impacts on hydrodynamics and sediment transport, *Ocean Dynamics* 66(10), pp. 1285–1299. doi:10.1007/s10236-016-0983-6.

Roc, T., Conley, D. C. and Greaves, D. (2013) Methodology for tidal turbine representation in ocean circulation model, *Renewable Energy*, 51, pp. 448–464. doi:10.1016/j.renene.2012.09.039.

Shives, M. and Crawford, C. (2016) Adapted two-equation turbulence closures for actuator disk RANS simulations of wind & tidal turbine wakes, *Renewable Energy*, 92, pp. 273–292. doi:10.1016/j.renene.2016.02.026.

Soediono, B. (1989) IHO standards for hydrographic surveys 5th edition, February 2008 Special Publication No. 44, *Journal of Chemical Information and Modeling*, 53, p. 160. doi:10.1017/CBO9781107415324.004.

Sun, X., Chick, J. P. and Bryden, I. G. (2008) Laboratory-scale simulation of energy extraction from tidal currents, *Renewable Energy*, 33(6), pp. 1267–1274. doi:10.1016/j.renene.2007.06.018.

Wessel, P. and Smith, W. H. F. (1996) A global, self-consistent, hierarchical, high-resolution shoreline database, *Journal of Geophysical Research: Solid Earth*, 101(B4), pp. 8741–8743. doi:10.1029/96JB00104.

Whitehouse, R. J. S., Harris, J. M., Sutherland, J. and Rees, J. (2011) The nature of scour development and scour protection at offshore windfarm foundations, *Marine Pollution Bulletin*. 62(1), pp. 73–88. doi:10.1016/j. marpolbul.2010.09.007.

World Energy Council (2013) World Energy Resources: 2013 survey, World Energy Council. Available at: http://www.worldenergy.org/wp-content/uploads/2013/09/Complete_WER_2013_Survey.pdf.

19 From Large Dams to Sustainable Hydropower

Katarzyna Anna Korys and Agnieszka Ewa Latawiec
International Institute for Sustainability

Agnieszka Ewa Latawiec
Pontifical Catholic University of Rio de Janeiro
and University of Agriculture in Krakow

CONTENTS

19.1 INTRODUCTION

The use of potential energy to drive a waterwheel and cogwheel gear dates to at least Roman engineer Marcus Vitruvius Pollio's year 13 BC grain mill (Raabe, 1987). But the world´s first hydropower plant began operation in 1882 on the Fox River, in Appleton, Wisconsin (Gurbuz, 2006). Today, the principle of its operation is based on converting the water potential energy into mechanical and electrical energy, using turbines (Ellaban et al., 2014).

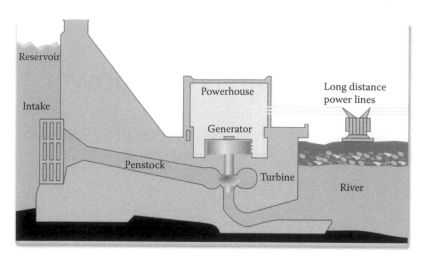

FIGURE 19.1   Simple scheme of a hydropower plant. (From Wikimedia commons.)

There are three main types of hydropower plants: run-of-river, storage hydro (reservoir), and pumped storage (Kaunda et al., 2012). The most common type stores the potential energy of inland water by building a dam on the river (impoundment facility). Water, which falls from a higher level, gains speed and flows through turbine blades (Figure 19.1). The generator converts the mechanical energy of water to electricity and sends it to the transformer, which in turn, regulates the voltage (Ellaban et al., 2014).

19.1.1   Difficulties in the Supply of Primary Energy Sources

Nearly 1.2 billion people (17% of the global population) have no access to electricity. Providing them electricity will augment global energy requirements, as will global economic and population growth (Brazil Energy Outlook, 2013). Electricity is a secondary energy source that requires a primary energy source to be converted. The supply of primary sources is going through a series of difficulties. Coal is still an important energy source worldwide. However, society demands energy, and if it is possible to obtain energy from other, less expensive and sustainable sources, it would be preferred (Höök et al., 2010). Conflict between Russia and Ukraine about natural gas—known as the "Natural gas crisis" and the "Cold War" that refers to a number of disputes between Ukrainian oil and gas company Naftohaz Ukrayiny and Russian gas supplier Gazprom (Shi, 2009) has raised concerns regarding the delivery of natural gas. The conflict included the series of events that began in 1992 when the USSR collapsed and the Ukrainians supported a pro-European government (Torres et al., 2009). Moreover, the situation in some regions of the Middle East, where more than half of the world's oil reserves is, located is unstable as the result of the military and political events that have begun from the Arab-Israeli War in 1973 (Anderson, 2000). In 1974, due to the aftermath of Middle East War crisis, the International Energy Agency was founded] and influenced by a number of factors such as demand and supply of natural resources. Nowadays the Organization of the Petroleum Exporting Countries, in view of its majority holdings of oil global production, can control the supply of natural resources and determine oil prices on the world´s market (Tomczonek, 2013), e.g. price for crude in Dubai in February 2017 was 54.17 US$ for a barrel (source: https://knoema.com/). Nuclear energy development receives much attention as an alternative to carbon-based energy sources; however, controversy emerged around nuclear power development, accentuated by the post-tsunami events at the Fukushima Daiichi nuclear power plant in Japan in March 2011 and continues. Although the disaster did not cause any immediate deaths, radioactive contamination from the Fukushima plant forced the evacuation of communities up to 25 miles away and affected up to 1,000,000 residents (Holt et al.,

2012). In the debates around nuclear power expansion, there is considerable attention to technical questions such as cost and design, but there is also attention to social issues such as risk assessment, social acceptability, public accountability, and public perception (Parkins and Haluza-DeLay, 2011). All these primary sources have, in addition, a negative influence on the environment, such as greenhouse gas emissions, air, water, and soil pollution among other impacts.

19.1.2 STATE OF GLOBAL HYDROPOWER

Hydropower contributes more than 16% of the global electricity generation and about 85% of the global renewable electricity. By 2011, over 160 countries had a compounded installed capacity of 11,000 hydropower stations and >930 GW. The leading generating countries are China, Brazil, Canada, the U.S., and Russia—these make up more than half the world's hydropower production (Ellaban et al., 2014). The largest plants are Three Gorges in China (22.5 GW) and Itaipu in Brazil/Paraguay (14.0 GW). According to data provided by Itaipu Binacional company, in 2016 Itaipu plant has produced a total of 103,098,366 MWh (https://www.itaipu.gov.br/en/energy/production-year-year).

19.1.3 HYDROPOWER IN BRAZIL

Brazil, Russia, India, and China are large, fast-growing, and vital to the world's economy. An important economic goal in Brazil seems to have been the attempted reduction of energy dependence via substitution of foreign energy supplies and development of domestic energy sources (Geller et al., 2004). In Brazil, hydropower is promoted as a renewable energy source in response to fast-growing energy demand and economic growth (Guatam et al., 2004).

Including hydropower, renewable sources provide 80% of the electricity in Brazil, one of the highest shares in the world (Brazil Energy Outlook, 2013). Hydropower supplies 70% of Brazil's electricity consumption (world average ~16%), and the current installed capacity in Brazil is ~114 GW with an estimated total potential of 260 GW (von Sperling, 2012). This is the third largest potential after Russia and China. The country has 176 large and 402 small hydro plants under operation, which makes Brazil one of the leading countries in the world regarding hydroelectrical electricity production. The forecast in the Ten-Year Energy Plan is that the country would have 71 new plants by 2017, with a potential output of 29,000 MW: 15 in the Amazon basin, 18 on the Paraná River, 13 in the Tocantins-Araguaia basin; and 8 on the Uruguay River; the 28 hydro plants planned in Amazon region would have a total installed capacity of 22,900 MW (von Sperling, 2012).

19.2 DAMS IN BRAZILIAN AMAZONIA: TECHNICAL AND SUSTAINABILITY ISSUES

The Amazon watershed holds one of the world's largest freshwater reserves. By May 2015, Brazilian states in the Amazon region had 15 large dams with reservoirs filled and 37 dams under construction or planned. The best sites have probably been exploited (Fearnside, 2015).

19.2.1 THE BALBINA POWERPLANT: HOW NOT TO BUILD A DAM

Concerns about the impact of the Balbina hydro plant on the Uatuamã River in Central Amazonia (Rosas et al., 2007) already existed at the planning stage. The project was criticized for potential forest loss, expensive construction, and displacement of the members of the Waimiri-Atroari tribe. The plant has five generators, an installed power of only 250 MW though the average power output is only 112.2 MW. Despite adverse local conditions (flat terrain and small size of the drainage basin), the dam was approved and began operation in February 1989. The project aimed at supplying Manaus with electricity (Fearnside, 1989), but during the construction work the city grew vastly, requiring other energy sources, which included transport from recently discovered oil and

FIGURE 19.2 Balbina Reservoir; described by Fearnside (1989) as "dead trees standing in the shallow water." (From Wikimedia commons.)

gas deposits or transmission from more distant dams. The flooded area (2360 square kilometers) was mostly covered by rainforest. To save time and workload, the forest was inundated without previous logging (Figure 19.2). Vegetation left to decompose in the shallow water contributed to the production of hydrogen sulfide—a gas with a characteristic rotten smell that also may cause corrosion of turbines due to oxygen-free, acidic water (Fearnside, 1989). Furthermore, detailed studies and measurements show significant emissions of greenhouse gases, like methane and carbon dioxide, from the reservoir and downstream (Kemenes et al., 2006, 2007, 2011).

Now, all the inundated area of Balbina on the left bank of Uatumã River is a protected area (Uatumã Biological Reserve) (Rosas et al., 2007); however, flooded habitats became an ecosystem of fragmented archipelagos of 3546 islands, which impede the free movement of many species of animals.

Balbina was considered one of the worst water uses in the country, in terms of electricity production and the environment. Early on, it seemed preferable to halt the project before filling the dam and use turbines and generators at more suitable sites, even if certain construction phases were already completed (Fearnside, 1989). The example of Balbina could have contributed to avoiding mistakes discussed in this chapter.

19.2.2 THE TUCURUÍ PLANT AND THE DEFOLIANT DISPUTE

The Tucuruí hydroelectric power plant was the first large-scale hydroelectric scheme completed in the Amazon region. The dam was part of a series of projects promoted by the Brazilian government to supply energy for the Grande Carajá's Project. The Carajás mine is the world's largest iron ore mine, with 7.2 billion metric tons of iron ore, as well as manganese, bauxite, copper, tin, aluminum, and gold. It has operated since 1970 by a government-created joint venture owned by the Vale Company and U.S. Steel. Brazil is currently the world's largest exporter of iron ore (mining technology.com). With a 7920 MW maximum capacity and 6.9-km long and 78 m high concrete wall (La Rovere and Mendes, 2000), it was one of 15 plants planned to harness the 20,645 MW potential of the Araguaia-Tocantins basin. The artificial lake formed by the dam covered 2435 square kilometers of tropical forest (Manyari and de Carvalho Jr., 2007), roughly the size of the Balbina reservoir, but Tucuruí generates nearly 8 times more power than Balbina. It has been defended by Eletronorte as the "example of successful hydroelectric development in the Amazon" (Fearnside, 2001). Still, the creation of the reservoir caused a variety of effects, like deforestation, resettlement, GHG emissions, and loss of species. However, in Tucuruí's case, the reason for the controversy was that during its construction toxic defoliants were used in an attempt to clear the forest.

Eletronorte was slandered for using harmful chemicals (Barham and Caufield, 1984). The decision not to clear a vast area of forest intended to be flooded caused a wave of criticism from environmentalists who argued that the consequences would equate to ecological disaster. In addition, decomposed trees would raise the acidity of the water and destroy the turbine by corrosion (at one Brazilian dam, corrosion of equipment cost the constructors U.S. $5 million, according to Barrow, 1998). Another issue was concern that waterweeds would spread over the surface of the reservoir and become a habitat for disease vectors, especially for malaria-transmitting mosquitoes. Under pressure from the studies' results, Eletronorte agreed to clear the parts of forest where the problems with decomposition would have been severe. The company decided to take action with only a few years left before the reservoir level started rising and there was further delay in choosing the contractor to clear the forest. Finally, the contract was awarded to CAPEMI, a company with no logging experience but connections with the Brazilian army.

Later, sources reported the contradictory information regarding the use of defoliants and participation of both companies in this operation. CAPEMI was accused of using herbicides in secret, due to their inability to keep the deadline. Both companies denied these allegations (Fearnside, 2001). Moreover, in 1982 Eletronorte asked the National Institute for Amazonian Research to conduct a study of several herbicides, including Tordon-101 and Tordon-155 (officially banned in 1977). The aim of the study was to estimate their potential usefulness in clearing the forest through aerial spraying.

The minister of the Interior, Mario Andreazza, had ordered Eletronorte not to proceed with plans of using herbicides (Andreazza was preparing to launch his presidential candidacy and decisions on herbicides could have had important public opinion repercussions) (Fearnside, 2001). The government nevertheless gave no assurance that the scheme would be abandoned. Furthermore, Barrow (1988) argues that some of the chemical defoliants leaked into the environment, resulting in deaths of people and livestock. The issue of herbicides became one of the polemical issues regarding Tucuruí.

19.2.3 THE BELO MONTE STRUGGLE

Belo Monte, on the Xingu River in the Brazilian Amazon, is one of the most contentious projects in recent years. It was strongly lobbied for by large corporations and the Brazilian government, despite cost estimates ranging $U.S. 13–27 billion. The gigantic construction is bound to be the world's third largest hydroelectric power plant, after Three Gorges Dam and Itaipu, with 11,233 MW capacity. The reservoir was filled in December 2015, the first turbines started operating in May 2016, and the scheme is to be fully operational in 2019 (Fearnside, 2017).

In 1987, Eletrobras released its "2010 Plan" of 297 dams to be built across Brazil, six of them on the Xingu River. However, ambitious plans concerning dams in the region of Xingu faced resolute opposition by the indigenous Kayapo people, whose lands bordered the river. Locals were terrified that the dam would flood their lands and affect the river's ecosystem. During the event known as "Altamira Gathering," the case against the project gained traction and international media were informed. The event received extraordinary publicity, and the Kayapo people gained support from representatives of the governments of other countries and public figures. The World Bank announced that it would refuse Brazil the loan for this purpose. As a result, out of the six dams to be built on the Xingu River, there remained only Belo Monte (Fearnside, 2017).

In the following years a battle took place among the government, opponents, and energy companies. Opposition came from residents of many regions of Brazil—especially indigenous people and settlers of the Volta Grande do Xingu region, where the effects of the hydroelectric dam would be greatest. There were also licensing issues. Nevertheless, Brazilian President Lula and Norte Energia S.A. signed the agreement for the construction of the Belo Monte dam. The construction of the hydroelectric power plant began in 2011 (Guatam et al., 2014).

This decision caused dismay and objections by many Brazilians and led to mass protests and street demonstrations across the country. In 2012, the construction was halted by the federal court in Brazil. However, after an appeal by the company, construction continued. Opponents argued that the dam would annihilate the surrounding rainforest ecosystem and contribute to the displacement of 20,000 to 40,000 people, destroying their livelihoods with little or no compensation. Also, the operation would be unprofitable: given the water regime of the Xingu River, the capacity factor (ratio of energy produced to the hypothetical maximum possible, in each period) would only be 40.7% (the average power production 4571 MW), as compared to Itaipu's 75%, Three Gorges' 46%, and 44% world average (Brazil Energy Outlook, 2013). The struggle is far from over (Fearnside, 2017).

19.2.4 Greenhouse Gas Emissions (CH_4, CO_2)

Inland waters—especially dam reservoirs—may have considerable impacts at the global scale (Gruca-Rokosz, 2012). Hydroelectric power stations are often perceived as "clean energy" because they lack smokestacks and seem not to cause emissions. This is an erroneous perception, because there are emissions of gases from the surface of reservoirs as well as downstream (Kahn et al., 2014).

Land covered by vegetation and then inundated creates perfect conditions for microbes to generate gases like carbon dioxide and methane. Organic matter accumulated in the reservoir is deposited in sediments and decomposed due to microbial activity. As long as oxygen is available, the gaseous product of microbial respiration is CO_2.

The production of greenhouse gases per unit of power generated (say, kWh) will depend on the amount of organic carbon flooded by the hydropower dam: one hectare of rainforest mean estimate of 157 tons of carbon which is equivalent to 576.6 tons of carbon dioxide (Fearnside, 1995, 2000). And the projected flooding of 12,000 km^2 of rainforest would yield 692 million metric tons of carbon dioxide (Kahn et al., 2014).

Concomitantly, CH_4 (with a global warming potential 23 times that of CO_2) is produced in a reservoir by the degradation of organic matter under anoxic conditions (Beaulieu et al., 2016) that occur in the sediment or the bottom layer of water (hypolimnion) when diffusion of oxygen in the deeper layers of the reservoir is hindered by stratification (Gruca-Rokosz, 2012). Methane can be released from a reservoir into the atmosphere by a sudden change in pressure and temperature, such as is caused by the release of water from the deep reservoir through a turbine (Fearnside, 2004). Diffusion also occurs as gas bubbles from the sediments are dissolved in the water column. After rising to the water surface, the bubbles are released to the atmosphere (Beaulieu et al., 2016). Ebullition contributed 60%–80% of total emissions from the surface of the reservoir (disregarding downstream emissions) and might be a major pathway in young hydroelectric reservoirs in the tropics (Deshmukh et al., 2014).

19.2.5 Deforestation and Land Losses

The Amazon River basin is the largest in the world with 70,000 MW hydropower potential (Manyari and de Carvalho, 2007). The Federal Government of Brazil has planned to build 58 hydroelectric dams in Brazilian Amazonia and an increase in hydroelectric power capacity of more than 35,000 MW; this does not include hundreds of dams proposed or under construction in other Amazonian countries (Kahn et al., 2014). According to these projects, over 12,000 square kilometers of rainforest would be flooded.

Losses would follow the Brazilian example of Sete Quedas (Seven Falls) National Park under the Itaipu reservoir (Fearnside, 2015), when the creation of the 1350 km^2 reservoir resulted in the disappearance of waterfalls, forests, and agricultural land (Ziober and Zanirato, 2014).

The losses would significantly affect the Amazon Basin, one of the world's most important ecological systems: it contains 40% of the remaining global tropical rainforest area (Aragão et al., 2014), one of Earth's greatest collections of plant, animal, and microbial biodiversity (Foley et al., 2007). The size and complexity of this ecosystem make it a key component of the global carbon, climate, and weather systems (Davidson et al., 2012). Moreover, the Amazon rainforests supply valuable products and materials, like spices, precious wood, minerals, and pharmaceutical commodities (Foley et al., 2007). Losses to hydropower projects would add to the forecasted 47% of the Brazilian Amazon that would be cleared by 2050 (Soares et al., 2006).

19.2.6 IMPACT ON BIODIVERSITY

Brazil is a megadiverse country; this is expressed in its ecosystems, biomes, landscapes, flora, and fauna (Zanirato, 2010). The Amazon Rainforest biome is a major part of Brazil's biological diversity and is under threat. There are no precise data on the exact number of extant species (Ziober and Zanirato, 2014), as unknown species of plants and animals are found every year in the Brazilian Amazon, while many become extinct before they are discovered (Costello et al., 2013).

Despite the fact that many of the species threatened in the Brazilian Amazon are strictly protected by Brazilian law (Law No. 9605/1998—Articles 29, 34 and 53), there are legal provisions to allow their complete extirpation by dam-building projects (Lees et al., 2016). Extirpation could be achieved in hydro projects directly as hunters and fishermen encroach on species to feed the construction workers or indirectly via regional impacts on biodiversity catalyzed by cascade effects (Lees et al., 2016). Fragmentation is a direct effect on habitat with protracted consequences on the reproduction of species. This why reservoir islands should not be used for species conservation as part of impact mitigation measures and should be counted as part of land impacted by dam creation (Jones et al., 2016).

Impacts on carnivorous fish can trigger such cascade effects: after Itaipú's reservoir was constructed, *Serrasalmus marginatus*, a piranha species, invaded the Upper Parana River (Agostinho, 2003) causing a reduction of the smaller, native species *S. spilopleura* (speckled piranha), important in commercial fisheries and the aquarium market (Luna et al., 2009). Upstream of Tucurui, the long-distance reproduction migrations of large catfishes (*Brachyplatystoma flavicans, B. filamentosum; Prochilodus nigricans, Anodus elongates*) were interrupted, which negatively affected populations of these species in the lower Tocantins, downstream of the dam (Ribeiro et al., 1995). In Belo Monte's reservoir, the 12 low-abundance species restricted to this section may well become extinct locally (Lyons et al., 2005). Meanwhile, modifications of the Xingu River hydrological cycles are likely to impact directly plant and animal communities adapted to annual fluctuations (Cunha and Ferreira, 2012).

19.2.7 SOCIAL IMPACTS

Large hydroelectric projects most often cause the relocation of considerable amounts of people and result in adverse social effects (Rosenberg et al., 1997). The construction of dams in Amazonia, home to indigenous tribes, has aroused controversy (Kahn et al., 2014). The Tucuruí dam is an example: parts of three indigenous reserves (Parakanã, Pucuruí, and Montanha) were flooded (Fearnside, 1999). The case of Belo Monte on the Xingu River has been widely covered by the media, due to the resistance posed by the local opponents. The indigenous groups throughout the Xingu River basin (the Kayapó, Arara, Juruna, and Xipai tribes) have adamantly and consistently spoken out against government plans to build the dam. The indigenous groups have been subjected to relocation, displacement, loss of homes and land, lack of compensation for losses incurred, and reckless migration plans. Conflicts between inhabitants and immigrant workers employed on the construction site are recurrent. Health effects have been noted—including mercury contamination and malaria, and fisheries that traditionally supported the population downstream of the dam have collapsed (Fearnside, 1999).

19.3 FROM RENEWABLE TO SUSTAINABLE HYDROPOWER

Greenhouse gas emissions, ecosystem, and social impacts shed doubts on claims of hydropower as clean energy, such as formulated by Brazil's largest power company, Eletrobras, the fourth largest "clean energy" company in the world. Still, there seems to be room for reductions in the socioenvironmental issues of hydropower, via more cost-effective technologies and adaptation to new environmental and social requirements (Ellaban et al., 2014; Renewable Energy Sources and Climate Change Mitigation, 2012).

Renewable energy sources tap nature's cyclical or long-lasting phenomena. As large hydropower clearly illustrates, the environmental and social impacts of renewable energy sources need to be placated for these sources to become sustainable. Hydropower supplies 70% of all renewable energy in

Brazil (Market study: Wind energy in Brazil, 2014) and a two-fold 2013–2035 hydropower capacity increase must be assumed in China and Brazil (from 246 GW and 70–430 and 151 GW, respectively; Brazil Energy Outlook, 2013). Impacts will be major unless technological changes take place (i.e., engineering, project management, and local involvement and overview). The engineering solutions exist, but a paradigm shift is needed to make local involvement happen at the design phase (Table 19.1; Figure 19.3).

TABLE 19.1
Lookup of Needed Innovations in the Hydropower Sector

Hydropower Advantages and Drawbacks	Possible Sustainable Innovation
One of the best conversion efficiencies of all energy sources (Ellaban et al., 2014).	Direct mechanical applications alongside electricity generation
However, all conversions entail losses (entropy increases) in the current transformation of potential energy to mechanical work, then to electricity, and back to mechanical work	
A renewable energy source that derives from very long-term hydrological cycles (Kaunda et al. 2012). However, reservoirs are prone to high evaporation rates	Smaller reservoirs that follow natural contours and benefit from tree or hill shades
Able to supply independent systems (domestic or industrial) and operate in isolation (da Silva Soito and Freitas, 2011)	Nongrid applications
Hydropower plants tend to have long economic lives. However large plants increase technological dependency or dependency from large corporate interests, especially where privatization or private-public corporations become a trend (like what happens in the water supply and sanitation industry). Large projects make it easier for corruption to go unnoticed.	Make hydro plants smaller, modular, and more adaptable and easier to maintain than larger components, subject to obsolescence, currently allow. Use of local materials, knowledge, and manpower, otherwise a single faulty part can take a whole plant out of operation. This decreases dependency. It is part of the appropriate technologies idea. Local input in the project phase and local overview in the construction and operation phases.
Reservoirs can host local industries: fisheries, water supply, irrigation, flood control, and recreation (da Silva Soito and Freitas, 2011; von Sperling, 2012). However, competition between activities and between users is common. Many immigrants have a reduced sense of place and less knowledge of local resources and tend to be more predatory and short term in relation to nature.	Combine prehydro plant and postplant activities. Involve preproject population during all project phases.
Increased national energy access and security. However, corporations are no longer bound to borders. Also, national interests most often favor large cities rather than the populations settled around a large project.	Local autonomy and resilience rather than national interests
Sustainable innovations can create economic opportunities. However, there is often separation of investment returns and socioeconomic benefits. Electricity benefits distant, not local, populations	Involve locals throughout the project
Power depends primarily on the flow and height difference in the river bed	
Potential energy (the primary energy) can be stored in the reservoir	
Fish	Applications that do not fragment habitat and do not involve fast-moving parts
Land	Applications without land flooding
Gases	Applications without land flooding

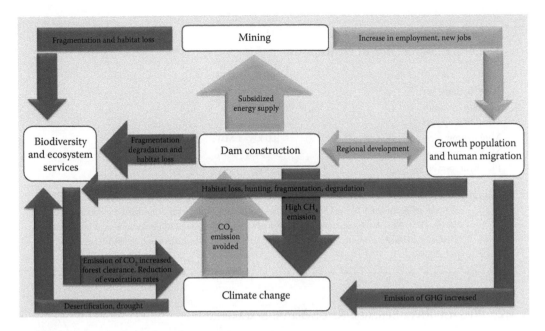

FIGURE 19.3 Scheme of interactions among dam construction, biodiversity, mining, climate change, and population growth. Dark gray arrows represent negative impacts, light gray arrows indicate positive impacts. (Modified from Lees, A.C. et al., *Biodivers. Conserv.*, 25, 451–466, 2016.)

19.4 SMALL-SCALE HYDROPOWER: TOWARD THE SUSTAINABLE TECHNOLOGY

Hydropower is primarily an alternative to producing energy while reducing the consumption of nonrenewable natural resources. It may also be the way to reduce environmental pollution caused by fossil fuels. Nevertheless, some of the great hydro investments proved to be ineffective, causing unfavorable and sometimes irreversible changes in the natural environment or leading to social conflicts at an international scale. By applying the safety measures and using the newest technologies, it may be possible to produce energy in a more sustainable and affordable manner in developing and rural areas. One of the promising alternative solutions for large-scale hydropower is small-scale hydropower (SHP).

19.4.1 DESCRIPTION AND CLASIFICATION

Due to the energy production capacity, expressed in megawatts, small hydropower plants (SHP) generally produce less than 10 MW (100 kW) (Paish, 2002); however, there is no international consensus on the definition of "small scale," and different countries have different criteria to classify hydropower plants. For example: Canada refers to upper limit capacities between 20 and 25 MW while some EU countries like Portugal, Spain, or Ireland accept 10 MW. Based on the data available (Law no. 9648/98.7 on December 9, 2003, with Resolution no. 652), small hydroelectric power plants in Brazil have an installed capacity limited to 30 MW and max 13 km² flooded (ANEEL, 2015) (Table 19.2).

SHP can be installed on already existing hydrological projects such as large-scale power plants or sawmills, small rivers, and streams. However, in contrast with large-scale hydropower systems, small-scale hydropower can be used to implement energy demands on rural areas or small urban centers with almost no impacts for environmental and ecosystem (da Silva et al., 2016) and is mainly "run-of-river" (Ferreira et al., 2015).

TABLE 19.2

General Classification of Hydropower

Type	Power Output
Large hydropower	>100 MW
Medium hydropower	10–100 MW
Small hydropower	1–10 MW
Mini hydropower	100 kW–1 MW
Micro hydropower	5–10 kW
Pico hydropower	<5 kW

Source: Singh, D., Micro hydropower, Recourse Assessment Handbook, Asian and Pacific Centre for Transfer of Technology of the United Nations—Economic and Social Commission for Asia and the Pacific (ESCAP). Available from http://www.sswm.info/sites/default/files/reference_attachments/SINGH%202009%20Micro%20Hydro%20Power%20Resource%20Assessment%20Handbook.pdf, 2009.

19.4.2 RUN-OF-RIVER SCHEMES: RESERVOIR VS. STREAMING

The performance and efficiency of run-of-river hydropower vary during the year depending on the state of water and rainfall, and these vary during the year. This type of power plant can work without interruption, and the amount of energy produced depends on the amount of water flowing in the river limited by the maximum permissible amount of water in m³/s flowing through the turbine. Water is diverted into the penstock and transmitted to turbines in a power plant at a lower elevation. The altitude difference between the inlet and the power station provides the kinetic energy that drives turbines and produces electricity (https://www.watershed-watch.org/publications/files/Run-of-River-long.pdf). In run-of-river projects, no dams are required. That eliminates issues associated with a reservoir, such as flooding large areas of land and fish migrating difficulty (Barthem et al., 1991); weir is the only construction built on the riverbed in order to divert requisite flow.

The classification of run-of-river is based on its capacity. It is important to note that some large-scale run-of-the-river plants with outputs of hundreds or thousands of MW are also operational (e.g., Belo Monte, Brazil).

There are several potential benefits of small-scale and run-of-river hydropower plants. Compared to the traditional construction, based on the dam's facilities, run-of-river projects are more cost efficient and can be constructed more quickly (Singh, 2014). Since no reservoirs are created, negligible amounts of mercury and greenhouse gases are emitted. Operating run-of-river facilities avoids other issues associated with reservoirs—such as sedimentation, resettlement, and land flooding; however, further environmental impacts should be considered as diversion in river flows may cause transformation in aquatic ecosystems. These and other drawbacks associated with damming rivers supported by the Amazon River basin were the subject of a significant number of scientific papers; nevertheless, several small communities of the Amazon region still have no access to electricity, which may have a negative impact on the economic development of the region and impede agricultural production. Electricity may significantly improve the lifestyle of the residents, developing regional food production by using agricultural machinery or production storage for business supply between harvests and generating jobs for the regional population (Blanco et al., 2008). For example: in the Amazon basin, where more than 70% of the remaining hydroelectric potential is found and the topography is not propitious for the creation of large reservoirs (da Silva et al., 2016), run-of-river projects could be a sustainable alternative to large hydro systems.

19.4.3 WHY SMALL HYDROPOWER (SHP) CAN BE A KEY ELEMENT FOR SUSTAINABLE DEVELOPMENT

Among the renewables, small hydropower is an attractive and environmentally friendly energy technology (Ferreira et al., 2015); thus, for the following reasons, it may be considered a significant factor for sustainable development (Nautiyal et al., 2011):

- Per the definition of renewable, SHP is a renewable source of energy because it uses the energy of following water; there is no depletion.
- No fossil fuels or other petroleum products are required for SHP to operate.
- Proper utilization of water resources provides energy from SHP.
- No big storage is formed (reservoirs).
- SHP is inflation free and effective: construction is relatively simple; no expensive equipment and construction work are required to establish and operate SHP.
- Low level of pollution: eliminating emissions of greenhouse gases (GHG) small hydropower can contribute to sustainable rural development.
- SHP development would provide electricity, communication links, transportation, and economy to rural and distant areas.
- SHP may provide other benefits like water supply, flood prevention, irrigation, fishing, and tourism.

19.4.4 SMALL-SCALE HYDROELECTRIC POWER PLANTS AND SUSTAINABLE DEVELOPMENT: THE BRAZILIAN CASE

Brazil´s hydropower potential is large, and the hydro energy sector in Brazil is relatively well developed. The way forward now is to promote more sustainable solutions in the hydropower section in Brazil and to extend the potential of SHPs (Reggiani, 2015). According to data provided by ANEEL (2015) the number of SHPs under operation was 475 and they corresponded to the generation capacity in the country of 4799 MW, representing 3.49% of the Brazilian energy grid. Currently, 37 SHP projects are under construction, which are expected to produce 428 MW of power; also, it is possible, that SHP could insert 1818 MW in power generation through an additional 128 stations granted in the following years.

In Brazil, the south and southeast/midwest have the highest concentrations of SHPs in operation, under construction, and granted and higher power generation capacity (Ferreira et al., 2015). Despite Brazil's enormous hydropower potential for small hydro schemes, only 18% of the known potential is used to generate electricity. Compared to other countries, which use much of their small-hydropower potential, that number is low (Ferreira et al., 2015). However, Brazil, with a SHP potential of 22,500 MW, strives to evolve its potential to about 6700 MW of its SHP installed capacity (ANEEL, 2014; Masera and Esser, 2013; Capik et al., 2012).

After the restructuring of the Brazilian electricity sector, which started in 1995, several resolutions, decrees, and laws were established, with the aim to regulate the sector and to attract investments (Table 19.3) (Leão, 2008). Since the regulation started, more than BRL 1 billion was invested by private investors in approximately 1000 engineering projects, totaling more than 9000 MW in projects. However, the number of SHP projects is having trouble growing due to the development of other renewables that may be competitive with SHP and internal policies that might cause obstacles to further evolution of small hydropower projects (Ferreira et al., 2015).

Compared to SHP, biomass and wind power have substantial benefits arising from their technological development, financial profit, and minimized cost of installation (Ferreira et al., 2015). It was estimated that in 2020 the generation capacity of biomass will increase by 74%, wind power by 1000%, and SHP power generation by 55% (Leão, 2008). To avoid difficulties with the development of SHP, Brazilian agencies supporting renewable sources and small hydropower such as the

TABLE 19.3

Regulations in the Brazilian Electricity Sector for Generation through SHP

Regulations for SHPs

Authorization no-cost to explore the hydraulic potential

Discounts superiors to 50% in the taxes of use of the transmission and distribution systems

Free commercialization of energy with customers whose load is less than 500 kW

Free commercialization of energy with customers, situated in isolated electrical system, whose load is less than 500 kW

Exemption related to the financial compensation for use of water resources

Participation in the division of fuel consumption account when replacing thermal oil generation in isolated systems

Exemption application, annually, of minimum 1% of operational net income in research and development of the electricity sector

Commercialization of energy generated by SHP with public concessionaires

Incentive Program for Alternative Sources of Energy established with the objective of increasing the participation of electricity produced by independent producers, designed based on SHP, wind power, and biomass

Review of the producers and criteria used in environmental licensing that determine the procedures and deadlines to be applied

Source: ANEEL, Agencia Nacional De Energia Elétrica, Guia do empreendedor de pequenas centrais hidroelétricas, Brasília, DF, 704 (In Portuguese), 2003.

Brazilian Association of Development of Small Hydroelectric Plants and the Brazilian Association of Clean Energy initiated relevant actions (inter alia: encouraging study on the theme PCHs Agencia Nacional De Energia Elétrica (ANEEL)—Portuguese: Pequena Central Hidrelétrica—Small Hydroelectric Plant, supporting companies in the sector, and deploying programs to simplify the ratification of studies and projects in ANEEL, etc.) (Ferreira et al., 2015). These actions might help to increase the investment potential of BRL 35 billion only in the SHP and increase to 5000 MW the small hydroelectric power installed capacity in the next 10 years (Lenzi, 2013).

Between 2001 and 2015, there was an increase of potential generation of SHP in Brazil: from 855 to 7799 MW (Ferreira et al., 2015). Regardless, financial issues greatly hinder realizing fully the potential of small-hydro projects in Brazil. Despite the country's enormous potential to become a world leader in clean energy (da Silva et al., 2016), the Brazilian government is currently buying more energy from nonrenewable energy sources, and only 1% of energy generated by SHP was bought in energy auctions in the last 8 years (Ferreira et al., 2015). Certain changes should be introduced in the financing of renewables projects (including SHP) to permit potential investors to conclude favorable financial agreements with the banks.

19.4.5 SMALL-SCALE (MICRO) HYDROPOWER UNDER A SUSTAINABLE DEVELOPMENT PERSPECTIVE IN THE AMAZON

For the small communities in the Amazon region setting micro-hydropower (MHP), typically run-of-river schemes would be practicable due to the grain harvest period that takes place between January and June and coincides with the wet season. The concurrence between the grain harvest (the peak of power demand) and floods (the peak of power production) may promote the energy production planning of MHP in the area. However, a small dam might be built (should be maximum 3 m high; this insures that MHP is still run-of-river construction). The height established by the weir plus the natural gross head: vertical distance from intake to turbine; the resulting pressure at the bottom when no water is flowing (Canyon Industries, 2013) is not sufficient to generate the required power (Blanco et al., 2008). Accomplishment in assembly and installation of MHP depends mainly on the local residents' commitment, and it starts with the energy demand, feasibility, construction, operation, and maintenance required. Therefore, using the hydropower resources of the small catchments of Amazonia, combined with reducing environmental impacts that may be caused by SHP and MHP, could cover local energy demand and be viable economically (generating income and creating jobs) and support the concept of sustainable development in isolated districts (Blanco et al., 2008).

19.4.6 SELECTED ISSUES OF SMALL HYDROPOWER PLANTS

Among all the renewable energy sources, SHP may be one of the most environmentally friendly and cost-effective technologies for electricity generation (Kumar and Singal, 2015); however, certain disadvantages should be considered before the implementation of SHP. The main negative effects are summarized below:

- Environmental issues: during the exploration of SHP, water is diverted away from part of the stream; caution should be exercised to ensure that it would not cause a damaging impact on the local environment and infrastructure. Also, the noise produced during SHP operation, caused by generators, turbines, and cooling ventilation, may be harmful for external environment (Singh, 2014). External noise levels can be decreased by improving the sound isolation of the turbine and power station and by controlling the vibration of the ventilation system (Singh, 2014).
- To fully utilize the electrical potential of a stream, the most appreciable localization is advisable. Main factors to consider are stream size, including flow rate, output and drop; distance between power station and location; and where the energy demand is needed and the availability of system components (e.g., transition lines, pipes, battery) (Ferreira et al., 2015).
- Some SHP may be problematic since emission of materials and dust from construction activity into water may result in the short-term increase of suspended particles in the aquatic environment. That could change the river qualities (transparency of the water) and affect aquatic species (Schwartz et al., 2005).
- Although the costs of maintenance and operating SHPs are low, periodic control is required to prevent the sudden failure of the hydro system (Schwartz et al., 2005).
- SHP projects have reduced or no storage capacity and cannot coordinate the output of power generation to adjust for consumer demand. Also, stream size fluctuates seasonally in many sites. To estimate whether energy requirements would be met, detailed study and planning are required before the implementation of any hydro project.

19.5 DISCUSSION

Water is a renewable, natural source and can produce relatively cheap and sustainable energy. Hydropower can be a solution for electricity demand, especially in areas where residents have limited access to the goods of civilization; it can significantly improve their quality of life. It can be used for those who have the option of replacing conventional energy sources like fossil fuels or gas. However, the construction of large hydroelectric power may have some negative effects, which were presented in the examples of misguided hydro plants in Brazil. Several enormous investments have brought significant negative environmental and social changes in the Amazon region by damming rivers. The Amazon forest is not only a Brazilian national treasure, but is also an important repository of natural resources and biodiversity. Every major decision must be made with due consideration for the future of the Amazonian ecosystem.

Small-scale, run-of-river projects could have a key role in the future, although certain conditions (including topography) are needed to run the power because the performance of hydropower depends primary on available water, high gradient in the riverbed, and flow. Hydro plants are not conductive to drought that may be caused by climate change. If hydrological conditions are unfavorable, hydropower can work only periodically, making negligible potential for the grid. However, if demand for power obtained by one source is not sufficient, the solution could be additional energy source provided e.g., small-scale hydropower and solar panels cooperating.

For developing countries (such as Brazil), significant increases in electricity consumption are projected in the upcoming years, and it is necessary to achieve a compromise between renewable sources to produce sufficient electricity supply (da Silva et al., 2016). Brazil has considerable potential to become a world leader in sustainable energy production, but the roles of solar, wind, and SHP should be evaluated in the country where electricity produced by hydropower supplies 70% of its electric power (da Silva et al., 2016).

Among the available technologies, which renewable energy would be better? All currently known renewables have advantages and disadvantages. When planning investments to obtain energy by using renewable energy sources, the following factors should be considered:

- Location of the investment planned
- Amount of power that can be obtained from each potential source
- Environmental factors on the site
- Invested cost per 1MW of the generation capacity

To achieve sustainable, environmentally friendly development of renewable technologies and make them competitive with traditional, nonrenewable energy sources, supporting policies and further investment in research and public consciousness are essential.

REFERENCES

ABRAPCH (2013). Associação Brasileira de Fomento as Pequenas Centrais Hidrelétricas. Relatório técnico 001/2013—Pequenas centrais hidroelétricas: Fundamentais para o desenvolvimento sustentável e diminuição da dependência de combustíveis fosseis do Brasil. Oficio DPR 01/13.

Agostinho, C.S., Hahn, N.S., and Marques, E.E. (2003). Patterns of food resource used by two congeneric species of piranhas. *Brazilian Journal of Biology.* 63(2): 177–182.

ANEEL (2003). Agencia Nacional De Energia Elétrica. Guia do empreendedor de pequenas centrais hidroelétricas. Brasília, DF; 2003. 704 (In Portuguese).

ANEEL (2014). Agencia Nacional de Energia Elétrica. Banco de Informações da Geração (BIG). ANEEL. Website, August; 2014 (In Portuguese).

ANEEL (2015). Agencia Nacional de Energia Elétrica. Resolução no. 652, de 9 de dezembro de 2003. Estabelece os critérios para o enquadramento de aproveitamento hidrelétrico na condição de Pequena Central Hidrelétrica e revoga a Resolução no. 394, de 04 de dezembro de 1998. Diário Oficial da União, Brasília, DF, 11 dez. Seção 1; 2003. 140 (In Portuguese).

Anderson, J.W. (2000). The Surge in Oil Prices: Anatomy of a Non-Crisis. Resources for the Future. Discussion Paper 00-17. Available from http://www.rff.org/.

Aragão, L. E. O. C., Poulter, B., Barlow, J. B., Anderson, L. O., Malhi, Y., Saatchi, S., Phillips, O. L., and Gloor, E. (2014). Environmental change and the carbon balance of Amazonian forests. *Biological Reviews*, 89, 913–931.

Barham, J. and Caufield, C. (1984). The problems that plague a Brazilian dam. *New Scientist*, 11, p. 10.

Barrow, C. (1988). The impact of hydroelectric development on the Amazonian environment: With particular reference to the Tucurui project. *Journal of Biogeography*, 15(1): 67–78.

Barthem, R. B., de Brito Ribeiro, M. C. L., and Petrere, M. (1991). Life strategies of some long-distance migratory catfish in relation to hydroelectric dams in the Amazon Basin. *Biological Conservation*, 55(3): 339–345.

Beaulieu, J. J., McManus, M. G., and Nietch C. T. (2016). Estimates of reservoir methane emissions based on a spatially balanced probabilistic-survey. *Limnology and Oceanography*, 61: S27–S40.

Blanco, C. J. C., Secretan, Y., and Mesquita, A. L. A. (2008). Decision support system for micro-hydro power plants in the Amazon region under a sustainable development perspective. *Energy for Sustainable Development*, 12(3): 25–33.

Brazil Energy Outlook. (2013). World Energy Outlook, International Energy Agency, Paris, France. ISBN: 978-92-64-20130-9. Available from http://www.iea.org/publications/freepublications/publication/WEO2013.pdf.

Canyon Industries (2013). Guide to Hydropower. An Introduction to Hydropower Concepts and Planning. Available from http://www.asociatiamhc.ro/wp-content/uploads/2013/11/Guide-to-Hydropower.pdf.

Capik, M., Yilmaz, A. O., and Cavusoglu, I. (2012). Hydropower for sustainable energy development in Turkey: The small hydropower case of the Eastern Black Sea Region. *Renewable and Sustainable Energy Review*, 16: 6160–6172.

Costello, M. J., May, R. M., and Stork, N. E. (2013). Can we name Earth's species before they go extinct? *Science*, 339(41): 413–416.

Cunha, D. A. and Ferreira, L. V. (2012). Impacts of the Belo Monte hydroelectric dam construction on pioneer vegetation formations along the Xingu River, Pará State, Brazil. *Brazilian Journal of Botany*, 35(2): 159–167.

da Silva Soito, J. L. and Freitas, M. A. V. (2011). Amazon and the expansion of hydropower in Brazil: Vulnerability, impacts and possibilities for adaptation to global climate change. *Renewable and Sustainable Energy Reviews*, 15: 3165–3177.

da Silva, R. C., de Marchi Neto, I., and Silva Seifert, S. (2016). Electricity supply and the future role of renewable energy sources in Brazil. *Renewable and Sustainable Energy Reviews*, 59: 328–341.

Davidson, E. A., de Arau´jo, A. C., Artaxo, P., Balch, J. K., Brown, I. F., Bustamante, M. M. C., Coe, M. T., DeFries, R. S., Keller, M., Longo, M., Munger, J. W., Schroeder, W., Soares-Filho, B. S., Souza Jr C. M., and Wofsy, S. W. (2012). The Amazon Basin in Transition. *Nature*, 481: 321–328.

Departamento Nacional de Água e Energia Elétrica [DNAEE] and Eletrobrás (1985). Manual de Micro centrais Hidrelétricas. Departamento Nacional de Água e Energia Elétrica e´ Eletrobrás, Brasília.

Deshmukh, C., Serça, D., Delon, C., Tardif, R., Demarty, M., Jarnot, C., Meyerfeld, Y., Chanudet, V., Guédant, P., Rode, W., Descloux, S., and Guérin, F. (2014). Physical controls on CH_4 emissions from a newly flooded subtropical freshwater hydroelectric reservoir: Nam Theun 2. *Biogeosciences*, 11: 4251–4269.

Ellaban, O., Abu-Rub, H., and Blaabjerg, F. (2014). Current status, future prospects and their enabling technology. *Renewable and Sustainable Energy Reviews*, 39: 748–764. Available from http://energyeducation.ca/encyclopedia/Main_Page.

Fearnside, P. M. (1989). Brazil's Balbina dam: Environmental versus the legacy of the Pharaohs in Amazonia. *Environmental Management* 13(4): 401–423.

Fearnside, P. M. (1995). Hydroelectric dams in the Brazilian Amazon as a source of greenhouse gases. *Environmental Conservation*, 22: 7–19.

Fearnside, P. M. (1999). Social impacts of Brazil's Tucuruı´ dam. *Environmental Management*, 24(4): 483–495.

Fearnside, P. M. (2000). Greenhouse gas emissions from land-use changes in Brazil's Amazon. *Global Climate Change and Tropical Ecosystems: Advances in Soil Science*; Lal, R., Kimble, J. R., Stewart, B. A., Eds; CRC Press, Boca Raton, FL, pp. 231–249.

Fearnside, P. M. (2001). Environmental impacts of Brazil's Tucuruı´ dam: Unlearned lessons for hydroelectric development in Amazonia. *Environmental Management*, 27(3): 377–396.

Fearnside, P. M. (2004). Greenhouse gas emissions from hydroelectric dams: Controversies provide a springboard for rethinking a supposedly 'clean' energy source. An editorial comment, *Climatic Change*, 66(1): 2.

Fearnside, P. M. (2015). Environmental and social impacts of hydroelectric dams in Brazilian Amazonia: Implications for the aluminium industry. *World Development* 77: 48–65.

Fearnside, P. M. (2017). Brazil's Belo Monte dam: Lessons of an Amazonian resource struggle. Contribution for Die Erde Special Issue "Resource Geographies: New perspectives from South America." ISSN 0013-9998. Available from http://www.die-erde.org/index.php/die-erde.

Ferreira, J. H. I., Camacho, J. R., Malagoli, J. A., and Guimarães Jr, S. C. (2015). Assessment of the potential of small hydropower development in Brazil. *Renewable and Sustainable Energy Reviews*, 56(2016): 380–387.

Foley, A. J., Asner, G. P., Costa, M. H., Coe, M. T., Defries, R., Gibbs, H. K., Howard, E. A., Olson, S., Patz, J., Ramankutty, N., and Synder, P. (2007). Amazonia revealed: Forest degradation and loss of ecosystem goods and services in the Amazon Basin. *Frontiers in Ecology and the Environment*, 5(1): 25–32.

Geller, H., Schaeffer, R., Szklo, A., and Tolmasquim, M. (2004). Policies for advancing energy efficiency and renewable energy use in Brazil. *Energy Policy*, 32: 1437–1450.

Guatam, A., Haubold, I., Pacey, V., Papirnik, D., Premjee, M., and Schlumpf, P. (2014). Brazil's Belo Monte: A cost-benefit analysis. *Energy and Energy Policy*, PBRO29000. Available from http://franke.uchicago.edu/.

Gruca-Rokosz, R. (2012). Reservoirs as a source of emission of greenhouse gases (published in Polish, original: Zbiorniki zaporowe jako zrodlo emisji gazow cieplarnianych). Inzynieria i Ochrona Srodowiska, t. 15, nr 1, 51–65.

Gurbuz, A. (2006). The role of hydropower in sustainable development. *European Water*, 13(14): 63–70.

Holt, M., Campbell, R. J., and Nikitin, M. B. (2012). Fukushima Nuclear Disaster. CRS Report for Congress. Congressional Research Service 7-5700, Available from http://www.crs.Gov; R41694.

Höök, M., Zittel, W., Schindler, J., and Alektell, K. (2010). Global coal production outlooks based on a logistic model. *Fuel*, 89(11): 3546–3558.

Itaipu Binacional. Available from https://www.itaipu.gov.br/en.

Jones, I. L., Bunnefeld, N., Jump, A. S., Peres, C. A., and Dent, D. H. (2016). Extinction debt on reservoir land-bridge islands. *Biological Conservation* 199: 75–83.

Kahn, J. R., Freitas, C. E., and Petrere, M. (2014). False shades of green: The case of Brazilian Amazonian hydropower. *Energies*, 7: 6063–6082.

Kaunda, C. S., Kimambo, C. Z., and Nielsen, T. K. (2012). Hydropower in the context of sustainable energy supply: A review of technologies and challenges. *Renewable Energy*, Article ID, 730631, 15 p. doi:10.5402/2012/730631.

Kemenes, A., Forsberg, B. R., and Melack, J. M. (2006). Gas Release from the Balbina Dam. Foz Do Iguaçu, Brazil, April 24–28, INPE pp. 663–667.

Kemenes, A., Forsberg, B.R., and Melack, J. M. (2007). Methane release below a tropical hydroelectric dam. *Geophysical Research Letters*, 34, L12809, 5 p. Available from http://www.bankinformationcenter.org/en/Document.102221.pdf.

Kemenes, A., Forsberg, B. R., and Melack, J. M. (2011). CO_2 emissions from a tropical hydroelectric reservoir (Balbina, Brazil). *Journal of Geophysical Research*, 116, G03004.

knoema. Available from https://knoema.com/.

Kumar, R. and Singal, S. K. (2015). Operation and maintenance problems in hydro turbine material in small hydro power plant. *4th International Conference on Materials Processing and Characterization*, 2(4–5): 2323–2331.

La Rovere, E. L. and Mendes, F. E. (2000). Tucuruì Hydropower Complex, Brazil, A WCD case study prepared as an input to the World Commission on Dams, Cape Town. Available from http://www.dams.org.

Leão, L. L. (2008). Considerações sobre impactos socioambientais de pequenas centrais hidrelétricas (PCHs)—modelagem e análise. Brasília: Centro do Desenvolvimento Sustentável, Universidade de Brasília; 2008 (In Portuguese).

Lees, A. C., Peres, C. A., Fearnside, P. M., Schneider, M., and Zuanon, J. A. S. (2016). Hydropower and the future of Amazonian biodiversity. *Biodiversity and Conservation*, 25: 451–466.

Lenzi, C. (2013). As PCHs no contexto energético futuro no Brasil. ABRAGEL—Associação Brasileira de Geração de Energia Limpa. Florianópolis—SC; 2013.

Luna, S. M., Torres, A. G., and Jegu, M. (2009). *Serrasalmus spiloplueura*, Speckled piranha. Retrieved May 4, 2009, Fish base Available from http://www.fishbase.org/Summary/SpeciesSummary.php?id=11973.

Lyons, K. G., Brigham, C. A, Traut, B. H, and Schwartz, M. W. (2005). Rare species and ecosystem functioning. *Conservation Biology*, 19: 1019–1024.

Manyari, W. V. and de Carvalho Jr., O. A. (2007). Environmental considerations in energy planning for the Amazon region: Downstream effects of dams. *Energy Policy* 35: 6526–6534.

Market Study: Wind Energy in Brazil. (2014). Ministry of Economic Affairs in The Netherlands. Larive International, 2014. Available from https://www.rvo.nl/sites/default/files/2014/08/Wind%20Study%20Brazil%202014.pdf.

Masera, L. H. and Esser, I., (2013). World Small Hydropower Development Report 2013. United Nations Industrial Development Organization: International Centre on Small Hydro Power. Available from http://www.smallhydroworld.org.

Nautiyal, H., Singal, S. K., Varun, G., Sharma, A. (2011). Small hydropower for sustainable energy development in India. *Renewable and Sustainable Energy Review*, 15: 2021–7.

Paish, O. (2002). Small hydro power: Technology and current status. *Renewable and Sustainable Energy Reviews*, 6(6): 537–556.

Parkins, J. R. and Haluza-DeLay, R. (2011). Social and ethical considerations of nuclear power development. Rural Economy, Staff Paper no 11-01. Available from http://ageconsearch.umn.edu/bitstream/103237/2/StaffPaper11-01.pdf.

Raabe, J. (1987). Great names and development of hydraulic machinery. *Hydraulics and Hydraulic Research: A Historical Review*, IAHR 1935–1985, G. Gabrecht, Ed., Balkerna, Rotterdam, The Netherlands, pp. 251–266.

Reggiani, M. C. P. (2015). Hydropower in Brazil—development and challenges. *ResearchGate*, doi:10.13140/ RG.2.1.3451.8808.

Renewable Energy Sources and Climate Change Mitigation (2012). Special Report of the Intergovernmental Panel on Climate Change. ISBN 978-1-107-02340-6. Available from https://www.ipcc.ch/pdf/special-reports/srren/SRREN_Full_Report.pdf.

Ribeiro, M. C. L. B., Petrere, M., Jr., Juras, A. A. (1995). Ecological integrity and fisheries ecology of the Araguaia-Tocantins River basin, Brazil. *River Research and Applications*, 11: 325–350.

Rosas, F. C. W., de Mattos, G. Y., and Mendes-Cabral, M. M. (2007). The use of hydroelectric lakes by giant otters *Pteronura brasiliensis*: Balbina Lake in central Amazonia, Brazil. *Oryx*, 41(4): 520–524.

Rosenberg, D. M., Berkes, B., Bodaly, R. A., Hecky, R. E., Kelly, C. A., and Rudd, J. W. M. (1997). Large-scale impacts of hydroelectric development. *Environmental Reviews*, 5: 27–54.

Schwartz, F., Pegallapati, R., and Shahidehpour, M. (2005). Small Hydro as a Green Power. 0-7803-9156-X/05/2005 IEEE. Available from http://www.iitmicrogrid.net/microgrid/pdf/papers/renewables/SmallHydro.pdf?iframe=true&width=980&height=780.

Shi, C. (2009). Perspective on natural gas crisis between Russia and Ukraine. *Review of European Studies*, 1(1). doi:10.5539/res.v1n1p56.

Singh, D. (2009). Micro hydro power. Recourse Assessment Handbook. Asian and Pacific Centre for Transfer of Technology of the United Nations—Economic and Social Commission for Asia and the Pacific (ESCAP). Available from http://www.sswm.info/sites/default/files/reference_attachments/SINGH%20 2009%20Micro%20Hydro%20Power%20Resource%20Assessment%20Handbook.pdf.

Singh, S. (2014). Study of Different Issues and Challenges of Small Hydro Power Plants Operations. International Conference on Advances in Energy Conversion Technologies 2014 (ICAECT).

Soares, B. S., Nepstad, D. C., Curran, L. M., Cerqueira, G. C., Garcia, A. R., Ramos, C. A., Voll, E., McDonald, M. A., Lefebvre, P., and Schlesinger, P. (2006). Modeling conservation in the Amazon basin. *Nature*, 440: 520–523.

Von Sperling, E. (2012). Hydropower in Brazil: Overview of positive and negative environmental aspects. *Energy Procedia*, 18: 110–118.

Tomczonek, Z. (2013). World oil market. Resources, consumption, directions of flows. Światowy rynek ropy naftowej- zasoby, konsumpcja, kierunki przeplywu. OPTIMUM. *Studia ekonomiczne* 64(4) Available from http://www.repozytorium.uwb.edu.pl

Torres, J. M., Alvarez, M., Laugé, A., and Sarriegi, J. M. (2009). Russian-Ukrainian Gas Conflict Case Study. Available from http://citeseerx.ist.psu.edu.

Zanirato, S. H. (2010). O patrimônio natural do Brasil. *Projeto História*. 40: 127–145.

Ziober, B. R. and Zanirato, S. H. (2014). Actions to safeguard biodiversity during the building of the Itaipu Binacional hydroelectric plant. *Ambiente & Sociedade n São Paulo v.*, XVII(1): 59–78.

20 Bioenergy Principles and Applications

Marina Islas-Espinoza
Mexico State University

Alejandro de las Heras
Independent Researcher

CONTENTS

20.1 FUNDAMENTALS

20.1.1 FREE ENERGY

The following can be credited to Gibbs and Helmholtz: the internal energy of a system is made up by reactants, where H is the energy contained by the number of chemical bonds in a given volume; G is the available energy to do work (movement, growth, maintenance, reproduction); S is entropy (energy loss); and T is temperature of a reaction (Gaudy and Gaudy, 1980). When a biochemical reaction takes place

$$\Delta H = \Delta G + T \Delta S \tag{20.1}$$

Where ΔG is the change in free energy, or useful energy, and T is proportional to the number of collisions between reactants.

20.1.2 Enzymes and Coenzymes

Consistent with Equation 20.1, biochemical reactions proceed from an initial energy level to a lower energy level, since entropy S augments any reaction. However, reactions are either exergonic (they release energy and proceed spontaneously) or endergonic (they require energy). The rate of reaction is the number of reacting molecules over time. In endergonic reactions, the number of reacting molecules is very low in any given time span, or it can take a very long time for the reaction to occur.

Enzymes are large biochemical compounds that accelerate the rate of reactions, by lowering the energy amount required by an endergonic reaction. Enzymes are energetically very efficient since few of them need to be synthesized, and they remain unmodified by the reaction (Cooper and Hausman, 2015).

Coenzymes are lighter molecules (often made up of vitamins) that also accelerate reaction rates. Again, they are unmodified by the reaction. Among coenzymes, adenosine triphosphate (ATP) stands out as being produced by all living beings.

20.1.3 Electron Carriers and ATP

Most biochemical reactions are redox reactions: electron exchanges between reactants or hydrogen removal by enzymes. Electron transfers between electron donors and acceptors, protons, and ATP are linked: ATP synthesis is driven by a proton electrochemical gradient (a difference of electric charge and chemical concentration) between two sides of a biomembrane. As shown by Mitchell in 1961, the "gradient is produced by the electron transfer and in opposite direction, the proton flow which frees energy to produce ATP" (Cooper and Hausman, 2015).

ATP is ubiquitous in biological energy reactions and diverse in the way different types of organisms obtain free energy: fermenters use organic compounds (nitrate and sulfate) as terminal electron acceptors, while organisms with respiratory metabolism (either aerobes or anaerobes) use oxygen.

In addition to electron donors and acceptors, there are coenzyme electron carriers and phosphate carriers (e.g., ATP). These coenzymes allow for coupled reactions (of the form A → B → C → D). In diverse redox exergonic reactions, energy is used to form ATP (Gaudy and Gaudy, 1980).

Although it is true that most chemical bonds are shared electrons, it is not in bonds that the energy is found, and it is not in bond breaking that energy is released; rather, it is in the rearrangement of electrons during reactions. This is consistent with Equation 20.1: energy G is released and entropy S augments (Cooper and Hausman, 2015).

20.1.4 Ion Pumps and Electricity

All living cells have membranes. Ions cannot penetrate, except through open pores and ion channels. Ions penetrate the membrane mostly through ion channels. Their opening and closing drives the transmission of electric signals (Cooper and Hausman, 2015). Ion pumps are responsible for maintaining internal ion concentrations and introducing a substance inside a membrane against its concentration gradient. Sometimes this requires membrane transport proteins, which can follow the concentration gradient (facilitated diffusion) or go against this gradient (active transport carried out by proteins (metabolic pumps) that use an extra source of energy). Ion pumps in particular use ATP.

The transport of calcium across muscle cell membranes uses ATP and results in much lower intracellular calcium concentrations, making the cell very sensitive to small increments in said concentration. Thus calcium has a key role in cell signaling, in particular in muscle contraction (Cooper and Hausman, 2015).

Also, the electron transfer from donor to acceptor induces an electric current. In fact, electric and chemical energy are equivalent:

$$\Delta G = -nF\Delta E_0 \tag{20.2}$$

with n electrons in the reaction, F Faraday (96500 Coulomb required to reduce or oxidize one mol), ΔE_0 the difference in electron potential in volts between two reactants (Gaudy and Gaudy, 1980).

Electric resistance is low when ion channels allow the passage of ions and electric current. When ion channels are closed, resistance is much higher.

20.1.5 MUSCLE

All cells are capable of movement. However, skeletal muscle cells are specialized in allowing animal movement. Each muscle is made of many muscle fibers. Each fiber is a large multinuclear cell. Each of these cells has a larger number of small fibers (myofibrils). Each myofibril is made of many aligned units (sarcomeres), each of which contracts thereby shortening the muscle (Figure 20.1). Sarcomere contraction is a sliding movement of actin protein filaments sliding along myosin protein filaments, according to the 1954 Huxley–Niedergerke–Hanson model. During the shortening of the sarcomere, myosin hydrolyzes ATP (Cooper and Hausman, 2015).

The regulation of muscle contraction starts with a signal from a motor neuron. This stimulation depolarizes the muscle cell membrane, leading to a calcium concentration increment associated with the shortening of the sarcomere. The strength of each muscle cell's contraction is invariable; what varies in muscular workload is the number of cells recruited. In cardiac and smooth muscle cells (located in hollow organs and tubes), contraction strength is modulated by adrenaline (a hormone and neurotransmitter). Adrenaline speeds up and strengthens cardiac contraction; it also relaxes vascular muscle cells. This increases cardiac flow and diminishes resistance in smooth vascular muscles.

The energy for contraction depends on the workload. With moderate ATP demand, supply is handled through oxidative metabolism. When contraction is exhausting, ATP is depleted in about one second, and creatine phosphate stored in muscle cells is used. Skeletal and cardiac muscle cells would also resort to anaerobic ATP generation with blood glucose as feed or feed provided by glycogen degradation (Colby, 1987).

Muscle contractions occur continually to maintain a body's posture; these tonic contractions do not exhaust muscle energy. Contrarily, muscle soreness takes place 24 h after exhausting exercise, because of diminutive tearing or swelling in the muscle. It was formerly believed that this delayed-onset muscle soreness was due to accumulated lactic acid released by anaerobic metabolism during strenuous effort (Hitchcock, 2007).

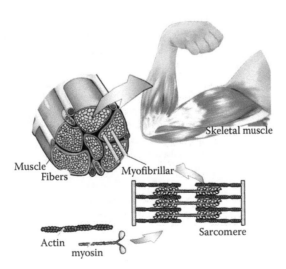

FIGURE 20.1 Human muscle anatomy. (From Tomoki Fukushima, Creative Commons 4.0. https://commons.wikimedia.org/wiki/File:筋肉英語版.png.)

20.1.6 WHAT DISTINGUISHES ORGANISMS

Organisms self-organize at the molecular level (including the genetic level), as well as at the ecological (or trophic) level. At the ecological level, organisms transform their habitat to reach equilibria with their physicochemical environment. They self-regulate, self-repair, self-detoxify (homeostasis), and establish feedback loops with the environment (one example being Table 20.1). Adaptation is transient (through learning) but also evolves irreversibly (through genetic information exchanges). Fitness (rather than efficiency) measures organismal or genetic success. Nondeleterious mutations also may enhance fitness. Enzymes allow work at lower temperatures than artificial chemical processes. Organisms, unlike machines, can work against thermodynamic gradients. Waste is unavoidable, but it is turned into a resource to feed on or mineralized and returned to biogeochemical cycles.

20.1.7 HYBRIDS, INCLUDING HUMANS

High levels of coupling, hybridization, and regulation prevail in organisms to maintain a continuous flow of energy through equilibria in ions, acidity, alkalinity, and redox potentials (Ponizovskiy, 2013).

Striking examples of hybridization are given by chloroplasts, mitochondria, and peroxisomes, reckoned to have a bacterial origin. Chloroplasts are responsible for photosynthesis. The transformation of solar to chemical energy occurs in the reaction center, where the absorbed excitation

TABLE 20.1
hiG Interactions with Geosphere, Biosphere, and DNA

Radiation Type (Energy)	Geosphere-Biosphere Interactions	Human Damages
Alpha (He nuclei, 4–9 MeV)	Keep Earth's iron core molten: vital for magnetosphere and avoid Earth's temperature cooling down. Natural greenhouse gases and solar radiation then bring temperature to an average 18°C, midway through the 5°C–40°C range of optimal enzymatic activity	Hazardous if inhaled/ ingested. Large exposures lead to cancer in lining of lung/stomach (IAEA, 2004)
Beta (electron or positron, 0.511 to 938.3 MeV)		Hazardous if inhaled or ingested. Large high-energy beta exposure can cause skin burns/ cancer (IAEA, 2004)
Gamma (photon, 0.1–3 MeV)	Filtered by magnetosphere. Magnetosphere deflects solar wind which would erase the O_3 layer	Penetrate the body producing single or double strand break, or base damage (Faraj et al., 2016)
X rays (photons < 25 MeV)	Filtered by ionosphere	
UV (< 24.8 eV)	UVC filtered by the O_3 and by atmospheric O_2 produced by photoautotrophs. O_2 was essential in emergence of O_3 layer. UV ionize the ionosphere, which absorbs extreme UV	UVA and UVB: distortion of the DNA structure avoiding replication. Linked to cancer mainly in skin and eyes, immune system problems (Kim et al., 2014)
Cosmic rays (9/10 H nuclei, 1/10 α particles, highly energetic charged heavy ions like iron nuclei, and β particles, > 100 MeV)	Deflected by the magnetic field. Cosmic rays ionize atmospheric N_2 and O_2 leading to O_3 layer depletion	Penetrate living tissue and are carcinogenic (Ghissassi et al., 2009)

Note: Energy = (1.24)103/λ. 1 eV = (1.602)10–19 J is the energy of a photon in the range visible to humans.

energy is converted into a stable charge-separated state by ultrafast electron transfer events (Romero et al., 2014). Mitochondrial oxidation is associated with the respiratory chain and the tricarboxylic cycle that leads to the production of ATP, while peroxisomal metabolism is associated with mechanisms of detoxification and the biosynthesis of specific fatty acids. Peroxisomes cooperate with mitochondria in lipid metabolism, oxidizing fatty acids, and in reactive oxygen species production (Demarquoy and Le Borgne, 2015).

In a human, around 10 billion mitochondria comprise 10% of the person's body weight (Perlmutter and Loberg, 2015). Human bodies are actually ecosystems whose cells contain symbiotic bacteria, collectively called the microbiome. The microbiome weighs around 2 kg and includes more genes than the whole human genome. Without the microbiome, mostly hosted in human intestines, many vital (immunity, digestion, physical, and microbiological barriers against pathogens, detoxification, enzyme, neurotransmission) functions would not be carried out in the human body. Mitochondria contain 5–10 copies of their own DNA (mitochondrial DNA transmitted in humans by the mother) and exert control over the nuclear DNA of the human cells (Perlmutter and Loberg, 2015).

20.1.8 NONEQUILIBRIUM THERMODYNAMICS

An organism keeps away from maximum entropy by continuously extracting energy and matter from its surrounding environment. Metabolism is what allows an organism to get rid of all the entropy it cannot help but produce while being alive (Schrödinger, 1944).

A metastable dissipative process is invariable not in the sense of maximum entropy but because new particles are constantly recycled in situ. A metastable process will settle in a stationary state of minimal entropy production, as close to equilibrium as possible (Prigogine, 1991). In particular, a metastable dissipative process is a system open to solar energy subsidies to keep life away from maximum entropy.

Organisms also circumvent entropic degradation through reproduction: the system replicates beyond individual death. Self-perpetuation takes place at the species, genetic, and molecular levels (in an autocatalytic cycle of chemical reactions A → B → C → D).

20.1.9 EXCESS ENERGY IN THE ANTHROPOCENE

Humans and domesticated animals consume increasing amounts of bioavailable energy in the geosphere (BEG):

$$BEG = hiG + loI + hiC \tag{20.3}$$

where hiG is the sum of the natural background radiation (Table 20.1; Figure 20.3) and anthropogenic high energy radiation emissions. loI are low intensity radiations (noise and vibrations as produced by machines), and infrared. hiC is a carbon compound. hiC is an essential component through which humans consume energy. Although it includes hydrocarbons, carbohydrates and fat are vital. Humans, like any other organism, are endowed with bodily mechanisms to dissipate excess energy. Now however, storage exceeds biological requirements, a process called the global nutrition transition (Popkin et al., 2013) of pandemic proportions due to the lack of physical activity and access to cheap dietary fat and sugar. Excess sugar is converted to fat, which is stored in adipose tissue, inflammatory and hormonally active (Frayn, 2010). Other changes occur in the microbiome, such as dominance of *Clostridium* bacteria in children addicted to sugar and fat, which induces craving for these compounds (Perlmutter and Loberg, 2015). Past a certain body mass index threshold, fat mass exceeds the cardiovascular and musculoskeletal capacity to sustain enough aerobic expenditure to process fat stores. All excess food energy intake further accumulates and at the same time requires energy to be maintained or metabolized. Dietary energy intakes must be based on energy expended, not on nutrient availability (see Chapter 22).

Insulation from loI noise and vibration has to an extent prevented deleterious effects. This has not been so for urban heat waves, which killed thousands of people in the U.S. and Europe in 1995 and 2002.

Of particular biological concern, in the discussion of sustainable energy technologies are the limits to hiG:

$$\lim hiG = geofilters + DNA\ protection + DNA\ repair \tag{20.4}$$

The geofilters are Earth's magnetic field, ionosphere, and atmosphere (Table 20.1; Figure 20.2).

An important component of hiG is the changes in Earth's balance of natural radiation (BNR) where

BNR = Earthbound extra-terrestrial radiation at the top of the magnetosphere

$$- absorption\ by\ geofilters + changes\ in\ biofilters + bodily\ ^{40}K\ and\ ^{14}C + lithospheric\ radiation \tag{20.5}$$

In the Anthropocene, three complicating sources of mutagenicity, genotoxicity, and cytotoxicity occur, explaining the difference between the total DNA-altering exposure (TDA) and BNR:

$$TDA = BNR + xenobiotics + pathogens + artificial\ radiation \tag{20.6}$$

so that

$$TDA > \lim hiG \tag{20.7}$$

which is possibly the tipping point at which human bodies find themselves in the Anthropocene. Xenobiotics are artificial chemical compounds. Pathogens here refer to those organisms that are linked to carcinogenesis in humans; in the Anthropocene, climate and land changes have modified the exposure of humans to pathogens (de las Heras et al., 2016). As to artificial radiation, it encompasses emissions from coal mining and burning, phosphate fertilizers, nuclear weapons, nuclear power plants and their waste, medical use of radioisotopes and X rays, airplane trips, and tanning devices.

Natural and anthropogenic exposure to radiation is a common cause of damage to the DNA. The double-strand helix of DNA has phosphorylated sugars in the exterior and four nucleotides or bases in the interior united by hydrogen bonds (two purines: A-adenine and G-guanine; and two pyrimidines: C-cytosine and T-thymine). When humans are exposed to radiation, the DNA structure is often broken or altered. Sometimes the DNA is repaired, but if the dose frequency or intensity exceeds the limits of the organism, the endpoints of damage are cancer, inheritable errors, or death. DNA repairs through (a) reversing the chemical reaction responsible for the damage or (b) eliminating damaged bases and restoring them with new synthesized DNA. Excision, recombination, and translesion synthesis are the main repair mechanism in humans. During excision, damaged bases are recognized and eliminated; the empty space is filled with new synthesized DNA

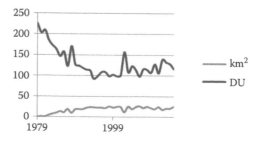

FIGURE 20.2 Extent and density of the Southern Hemisphere ozone layer. Ozone hole mean 7 September-13 October area (million km²). Minimum 21 September-16 October ozone density (Dobson Units, DU). By definition, DU<220 in the hole; such values did not exist in the historical record. (From http://ozonewatch.gsfc.nasa.gov/meteorology/annual_data.txt.)

using the nondamaged complement strand as a template. Recombination reestablishes the sequence of the broken double strands with homologue sequences of an intact chromosome after replication, while the new synthesized chromatids remain united or reattaches the two broken extremes of each strand, but the latter mechanism is associated with mismatches through deletion of bases in the damaged point. Without a template, cells use the error-prone recovery mechanism known as translesion (Cooper and Hausman, 2015). Also, a cell cycle may stop progress from one phase of the cycle to the next (cell-cycle checkpoints) or go to apoptosis (programmed cell death), protecting the organism at the expense of the individual cell (Sinha and Häder, 2002).

Difficulties remain in compounding data on labor, domestic, and environmental exposures, and in relating exposure to damages. But there are at least two trends in damages to future generations of humans: first, lessened quality of human sperm has been documented in the last decades (Sharpe, 2012). Cellular phone radiation is among the factors implicated in this trend (Agarwal et al., 2009). Second, an increased trend in cancers has been observed, even as most of them are preventable (IARC, 2014).

20.1.10 Natural and Anthropogenic Alterations to Geophysical Shielding against Extraterrestrial Radiations

The biogenically derived stratospheric ozone layer protects life on Earth from UV. Starting in 1979, however, anthropogenic chlorofluorocarbon (CFC) gases seasonally have depleted the ozone layer, mainly in a vast area around the South Pole. Although the depletion process seems to have stabilized, there are no clear signs of recovery (Figure 20.2); in 2014, a record ozone hole reached South America. Furthermore, dichloromethane, a very short-lived substance whose concentrations had diminished in the 1990s, is being used in industrial processes and increasing in abundance in the lower stratosphere. Ironically, it is used in the manufacture of "ozone-friendly" hydrofluorocarbons, which had replaced CFCs. Dichloromethane has more ozone depletion efficiency and a more powerful influence on climate than CFCs (Hossaini et al., 2015).

In addition, Earth's magnetic field has been weakening since at least 1840 (Zhong et al., 2014). Consequences are, first, a lessened filtration of alpha, beta, and gamma extraterrestrial radiation and cosmic rays (Table 20.1 and Figure 20.3). Second, this decaying magnetic dipole will let the solar wind buffet the ozone layer with an ensuing reduction of UV filtration. Third, oxygen will escape Earth with more ease (Wei et al., 2014) possibly leading to higher exposure to UV on Earth's surface.

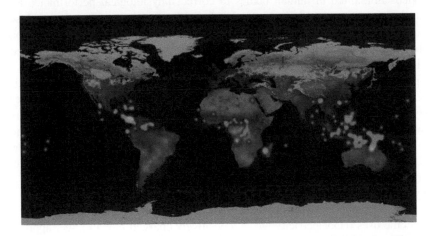

FIGURE 20.3 Distribution of gamma rays around Earth's equator. The red dots show the ~500 terrestrial gamma-ray flashes daily detected by the Fermi Gamma-ray Space Telescope through 2010. (From Goddard Multimedia, NASA/Goddard Space Flight Center. https://en.wikipedia.org/wiki/File:Antimatter_Explosions. ogv.)

20.2 APPLICATIONS

Living organisms can help generate energy sustainably based on their capabilities such as catalysis, energy capture and storage, and fuel synthesis. They can also help save energy in green chemistry via CO_2 fixation, chemosynthesis, and biorefining rare materials. They are also endowed with self-repair, self-replication capabilities, applicable to long-term bioremediation of waste and waste sites. These capacities afford technologies at lower economic and environmental impact costs (Adesina et al., 2017).

Applications with live organisms can further be hybridized with life-inspired (biomimetic) devices: when lighting devices track solar light actively—following the model of sunflower phototropism—collect outdoor light and transmit it inwards, improvement in performance can be 30%–36% over passive light collection technologies (Yuan et al., 2017).

Two sets of applications stand out in their ability to solve sustainability issues in all their conventional environmental/economic/social classes. These are biomechanical and biomass waste recovery applications. Their importance stems from their storability, portability, and dispatchability (ease of use on demand). These applications pertain to self-organizing systems, so rather than large investment and maintenance costs, they require hybridization to close an organic matter cycle. Such a system should aim for self-sustainability: maximal resort to natural processes and minimal or null artificial inputs. In particular, since biomass waste has recently captured atmospheric CO_2, the combustion of biomass-waste-based biomethane is carbon neutral (Chong, 2008). As to biomechanical applications using human power, they are needed to face today's excess energy intake by an increasing number of humans. Animal power should increase too, while meat consumption should decrease due to the large environmental hoofprint of animal husbandry.

20.2.1 BIOMECHANICAL APPLICATIONS

These applications fulfill essential agricultural, medical, and transport goals. Horsepower is a shift to local flow-limited renewable sources, away from a technology supported by nonlocal processes, driven by nonrenewable sources, and beyond the control of the farmer. The horse generates traction and emotional comfort, but also meat, leather, horsehair, and manure (a soil amender after it has gone through biomethane production). Horse inputs are more renewable (60%) than tractor inputs (9%). The farmer is in control of the information needed to help the horse manage its own life (Rydberg and Jansén, 2002).

Most devices implanted in human bodies to correct physiological performance are powered by batteries. But conversion of glucose into electrical energy using implantable glucose fuel cells in the blood vessels could supply energy to small electronic devices such as cardiac pacemakers, implanted biosensors, or artificial urinary sphincters. For this, glucose biofuel cells need improved power and lifetime (Cosnier et al., 2014).

Human-powered nebulizers to treat respiratory diseases (Hallberg et al., 2014) require only the manual energy of a health worker or caregiver to generate an airflow while the patient is being treated; they are virtually fail proof. A biomechanical energy harvester allows a recharge of prosthetics or medical devices, with a little extra effort (5W) when walking. The use of negative muscle work resembles the use of hybrid car brakes that recharge a battery. The electricity generated is 10 times that of devices installed in shoes. The additional metabolic cost is 1 W (1/8 of conventional human energy generation) (Donelan et al., 2008).

Even tiny, continued efforts can recharge portable electronic devices with the energy, otherwise dissipated, of another energy harvester using a computer keyboard and typing > 100 characters/min. This scheme uses triboelectricity: when a positively charged human finger approaches a key, free electrons flow toward the upper electrode. Then, as the finger lifts, it produces another current, from the lower to the upper electrode (Chen et al., 2015). Other technologies to harvest energy from human motion include, but are not limited to, piezoelectric materials, electromagnetic generators, and dielectric elastomers (Partridge, 2014).

In human-powered vehicles, improved aerodynamics and reduced friction are paramount. The effective frontal area is the coefficient of drag times the frontal surface. Effort efficiency depends on the vehicle (i) holding the body by the saddle, preventing the muscles from making an effort to maintain posture and balance, (ii) allowing the muscles to always operate in the optimal direction (iii). Cycling is particularly efficient in generating energy, since it draws on the most powerful muscles of the body. Rowing in turn uses a kinematic chain that involves the whole body, and not just the legs. It is therefore not surprising that cycling and rowing have allowed flight across the British Channel or ocean circumnavigating (Figures 20.4 and 20.5) using only human energy: the Moksha boat crossed several oceans in 1998–2007.

20.2.2 Waste Recovery Applications

20.2.2.1 Biorefineries

Biorefineries are facilities for fractioning and refining biomass to increase its value through conversion from several substrates to several products, preferably using all resource components. The most common substrates are still sugar, straw, wood, and starch for ethanol, lignin, and organic acids; plant and algae oil; cellulose and lignocellulose (Al-Kaidy et al., 2015). This concept goes beyond the exhaustion of biomass into a range of products based on four principles: sustainability, cascading, nonconflict with food, and neutral carbon footprint (Escamilla-Alvarado, 2017).

Living systems manage their chemistry more efficiently than manmade chemical refineries, and most of the wastes they generate are recyclable or biodegradable, operate at lower temperatures, and produce less toxic waste and fewer emissions than conventional chemical processes to produce energy (Erickson et al., 2012).

Biorefinery might develop the use of lignin, celluloses, and hemicelluloses that do not compete with food or land and are massive agricultural, forestry, and municipal residues, as substrates for the synthesis of biofuels with the help of novel enzymes improved by metabolic engineering, the implementation of pretreatments, recombinant technology, biocatalyst design, and reaction engineering and chemicals (Seibel et al., 2014).

FIGURE 20.4 The Gossamer Albatros crossed the Channel in April 1979. (From NASA on The Commons. From https://commons.wikimedia.org/wiki/File:Solar-powered_Gossamer_Penguin_in_flight.jpg.)

FIGURE 20.5 Hybrid pedal/row kayak/sail/boat. (From "Mike" Michael L. Baird. Creative Commons Attribution 2.0 Generic license. https://commons.wikimedia.org/wiki/File:Hobie_Mirage_Adventure_Island_Trimaran_sail_yak.jpg.)

Engineered metabolic pathways have been developed to produce butanol, oleaginous fuels, branched-chain alcohols, medium-chain fatty acids, alkanes, gasoline-like molecules, biodiesels, and aviation fuels (Adesina et al., 2017). The use of algae, microalgae, and aquatic biomass as feedstock for biofuel production is an interesting option due to fast-growth rates, efficient CO_2 capture, a short harvesting cycle, absence of lignin, low hemicellulose content, and growth in areas unsuitable for agricultural purposes, which may be cultured with nutrients emitted from marine animal aquaculture in manmade open ponds and closed systems to exclude the heterotrophs that graze on the algae (Harish et al., 2015).

Mineralization of substrates by microorganisms can construct nanostructured porous materials for electrodes of batteries. Electroactive microbes can transfer electrons over long distances between the cell surface and external substrates by using conductive pili (bacterial nanowires). The use of nanostructured electrode materials is also attractive to improve the capacity, cycling life, and safety of batteries. Viruses have the potential for templating carbon nanotube electrodes at ambient temperatures, and bacterial biofilms can be used for the design of nanofibers (Adesina et al., 2017).

New applications with networks made of nucleic acids have been used to program, calculate, and store events such as molecular interactions and facilitate regulatory circuits; these have been combined with logic-gated nanorobots, solid-state nanochannels, or switchable nanovalves and artificial aptamer-lipid-receptors (Al-Kaidy et al., 2015). In situ use of nanomaterials seems to be a must, lest nanoparticles in the environment continue to accumulate.

20.2.2.2 Biomethane

Organic residues and nonfood agricultural and forestry residues are recommended when converting biomass to energy. Certain microalgal species seem to be good substrates for anaerobic fermentation, resulting in the production of biogas with relatively high methane content (Mussgnug et al., 2010).

Biomethane seems to be cost-effective, less toxic, and more efficient than other biofuels in life cycle assessment studies (Islas-Espinoza and Weber, 2014). Methane combustion is carbon neutral since it emits only one molecule of CO_2 per molecule of CH_4, compared to other fossil fuels. Details about methane generation and purification are explained in Chapter 21.

20.2.2.3 Biohydrogen

Fuel combustion for water electrolysis aiming at hydrogen production contaminates the air, is toxic, and is difficult to store. Hydrogen as a fuel does not seem to have significant impacts on climate at this time, but its large-scale use is suggested to lead to increased leakage of hydrogen into the atmosphere. Currently, the atmospheric lifetime of hydrogen appears to be controlled by soils. Hydrogen is an indirect greenhouse gas that would affect the lifetime of methane in the atmosphere, and if concentrations in the atmosphere increase, then migration to the stratosphere could enhance ozone depletion (Schauer, 2015).

Recent research has been carried out to imitate nature: the chlorophyll of plants absorbs solar light to produce H_2O and O_2, but instead of chlorophyll, nanoparticles of rhodium and ruthenium are being evaluated to produce hydrogen from water (Takeuchi, 2011). There are three methods of biohydrogen production: splitting water to hydrogen and oxygen by green algae and cyanobacteria (direct biophotolysis); photodecomposition of the accumulated biomass enriched by carbohydrates during the process of photofermentation by photosynthetic bacteria (indirect biophotolysis); and dark fermentation of organic compounds (Voloshin et al., 2016).

In dark fermentation, hydrogen is produced by anaerobic bacteria on carbohydrate-rich substrates. The effluent from the dark fermentation is usually rich in organic acids and can be used for photofermentative biohydrogen production using photosynthetic bacteria (Harish et al., 2015).

20.2.2.4 Beyond Hybridizing: Designing Human-Natural Cycles

Considerable heat is dissipated by transport (see Figure 1.2, p. 4), and transportation of food from rural breadbaskets to urban markets is a major component of this issue. This has inspired the local food movement and more recently sparked interest in urban food production in green roofs and reclaimed empty lots. The city of Detroit after the demise of the car manufacturing industry is exemplary in this respect.

Those innovations leave plenty of room for improvement, along two converging hybridization lines: on the one hand, the microbiological and molecular level represented by biorefineries and biomethane production. On the other hand, the ecosystem-level transition where isolated water bodies and islands of greenery in a sea of asphalt give way to green-and-blue corridors and then to mixed rural and urban spaces with maximal solar gain by natural and human communities.

Hybridization starts with standalone devices, in particular devices using the sun and waste to produce energy and recycling heat (Figure 20.6); at this early stage of hybridization, the goal is to recycle all waste, which then becomes a valuable resource, following the principles of biorefinery. Wastewater for instance becomes—through (an)aerobic bioremediation—biomethane, soil amender, while water is reclaimed via hybrid solar-biomethane disinfection (Figure 20.6). Urban environments where this started taking place included Hammarby in the vicinity of Stockholm; the idea then diffused to the Bo01 neighborhood of Malmö, also in Sweden.

The hybridization of devices then has to become part of a cycle where primary production by vegetables is promoted by the use of solar and waste-based energy (Figure 20.7). Solar energy, even at the latitude of the Netherlands is more than enough for greenhouse vegetable production, as shown more than a decade ago by Zonneterp, which recirculated excess greenhouse heat into a housing block. Zonneterp's idea can be expanded in many ways. For instance, two half-spheres (a triple-glazed greenhouse at the top and a parabolic solar concentrator at the bottom) make up the outer envelope of an anaerobic reactor. The apparatus is partially belowground to minimize advective wind cooling and radiation heat losses to the cooler atmosphere. The apparatus is further protected on the less sunny sides by a building or orchard trees. Below the solar concentrator, heat-trapping materials (like zeolites) slowly release heat at night. Hot (55°C) effluent from the reactor flows through a drained field that fertirrigates the orchard or the nursery trees in a minitunnel greenhouse. Dissolved methane in the effluent water is trapped by the soil and acts as greenhouse

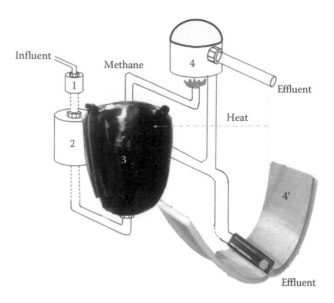

FIGURE 20.6 Hybrid system for biomethane production using food/animal wastes or wastewater. (1) Primary filtration (zeolite), (2) Secondary filtration (biofilm), (3) Anaerobic digestion, (4) Methane heat generation, (4′) Solar heat and waste disinfection.

gas, heating the soil of the orchard. Methane combustion is done belowground so that carbon dioxide emissions also act as greenhouse gases in the soil and promote microbial life. In this scheme, agro waste and wastewater are recirculated (Figure 20.7). Efficiency enhancements can proceed along several paths, such as the use of a dome-shaped Fresnel Köhler concentrator instead of a greenhouse, in colder climes.

The evolution of hybrids then takes over whole landscapes (Figure 20.8), so that human-dominated ecosystems transition toward nature-dominated cycles. Urban heat islands disappear as rooftop and balcony gardens dominate and CO_2 fertilizes plants. Entropy in these cycles is reduced as heat is absorbed by plants and soil and landscapes, and transportation of produce to urban markets is forgone. Energy gain is maximized via photosynthesis. An application of these ideas might follow the model of aquaponics in Inle Lake in Myanmar, where crops are grown on a floating vegetal mat, underneath which a subaquatic ecosystem feeds on detritus. This ecosystem can also harbor methane production in the sediment: rather than letting it diffuse to the atmosphere, the trophic chain started by methanotrophs is used in biorefinery to make fish feed and further increase the productivity of the ecosystem.

FIGURE 20.7 Waste-based energy for high-latitude permaculture.

FIGURE 20.8 Urban-Rural hybridization. (A) Urban-Rural mixed uses: energy farm and agriculture close by (web-based) market or interchange, (B) Underwater aquaponics, (C) Sediment, (D) Fish, (E) Methanogenic activity, (F) Detritus, (G) Methanotrophs, (H) Fish feed. (Adapted from Hora dara. Creative Commons 3.0. https://commons.wikimedia.org/wiki/File:The_Social_Ecology.jpg.)

REFERENCES

Adesina, O, IA Anzai, JL Avalos, and B Barstow. 2017. Embracing biological solutions to the sustainable energy challenge. *Chem* 2: 20–51. doi:10.1016/j.chempr.2016.12.009.

Agarwal, A, NR Desai, K Makker, A Varghese, R Mouradi, E Sabanegh, and R Sharma. 2009. Effects of radiofrequency electromagnetic waves (RF-EMW) from cellular phones on human ejaculated semen: An in vitro pilot study. *Fertility and Sterility* 92: Elsevier Ltd: 1318–25. doi:10.1016/j.fertnstert.2008.08.022.

Al-kaidy, H, A Duwe, M Huster, K Muffler, C Schlegel, T Sicker, R Stadtmüller, N Tippkötter, and R Ulber. 2015. Biotechnology and bioprocess engineering – From the first Ullmann' S article to recent trends. *ChemBioEng Reviews* 2 (3): 175–84. doi:10.1002/cben.201500008.

Chen, J, G Zhu, J Yang, Q Jing, P Bai, W Yang, X Qi, Y Su, and ZLWang. 2015. Personalized keystroke dynamics for self-powered human-machine interfacing. *ACS Nano* 9: 105–16.

Chong, J. 2008. Methane: A natural gas. *Microbiology Today* 35 (3): 124–27.

Colby, DS. 1987. *Compendion de Bioquimica*. First. Mexico City: El manual moderno.

Cooper, GM, and RE Hausman. 2015. *The Cell: A Molecular Approach*. Seventh. Sunderland, MA: Sinauer.

Cosnier, S, AL Goff, and M Holzinger. 2014. Towards glucose biofuel cells implanted in human body for powering artificial organs : Review. *Electrochemistry Communications* 38. Elsevier B.V: 19–23. doi:10.1016/j.elecom.2013.09.021.

de las Heras, A, M Islas-Espinoza, and A Amaya-Chavez. 2016. Pollution: The pathogenic and xenobiotic exposome of humans and the need for technological change. *Encyclopedia of Environmental Management*, 1–10. doi:10.1081/E-EEM-120052926.

Demarquoy, J, and F Le Borgne. 2015. Crosstalk between mitochondria and peroxisomes. *World Journal of Biological Chemistry* 6 (4): 301–9. doi:10.4331/wjbc.v6.i4.301.

Donelan, JM, Q Li, V Naing, JA Hoffer, DJ Weber, and AD Kuo. 2008. Biomechanical energy harvesting. *Science* 319: 807–10. doi:10.1126/science.1149860.

Erickson, B, JE Nelson, and P Winters. 2012. Perspective on opportunities in industrial biotechnology in renewable chemicals. *Biotechnology Journal* 7: 176–85. doi:10.1002/biot.201100069.

Escamilla-Alvarado, C. 2017. An overview of the enzyme potential in bioenergy-producing biorefineries. *Journal of Chemical Technology and Biotechnology* 96 (5): 906–24. doi:10.1002/jctb.5088.

Faraj, KA, MM Elias, AH Al-Mashhadani, and S Baatout. 2016. Effect of x- and gamma rays on DNA in human cells effect of x- and gamma rays on DNA in human cells. *European Journal of Scientific Research* 53: 470–76.

Frayn, NK. 2010. *Metabolic Regulation: A Human Perspective*. Third. Oxford, UK: Wiley-Blackwell.

Gaudy, AF, and ET Gaudy. 1980. *Microbilogy for Environmental Scientists and Engineers*. First. New York: McGraw-Hill.

Ghissassi, FE, R Baan, K Straif, Y Grosse, B Secretan, V Bouvard, L Benbrahim-tallaa et al. 2009. Special report : Policy a review of human carcinogens—Part D : Radiation. *Lancet Oncology* 10 (8): 751–52. doi:10.1016/S1470-2045(09)70213-X.

Hallberg, CJ, MT Lysaught, CE Zmudka, WK Kopesky, and LE Olson. 2014. Characterization of a human powered nebulizer compressor for resource poor settings. *BioMedical Engineering OnLine* 13: 1–11.

Harish, BS, MJ Ramaiah, and KB Uppuluri. 2015. Bioengineering strategies on catalysis for the effective production of renewable and sustainable energy. *Renewable and Sustainable Energy Reviews* 51. Elsevier: 533–47. doi:10.1016/j.rser.2015.06.030.

Hitchcock, ST. 2007. Skeletal muscle. In *Body: The Complete Human*, edited by ST Hitchcock, 106–23. Washington, DC: National Geographic Society.

Hossaini, R, MP Chipperfield, SA Montzka, A Rap, S Dhomse, and W Feng. 2015. Efficiency of short-lived halogens at influencing climate through depletion of stratospheric ozone. *Nature Geoscience* 8 (3): 186–90. doi:10.1038/ngeo2363.

IAEA. 2004. *Radiation, People and the Environment*. Edited by J Ford. Austria: IAEA.

IARC. 2014. *World Cancer Report 2014*. Edited by BW Stewart and CP Wild. Lyon, France: IARC.

Islas-Espinoza, M, and B Weber. 2014. Bioenergy solutions. In *Sustainability Science and Technology: An Introduction*, edited by Alejandro de las Heras, 157–71. Boca Raton, FL: CRC Press.

Kim, S, S Jin, and GP Pfeifer. 2014. Formation of cyclobutane pyrimidine dimers at dipyrimidines containing 5-hydroxymethylcytosine. *Photochemical and Photobiological Sciences* 12 (8): 1409–15. doi:10.1039/c3pp50037c.Formation.

Mussgnug, JH, V Klassen, A Schlüter, and O Kruse. 2010. Microalgae as substrates for fermentative biogas production in a combined biorefinery concept. *Journal of Biotechnology* 150. Elsevier B.V.: 51–56. doi:10.1016/j.jbiotec.2010.07.030.

Partridge, JS. 2014. Human energy harvesting in the urban environment. EngD Thesis. University College London Faculty of Engineering, London.

Perlmutter, D, and K Loberg. 2015. *Brain Maker : The Power of Gut Microbes to Heal and Protect Your Brain-for Life*. First. New York: Little Brown and Company.

Ponizovskiy, MR. 2013. The central regulation of all biophysical and biochemical processes as the mechanism of maintenance stability of internal energy and internal medium both in a human organism and in cells of an organism. *Modern Chemistry and Application* 1 (1): 1–2. doi:10.4172/2329-6798.

Popkin, BM, LS Adair, and SW Ng. 2013. Now and then: The global nutrition transition: The pandemic of obesity in developing countries. *Nutrition Reviews* 70: 3–21. doi:10.1111/j.1753-4887.2011.00456.x.NOW.

Prigogine, I. 1991. The behavior of matter under nonequilibrium conditions: Fundamental aspects and applications. Austin, TX.

Romero, E, R Augulis, VI Novoderezhkin, M Ferretti, J Thieme, D Zigmantas, and R Van Grondelle. 2014. Quantum coherence in photosynthesis for efficient solar-energy conversion. *Nature Physics* 10: 1–7. doi:10.1038/NPHYS3017.

Rydberg, T, and J Jansén. 2002. Comparison of horse and tractor traction using emergy analysis. *Ecological Engineering* 19: 13–28.

Schauer, JJ. 2015. Design criteria for future fuels and related power systems addressing the impacts of non-CO_2 pollutants on human health and climate change. *Annual Review of Chemical and Biomolecular Engineering* 6: 101–20. doi:10.1146/annurev-chembioeng-061114-123337.

Schrödinger, E. 1944. *What Is Life – The Physical Aspect of the Living Cell*. Cambridge, MA: Cambridge University Press.

Seibel, J, U Bornscheuer, and K Buchholz. 2014. Enzymatic degradation of (ligno) cellulose. *Angewandte Reviews* 53: 2–20. doi:10.1002/anie.201309953.

Sharpe, RM. 2012. Sperm counts and fertility in men: A rocky road ahead. *EMBO Reports* 13 (5). Nature Publishing Group: 398–403. doi:10.1038/embor.2012.50.

Sinha, RP, and D-P Häder. 2002. UV-induced DNA damage and repair : A review UV-induced DNA damage and repair : A review. *Photochemical and Photobiological Sciences* 1: 225–36. doi:10.1039/B201230H.

Takeuchi, N. 2011. *Nanociencia Y Nanotecnología: La Construcción de Un Mundo Mejor Átomo Por Átomo*. First. Mexico City: FCE, CN y N-UNAM, SEP, CONACYT.

Voloshin, RA, MV Rodionova, SK Zharmukhamedov, and T Nejat Veziroglu. 2016. Review: Biofuel production from plant and algal biomass. *International Journal of Hydrogen Energy* 41 (39). Elsevier Ltd: 17257–273. doi:10.1016/j.ijhydene.2016.07.084.

Wei, Y, Z Pu, Q Zong, W Wan, Z Ren, M Fraenz, E Dubinin et al. 2014. Oxygen escape from the earth during geomagnetic reversals : Implications to mass extinction. *Earth and Planetary Science Letters* 394. Elsevier B.V.: 94–98. doi:10.1016/j.epsl.2014.03.018.

Yuan, Y, X Yu, X Yang, Y Xiao, B Xiang, and Y Wang. 2017. Bionic building energy efficiency and bionic green architecture : A review 74 (November 2016): 771–87. doi:10.1016/j.rser.2017.03.004.

Zhong, J, WX Wan, Y Wei, SY Fu, WX Jiao, ZJ Rong, LH Chai, and XH Han. 2014. Increasing exposure of geosynchronous orbit in solar wind due to decay of earth's dipole field. *Journal of Geophysical Research : Space Physics* 119: 9816–22. doi:10.1002/2014JA020549.Received.

21 Advanced Biomethane Processes

Sevcan Aydin and Bahar Yavuzturk Gul
Istanbul Technical University

Aiyoub Shahi
University of Tabriz

CONTENTS

21.1 INTRODUCTION

Eco-friendly and economical energy production gains importance every day in the world due to overpopulation and quickly developing industry. Because renewable energy reduces poverty and leads to sustainable development, recently many countries have started to use renewable energy. Reduction in the reserves of nonrenewable energy sources and the role of these resources in climate change is another reason for the increasing importance of renewable energy (Pehnt, 2006). Concordantly, biogas (produced by processing various renewable energy sources) has an spatial role in the succession of fossil fuels, which cause the increase of greenhouse gases emissions, global warming, and climate change effects, in power and heat production. Alternatives like biomass, solar, geothermal, and hydraulic resources can be considered potential renewable energy sources (Hosseini and Wahid, 2016).

Biomass, one of the most important renewable sources of energy, is defined as the organic matter produced by the photosynthetic conversion of solar energy to chemical form (Herbert and Krishnan, 2016). Thus, biomass can be used as fuel to provide green power as renewable and sustainable energy. The process of breaking down the biological materials in an anaerobic environment generates biogas. After that, biomethane is procured owing to biogas purification process. Although there are many sources of biomass such as municipal solid waste, agricultural residue, forestry crops and residue, sewage, industrial residue, and animal residue, biomass includes lignocellulosic compounds, which are more favorable for biogas production (Aydin, 2016a; Aydin et al., 2017).

Anaerobic digestion is a series of biological process providing catabolism of organic materials under anaerobic conditions by microorganisms. Biogas formation (i.e., methane and carbon dioxide) as a renewable energy is obtained thanks to anaerobic digestion by microbial biomass (Aydin, 2016a; Cavinato et al., 2017). Thereby, through this process biomass is converted to energy.

Using anaerobic digestion technologies, municipal solid waste, animal manure, food waste, industrial wastewater, and other organic origin wastes are converted into biogas. Because this biogas is environmentally friendly and economically beneficial, it is considered that the biogas production during anaerobic digestion is one of the most effective ways of producing renewable energy production. Especially methane (one of the end products of anaerobic digestion) is a quite important biogas because it provides a renewable alternative for heat and power. In addition to an important renewable energy source, fertilizer production, pathogen removal, pollution control, waste stabilization, and odor reduction can be obtained due to anaerobic digestion (Si et al., 2016).

The aim of this chapter is to investigate the advantages and difficulties of biomethane production using an anaerobic digestion process. The anaerobic biodegradation steps involved in the digestion process and anaerobic microbial relationships during biomethane production will be explained.

21.2   ESSENTIALS OF THE ANAEROBIC DIGESTION PROCESS

Several groups of facultative and anaerobic microorganisms take part in the stages of the anaerobic digestion process to degrade organic material. The synergetic community of microorganisms found in an anaerobic digester conducts the process of fermenting organic matter into methane. Anaerobic digestion is mediated during the stages of hydrolisis, acidogenesis, acetogenesis, and methanogenesis by anaerobic microorganisms. Mentioned stages are shown in Figure 21.1 (Aydin et al., 2015). Complex particulate materials such as lipids, carbohydrates, and proteins must be hydrolyzed to soluble organic matter that can be absorbed by microbial cells. This hydrolysis step is done by specific extracellular enzymes that are produced by hydrolytic fermentative bacteria under anaerobic conditions in anaerobic digesters. pH, temperature, substrate composition, cell residence time, and the by-products of hydrolytic bacteria are important factors affected by the reaction rates of extracellular enzymes in hydrolysis. The microbial community of the hydrolysis stage is considerably heterogenic. It was found that the compounds containing cellulose are degraded by *Clostridium* spp., but, *Bacillus* spp. are responsible for the degradation of protein and fats. The most widespread hydrolytic microorganisms are classified as cellulolytic (*Clostridium thermocellum*), proteoytic (*Clostridium bifermentas*,

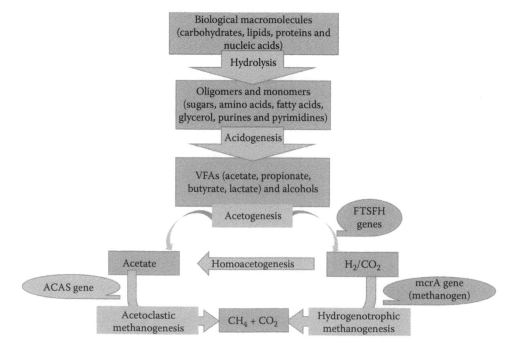

FIGURE 21.1   Anaerobic process and targeting metabolic genes.

Peptococcus spp.), lipolytic (genera of *Clostridia* and *Micrococci*) and aminolytic (*Clostridium butyricum, Bacillus subtilis*) bacteria. It was found that several anaerobic fungi also degrade cellulose and hemicelluloses (Yuan and Zhu, 2016). The hydrolytic microorganisms are also capable of degradation of some intermediate products to simple volatile fatty acids (VFAs), lactic acid, carbon dioxide, hydrogen, ethanol, ammonia, and hydrogen sulfide. After hydrolysis, the soluble monomers that are generated are broken down to short-chain organic acids, alcohols, hydrogen, and carbon dioxide by facultative and obligatory anaerobic bacteria in the second step, defined as acidogenesis. The concentration of hydrogen ions is important for determining the type of end products that will be generated. The partial pressure of hydrogen must be high in order to form acetate. Acidogenic or fermentative bacteria can metabolize amino acids and sugars to intermediary products like acetate and hydrogen. While single amino acids are produced by *Clostridia, Mycoplasmas* and *Streptococci*, butanol, butyric acid, acetone, and iso-propanol are usually produced by *Clostridum* sp. Butyrate is produced by *Butyribacterium*; acetone and butanol are produced *by Clostridium acetobutylicum. Clostridium butylicum* also produces butanol, hydrogen, carbon dioxide, and iso-propanol. During these reactions, different pathways are used. In the degradation of carbohydrates, propionic acid is formed by the succinate pathway and the acrylic pathway. Butyric acid and fatty acids are degraded by the beta oxidation reaction. Proteins are degraded by the Stickland reaction, and if cysteine is degraded, hydrogen sulfide can be formed (Zahedi et al., 2016).

In the third step of anaerobic digestion (acetogenesis) the products of acidogenesis are converted into acetate, hydrogen, and carbon dioxide by acetogenic bacteria from which methane can be obtained. Because acetate is the most common and significant precursor of methane production, acetogenic bacteria have an important role for methanogenic microorganisms. Hydrogen-producing or hydrogen-consuming acetogenetic bacteria play a role in the conversion acidogenesis end products of acetate (Aydin et al., 2015). There are two different types of acetogenic mechanisms: acetogenic hydrogenation and acetogenic dehydrogenation. Acetogenic hydrogenation includes the production of acetate from fermentation hexoses or from CO_2 and H_2. Acetogenic dehydrogenation refers to the anaerobic oxidation of long and short chain volatile fatty acids by obligate hydrogen-producing or obligate proton-reducing bacteria. Acetic acid-producing bacteria are *Methanobacterium bryantii, Desulfovibrio Syntrophobacter wolinii, Syntrophomonas wofei,* and *Syntrophus buswellii,* which are the most common acetic acid-producing bacteria (Zahedi et al., 2016).

Methanogenesis (also called methane fermentation) is the final step in anaerobic digestion where products of acetogenesis are transformed to methane and carbon dioxide by methanogenic *Archaea,* which consists of strictly anaerobic microorganisms belonging to *Euryarchaeota.* Phylogenetically, methanogens are classified in domain Archaea. Archaeal cells have unique properties separating them from the other two domains of Bacteria and Eukaryota. These three domains are separated based on rRNA analysis. Methanogens are grouped into five orders within the kingdom *Archaeobacteria: Methanobacteriales, Methanococcales, Methanomicrobiales, Methanosarcinales,* and *Methanopyrales.* Most methanogenic archaea can produce methane (CH_4) and carbon dioxide from carbon monoxide, formate, and a few alcohols. Methyl groups can also be reduced to methane by methanogens. Only a limited number of substrates can be used by methanogens to produce methane. They can be classified into three main groups according to their affinity for different substrates (Madigan et al., 2002). In the CO_2 type of substrate, methane is produced from carbon dioxide and hydrogen by hydrogenotrophic methanogens including *Methanobacteriales, Methanomicrobiales, Methanococcales,* and *Methanosarcinaceae.* In the methyl group of substrate, methane is produced from methyl compounds by methylotrophic methanogens via two different pathways. While methane is formed by reducing methyl group substances with an external electron donor such as H_2 in the first mechanism, methyl compounds are converted to methane without H_2 in the second mechanisms. In the acetate type of substrate, methane is produced from acetate via acetoclastic methanogens containing *Methanosarcina* and *Methanosaeta.* Because acetate is the major product of fermentation, it is generally dominant in the anaerobic digester systems. Almost 70% of all methane production is provided by acetoclastic methanogens. *Methanosarcina* have a higher methane yield than *Methanosaeta* (Aydin et al., 2015; Zahedi et al., 2016).

21.3 ENVIRONMENTAL AND OPERATIONAL FACTORS AFFECTING ANAEROBIC DIGESTION

Various parameters affect the performance of anaerobic digestion, such as temperature, hydraulic retention time (HRT), solid retention time (SRT), organic loading rate (OLR), pH, alkalinity, and micro- and macronutrients. Temperature is one of the most significant factors affecting the performance of anaerobic digestion in terms of ionization equilibrium, solubility of substrates, substrate removal rate, specific growth rate, decay biomass yield, and half saturation constant. Stover et al. (1994) demonstrated that anaerobic digestion processes, especially conversion of methane from acetate, are relatively sensitive to the temperature variations. Biogas production can be influenced adversely by temperature fluctuations. Anaerobic digestion can operate under psychrophilic (<25°C), mesophilic (25°C–40°C), and thermophilic (>45°C) conditions; however, mesophilic (25°C–40°C) and thermophilic (45°C–60°C) conditions are more promising. Thermophilic conditions provide many advantages like higher specific growth rates and higher metabolic rates, so, processes can be faster and more effective. On the other hand, lower stability and energy requirements are important downsides of thermophilic conditions. Thus, anaerobic digestion is usually processed under mesophilic conditions, mostly between 35°C and 42°C for lower stability and higher susceptibility to changes in environmental and operational conditions of thermophilic conditions. Nevertheless, temperature effects mostly rely on the kind of microorganisms in anaerobic digestion processes (Chae et al., 2008). Methanogenesis can be done in both mesophilic and thermophilic temperature conditions. Although bacteria are not usually sensitive to temperature fluctuations until the metanogenesis step, methanogenic Archaea is highly sensitive to even small temperature changes (Aydin et al., 2015).

HRT, described as a criterion for biogas production and waste stabilization, is another factor affecting anaerobic digestion. Because optimum HRT depends on substrate characterization and temperature, it can vary for different substrates and temperature conditions. However, it should be sufficiently long to provide microbial growth for processes of anaerobic digestion. If the HRT is too short, the organic material cannot be completely degraded, and it leads to low biogas production, washout of the microorganisms, and inhibition of the process.

The SRT relating to the growth rate of microorganisms and to anaerobic digester volume is also a significant parameter in the anaerobic digestion process. It is the same with HRT if there is no recycling or supernatant withdrawal. One must properly choose SRT and volume of digester because the digestion procedure is a function of time required by microorganisms to digest the organic material. It was found that the optimum SRT in digesters is about 30 days for mesophilic digestion and longer for low-temperature digestion (Metcalf & Eddy, Inc., 2003).

Another important parameter that affects anaerobic digestion performance is the OLR, a certain amount of organic matter that is fed daily per cubic meter of digester working volume and generally described as volatile solids. If nutrients in the digester can be easily degraded, anaerobic digestion process can be affected because of the acidification phase, which has more end-products. Thus, OLR is for methanogenic activity and biogas production (Rincón et al., 2008).

pH is another significant factor affecting anaerobic digestion systems. It is a pivotal parameter because of the solubility of matters and the reaction potential of microorganisms; so all digestion performances are directly influenced by pH. It was investigated that the optimal range of pH to provide maximum biogas efficiency in anaerobic digestion is 6.5–7.5; however, the range of pH can be relatively wide due to different types of substrate and digestion techniques. Microorganisms in anaerobic digesters are also important factors for determining the optimum pH. While methanogen function mostly occurs in a pH between 6.5 and 7.5, anaerobic bacteria can generally grow in pH between 6 and 8. Although anaerobic fungi can tolerate a wide range of pH, the optimum pH conditions are acidic (Orpin and Joblin, 1997). However, fluctuations from an optimum range of pH can cause excessive production and aggregation of acidic or basic products like organic fatty acids or ammonia. On the other hand, the accumulation of VFA cannot usually lead to a pH drop because of the buffering capacity of the substrate. In addition, if alkalinity is not sufficiently high, organic acids produced by acidogenic bacteria leads to a decrease in pH. However, the bicarbonate produced by methanogens can buffer the reduction of pH under normal conditions. If buffering capacity cannot be enough, especially in unfavorable environmental conditions, acidity can lead to an inhibitory effect on methanogens. Yet, acidogenic bacteria and anaerobic fungi are more resistant to acidity (Metcalf & Eddy, Inc., 2003).

Micronutrients, which are also called trace elements, such as iron, nickel, cobalt, selenium, molybdenum, and tungsten, and macronutrients such as carbon, nitrogen, and phosphorus are other important factors in anaerobic digestion due to microbial growth and survival. For methanogenesis, iron, nickel, magnesium, calcium, sodium, barium, tungstate, molybdate, selenium, and cobalt are considerably important. Selenium, tungsten, and nickel are required for the enzyme systems of acetogenesis and methanogeneis. The addition of macronutrients also provides positive impact on the anaerobic digestion process and so on biogas potential. Yen and Brune (2007) proved that the addition of carbon-rich waste papers as a macronutrient into an algal biomass digester leads to enhance methane yield owing to ensuring the balance between carbon and nitrogen in the feed.

Light metal ions and heavy metals are also required for the growth of microorganisms. Sodium, potassium, magnesium, and calcium as the most significant light metal ions stimulate microbial growth in anaerobic systems; however, their excess quantity causes the deceleration of the growth and causes severe inhibitions or toxicity. Heavy metals such as chromium, iron, cobalt, copper, zinc, cadmium, and nickel can be also found in important concentrations in anaerobic digesters. However, they can induce toxic effects on anaerobic processes because they cannot be biodegraded and they can accumulate on the system. Thus, heavy metals should be present at trace amounts for microbial activity and avoiding potential toxicity (Madigan et al., 2002; Metcalf & Eddy, Inc., 2003).

Other inhibitory substances are also important because they cause severe failures of anaerobic digestion processes. These materials lead to changes in the microbial population structure or inhibit bacterial growth. Fundamental indicators for inhibition are the accumulation of organic acids and the decrease of the biogas production rate. Although it has long been known that the inhibition can impact all groups of microorganisms in the anaerobic digestion processes, such as bacteria, archaea and anaerobic fungi, methanogens are the most sensitive groups to inhibitory or toxic material (Chen et al., 2008).

Ammonia, which is produced by the breakdown of nitrogenous compounds, is one of the important inhibitory factors affecting anaerobic digestion performance. Although ammonia inhibition was observed to start at concentrations of 1500–2500 mg-N/L, adaptation of anaerobic consortia involved in the biogas production process to ammonia and their tolerance to 4 g-N/L total ammonia was determined. However, 3000 mg/L of ammonia may have inhibitory effects on methanogens because they are the least resistant microorganisms to ammonia inhibition. On the other hand, it was generally asserted that 50–200 mg/L of ammonia is beneficial, 200–1000 mg/L of ammonia has no adverse effect, 1500–3000 mg/L of ammonia is an inhibitor for pH > 7.4–7.6, and above 3000 mg/L of ammonia is toxic on anaerobic processes. Inorganic nitrogen is found in the forms of ammonium (NH^{4+}) and free ammonia (NH_3) in anaerobic digesters. The free ammonia concentration depends mostly on total ammonia concentration, temperature, and pH. It was shown that thermophilic temperatures can more easily inhibit the methane fermentation of high ammonia-containing anaerobic digesters (Angelidaki and Ahring, 1994).

Another inhibitory parameter that affects the anaerobic digestion process is sulfate because of H_2S, which is the toxic form of sulfide. It is reduced to sulfide by the sulfate reducing bacteria (SRB) in anaerobic digesters and H_2S penetrates cells. Therefore, sulfate can inhibit metanogenesis owing to the competition for acetate and hydrogen by SRBs. In addition, sulfides can be produced by sulfur containing inorganic compounds during the biological production in the anaerobic digestion. If the concentration of soluble sulfide is less than 100 mg/L, it can be tolerated. However, Stronach et al.

(1986) demonstrated that higher than 200 mg/L of sulfate has an inhibitory impact on anaerobic digestion systems.

An additional example of the inhibition in anaerobic digestion systems is organic chemicals. Because they cannot be sufficiently dissolved in the water, they are absorbed by surfaces of solids. Thus, organic chemicals accumulate and cause the membranes of bacteria to swell and leak, disrupting ion gradients and finally causing cell lysis. A concentration of toxic materials, concentration of biomass, toxicant exposure time, sludge age, feeding, acclimation, and temperature are the most significant parameters influencing the inhibition of organic chemicals.

The concentration of VFAs, which are produced from complex organic material by acidogenic microorganisms in anaerobic digesters, is another significant consideration for efficient performance of digesters (Béline et al., 2017). Because VFAs are considerably dependent on the changes in pH, alkalinity, and the activity of methanogens, they are one of the most sensitive indicators to measure the performance of anaerobic digesters. While acetic acid, propionic acid, butyric acid, valeric acid, caproic acid, and enanthic acid are main groups of VFAs, acetate and propionate are the predominant VFAs. Acetate, carbon dioxide, and hydrogen, which are used by methanogens for the generation of methane, are produced because of the oxidation of the VFAs. Thus, VFAs are significant intermediary products in the metabolic pathway of methane production. If they can cause inhibitory effect on anaerobic digestion processes, system failure can arise and biogas production can be negatively affected. Accumulation of VFA production can lead to the inhibition of methane production. Moreover, microbial activity balance in anaerobic digesters can be readily disturbed by increasing VFA and decreasing methane production. In confirmation of this, Ianotti and Fischer (1983) reported that 35 mg/L of acetic acid, higher than 3000 mg/L of propionic acid, and 1000 mg/L of butyrate concentration inhibit microbial growth. Labatut and Gooch (2012) also reported that biogas production can be limited at VFA concentrations above 1500–2000 mg/L.

In addition to the abovementioned parameters, some operational factors such as mixing and types of digester are also considerably important for anaerobic digestion performance. Because mixing provides the complete contact between the reactor contents and the biomass, it is particularly significant for anaerobic digesters operating with particulate substrates. The possible inhibitory impacts of local VFA accumulations and other digestion products can be reduced by mixing. Mechanical mixers, biogas recirculation, or slurry recirculation can be used to accomplish the mixing. While mixing can vary between 20 and 100 rpm in lab-scale anaerobic digesters, mixing in high rpms is difficult in full-scale digesters (Wu et al., 2010; Cavinato et al., 2017).

The design of a digester is another conspicuous operational factor because of composition, homogeneity, and the dry matter content of the anaerobic digester system. For animal manure and algal biomass, which is rich in solid materials, high-rate reactors are not quite suitable due to granule formation causing coagulation. Anaerobic digestion systems can be performed in batch-wise, semi-continuous or continuous mode. While there isn't any addition of wastes during anaerobic digestion process in a batch system, quantities of waste are periodically added and removed to a digester leading to a de facto semi-continuous system. The raw waste is fed regularly into a digester, displacing an equal volume of digested material in the continuous-flow tank reactor systems (Wu et al., 2010; Béline et al., 2017).

21.4 BIOGAS FROM AN ANAEROBIC DIGESTION SYSTEM

Today, the use of renewable energy sources has increased due to fossil fuels' pressures on the global environment and generation of greenhouse gases accumulating in the atmosphere and causing climate change. Recently, in order to decrease the greenhouse gases in the atmosphere and prevent climate change, alternative renewable energy sources were suggested in the Kyoto protocol. Biomass is one of the most important potential sources of renewable energy because it can be found almost everywhere and can be stored. Thus, it can be said that biomass is a nonstop energy source compared to other sources. Although the initial applications of anaerobic digesters were for the stabilization and treatment of waste sludge, anaerobic digestion systems are also a source of renewable energy and the most substantial part of the production of biogas. These systems are natural phenomena in which organic matter is converted by various microorganisms in an oxygen-free environment into biogas, including methane and carbon dioxide (Aydin et al., 2015).

Biogas is utilized by the main conversion processes of anaerobic digesters. Organic materials such as manure, food scraps, crop residue, algae, or wastewater sludge are fed into the digester and stirred for 30–60 days, slowly producing a combination of methane, carbon dioxide, and other gases, which are known as biogas. Biogas can be used for power generation, heating, electricity, and cooling needs or piped into the natural gas grid. After the completion of biogas production, high-quality fertilizer called digestate is produced from the wastes, and all processes start again. Therefore, biogas can be used for various purposes such as household, commercial, and industrial applications (Yuan and Zhu, 2016).

21.5 FEEDSTOCKS FOR BIOGAS PRODUCTION

Biomass materials supply feedstocks for several biogases, end-products, and end-uses. Although all kinds of organic waste containing carbohydrates, fats, lipids, cellulose, and hemicelluloses can be used as a substrate for the anaerobic metabolism process, the majority of biomass for biogas feedstocks is classified in three main sources, such as forests, agriculture, and waste. In addition, nonforest lands including grasslands, savannahs, and algaculture (cultivation of algae) are potential sources of biogas feedstocks. On the other hand, biomass sources can be divided into two extensive groups: woody and nonwoody. While forests provide just woody materials, agriculture wastes provide both woody and nonwoody biomass for biogas production as seen in Figure 21.2 (Williams, 2011). In addition to types of feedstocks, retention time and the digestion system are important for the composition of the biogas and methane yield. The theoretical biogas yield depends on the carbohydrate, protein, and fat content of the substrate. Agricultural crop wastes and residue, municipal solid waste, sewage, forestry crops and residue, industrial residue, and animal residue have been mostly used as a biomass source in anaerobic digestion processes. Moreover, these substrates can be used with the additional cosubstrates to increase the substances of organic material for efficient biogas production (Fliegerova et al., 2012; Aydin 2016a).

After treatment of municipal wastewater, a large amount of sewage sludge is produced. Sewage sludge, also called wastewater sludge, contains dry matter consisting of nontoxic organic compounds and microbiological sludge. Therefore, the use of sewage in anaerobic digesters is one of the most significant choices for biogas production. Wastewater sludge can be stabilized, converted into volatile compounds and biogas by anaerobic digestion; then, the biogas can be used as an energy source at wastewater treatment plants or somewhere else. Production of biogas from sewage sludge is already carried out worldwide on small, medium, and large scales (Kamali et al., 2016; Yuan et al., 2016).

Industrial wastes such as pharmaceuticals, personal care products, steroid hormones, surfactants, some industrial chemicals, pesticides, and biopesticides can also be considered an important feedstock for biogas production. Because industrial wastes have wide ranges of organic matters and high chemical oxygen demand (COD) levels, they are favorable alternatives for anaerobic biogas production. Moreover, anaerobic digestion is a very useful process for high COD containing wastewaters. Because most of these compounds are resistant, they cannot be completely removed by conventional wastewater treatment plants. Thus, the use of industrial waste in anaerobic digesters as a feedstock is comparatively important to induce undesirable ecological effects on water quality and the environment (Ennouri et al., 2016; Suksong et al., 2016).

Another common source of biomass for biogas production is forestry crops and residues. There are many sources of crop and forest residues, and they have different amounts of biomass remaining

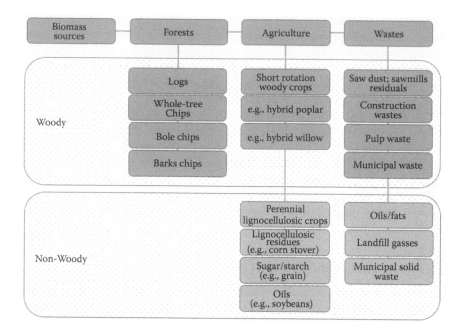

FIGURE 21.2 Sources and types of biomass materials for conversion into biogas.

after harvest. Small trees, branches, tops, and unmerchantable wood are usually considered the main sources of forest residues. They can be considered feasible substrates for biogas, which is one of the high value products. After timber harvesting, they can be collected and used for energy production. Conversion of forestry residues, an important lignocellulosic biomass to biogas, is quite important for environment. Since biogases are cleaner-burning than fossil fuels, they can enhance energy security and reduce greenhouse emission (Rudie et al., 2016).

Codigestion has environmental, economic, and technological advantages (Croxatto Vega et al., 2014). Reduction in the concentration of toxic compounds, the ability to set the necessary nutrients and their amounts, the synergistic growth effect among anaerobic bacteria, the increase in biogas production, and the higher rate of digestion are some benefits of the synergistic impacts of different organic substrates (Sosnowski et al., 2003). Sugar beet, maize silage, beet residues, fodder beet, molasses, and glycerine have been recommended by Leuenberger and Jungbluth (2009) as cosubstrates. Substrates should be free of harmful materials such as pathogens or antibiotics, which could reduce the digestion efficiency (Deublein and Steinhauser, 2010). Harmful substances could also restrict the applicability of the fermentation residue as a fertilizer. Organic content should match the fermentation process chosen, and nutritional values should be high enough to ensure high gas generation (Deublein and Steinhauser, 2010).

Organic materials from the sea are new substrates that have recently attracted the interest of scientists. Micro and macro algae together with Cyanobacteria are important aquatic carbon sources (Dębowski et al., 2013). Algae produce large amounts of biomass in a short time under favorable ecological conditions. Having less cellulose than plants and no lignin makes them easier to digest. Algae for biogas purposes can be cultivated or harvested from the sea (Hansson et al., 2012). Common reeds have been recommended as a new substrate for biogas production, through a pilot and full-scale study in Sweden (Risén et al., 2013). The waste and wastewater from the fish industry can be a source of biogas production but because of a high amount of proteins and fats the methane production process is combined with risks of ammonia and long chain fatty acid accumulation inhibiting the process (Palatsi et al., 2009; Nges et al., 2012). Codigestion with a carbohydrate-rich substrate overcomes this limitation (Cuetos et al., 2008).

21.6 ENHANCEMENT OF BIOGAS PRODUCTION ON ANAEROBIC DIGESTERS

Lignocelluloses contain a large range of biomass, especially animal manures, municipal solid waste, crop residues, forest residues, energy crops, and algal biomass (Williams, 2011; Rudie et al., 2016). Hemicellulose, lignin, extractives, and several inorganic materials are found in lignocelluloses. Cellulose, which is also called β-1-4-glucan, is a linear polysaccharide polymer of glucose made from cellobiose units. Hydrogen bond found between the cellulose chain and this structure is defined as "elementary and microfibrils." These fibrils are attached to each other by hemicelluloses, amorphous polymers of different sugars, pectin, and covered by lignin. This special and complicated structure makes cellulose resistant to conventional treatment methods. Biodegradation of lignin is particularly difficult because it is a considerably complex molecule made of phenylpropane units linked in a three-dimensional structure. If a molecule has a high proportion of lignin, this molecule is highly resistant to chemical and enzymatic degradation. Lignin, hemicellulose, and cellulose are linked with chemical bonds that are quite difficult to break (Watanabe et al., 2016).

Pretreatment methods are the most commonly used techniques for increasing biogas production from woody biomass sources in anaerobic digesters (Nitsos et al., 2016; Ennouri et al., 2016). Hydrolysis performance can be improved by pretreatment methods rates owing to changes in the chemical and physical structures of the lignocellulosic materials. Pretreatment, a well-investigated process for biogas production from lignocellulosic materials, can be physical, chemical, or biological. In general, all pretreatment methods pretreat waste materials and improve the biodigestibility of the wastes for biogas production and accessibility of the enzymes to the materials (Nitsos et al., 2016). As a consequence of pretreatment, adverse effects of difficult biodegradable lignocellulosic compounds can be prevented and biogas production on anaerobic digesters can be improved. In addition to biogas, bioethanol production can be obtained during pretreatment process. In the literature, there are some physical pretreatment methods such as pyrolysis, mechanical comminution, and chemical pretreatment methods like ozonolysis and acid hydrolysis. Despite the fact that these pretreatment methods could improve biogas, especially methane yield, the energy cost of these pretreatment technologies was quite high. Moreover, if the thermochemical pretreatment methods are used, they can caused a probable configuration of inhibitory substances. However, biological pretreatment methods that contain enzymes and microorganisms that naturally convert lignocellulosic compounds to their natural habitat and under physiological conditions are notable alternatives for improving the biogas potential of anaerobic digester (Ennouri et al., 2016; Sun and Schnürer, 2016; Yildirim et al., 2017).

Bioaugmentation, a method of the addition of specific microorganisms, selected strains, or mixed cultures, is a considerably important biological pretreatment method (Hu et al., 2016). Bioaugmentation is used in anaerobic digesters to enhance the hydrolysis yield, nutrient recovery, and biogas production from woody biomass (Aydin, 2016b). It can be an efficient alternative process for overcoming troubles and improving the performance of biogas production in anaerobic biological treatments. For example, Angelidaki and Ahring (2000) discovered that the biogas potential of animal manure is increased by bioaugmentation with the hemicellulose degrading bacterium B4. It was shown that there is an increase of approximately 30% in methane potential over controls thanks to biological treatment of animal manure. Aydin (2016a) investigated how bioaugmentation with the bacterium *Clostridium thermocellum* at various inoculum ratios affects the CH_4 production from microalgae. This study showed that bioaugmentation with *Clostridium thermocellum* provided increased microalgae biomass and caused an 18%–38% increase in methane production due to increased cell disruption. Although anaerobic digesters are bioaugmented with various kinds of microorganisms, rumen microorganisms are considered one of the most important biological pretreatment options for the improvement of biogas production in recent years.

Despite the fact that some studies have been conducted to improve biogas production from macroalgae, there is no study about biological pretreatment or bioaugmentation. Nielsen and Heiske (2011) researched the effects of mechanical pretreatment in the form of maceration on the methane yield of *Ulca lactuca*. Montingelli et al. (2015) demonstrated that improvement of methane yield in anaerobic digesters fed with macroalgae depends upon the type of pretreatment and algal species. Although this study showed the effects of physical pretreatment on anaerobic digestion, it is also revealed that the effects of different pretreatments under optimal AD parameters must be investigated for enhancement methane production from macroalgae.

Rumen is a natural cellulose-degrading system in mammalian animals. It is like a large fermentation room in which the microbial population helps to digest the herbivorous' diet. Budiyono et al. (2014) studied the effects of animal ruminal fluid on improving the biogas production rate from cattle manure at mesophilic conditions. A series of laboratory experiments was carried out in batch digesters. Because of the study, it was observed that the ruminal fluid inoculated to biodigesters considerably affected the biogas production. When compared to manure substrate that was not bioaugmented with ruminal fluid, it was proven that ruminal fluid inoculums induce the biogas production rate and efficiency increases more than two times. Jin et al. (2014) studied anaerobic fermentation of biogas liquid pretreated maize straw by rumen microorganisms in vitro. They found that rumen microorganisms have feasible and efficient influence on anaerobic hydrolytic acidification of biogas liquid pretreated maize straw. Some studies have evaluated the bioaugmentation of anaerobic digestion processes with rumen anaerobic bacteria. Cirne et al. (2006) indicated that bioaugumentation with an anaerobic rumen bacterium on anaerobic digestion improved the hydrolysis of the lipid fraction. The bioaugmenting lipolytic bacterium strain (*Clostridium lundense*) was isolated from bovine rumen fluid. It was showed that anaerobic digesters, which are bioaugmented with anaerobic rumen strain, provided an increase in the methane production rate and accordingly, a reduction in the digestion period required to achieve the same methane yield as the control. Fliegerova et al. (2012) worked on potential influence of anaerobic fungi on biogas production. Batch, semi continuous, and continuous reactors fed with anaerobic slurry, grass, and maize silage were bioaugmented with anaerobic rumen fungi. This study showed that there is a positive effect of rumen anaerobic fungi on biogas amount and quality. The biogas amount was improved by anaerobic fungi by 9% up to 18%; the methane ratio in biogas was higher by about 2.5% depending on the used substrate and species of rumen fungi. It was considered that efficient hydrolysis due to rumen fungi caused improvement in the degradation of substrates and provided the highly effective biogas yield and quality.

21.7 BIOGAS PURIFICATION

The amount of biogas produced in an anaerobic digestion unit burns in cogeneration units for supply heat and electricity demand of the process. The remaining amount can be purified to biomethane. Produced methane then can be fed into the natural gas grid or pumping stations. In addition to other benefits, purified biomethane causes a reduction in greenhouse gas emission when used as vehicle fuel (Holm-Nielsen et al., 2009). It emits less nitrogen oxide, hydrocarbon, and carbon monoxide than gasoline and diesel (Zhao et al., 2010). Biogas can be purified using different techniques.

Scrubbing is a simple and effective technique for the removal of both CO_2 and H_2S gases. Water scrubbing is a process that benefits from the higher solubility of CO_2 and H_2S to remove these gases from the biogas (Lien et al., 2014). Within a packed column in which biogas is pressurized and fed to the bottom and water is fed on the top, a purely physical adsorption process happens countercurrently. Selective removal of H_2S can also take place via water scrubbing since the H_2S is more soluble than carbon dioxide in water. Scrubbing with polyethylene glycol also happens with a physical adsorption process. CO_2 and H_2S are more soluble than methane in this solvent (Biernat and Samson-Bręk, 2011). Because of more solubility of CO_2 and H_2S in the solvent than water, a lower solvent demand and pumping is needed. A disadvantage of using water in the scrubbing process is the decline in pH as a result of CO_2 dissolution that leads to equipment corrosion caused by H_2S (Naegele et al., 2013).

Chemical adsorption as another purification method is based on the formation of reversible bonds between the solution and solvent (Zulkeflia et al., 2016). Aqueous solutions of amines (i.e., mono-, di- or tri-ethanolamine) and alkaline salts (e.g., NaOH, KOH, and $Ca(OH)_2$) are the most common solvents with negligible CH_4 loss. During the process, CO_2 is emitted to the atmosphere. Amino solution used in the process must be replaced a few times a year and thus it needs to be managed as a waste.

Pressure swing adsorption (PSA) is one of the most common technologies of biogas purification and can enrich methane up to 97% (Patterson et al., 2011). PSA is used to separate gas species from a gas mixture using pressure proportional to the molecular characteristics of the species and their affinity to the used adsorbent materials (e.g., zeolites and granular activated carbon) (Grande, 2012). The membrane separation technique; the biological process for H_2S removal using chemotropic bacterial species, microalgae cultures and anaerobic phototrophic bacteria; and the cryogenic separation method are other purification methods used for the enrichment of the produced methane in biogas production sites.

21.8 BIOGAS, FUTURE, AND SUSTAINABLE DEVELOPMENT

The environment and social-economic stability should be considered and balanced well with energy issues. Developing countries hope to supply most of their energy demands from renewable energy systems. The EU energy policies have been fixed to supply 20% of the union energy demand from renewable energy by the year 2020 (Holm-Nielsen et al., 2009). Biogas can be used to produce heat and electricity and fed into a natural gas grid or used as a transport fuel. Research, development, and implementation programs have been actively initiated to increase the use of biogas. Early work in this field focused on suitable feedstock and full-scale biogas purifying technologies. Recently, using energy crops and algae as feedstock and using the thermochemical process to create renewable natural gas have developed new interests. Biogas production is supportive of agriculture; with new adaptations in technology, it can improve the status of farmers (Minde et al., 2013).

Methane as a second-generation biofuel has found a special position in the future of biofuels (Biernat and Samson-Bręk, 2011). Therefore, scientific research is needed to develop effective technology for different operational conditions using different feedstock. Future research directions on biogas technology should include different aspects of this technology such as the most effective biogas cleaning technology, low-cost energy crops, cosubstrate systems, biogas market, and scientific innovations. Future energy crops can make the potential availability of biofuel almost unlimited. The future of biofuel will also depend on the development of markets for renewable resources. Because of the high price of oil, renewable natural gas is a low-cost option for the transportation sector, and this sector will play a main role in the future of biofuel production. Finally, technology, market advances, and energy policy determine the success and future of the biogas sector.

21.9 CONCLUSION

Environmentally friendly and low-cost energy production has gained importance in the world. Thus, many countries have begun to use renewable energy in recent years. Utilization of biogas from biomass, which is one of the most important alternatives in renewable energy sources, became an emerging application around the world. A conspicuous benefit of anaerobic digestion is the production of methane-rich biogas. However, a major problem in energy production from cellulosic biomass is the negative effect on the performance of anaerobic digesters and reduction in yield. Therefore, various pretreatment techniques have attracted attention in improving the biogas potential of anaerobic digester systems. Although effects of different bacterial species on biogas production have been studied, effects on anaerobic fungi in the process are still unknown.

REFERENCES

Angelidaki, I. and Ahring, B. K. 2000. Methods for increasing the biogas potential from the recalcitrant organic matter contained in manure. *Water Science and Technology* 41: 189–194.

Angelidaki, I. and Ahring, B. K. 1994. Anaerobic thermophilic digestion of manure at different ammonia loads: effect of temperature. *Water Research* 28: 727–731.

Aydin, S., Ince, B., and Ince, O. 2015. Application of real-time PCR to determination of combined effect of antibiotics on Bacteria, Methanogenic Archaea, Archaea in anaerobic sequencing batch reactors. *Water Research* 76: 88–98.

Aydin, S. 2016a. Enhancement of microbial diversity and methane yield by bacterial bioaugmentation through the anaerobic digestion of *Haematococcus pluvialis*. *Applied Microbiology and Biotechnology* 100: 5631–5637.

Aydin, S. 2016b. Microbial sequencing methods for monitoring of anaerobic treatment of antibiotics to optimize performance and prevent system failure. *Applied Microbiology and Biotechnology* 100: 5313–5321.

Aydin, S., Yıldırım, E., Ince, O., and Ince, B. 2017. Rumen anaerobic fungi create new opportunities or enhanced methane production from microalgae biomass. *Algal Research* 23: 150–160.

Béline, F., Rodriguez-Mendez, R., Girault, R., Le Bihan, Y., and Lessard, P. 2017. Comparison of existing models to simulate anaerobic digestion of lipid-rich waste. *Bioresource Technology* 226: 99–107.

Biernat, K. and Samson-Bręk, I. 2011. Review of technology for cleaning biogas to natural gas quality. *Chemik* 65(5): 435–444.

Budiyono, B., Widiasa, I. N., Johari, S., and Sunarso, S. 2014. Increasing biogas production rate from cattle manure using rumen fluid as inoculums. *International Journal of Science and Engineering* 6: 31–38.

Croxatto Vega, G. C., ten Hoeve, M., Birkved, M., Sommer, S. G., and Bruun, S. 2014. Bioresource technology choosing co-substrates to supplement biogas production from animal slurry: a life cycle assessment of the environmental consequences. *Bioresource Technology* 171: 410–420.

Cavinato, C., Da Ros, C., Pavan, P., and Bolzonella, D. 2017. Influence of temperature and hydraulic retention on the production of volatile fatty acids during anaerobic fermentation of cow manure and maize silage. *Bioresource Technology* 223: 59–64.

Chen, Y., Cheng, J. J., Creamer, K. S. 2008. Inhibition of anaerobic digestion process: a review. *Bioresource Technology* 99: 4044–4064.

Chae, K. J., Jang, A. M., Yim, S. K., and Kim, I. S. 2008. The effects of digestion temperature and temperature shock on the biogas yields from the mesophilic anaerobic digestion of swine manure. *Bioresource Technology* 99(1), 1–6.

Cirne, D. G., Björnsson, L., Alves, M., and Mattiasson, B. 2006. Effects of bioaugmentation by an anaerobic lipolytic bacterium on anaerobic digestion of lipid-rich waste. *Journal of Chemical Technology and Biotechnology* 8: 1745–1752.

Cuetos, M. J., Gómez, X., Otero, M., and Morán, A. 2008. Anaerobic digestion of solid slaughterhouse waste (SHW) at laboratory scale: Influence of co-digestion with the organic fraction of municipal solid waste (OFMSW). *Biochemical Engineering Journal* 40:99–106.

Dębowski, M., Zieliński, M., Grala, A., Dudek, M. 2013. Algae biomass as an alternative substrate in biogas production technologies-review. *Renewable and Sustainable Energy Reviews* 27:596–604.

Deublein, D., Steinhauser, A. 2010. Biogas from Waste and Renewable Resources: An Introduction, 2nd, Revised and Expanded Edition. Wiley-VCH Verlag GmbH & Co. kGaA. ISBN: 978-3-527-32798-0.

Ennouri, H., Miladi, B., Diaz, S. Z., Güelfo, L. A. F., Solera, R., Hamdi, M., and Bouallagui, H. 2016. Effect of thermal pretreatment on the biogas production and microbial communities balance during anaerobic digestion of urban and industrial waste activated sludge. *Bioresource Technology* 214: 184–191.

Fliegerova, K., Prochazka, J., Mrazek, J., Novotna, Z., Strosová, L., and Dohanyos, M., 2012. *Biogas and Rumen Fungi. Biogas: Production, Consumption and Applications*, Nova Science Publishers Inc., Hauppauge, pp. 161–180.

Grande, C. A. 2012. Advances in pressure swing adsorption for gas separation, *ISRN Chemical Engineering* 2012: 100–102.

Hansson, A., Tjernström, E., Gardin, M., and Finnis, P. 2012. Wetlands algae biogas – A southern Baltic Sea eutrophication counter act project. Municipality of Trelleborg, Trelleborg, Sweden.

Herbert, G. J., Krishnan, A. U. 2016. Quantifying environmental performance of biomass energy. *Renewable and Sustainable Energy Reviews* 59: 292–308.

Holm-Nielsen, J. B., Al Seadi, T., and Oleskowicz-popiel, P. 2009. The future of anaerobic digestion and biogas utilization. *Bioresource Technology* 100(22): 5478–5484.

Hosseini, S. E. and Wahid, M. A. 2016. Hydrogen production from renewable and sustainable energy resources: promising green energy carrier for clean development. *Renewable and Sustainable Energy Reviews* 57: 850–866.

Hu, Y., Hao, X., Wang, J., and Cao, Y. 2016. Enhancing anaerobic digestion of lignocellulosic materials in excess sludge by bioaugmentation and pre-treatment. *Waste Management* 49: 55–63.

Ianotti, E. L. and Fischer, J. R. 1983. *Effects of Ammonia, Volatile Acids, pH and Sodium on Growth of Bacteria Isolated from a Swine Manure Digester, Proceedings of the 40th General Meeting Society of Industrial Microbiology*, Orlando, FL.

Jin, W., Xu, X., Gao, Y., Yang, F., and Wang, G. 2014. Anaerobic fermentation of biogas liquid pretreated maize straw by rumen microorganisms in vitro. *Bioresource Technology* 153: 8–14.

Kamali, M., Gameiro, T., Costa, M. E. V., and Capela, I. 2016. Anaerobic digestion of pulp and paper mill wastes–An overview of the developments and improvement opportunities. *Chemical Engineering Journal* 298: 162–182.

Labatut, R. A. and Gooch, C. A. 2012. Monitoring of anaerobic digestion process to optimize performance and prevent system failure. Proceedings of got manure? *Enhancing Environmental and Economic Sustainability* 14: 209–225.

Leuenberger, M. and Jungbluth, N. 2009. *LCA of Biogas Production from Different Substrates-Pre-Evaluation of Substrates*. ESU-services im Auftrag des Bundesamts für Energie BFE, Uster, CH.

Lien, C., Lin, J. and Ting, C. 2014. Water scrubbing for removal of hydrogen sulfide (H2S) inbiogas from hog farms, *Journal of Agricultural Chemistry and Environment* 3:1–6.

Madigan, M. T., Martinko, J. M., and Parker, J. 2002. *Brock Biology of Microorganisms*, Prentice Hall, Inc., Upper Saddle River, NJ.

Metcalf & Eddy, Inc.. 2003. *Wastewater Engineering, Treatment and Reuse*. McGraw-Hill, New York.

Minde, G. P., Magdum, S. S., and Kalyanraman, V. (2013). Biogas as a sustainable alternative for current energy need of India, *Journal of Sustainable Energy & Environment* 4: 121–132.

Montingelli, M. E., Tedesco, S., and Olabi, A. G. 2015. Biogas production from algal biomass: a review. *Renewable and Sustainable Energy Reviews* 43: 961–972.

Naegele, H., Lindner, J., Merkle, W., Lemmer, A., Jungbluth, T., and Bogenrieder, C. 2013. Effects of temperature, pH and O_2 on the removal of hydrogen sulfide from biogas by external biological desulfurization in a full scale fixed-bed trickling bioreactor (FBTB). *International Journal of Agricultural and Biological Engineering* 6(1): 69–81.

Nielsen, H. B. and Heiske, S. 2011. Anaerobic digestion of macroalgae: methane potentials, pre-treatment, inhibition and co-digestion. *Water Science and Technology* 64: 1723–1729.

Nitsos, C. K., Choli-Papadopoulou, T., Matis, K. A., and Triantafyllidis, K. S. 2016. Optimization of hydrothermal pretreatment of hardwood and softwood lignocellulosic residues for selective hemicellulose recovery and improved cellulose enzymatic hydrolysis. *ACS Sustainable Chemistry and Engineering* 4: 4529–4544.

Nges, I. A., Mbatia, B., and Björnsson, L. 2012. Improved utilization of fish waste by anaerobic digestion following omega-3 fatty acids extraction. *Journal of Environmental Management* 110: 159–165.

Orpin, C. G. and Joblin, K. N. 1997. The rumen anaerobic fungi. In: Hobson P.N., Stewart C.S. (eds) *The Rumen Microbial Ecosystem*. Springer, Dordrecht, The Netherlands, pp. 140–195.

Palatsi, J., Laureni, M., Andrés, M. V., Flotats, X., Nielsen, H. B. and Angelidaki, I. 2009. Strategies for recovering inhibition caused by long chain fatty acids on anaerobic thermophilic biogas reactors. *Bioresource Technology* 100: 4588–4596.

Patterson, T., Esteves, S., Dinsdale, R., and Guwy, A. 2011. An evaluation of the policy and techno-economic factors affecting the potential for biogas upgrading for transport fuel use in the UK. *Energy Policy* 39(3): 1806–1816.

Pehnt, M. 2006. Dynamic life cycle assessment (LCA) of renewable energy technologies. *Renewable Energy* 31: 55–71.

Rincón, B., Borja, R., González, J. M., Portillo, M. C. and Sáiz-Jiménez, C. 2008. Influence of organic loading rate and hydraulic retention time on the performance, stability and microbial communities of one-stage anaerobic digestion of two-phase olive mill solid residue. *Biochemical Engineering Journal* 40(2): 253–261.

Risén, E., Gregeby, E., Tatarchenko, O., Blidberg, E., Malmström, M. E, Welander, U. and Gröndahl, F. 2013. Assessment of biomethane production from maritime common reed. *Journal of Cleaner Production* 53:186–194.

Rudie, A. W., Houtman, C. J., Groom, L. H., Nicholls, D. L. and Zhu, J. Y. 2016. A survey of bioenergy research in forest service research and development. *Bioenergy Research* 9: 534–547.

Si, B. C., Li, J. M., Zhu, Z. B., Zhang, Y. H., Lu, J. W., Shen, R. X. and Liu, Z. 2016. Continuous production of biohythane from hydrothermal liquefied cornstalk biomass via two-stage high-rate anaerobic reactors. *Biotechnology for Biofuels* 9: 254.

Sosnowski, P., Wieczorek, A. and Ledakowicz, S. 2003. Anaerobic codigestion of sewage sludge and organic fraction of municipal solid wastes. *Advances in Environmental Research* 7:609–616. doi:10.1016/S1093-0191(02)00049-7.

Stover, E. L., Brooks, S., and Munirathinam, K. 1994. Control of biogas H_2S concentrations during anaerobic treatment. American Institute for Chemical Engineers Symposium Series No. 300, p. 90.

Stronach, S. M., Rudd, T., and Lester, J. N. 1986. *Anaerobic Digestion Process in Industrial Wastewater Treatment*, Springer, Berlin.

Suksong, W., Kongjan, P., Prasertsan, P., Imai, T., and Sompong, O. 2016. Optimization and microbial community analysis for production of biogas from solid waste residues of palm oil mill industry by solid-state anaerobic digestion. *Bioresource Technology* 214: 166–174.

Sun, L. and Schnürer, A. 2016. Draft genome sequence of the cellulolytic strain *Clostridium* sp. Bc-iso-3 isolated from an industrial-scale anaerobic digester. *Genome Announcements* 4: e01188-16.

Watanabe, R., Nie, Y., Takahashi, S., Wakahara, S., and Li, Y. Y. 2016. Efficient performance and the microbial community changes of submerged anaerobic membrane bioreactor in treatment of sewage containing cellulose suspended solid at 25°C. *Bioresource Technology* 216: 128–134.

Williams, C. L. 2011. *Wisconsin Grassland Bioenergy Network*, Sources of Biomass.

Wu, X., Yao, W., Zhu, J., and Miller, C. 2010. Biogas and CH_4 productivity by co-digesting swine manure with three crop residues as an external carbon source. *Bioresource Technology* 101: 4042–4047.

Yen, H. W. and Brune, D. E. 2007. Anaerobic co-digestion of algal sludge and waste paper to produce methane. *Bioresource Technology* 98: 130–134.

Yuan, H. and Zhu, N. 2016. Progress in inhibition mechanisms and process control of intermediates and by-products in sewage sludge anaerobic digestion. *Renewable and Sustainable Energy Reviews* 58: 429–438.

Yıldırım, E., Aydin, S., Ince, O., and Ince, B. 2017. Improvement of biogas potential of anaerobic digesters using rumen fungi. *Renewable Energy*, doi: 10.1016/j.renene.2017.03.021.

Zahedi, S., Solera, R., Micolucci, F., Cavinato, C. and Bolzonella, D. 2016. Changes in microbial community during hydrogen and methane production in two-stage thermophilic anaerobic co-digestion process from biowaste. *Waste Management* 49: 40–46.

Zhao, Q., Leonhardt, E., MacConnell, C., Frear, C., and Chen, S. 2010. *Purification Technologies for Biogas Generated by Anaerobic Digestion*, CSANR Research Report-001, 2010.

Zulkeflia, N. N., Masdar, M. S., Jahim, J., and Majlan, E. H. 2016. Overview of H_2S removal technologies from biogas production, *International Journal of Applied Engineering Research* 11(20): 10060–10066.

22 Human Energy
System Integration, Efficiency, Recovery

José Antonio Aguilar Becerril, Diana Gabriela Pinedo Catalán, and Paola Yazmín Jiménez Colín
Mexico State University

Jaime Manuel Aguilar Becerril
Mexican Social Security Institute

CONTENTS

22.1 ENERGY SYSTEMS IN THE HUMAN BODY

During physical work, all systems and organs of the human body are involved: the muscular system is the effector of the central nervous system, which requires the other systems (cardiovascular, pulmonary, endocrine, renal, and others) to supply muscle tissue with energy to maintain motor activity.

Muscle contraction during physical exercise is possible due to energy transformations from chemical energy stored in the bonds of molecules in different metabolic substrates into mechanical energy. Adenosine triphosphate (ATP) is an intermediary molecule throughout these processes. In this transformation, much energy is released as thermal energy. The increase in temperature in different metabolic reactions, mediated by enzymatic complexes, makes these reactions more efficient energy wise. In practical terms, this justifies warmup before training. The metabolic substrates that allow the production of ATP come from bodily reserves or daily food intake. The substrates most used in the different metabolic pathways during physical exercise are carbohydrates and fatty acids.

Three energy systems in which ATP resynthesis takes place depend on the intensity and duration of physical exercise. These are the phosphagen, glycolysis, and aerobic systems.

22.1.1 THE PHOSPHAGEN ANAEROBIC SYSTEM

The phosphanenic system or alactical anaerobic system provides energy at the beginning of any physical activity and for activities of very high intensity and short duration. The most important substrates are ATP and phosphocreatine (PC); others are adenosine diphosphate (ADP), adenosine monophosphate (AMP) guanosine 5′ triphosphate (GTP), and uridine 5′ triphosphate (UTP). All have high-energy phosphate bonds.

ATP is hydrolyzed by the enzyme ATPase located in the myosin heads of the muscle; these trigger the displacement of actin that results in contraction. The energy released in the hydrolysis of one ATP molecule during exercise is approximately 7300 calories (this varies depending on temperature and muscle pH). ATP hydrolysis

$$ATP + H_2O = ADP + P$$

releases energy for muscle work, metabolic syntheses, and other cellular functions. Cell reserves will deplete in 1 s during physical exertion. PC allows for fast resynthesis of ATP after its use, catalyzed by creatine kinase, which is activated with increasing ADP concentration:

$$\uparrow ADP + PC + H^+ = ATP + C$$

PC in muscle cells is depleted in 2 s during very intense exercise if the cell has only this substrate to maintain an effort.

22.1.2 ANAEROBIC GLYCOLYSIS

Through this system, only carbohydrates can be metabolized in the cytosol of a muscle cell. Energy is obtained without involving oxygen directly, and 2 ATP can be resynthetized per molecule of glucose. It provides energy for an effort lasting a few seconds and up to 1 min. The transport of glucose into the cell is facilitated (facilitated diffusion) by a membrane transporter called glucose transporter 4. In the later stages of prolonged exercise, when muscle glycogen and glycaemia are low, it seems that increases in free fatty acids limit the consumption of glucose. The step from glucose to glucose 6 phosphate in the muscle cell is irreversible.

During glucose catabolism to pyruvate in the cytoplasm, the net energy yield equals the synthesis of 6 molecules of ATP; 2 ATP are formed in the cytosol (by anaerobic glycolysis) and 4 ATP in mitochondria by the reoxidation of nicotinamide adenine dinucleotide (NADH). If this NADH reoxidation is not possible, pyruvate intervenes, reducing to lactic acid:

$$\text{Pyruvic acid} + \text{NADH} + \text{H}^+ = \text{Lactic acid} + \text{NAD}^+$$

Thus, through anaerobic glycolysis, only 2 molecules of ATP and 2 molecules of lactic acid are formed that cause states of metabolic acidosis, the metabolic consequence of which is muscle fatigue.

Lactic acid completely dissociates to the normal pH of the muscle cell, giving rise to lactate and protons. Protons must be buffered into the cell to maintain the acid–base state. Bicarbonate is the most widely used compound, so when hydrogen ion is added, it increases the production of carbon dioxide (CO_2) during intense exercise.

22.1.3 AEROBIC SYSTEM

Carbohydrates, fats, and to a lesser extent proteins can be used to obtain energy through the Krebs cycle. Energy is much greater than that obtained by glycolysis. In the Krebs cycle, ATP, CO_2, and hydrogen are formed, electrons are transferred to the mitochondrial respiratory chain, where they react with O_2 in order to form H_2O and generate a greater amount of energy by the coupling between oxidation and reduction reactions. Endurance training largely augments the activity of several enzymes in the Krebs cycle and respiratory chain; it also increases the number and size of mitochondria.

22.1.3.1 Carbohydrates (Oxidation of Pyruvate)

Pyruvate formed in glycolysis goes into the mitochondria and is transformed to acetyl Co-A by pyruvate dehydrogenase, thus entering the Krebs cycle. The most important function of this cycle is to generate electrons to pass through the respiratory chain, where a large amount of ATP is resynthesized through oxidative phosphorylation. The limiting enzyme is isocitrate dehydrogenase, which is inhibited by ATP and stimulated by ADP. In addition, both ADP and ATP stimulate and inhibit, respectively, the transport of the electron chain. The net energy yield of this aerobic metabolism is 36 ATP versus the 2 ATP obtained in anaerobic glycolysis.

In rest phases, glucose is stored in the body after phosphorylation into glycogen through glycogen synthetase. When performing exercises, it is necessary to obtain glucose; a process called glycogenolysis synthesizes 1 molecule of ATP, which is why the net energy efficiency is 37 ATP.

In addition to these mechanisms, gluconeogenesis is the synthesis of glucose from amino acids, glycerol; and lactate; glycogenesis is the synthesis of glucose from pyruvate, of which the former can represent up to 45% of the hepatic glucose production during exercise.

22.1.3.2 Lipids

Lipids provide a vast source of energy during exercise; their use increases with the duration of exercise. Their metabolism is purely aerobic and as energy substrate save carbohydrates whose exhaustion is related to muscle fatigue in long-duration exercises. Triglycerides from adipocytes are broken by the action of lipase (lipolysis) on glycerol and fatty acids (FA). The former acts as gluconeogenic precursor while FAs are transported to the muscle cell where after suffering a series of changes in the cytoplasm, they enter the mitochondria through a carrier, carnitine, and there beta-oxidation occurs, which results in the formation of acetyl Co-A molecules entering the Krebs cycle with a yield of 12 ATP each.

During exercise, there is an increase of adrenal sympathetic activity and decrease of insulin stimulate lipolysis. The consumption of FA depends on several factors: muscle blood flow (most important), intensity and duration of the exercise, and degree of training. Endurance training causes an increase of mitochondrial mass and increased carnitine activity.

An overall improvement of FA entry into the mitochondrial matrix depends on ketone bodies used as an energy source; enzymes involved in the use of ketones are increased in trained subjects.

22.1.3.3 Proteins

Proteins contribute 4%–15% of the total energy in long-duration exercises (>60 min) associated with increases in blood levels of the amino acids leucine and alanine, reflecting an increase in proteolytic processes at the hepatic and muscular levels. NH2 is a functional group derived from ammonia or any of its alkylated derivates by the removal of one of its hydrogen atoms. NH_2 groups are converted to urea whereas their structural carbons are transformed into pyruvate, acetyl Co-A (the acetylated form of coenzyme A, formed as an intermediate in the oxidation of carbohydrates, fats and protein in human metabolism) or some of the intermediates of the Krebs cycle. Exercise modifies three important protein metabolism processes:

1. It increases ammonium production from ATP deamination that occurs when the rate of ATP production exceeds that of formation.
2. It increases urea production in the liver in long-duration exercises, which is eliminated by urine.
3. It increases the oxidation of amino acids with negative nitrogen balance, especially branched chain ones (e.g., leucine), which are catabolized in skeletal muscle, their carbons are oxidized and the nitrogen residues participate in the formation of alanine, which acts as gluconeogenic substrate in the liver (alanine–glucose cycle).

22.2 THE MEASUREMENT OF PERFORMANCE: OXYGEN TRANSPORT AND ANAEROBIC POWER

Maximum exercise tolerance is related to depletion or accumulation of metabolites, which interfere with efficient chemomechanical energy transduction. Stimulated afferences result in an intolerably high-perceived effort. Oxygen flux from the atmosphere to its site of utilization at mitochondrion plays a prominent role in minimizing the rate at which this limitation occurs, therefore transfer inefficiencies at any site in the energy conduction chain reduce tolerance to exercise and/or maximum oxygen uptake (VO_2 max); healthy subjects become unable to continue an exercise at a point at which these stressors reach intolerable levels (Figure 22.1). VO_2 max is a quantitative measure of the capacity to resynthesize ATP aerobically: it is an important indicator of a subject's to maintain intense physical effort. VO_2 phases are the cardio dynamic phase, where circulatory transit delay from muscles to lungs occurs. In the mono exponential increase phase, VO_2 adjusts to skeletal muscle activity. In the steady state phase, VO_2 peaks. Another phase occurs during moderate exercise intensities.

22.2.1 ERGOMETRY

Ergometry is a functional overload test (Figure 22.2) that evaluates the aerobic capacity (VO_2 max). Ergometry directly uses a clinical gas analyzer; it can also use ancillary equations (taking into

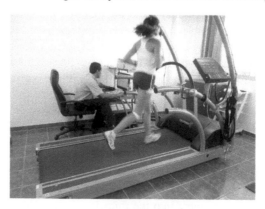

FIGURE 22.1 Velocity athlete performing endless stress test.

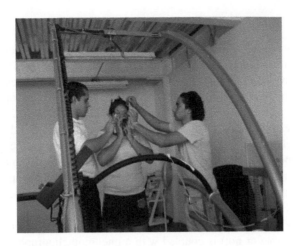

FIGURE 22.2   Evaluation of cardiorespiratory capacity in weightlifting athletes.

consideration age, gender, height, or weight). Ergometers of various kinds assess the upper or lower extremities or both. Ergometers can be automatic or manual; the latter require more operators.

Before ergometry, a medical examination is mandatory. Emergency equipment including a defibrillator and environmental control equipment (thermometer, barometer, and hygrometer) are compulsory. Adequate air movement is also indispensable. Uncontrolled environmental conditions modify cardiovascular responses and the maximum performance may decrease. Optimal conditions need to be met in the test room: humidity should be 50%, to enhance cutaneous heat loss, and temperature 22°C–26°C.

Progressive ramp tests provide an array of physiological information, reflective of work efficiency, e.g., estimated lactate threshold, VO_2 gain, mean response time. A ramp test is typically continued to the limit of subject tolerance: peak VO_2 is obtained as a proxy of VO_2 max. A treadmill (mobile platform with regulated speed and inclination and lateral support) is the most widely used apparatus, because it simulates the natural walk movement. It can be used both in athletes and in the general population. Stationary bicycles are ergometers with mechanical, electric, pneumatic, or hydraulic resistance. The physiological response to exercise on the cycle is 5%–20% lower than on the treadmill. Arm ergometers are used in subjects with upper body activity; 60–75 rpm must be maintained. Electrocardiographic recording systems allow for continuous monitoring of the heart and evaluate changes during exercise and recovery. The blood pressure response is also very important, in asymptomatic and diseased persons.

Gas analyzers are devices made up of a mask (open circuit spirometry) and a conduit that is connected to the computer; they analyze both environmental air and expired air. The size of the mask may interfere with maximum performance.

Cardiorespiratory functional capacity inaccuracies (see Figure 22.2) are associated with sedentary healthy adults and heart disease patients reaching fatigue and discomfort far below their physiologic maximum, i.e., far below the physiologic VO_2 max. Different protocols with distinct aims are used for ischemic and non-ischemic heart disease patients, in healthy people, athletes, and sedentary and disabled persons. Tests also differ for each ergometric system. Tests can be continuous or discontinuous, maximal (100% heart rate) or submaximal (85% heart rate).

The Bruce and Pugh tests are widely used. Bruce's protocol is a test for treadmill, with changes in speed and slope every 3 min. It is frequently used to diagnose and evaluate coronary heart disease and is an indirect mode to determine functional capability in sedentary people. The modified Bruce protocol develops lighter stages, for those with lower functional capability. Pugh's protocol is constant in slope, with speed increases every three minutes, frequently used in trained persons.

22.2.2 SPIROMETRY

Lung function is evaluated by spirometry; it is a general respiratory health screening. This is a physiological test that measures volumes and flows of air in a time frame, assessing the patient's capacity to expel air from the lungs after a maximal inspiration. Much can be learned about the mechanical properties of the lungs by measuring forced maximal expiration and inspiration.

The most important aspects to measure are forced vital capacity (FVC) and forced expiratory volume (FEV1) in L per second. The forced vital capacity is the maximal volume of air exhaled with forced and complete effort, starting from full inspiration. The forced expiratory volume in 1 s is the volume of air expelled in the first second of a forced expiration from a maximal inspiration. This test is mostly dependent on the correct operation and accuracy of the spirometer and the use of relevant predicted normal values, as well as the staff who is performing the spirometry (should be trained personnel) and the correct breathing maneuver.

Spirometers measure volumes only, flows only, or both (Figure 22.3). Some include an oximeter. The mouthpiece (disposable or not) is coupled with a pneumotachograph connected to a monitor displaying graphic results.

Most spirometers calculate normal values, from patient's data (gender, age, height, and weight). The simplest handheld spirometers produce FEV1 and forced expiratory vital capacity (FEVC) readings, while advanced spirometers produce traces of exhaled air's volume and a volume–time curve. Electronic devices also display flow–volume curves. There are other settings to measure, such as vital capacity, which is the maximum volume of air exhaled or inhaled during either a maximal LY forced (FVC) or a slow (VC) maneuver; if FVC and VC are equal at least an airflow obstruction is present. The relation FEV1/VC is forced expiratory volume in 1 s expressed as a percentage of the VC or FVC express airflow limitation. Peak expiratory flow (PEF) is the maximal expiratory flow rate achieved.

Forced expiratory flow (FEF) (Figure 22.4), the average (25%–75%) exhaled flow beyond the FVC maneuver, is less reproducible than FEV1. FEF50% and FEF75% are the maximal expiratory flow measured when 50% and 75% of the FVC has been expired.

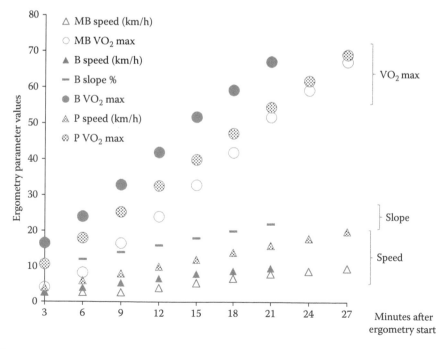

FIGURE 22.3 Ergometry tests and speed, slope, and VO_2 max parameters.

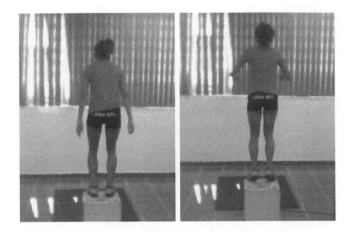

FIGURE 22.4 Resistance athlete performing power test on anaerobic platform.

Normal predicted values for ventilatory function depend on gender, age, height, and ethnic origin. Males have a larger FEV1, FVC, FEF25%–75%, and PEF, but a slightly lower FEV1/FVC. With age, FEV1, FVC, FEF25%–75%, and PEF increase, while FEV1/FVC decreases, until the age of 20 in females and 25 years in males, so all values gradually fall. All indices other than FEV1/FVC increase with standing height. Caucasians have the largest FEV1 and FVC and, of the various ethnic groups, Polynesians have among the lowest. The values for black Africans are 10%–15% lower than for Caucasians (their thorax is shorter).

22.2.3 ANAEROBIC POWER

The ability of ATP and phosphocreatine energy pathways to produce energy for muscle contraction per unit of time is known as anaerobic power (Figure 22.5). This system is used in short bursts of intense power output, inasmuch as it is depleted quickly. Anaerobic peak power is the highest mechanical power generated during any 3–5 s interval of the test. Anaerobic capacity is the total amount of work accomplished over a 30-s period.

The maximum anaerobic energy that can be used is proportional to the quantity of phosphocreatine and lactate the body can gather. Lactate accumulation is the result of an anaerobic metabolism during activity using glycogen as substrate and is not accumulated in all types and intensities of activity. The blood lactate concentration in the lactate pool is the result of lactate resulting from working muscles and various tissues and lactate disappearing in the skeletal muscles, the heart, the liver, and the kidney cortex.

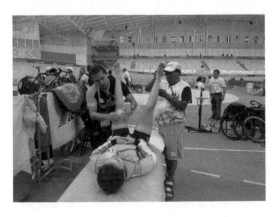

FIGURE 22.5 Recovery's work by means of thermoregulation.

The rate of ATP production by oxidative sources becomes insufficient when exercise intensity increases; anaerobic ATP production is needed, then ATP is produced by glycolysis or glycogenolysis. The endpoint is pyruvate; it can be reduced to lactate or oxidized to CO_2 or H_2O. During steady state, lactate production is equal to lactate removal; as a result, the lactate concentration in the lactate pool stays constant, and the rate of oxygen consumption is a measure of the whole body energy expenditure, despite the magnitude of lactate production and removal or the absolute blood lactate concentration. The maximal lactate steady state is the highest level of intensity where a steady state condition of lactate can be obtained, which is also referred to as lactate threshold.

Anaerobic tests are divided into those measuring anaerobic power and those measuring anaerobic capacity. The values of maximal anaerobic power obtained with different protocols (force-velocity tests, vertical jump tests, staircase tests, and cycle ergometer tests) are different but generally well correlated.

Existing equipment widely used for testing muscular strength, power, and jump height include contact mats, infrared mats, force platforms, and position transducers, with a system for measuring neuromuscular qualities. These devices have four principal parts: a contact mat with electronic sensors connected to a computer; a chronometer to measure jump partial and contact time, expressed into a graphic through mathematical formula; a strobe light; and sound equipment. The analysis is based on a shortening-stretching cycle method, which evaluates different components of the anaerobic capacity, such as contact time, flight time, jump height, reactive strength index, peak power output, and power.

Contact mats are widely used in laboratories; there are also portable devices especially designed to be used in the field. It is important to analyze the validity and reliability of all equipment used during these assessments, before using them to test muscular strength, power, and jump height.

22.3 ENERGY RECOVERY: NUTRITION

Energy and nutrient requirements increase with exercise and must be met to maintain an adequate body weight to develop optimal performance. Factors that determine the caloric requirements include intensity and type of activity; duration of exercise; age, sex, and body composition; room temperature; and degree of training.

22.3.1 CARBOHYDRATES

Carbohydrates fulfill a fundamental energy function. One gram of carbohydrates provides ~4 kcal. They are the main fuel for the muscle during physical activity; a carbohydrates-rich diet is essential for athletes, equivalent to 60%–65% of the total energy for the day. With these quantities, the reserves (in the form of glycogen) necessary for muscle contraction can be maintained.

There are two different types of carbohydrates: Simple or quick absorption monosaccharides and disaccharides are found in fruits, jams, candies, and milk (lactose). Complex or slow absorption carbohydrates are found in cereals and their derivatives (flour, pasta, rice, bread, corn, and oats), legumes (beans, lentils and chickpeas), and potatoes.

22.3.2 FAT

One gram of fat provides up to 9 kcal. A fat-rich diet (>35% of the total energy required) will be low in carbohydrates, so that glycogen storage will be insufficient. A suitable diet for an athlete should contribute 20%–25% of the total calories (exceptions are extreme cold conditions). Excess fat, especially animal fat, can increase the blood cholesterol, with negative health consequences. If fat content is low (<15%), there is a risk of deficiencies in fat-soluble vitamins (A, D, E, K) and essential fatty acids (see Table 22.1).

TABLE 22.1

Daily Vitamin and Mineral Needs by Age

		11–18 Years Male	+18 Years Male	11–15 Years Female	+15 Years Female
	Weight (kg)	45–60	70	45	55
	Protein (g)	45	56	46	44
Vitamins and folic acid	A (mg)	1	1	0.8	0.8
	D (μg)	10	6	10	6
	E (mg)	8	10	8	8
	C (mg)	50	60	50	60
	B1 (mg)	14	1.2	1.1	1
	B2 (mg)	1.6	1.4	1.3	1.2
	B3 (mg)	18	16	15	13
	B6 (mg)	1.8	2.2	1.8	2
	Folic Acid (mg)	0.4	0.4	0.4	0.4
	B12 (μg)	3	3	3	3
Main minerals	Calcium (mg)	1200	800	1200	800
	Phosphorus (mg)	1200	800	1200	800
	Magnesium (mg)	350	350	300	300
	Iron (mg)	18	10	18	10
	Zinc (mg)	15	15	15	15
	Iodo (mg)	150	150	150	150

Source: Alcocer Díaz, L.F, *Manual de sustancias prohibidas en el deporte y complementos alimenticios,* 3era edición, México, 2011.

Vegetable oil (other than palm and coconut), blue fish, and nuts have a better lipid profile. In exercise, the importance of fat as the substrate that provides energy is limited to the aerobic energy metabolism. The contribution of fat as fuel for muscle increases as the exercise duration increases and intensity decreases.

22.3.3 PROTEINS

Proteins are the structural basis of our organism. They are made of 20 different amino acids, divided into two groups: The essential amino acids (cannot be synthesized by the human body and must be provided by food) and nonessential amino acids (can be produced by our body). Good-quality proteins are rich in essential amino acids. To achieve quality animal protein, proteins from different plants should be combined. It is recommended that proteins provide around 12%–15% of the total energy of the diet. During sports training, 1.2 to 1.6 g protein/kg body weight is required. In some disciplines, athletes anxious for muscle development greatly exceed intake recommendations through supplements. An excess of proteins can cause an accumulation of toxic substances and harmful effects.

22.3.3.1 Eating and Drinking before Competition

A week before a competition two main objectives are to optimize the store of carbohydrates in muscles and liver (as glycogen) to maximize the energy reserve and to be well hydrated. The days leading up to the event are very important, and diet should be based on a high intake (65%–75%) of carbohydrates, 15%–20% fat, and 10%–12% protein. Drink preparation is determined by the type and frequency of competition.

22.3.3.2 Eating on the Day of the Competition

A meal rich in carbohydrates taken hours before the competition can help to complete the glycogen storage in the body. The liver is responsible for maintaining the plasma glucose, so it requires constant food consumption, and athletes who fast before competition (little dinner and no breakfast) and do not consume carbohydrates during competition, are at higher risk of hypoglycemia during the performance.

The intake before the competition will be high in carbohydrates and low in fat, protein, and fiber. To allow for digestion, this meal should take place 3–4 h before the performance. In the hour before competition, food in liquid form is easier and quicker to assimilate and so is highly recommended.

22.3.3.3 Eating during Exercise

During long-duration performances (>60 min), intake is based on carbohydrates. The goal is a rate of 40–60 g per hour, as carbohydrates help to delay the onset of fatigue and maintain performance in the last phases of physical effort. In sports like cycling or sailing, it is possible to take solid foods in the form of dry fruit, bananas, seeds, and so forth. Sports drinks help replace the electrolytes and liquids lost by sweating, prevent dehydration, and provide carbohydrates.

22.3.3.4 Feeding after Exercise

Following the exercise sports drinks, proteins and carbohydrate-rich food are recommended (1 g carbohydrates/kg body weight). The immediate objective is to restore hepatic and muscular reserves of glycogen and liquid losses. The important point is to choose food with a moderate to high glycemic index for fast refueling. Suitable foods include pasta, noodles, rice, and cooked or roasted potatoes; as far as possible, fatty foods (fried foods, breaded, braised) should be avoided, since they slow carbohydrate replenishment and can produce gastrointestinal discomfort.

22.4 ENERGY RECOVERY: QUALITY OF SLEEP

While sleeping, the organism rests, recovers energy, and assimilates training loads. The heart benefits from growth hormones and melatonin. Material conditions let the athlete enjoy and maximally benefit from a few hours of quality sleep. Some of the benefits of sleep are

Strengthened immune system
Lessened heart and circulatory effort, lower blood pressure
Total relaxation of the locomotor system: muscles, joints, and back are released from tensions
Decreased energy expenditure, saving energy for the day
Slower and deeper breathing, improving oxygenation of all cells in preparation for effort
Decreased heart rate; coronary cells benefit from the restorative action of growth hormone and melatonin.

The restriction of nighttime sleep to <6 h for a period of more than four consecutive days has been shown to be detrimental in terms of cognitive function, performance and mood, appetite regulation, glucose metabolism, and immune function. Current recommendations are that adults sleep 8 h per night to avoid neurobehavioral alterations.

Polysomnography (PSG) measures body functions such as brain electric and cardiac activities, eye movements, and muscle activity. Thus the PSG provides information on the stages of sleep and is considered the reference standard to measure quantity and quality of sleep. In addition, the Insomnia Severity Index identifies the absence of clinical insomnia or levels of insomnia. The Pittsburgh Sleep Quality Index evaluates sleep quality through seven components: subjective quality of sleep, latency (amount of time taken to fall asleep), duration, efficiency, alterations, use of hypnotic medication, and daytime dysfunction.

TABLE 22.2
Impacts of the Duration of Sleep Deprivation

Duration (h)	Impacts
24	Decreased aerobic performance, as repetitive or sustained efforts are more affected than maximum effort
	Altered mood state profile: vigor, fatigue and general mood
30	Decreased anaerobic performance and sprinting time
36	Decreased maximum power

22.4.1 EFFECTS OF POOR SLEEP ON PERFORMANCE

Although the mechanism of performance reduction is unclear, increased perception of effort seems to be a major cause. It is also known that most athletes experience acute bouts of partial sleep deprivation, where sleep is reduced for several consecutive hours. Partial sleep deprivation inhibits psychomotor functions and significantly affects maximal and submaximal tests but does not appear to restrict gross motor functions (muscle strength, pulmonary power, and resistance) (Table 22.2).

22.4.2 EFFECTS OF INCREASED SLEEP AND NAPS

It has been observed that by obtaining as much extra sleep as possible, faster times have been recorded in sprint tests and more accuracy in throws. The benefits from a brief nap are particularly evident if evening training sessions are held. After a 30-min nap, sprinting times and wakefulness improve.

Naps can reduce drowsiness and be beneficial when learning skills, strategies, or tactics in athletes with lack of sleep. Athletes who get up early can benefit from naps too. This is because even with 8h in bed, latency can increase and efficiency of sleep decrease, at some stages of training. During competitions, up to 80% of athletes may have increased latency, wake up early, wake up at night, and present nervousness and anxious thoughts. Poor sleeping habits (watching television in bed, caffeine misuse) can worsen these situations.

Recommendations to maximize sleep efficiency include:

A. Have a cool, dark, and quiet bedroom. During trips, use eye masks and earplugs
B. Develop a good sleep routine, going to bed and getting up at a usual time
C. Avoid watching TV in bed or using computers or cell phones
D. Avoid caffeine 4–5h before bedtime
E. Take naps (<30–45min) well before bedtime

22.5 HYDRATION

Water gain comes from consumption (liquids and food) and production (metabolic water); water loss occurs through respiration, gastrointestinal and renal excretion, and sweat. Water balance is directly related to heat balance. Heat production is essential to metabolism. During physical activity, a high amount of heat is a byproduct of energy generation. Thermoregulation is in complete homeostasis (temperature varies very little); it is controlled at the hypothalamic level with information from thermoreceptors in different parts of the body.

Heat losses follow four mechanisms:

1. Radiation in the form of infrared rays (≤60% loss)
2. Evaporation of water by sweat (>20% loss)

3. Convection when air and/or water come in contact with the body (≤12% loss)
4. Thermal conduction when solid objects touch the body (3% loss)

Monitoring and simple indicators help determine the level of hydration and balance of liquids during speed and resistance training. Urinary density is a quantification of fluids ingested during training as well as body weight before and after workouts. Body weight is a good indicator of water status, since 1% body weight loss causes a 2.5% decrease in plasma volume, i.e., slight dehydration (one of the main causes of performance reduction through compromised physiological and cognitive functions). It is important not to exceed 2% of body mass loss, otherwise the mental and psychomotor capacities deteriorate, mood changes, and depression feelings appear in young people, when liquid intake is voluntarily suspended. A 2.1% weight loss in swimmers is a limit value at which performance begins to be affected. Body weight is stable enough to monitor the daily balance of liquids, even over long periods involving intense exercise and sharp changes in fluid intake.

During prolonged exercise in hot conditions, losses ≤1–2 L water/h are observed, with increases in water deficit, physiological tension, central temperature, heart rate, and perception of exertion.

Another variable to consider is the rate of maximum sweating, ≤2–3 L/h or 2%–4% body mass loss, with consequent performance decrease. Interindividual variation is large; some subjects are more prone to dehydration than others.

A high rate of sweating (calculated as body weight loss, fluid consumption, or urine excreted during exercise) results in decreased blood volume, with difficulties in satisfying the demand of muscle for substrates and heat transfer through the skin.

22.5.1 WATER RECOVERY

Post-exercise recovery is part of the preparation for the next workout, and replacing water and salt lost through transpiration is essential. Around 1.2–1.5 L fluid/kg weight loss should be drunk during training or competition. Hydration and energy recovery are linked to physical performance: sports drinks allow for simultaneous carbohydrate (4–8 g/100 mL) and fluid recovery. Milk intake increases the synthesis of protein, and this leads to a better net muscular protein balance. Milk has a high density of nutrients and vitamins for endurance athletes compared to traditional sports drinks. Also, taste is the determining factor in the choice of drink after exercise. The International Federation of Sports Medicine recommends fluid intake before, during, and after prolonged submaximal exercises as detailed in Table 22.3.

22.5.2 THERMOREGULATION IN SPORTSWEAR

Were it not for thermoregulation during physical activity, temperature would increase by one degree every 5 min, with lethal consequences. Ice has been a traditional aid in thermoregulation after exercise (Figure 22.6). More recently, body-mapping studies have identified the areas with highest temperature during physical activity, which require more ventilation to help the athlete stay fresh, comfortable, and protected during activity. Fanger's comfort equation evaluates the energy balance of thermal comfort in a given environment.

Materials such as Outlast were designed to maintain a constant temperature, by means of microcapsules containing phase change materials. Microfibers absorb excess heat and transmit it back, when the body temperature descends. Such materials adapt to the body temperature and reduce overheating. They help to reduce perspiration and increase comfort (i.e., the balance between the heat produced by the metabolism and the amount of heat dissipated by the body) by avoiding excess heat. Ionized clothing (Ion X) claims body ionization through a negatively charged electromagnetic field, which aims at enhanced delivery of blood oxygen to muscles, yield increases of 2.7%, and speeding of muscle recovery. The supply of oxygen delays the intervention of the anaerobic energy

TABLE 22.3
Recommendations for Ingestion of Fluid

		Exercise duration (and intensity, % VO$_2$ max)		
		<60 mn	60–180 min	>180 mn
		(80%–130%)	(60%–90%)	(30%–70%)
Pre-event	Volume (mL)	300–500		
	Content	30–50 g CHO		
Event	volume (mL)	500–1000 mL		
	<60 min	6% CHO	6% CHO	
			10–20 mEq Na⁺/L + 10–20 mEq Cl⁻/L	
	>60 min drink		*8–12% CHO*	
Recovery	Volume (mL)	500–1000 mL/h		
	<120 min	10–20 mEq Na⁺/L + 10–20 mEq Cl⁻/L	0.7 g/kg/h	
			10–20 mEq Na⁺/L + 10–20 mEq Cl⁻/L	
		6–12% CHO	*8–12% CHO*	
	>120 min	6% CHO		
		10–20 mEq Na⁺/L + 10–20 mEq Cl⁻/L		

Source: Gisolfi, C.V., Duchman S.M., *Med. Sci. Sports Exerc.*, 24, 679, 1992.

system, thus reducing lactic acid concentrations. Sphere React provides and manages thermal and moisture comfort. "Skins" are garments that help to compress muscles in certain areas and might reduce lactic acid, fatigue, and recovery time, thus improving performance. Tencel, an outer garment layer, is composed of crystalline nanofibrils that absorb water, helping to remove sweat and moisture, with a cooling effect by evaporation, avoiding overheating.

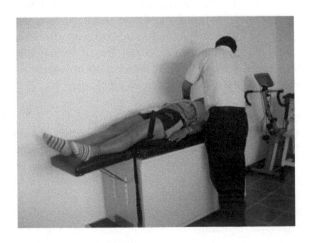

FIGURE 22.6 Athlete with fatigue symptomatology.

22.6   ERGOGENIC AIDS

Ergogenic aids are any maneuver or method (physical, psychological, nutritional, or pharmacological) increasing the capacity to perform physical work and improving performance. These aids can improve the preparation for exercise, efficiency, and post-exercise recovery. They can allow for greater tolerance to exercise or higher-intensity training. They can also promote and maintain the athlete's health. These aids are not in the list of doping substances of the International Olympic Committee and the World Anti-Doping Agency.

Under U.S. federal regulations, supplements of a dietary nature with one or more nutrients, including amino acids, vitamins, minerals, or botanical substances, are not regulated by the Food and Drug Administration. As to complements, they contribute nutrients when food intake does not meet bodily needs. Examples are carbohydrate and/or protein energy shakes when energy needs are so high that additional, easily assimilated, concentrated energy sources are required. The potential biochemical or psychological benefits can be acute and chronic. The former occur when ergogenic aids induce physiological or psychological changes in sports performance, immediately after use. Chronic effects, on the other hand, occur when such an ergogenic element is required over a particular training phase.

Currently there are contradictory criteria related to the use of ergogenic aids in the practice of sports and physical activity. The manufacture of nutritional aids has increased, but not all have been evaluated or validated under scientific and nutritional criteria. Some of them have been positively evaluated:

Antioxidants of the vitamin A, C and E, whose function is to reduce the damage caused by the free radicals produced by the metabolism (without overdosing since this would cause oxidative stress damaging the cell membrane). Exercise is a factor that increases free radicals, but, with antioxidants there is no evidence of improved sports performance; their main benefit lies in the strengthening of the immune system. Regular training also promotes the increase of endogenous antioxidants. Doses of 500 mg/day for vitamin C, 500 IU for vitamin D and E, and 3000 IU for vitamin A are recommended. These vitamins are suggested to athletes who start a period of intense training, are acclimatizing to warm temperatures, or move to a higher altitude for training or competition.

Bicarbonate and citrate are the most important acid–base buffers in the body, which decrease the excess of muscle hydrogen ions when anaerobic efforts are made. Acute doses (300 mg/kg 2 h before exercise) and chronic doses (500 mg/kg divided into 5 doses per day for 5 days) are recommended in high-intensity performances to improve the adaptation during training.

Caffeine appears to mobilize fatty deposits in the body, modify muscle contractility, change perception of exertion and fatigue, and release adrenaline. For this reason, it is used in exercises of high intensity, high resistance, and brief duration. It has shown no effects on muscle strength. The recommended dose is 1–3 mg/kg body weight before, during, and even after exercise. If it is consumed above this dose, it can cause tachycardia and loss of coordination and precision in movements.

Calcium is required for growth in children and adolescents or women with menstruation or any gynecological or metabolic condition that requires its administration. The recommended dose is 600 mg/day.

Creatine is an energy source that contributes phosphate groups to ADP and then converts it to ATP, favoring the phosphagen pathway (ATP-CP) during high intensity physical work with intermittent recoveries, such as sprints or short explosive efforts. These movements increase work capacity and indirectly muscle mass (for athletes who require a body mass increase of 1–2 kg). The usual dose is 70 mg/kg body weight, divided into 4 or 5 doses, the first week of each month. It should be ingested with drinks rich in carbohydrates with a high glycemic index.

Electrolytes help to maintain constant body temperature and, with body water, keep ionic homeostasis at the cellular level, avoiding symptoms of fatigue even up to hypovolemic shock. The usual dose is 0.5 grams of sodium per liter (in competitions > 1 h).

Vitamins and minerals are contained in food, helping metabolic processes (cofactors). They are used by athletes with risk factors of poor diet and women with polimenorrea and during the prescription of low calorie diets. Doses depend on age and type of vitamin or mineral (see Table 22.1).

Chondroitin may help joint protection and slow cartilage wear. However, it seems that the benefit of this substance is minimal or null.

Glucosamine is used for cartilage maintenance of the joints, preventing deterioration (does not regenerate them and is usually taken in conjunction with chondroitin). Glucosamine sulfate at a dose of 500 mg every 8 h may relieve moderate to severe arthritis pain.

Citrulline malate is used for the treatment of asthenia or fatigue in patients with infectious and postsurgical diseases. These effects are achieved after 6–14 days of antifatigue therapy with a usual dose of 3 g/day, improving aerobic function in the muscles of patients with asthenia. Doses of 8 g/day confer benefits on postexercise muscle recovery and increase performance in athletes with high intensity loads of anaerobic exercise where there is a considerable increase in lactacidemia, ammonia, and acidosis.

It has been shown that very few substances can help increase the state of physical performance, especially in athletes who have already reached their peak physical performance. These substances do not replace physical work or a balanced diet and may expose the athlete to health risks or doping results. This is why the best recommendation is to abandon the self-administration of drugs or food supplements and leave this task to trained personnel.

22.6.1 Fatigue

Fatigue is the qualitative and reversible feeling of tiredness. Reduced muscle performance or inability to maintain strength sometimes produces irreversible injuries to muscles and other organs appearing in one or more places. Fatigue tends to be multifactorial depending on the subject, the environment, the duration of the intensity, and the type of exercise. During intense loads (intervallic or repetitive anaerobic exercises) in hot climates, fatigue occurs rapidly in muscle fibers, which accumulate large amounts of lactate. Acidosis modifies the permeability of the wall of muscle cells and decreases their functions, causing an increase in urea concentration.

Cellular restoration takes between 1 and 4 days and can be accelerated through light exercise recovery sessions, as this eliminates the lactate found in the muscle. The most intense elimination of lactate (8 mg/100 mL/min) is obtained with a high oxygen consumption of 40%–70% of the maximum oxygen consumption (VO_2 max).

High lactate concentration in muscles and blood damages coordination of movement. A high level of lactate in the blood (above 7–8 mmol/L) increases the risk of microlaceration in muscle tissue and can cause in the long run, more severe injuries. Fatigue can be classified according to the duration of training. In a macrocycle, it would be "acute fatigue," in a microcycle, it would be "sub-acute fatigue," and in a mesocycle, it would be "chronic fatigue or overtraining."

According to the site in the body, fatigue is classified as

1. Central fatigue, which occurs in the central nervous system due to altered communication between the brain and muscle fibers. It appears when acetylcholine is produced in less quantity, hindering normal stimulation of the motor plaque. Other factors involved in this type of fatigue are motivation, anxiety, and stress (Figure 22.7).
2. Peripheral fatigue, found in the contractile mechanisms of muscle cells (Figure 22.8), is caused by the overworking of a specific muscle or several muscles. Other factors involved in this type of fatigue are excessive intensity of training sessions, inadequate energy supply (macronutrients), alterations of other energy systems (mainly cardiovascular), alterations of endocrine functions, previous damage to muscle or motor plate, and climatic factors (humidity, altitude, and temperature).
3. Metabolic fatigue, related to a decrease in the concentration of energy substrates or inhibition of enzymatic processes (Figure 22.8), due to the intensity of the exercise.

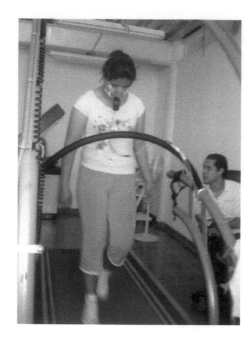

FIGURE 22.7   Athlete performing a stress test reaching musculoskeletal fatigue.

FIGURE 22.8   Athlete performing anaerobic power and subsequent metabolic fatigue.

RECOMMENDED REFERENCES

Alvarado G. et al. Determination of the anaerobic capacity in professional tennis players by using a strength platform base on the MICHECEVI® method. *British Journal of Sports Medicine* 2016; 50: A85.1–A85.10.

American College of Cardiology (ACC)/American Heart Association (AHA). ACC/AHA 2002 Guideline update for exercise testing: summary article. *Circulation* 2002; 106: 1883–1892.

Baker L. B. et al. Normative data for regional sweat sodium concentration and whole-body sweating rate in athletes. *Journal of Sports Sciences* 2016; 34: 358–368.

Bellamy D. et al. Spirometry in Practice. A practical guide to using Spirometry in primary care. British Thoracic Society (BTS) COPD Consortium. 2005.

Blumer P. et al. The effects of twenty-four hours of sleep are the performance of national-caliber male collegiate weightlifters. *Journal of Strength and Conditioning Research* 2010; 21: 1146–1154.

Burton D. et al. *Spirometer Users' and Buyers' Guide*. Melbourne: National Asthma Council Australia, 2015.

Dardouri W. et al. Relationship between repeated sprint performance and both aerobic and anaerobic fitness. *Journal of Human Kinetics* 2014; 40: 139–148.

Day J. R. et al. The maximally attainable $V^{\cdot}O_2$ during exercise in humans: the peak vs. maximum issue. *Applied Physiology* 2003; 95: 1901–1907.

Duran A. S. et al. Sleep quality, excessive daytime sleepiness and insomnia in Chilean paralympic athletes. *Nutricion Hospitalaria* 2015; 32: 2832–2837.

Erlacher D. et al. Sleep habits in German athletes before important competitions or games. *Journal Sports Sciences* 2011; 29: 859–866.

Fan J. and Tsang H. W. K. Effect of clothing thermal properties on the thermal comfort sensation during active sports. *Textile Research Journal* 2008; 78: 111–118.

Fields J. and Turner J. L. Performance-enhancing sports suplements. In R. M. Buschbacher, N. D. Prahlow, and S. J. Dave (Eds.), *Sport Medecine and Rehabilitation: A sport-Specific Approach* (2nd ed., pp. 39–49), Philadelphia; Lippincott Williams & Wilkins, 2009.

Flanagan E. P. and Comyns T. M. The use of contact time and the reactive strength index to optimize fast stretch-shortening cycle training. *Strength and Conditioning Journal* 2008; 30.5: 32–38.

Fluis A. C. and Electrolyte Balance. In: M. G. Wohl, R. S. Goodhart, Ed. Modern nutrition in health and disease (pp. 404–424). Philadelphia: Lea & Fegiber, 1968.

Gastin P. B. Victorian Institute of Sport, Melbourne, Australia energy system interaction and relative contribution during maximal exercise. *Sports Medicine* 2001; 31: 725–741.

Gibbons R. J. et al. ACC/AHA guidelines for exercise testing update: a report of the American College of Cardiology/American Heart Association Task Force on Practice Guidelines (Committee on Exercise Testing). *Journal of the American College of Cardiology* 1997; 30(1): 260–311.

Gisolfi C. V. and Duchman S. M. Guidelines for optimal replacement beverages athletic events. *Medical and Science in Sports and Exercise* 1992; 24: 679–687.

Johns D. P. and Pierce R. *Pocket Guide to Spirometry*, 2nd edition. McGraw-Hill Australia, 2007.

Marian V. et al. Improved maximum strength, vertical jump and sprint performance after 8 weeks of jump squat training with individualized loads. *Journal of Sports Science and Medicine* 2016; 15, 492–500.

Miller M. R. et al. General considerations for lung function testing. *European Respiratory Journal* 2005; 26: 153–161.

Miller M. R. et al. Standardisation of spirometry. *European Respiratory Journal* 2005; 26: 319–338.

Moxnes J. F. et al. On the kinetics of anaerobic power. *Theoretical Biology and Medical Modelling* 2012 Jul 25; 9: 29.

National Asthma Council Australia. *Australian Asthma Handbook, Version 1.1 [website]*. Melbourne: National Asthma Council Australia, 2015.

National Asthma council Australia. *Spirometry Quick Reference Guide*. A guide to performing high-quality Spirometry. 2012.

Pellegrino R. et al. Interpretive strategies for lung function tests. *European Respiratory Journal* 2005; 26: 948–968.

Pérez-Guisado J. and Jakeman P. M. Citruline malate enhances athletic anaerobic performance and relieves muscle soreness. *Journal of Strength and Conditioning Research* 2010; 24: 1215–1222.

Périard J. D. et al. Adpatations and mechanisms of human heat acclimation: Applications for competitive athletes and sports. *Scandinavian Journal of Medicine and Science in Sports* 2015; 25(Suppl. 1): 20–38.

Petri N. M. et al. Effects of voluntary fluid intake deprivation on mental and psychomotor performance. *Croatian Medical Journal* 2006; 47: 855–861.

Pina I. L. et al. Guidelines for clinical exercise testing laboratories: a statement for healthcare professionals from the Committee on Exercise and Cardiac Rehabilitation, American Heart Association. *Circulation*. 1995; 91: 912–921.

Postolache T. et al. Sports chronobiology consultation: from the lab to the arena. *Clinical Sports Medicine* 2005; 24: 415–456.

Pugh L. Oxygen intake in track and treadmill running with observations on the effect of the air resistance. *Journal of Physiology* 1970; 207: 823–835.

Reichenbach S. et al. Meta-analysis: chondroitin for osteoarthritis of knee or hip. *Annals of Internal Medicine* 2007; 146: 580–590.

Roy B. D., Milk: the new sportsdrink? A review, *Journal of the International Society of Sports Nutrition* 2008; 5: 15.

Skein M. et al. Intermittent-sprint performance and muscle glycogen after 30 h of sleep deprivation. *Medical and Science in Sports and Exercise* 2011; 43: 1301–1311.

Tanaka H. et al. Age-predicted maximal heart rate revisited. *Journal of American College of Cardiology* 2001; 37: 153–156.

Terrados N. et al. Physiologial and medical strategies in post-competition recovery-practical implications based on scientific evidence. *Serbian Journal of Sport Sciences* 2009; 3: 29–37.

Titchennal C. A. Chapter 15: Ergogenic aids. In M. A. Clark and S. C. Lucett (Eds.), *NASM′s Essentials of Sport Performance Training* (pp. 415–428) Philadelphia; Lippincott Williams & Wilkins, 2010.

Vandewalle H. et al. Standard anaerobic exercise tests. *Sports Medicine* 1987; 4: 268–289.

Waterhouse J. The role of a short post-lunch nap in improving cognitive, motor, and sprint performance in participants with partial sleep deprivation. *Journal of Sports Science* 2007; 25: 1557–1566.

Wilmore J. H. et al. Role of taste preference on fluid intake during and after 90 min of running at 60% VO_2 max in heat. *Medicine and Science in Sports and Exercise* 1998; 30, 587–595.

Yang K. et al. Analysis and prediction of the dynamic heat-moisture comfort property of fabric. Fibres and Textiles in Eastern Europe 2008; 16–3 (68): 51–55.

23 Sustainable Energy for Houses

David Morillón Gálvez and Francisco Javier Ceballos Ochoa
National Autonomous University of Mexico

CONTENTS

23.1 ENERGY SUSTAINABILITY

In the European Union in 2010, buildings were responsible for consuming 40% of the energy produced and were the source of 36% of CO_2 emissions. Therefore, energy efficiency and the autonomous generation of energy in buildings are priority targets for combatting climate change and reducing dependence on fossil fuels. As urban populations continue to grow worldwide, the search for energy sustainability in urban settings depends on reduced and efficient energy use and renewable energy technologies.

Current design tools and methods for sustainable housing include the ICC and LEED in the U.S. and ISO, BREEAM in the UK, Passivhaus in Germany. More recently, the concept of Net Zero Energy Buildings (ZEB) has emerged to mitigate CO_2 emissions and achieve energy sustainability. The growing number of examples built and international research highlight the growing attention to buildings of this type. However, beyond documented ZEB case studies, there is room for improvement of methods: a common and unambiguous definition, criteria for energy balance (e.g., primary energy, carbon emissions), efficiency parameters, or indicators specific to these types of buildings.

This chapter highlights zero energy buildings (ZEB) methods and results. It is posited that net zero energy (NZE) can be easier to achieve in bioclimatic designs with efficient energy use, no use of air conditioning, and more solar insolation. This can sometimes lead to net generation of energy.

We verified this approach in three case studies representative of climates of the world: subtropical high plateau, desert, and coastal tropical forest. Results of the case studies are compared in terms of thermal behavior, the final use of energy, and energy generated by a photovoltaic system. The impact of NZE housing, environmental benefits, and impact on energy networks is discussed, and some recommendations on technologies are made.

23.2 METHOD

The method sequence is as follows:

Bioclimatic diagnosis

Energy-relevant design recommendations

In situ thermal evaluation prior to and after improvements on final consumption of energy

Measurement of final use of energy in the house to reduce it

Identification of applicable and available efficient energy technologies for energy savings and generation

Estimate of the energy balance achieved with renewable sources integrated to the dwelling (Figure 23.1).

All calculated parameters were applied to a prototype architectural housing design. The guidelines of the Official Mexican standard on energy efficiency and the housing code (NOM-020-ENER-2011), the criteria and indicators for sustainable housing developments, the green mortgage program (Morillón, 2007), the sustainable building certification program in Mexico City, and the Mexican standard for sustainable building were complied with. The following average energy consumption data in a Mexican house were used: cooking (52%), water heating (29%), and appliances and lighting (19%). Of the last, refrigerators consume 35%, air conditioners 26%, televisions 24%, irons 8%, washing machines 6%, and heaters 1% (Rosas et al., 2011).

The bioclimatic design included climatic data and input in a hygrothermal comfort evaluation. The maximum and minimum average temperatures were used to assign a bioclimate appropriate for the location under study (Morillón, 2004). The climatological data were taken from weather station

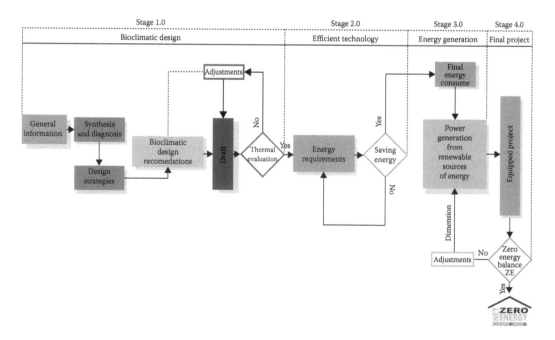

FIGURE 23.1 Methodology for ZEB.

9047 in Mexico City (DF) (19°27′ N, 99°11′ W, 2255 masl). Relative humidity and prevailing wind data were used.

Monthly conditions of thermal sensation, using the bioclimatic diagram of Olgyay and the bioclimatic comfort diagram of Givoni, were obtained using hourly temperature and relative humidity, as well as the months of highest and minimal comfort temperature, May (T_p=23.9°C) and January (T_p=22.2°C), respectively. Analysis by means of the diagram of isorequirements (Table 23.1), shows yearly variations.

In the hygrothermal comfort evaluation, the center of the comfort zone or thermopreference (T_p) as a function of the average ambient temperature (T_{amb}) for each month was (Aliciems equation)

$$T_p(°C) = 17.6 + 0.31\, T_{amb} \tag{23.1}$$

In order to know the thermal behavior of the preproject and the energy efficiency, the preliminary design was submitted to a thermal simulation of the heat gain in the envelope

$$\Phi_p = \Phi_{pc} + \Phi_{ps} \tag{23.2}$$

where the thermal gain by conduction (Φ_{pc}) plus the radiation gain (Φ_{ps}) equals the total heat gain (Φ_p).

TABLE 23.1

Diagram of Isorequirements for the Temperate Bioclimate (DF), Based on Olgyay Monthly Bioclimates

Hour	Jan	Feb	Mar	Apr	May	Jun	Jul	Aug	Sep	Oct	Nov	Dec
06:00												
07:00											Cold	
08:00												
09:00												
10:00												
11:00												
12:00												
13:00					Hot				Comfort			
14:00												
15:00												
16:00												
17:00												
18:00												
19:00												
20:00												
21:00												
22:00												
23:00		Cold										
00:00												
01:00												
02:00												
03:00												
04:00												
05:00												

The design and evaluation model of solar control in buildings was determined according to the number of hours of cold and the number of hours of comfort and heat per month (6:00 am through 6:00 pm). For example, for February, the number of cold hours is five, from 6:00 am to 11:00 am, and there are eight hours of comfort, from 11:00 am to 7:00 pm. The number of hours obtained is multiplied by the number of days in the month: 5h of cold times 28 days in the month equals 140. The same is applied for hours of comfort and warmth. Finally, each row is added to obtain the hours in which sunshine ($T1$) and sun protection ($T2$) are required for each semester (Tables 23.2 and 23.3).

The thermal evaluation helped determine whether the applied recommendations were optimal in the local bioclimate. The energy consumption and energy generation requirements using renewable sources were then input for the energy balance and assessing whether the dwelling had achieved NZE grade. The information obtained was transferred to stereographic solar charts, to obtain shading hours required per semester ($T2$). The hours that require shade ($T2$) that are outside the eaves (K) and those that require sunning ($T1$) that are outside the eaves (L) defined the efficiency of the eaves with respect to the orientation

$$EPC = 1 - (K / T2) \text{ and} \tag{23.3}$$

$$EPF = L / T1 \tag{23.4}$$

The zero energy (ZE) balance is (Sartori et al., 2012):

$$EENcR = \text{exported energy} - \text{imported energy} \geq 0 \tag{23.5}$$

$EENcR$ first has to account for the architectural object itself (constructive system in values of transmittance and thermal resistance, orientations, and ventilation). Second, $EENcR$ has to be achieved in situ via solar photovoltaic, solar water heating, wind turbines, and possibly connection to a local grid. For the design of a ventilation system, four steps were identified:

1. Clear image of the directional wind range, throughout the annual cycle
2. Determination of cooling needs (diurnal and seasonal) for thermal comfort
3. Evaluation of the shelter of neighboring structures or topographies that alter the flow of the wind
4. A system of windows whose functional characteristics correspond to both wind and required thermal comfort

TABLE 23.2
Number of Hours Requiring Sun ($T1$) and Solar Protection ($T2$), January–June

	Jan	Feb	Mar	Apr	May	Jun	Total
$T1$	186	140	124	120	93	90	753
$T2$	217	224	279	270	310	300	1600

TABLE 23.3
Number of Hours Requiring Sun ($T1$) and Solar Protection ($T2$), July–December

	Jul	Aug	Sep	Oct	Nov	Dec	Total
$T1$	124	124	150	186	180	186	950
$T2$	279	279	240	217	210	217	1442

23.3 RESULTS

23.3.1 CLIMATE

The annual average temperature is 17.7°C, with 28.1°C summer maximum and 6.5°C winter minimum.

Mexico City has a temperate bioclimate, with average and minimum temperatures below comfort range; the maximum slightly exceeds comfort; daily oscillation is 10°C–15°C. Average and maximum relative humidity are within the comfort range. Rainfall is ~900 mm. Average relative humidity (7:00 am) is 70%–85% and 40%–45% (4:00 pm). The highest relative humidity is 48%–88% (July, rainy season) and lowest 32%–74% (March).

The winds are cold in winter at night. The average annual wind speed is 0.8 m/s predominantly northwesterly (341°). Daytime wind periods are: 1:00–8:00 am (307°–327°, 0.6–1.1 m/s); 9:00 am–3:00 pm (2°–29°southwesterly, 0.7–1.2 m/s); 4:00 pm–12:00 am (0.7–1.2 m/s (south) westerly).

Solar radiation fluctuates around 5.61 kWh/m²day, with highest value in April and 4.15 kWh/m² day in November. Highest values are 10:00 am through 2:00 pm.

23.3.2 BIOCLIMATE

Annual average temperature T_p was ~14.5°C with a 3°C amplitude of the comfort zone. Temperatures outside this range indicate overheating and cold periods. T_p and comfort zone center do not show extreme changes during the year. Minimum temperatures are below the comfort zone while maximum temperatures exceed the comfort zone only in spring.

The month of highest comfort temperature is May (T_p=23.9°C). Comfort temperature times are between 9:00 am and 8:00 pm (Table 23.1). Midday temperature requires a 0.3 m/s ventilation. Between dusk and dawn, temperatures are below the comfort range, with a very humid environment. A cold sensation is generally felt from 8:00 pm through 9:00 am. The recommendation of Olgyay's climate chart is radiation between 70 and 350 W/m², to return to the comfort zone.

In January, the coldest month, temperatures are below the comfort zone and reach thermal sensations close to numbness between 6:00 pm and 10:00 am. So the recommendation is to use solar radiation with levels above 420 to 70 W/m².

The isorequirements diagram (Table 23.1) shows that the comfort temperature in the case study (DF) does not have large variations during most of the year and is constant between 11:00 am and 5:00 pm. The months with hot temperatures are March, April, and May (spring). During the autumn (September–November) and winter (December–February), the morning and afternoon are cold.

23.3.3 SOLAR BEHAVIOR ANALYSIS

The eave angles and the weighted efficiency values suggest that the west orientation was the least favorable because of its high solar incidence during the day in relation to the air conditioning requirements. The northern orientation was the most favorable; it received direct radiation throughout the day. Accordingly, the winter sensations of cold needed attenuation through passive heating. In contrast, spring temperatures above the comfort range required curtains on the west side.

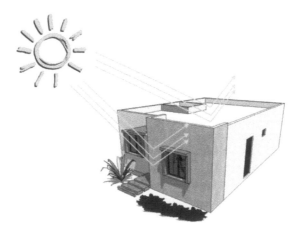

FIGURE 23.2 The use of color high reflectance ceilings and walls in hot climates, eaves on windows for solar control.

23.3.4 DESIGN STRATEGIES

23.3.4.1 Passive Heating

During the months of cold (October–February), solar passive heating was suggested, since geo-thermal energy was deemed more complicated to come by. Factors to be considered are (Morillón, 2011) orientation of the facades, solar elevation, and depth of the buildings spaces. Passive heating systems include direct absorption onto internal surfaces, indirect absorption onto outer envelope, and heating via heat transmission from an outside surface. During the months of cold, passive solar heating, direct heating in the morning on the southeast facades, and indirect heating in the evenings were suggested. This would also avoid heat loss from windows. Transition spaces between the exterior and the interior were recommended.

During the months of comfort, storage of heat in floors, ceilings, and walls of the west and south facades, as well as ventilation through the windows, was recommended.

During the months of heat (March–May), ventilation of 0.3 m/s was required. Ventilation through the windows should focus on the management of air flowing toward the occupants rather than the structure (Morillón, 2011). Heat gain through the envelope was to be mitigated by high reflectance colors on the envelope (Figure 23.2).

23.3.4.2 Bioclimatic Architectural Recommendations

Our case study was a 62.5–97.5 m² social housing building (Figure 23.3) with flat roof, 2.40 m floor-to-ceiling height, and constructive layers described in Table 23.4. Rough texture on walls and unilateral ventilation for protection from cold winter winds promoted winter heat gain. Southeastern and southwestern windows should have >80% of the surface gaining heat from direct radiation. The 1.20×1.20 m windows were to have solar control devices with optimum angles. Deciduous trees on the southern side captured solar radiation in winter (Figure 23.4). The interior patios acted as a greenhouse for thermal control, in addition to having natural ventilation in spring and summer.

23.3.4.3 Thermal Evaluation

The heat gain in the envelope was much lower than a reference building complying with the Mexican building envelope guidelines. This was largely thanks to the solar control devices with which it was possible to reduce the gain of heat by radiation. The project was considered energy efficient from the point of view of architectural design and optimal in terms of thermal comfort of the user. In hot climates, this efficiency translated into considerable savings of electric power through a decreased use of cooling systems.

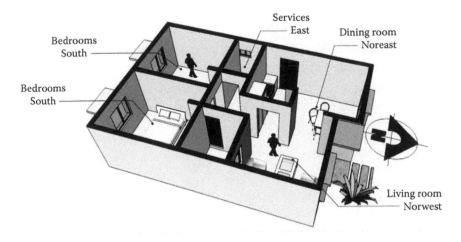

FIGURE 23.3 Architectural project with space localization as recommended for the city of Chihuahua, Chihuahua.

TABLE 23.4
Materials Used in the Architectural Project and Their Thermal Properties

Compound	Material	Thickness (m)	Thermal Conductivity (λ) W/m °C	Thermal Resistance m² °C/W	Transmittance W/m² °C
Wall	Cement block	0.18	0.998	0.476	2.099
	Plaster	0.013	0.372		
	White lime-based stucco	0.010	0.698		
Window	White glass	0.003	0.930	0.204	4.911
Door	Plywood (pine, two layers)	0.006	—	0.492	2.03
Roof	Waterproofing	0.005	0.17	1.001	0.999
	Warped concrete	0.04	0.698		
	Tezontle	0.12	0.186		
	Concrete slab	0.10	1.74		

FIGURE 23.4 Eaves in the windows with the optimal angles for the locality and deciduous leaf trees in the southern facade. (From Morillón, G. D. et al., Human bioclimatic atlas for México. *Solar Energy J.,* 76,781–792, 2004.).

TABLE 23.5
End Use of Electric Energy for a Conventional House in Temperate Climate

Concept	Consumption (kWh/year)	(%)
Illumination	480.0	40
Refrigerator	348.0	29
Television	156.0	13
Consumer durables	84.0	7
Ironing	72.0	6
Washing machine	60.0	5
Total (kWh/year)	1200.0	100
LP gas (kg/year)	388.8	—

23.3.4.4 Energy Requirements and Efficient Technology

The baseline of energy consumption in the temperate bioclimate was 1.20 MWh/year (Morillón, 2009), out of which the highest percentage was lighting, refrigeration, and television use (Table 23.5). The final consumption of electric energy through using efficient technologies saving was 30.69 kWh/month (replacement of incandescent bulbs and refrigerator). This resulted in a final consumption of 831.72 kWh/year. The final consumption of gas was 26.84 kg/month, by replacing the conventional heater with a solar heater. Savings in electric energy consumption were 31%, and gas savings were 83% of the total baseline.

23.3.4.5 Generation of Renewable Energy

Renewable generation of energy is fundamental for a ZEB house. A photovoltaic system was to be interconnected to the grid; its size depended on

- Latitude, longitude, and altitude as well as monthly solar radiation (Almanza, 2005)
- Daily electrical energy consumption
- Daily electricity consumption including a 10% factor for peak uses
- Photovoltaic panel type (here, a crystalline silicon photovoltaic module, 36 cells connected in series with block diodes, $P_{mod}=210$ W, $V_{max}=18.30$ V, $I_{max}=11.50$ A, $I_{cc}=12.10$ A, $V_{ca}=22.80$ V)
- Consideration for the least favorable month: number of modules ≥ requirement of the least favorable month/total power of the solar array: four modules ≥3.20=671.75 W/210 W, with total power of the photovoltaic (PV) system corresponds to an array of 840 W
- Net energy generated by PV: total PV power times the daily radiation times energy generated in a day in the most unfavorable month times losses factor: 4.24 kWh/day=[(0.840 kW) * (5.61 kW/m²day)] * (0.90), with a net energy generated by the PV system annually of 1331.70 kWh/year.

23.3.4.6 Equipped House and Energy Balance

At this stage, the architectural project was energetically efficient from the viewpoints of bioclimatic design, use of efficient technology, and dimension of the PV system. However, ZEB balance was to be achieved.

Considering 4 PV modules the energy balance was negative and 315.41 kWh/year needed to be imported from the electricity grid. After adding a fifth 210 W PV module, the total power of the PV array was 1050.0 W, generating 1664.63 kWh/year, and the surplus was 17.44 kWh/year, bringing the house beyond the NZE grade.

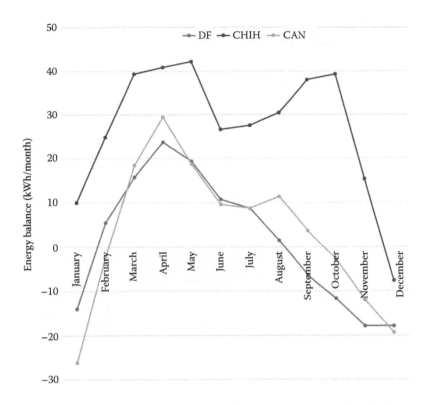

FIGURE 23.5 Energy balance between energy consumed and energy generated for the three case studies.

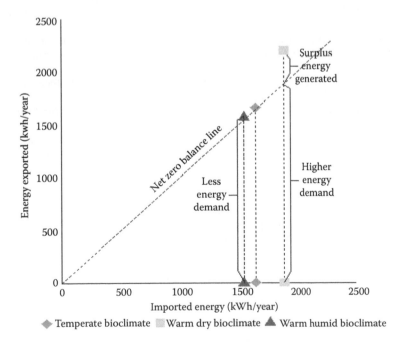

FIGURE 23.6 Graphic representation of ZEB for the three case studies.

23.4   ENERGY DIAGNOSIS OF CONVENTIONAL AND ZERO ENERGY HOUSING

The foregoing results for a temperate bioclimate were also obtained for warm dry and hot humid bio-climates (Figure 23.5). They were compared to the baseline electric energy consumption (Morillón, 2009). The warm dry climate had the highest energy consumption (2.29 MWh/year), followed by the hot humid climate (2.27 MWh/year) and the temperate climate (1.20 MWh/year). Solar radiation marked the difference in the amount of energy generated annually. The dry hot bioclimate had a large energy surplus (326.93 kWh/year), compared to the humid warm (15.44 kWh/year) and temperate (37.60 kWh/year) surpluses. The three case studies reached Net Zero Energy grade (Figure 23.6).

The hot humid climate, owing to the number of overcast days, did not have a large potential for PV energy surpluses. The temperate climate had a lower energy demand and it was easier to achieve the NZE grade.

23.5   DISCUSSION

Paramount to energy sustainability is coupling efficient design and efficient technology. This means avoiding conventional elements for air conditioning and reducing energy demand. NZE methodology and technologies rely on bioclimatic design, energy-efficient devices, reduced energy consumption, and generation of energy by renewable sources.

The energy savings represent important environmental benefits, since for each saved kWh, the emission of $0.60\,kg\ CO_2$ is reduced, preventing a total emission of $220.96\,kg\ CO_2$/year for the temperate bioclimate, $623.44\,kg\ CO_2$/year for the dry warm bioclimate, and $855\,kg\ CO_2$/year for the wet warm bioclimate, totaling $1700.14\,kg\ CO_2$/year to the atmosphere.

REFERENCES

Estrada-Cajigal Ramírez, V., Almanza Salgado, R., Irradiaciones global, directa y difusa, en superficies hori-zontales e inclinadas, así como irradiación directa normal, en la República Mexicana, Ed. II-UNAM, México, 2005.
Morillón, G. D., *Bases para una hipoteca verde en Mexico, camino a la vivienda sustentable,* en Estudios de Arquitectura Bioclimática Anuario 2007 Vol. IX, Fuentes Freixanet Víctor A., Edit. Limusa, México, 2007.
Morillón, G. D., *Línea base para la vivienda sustentable en México: GEI,* Informe Técnico, Banco Mundial, México, 2009.
Morillón, G. D., Saldaña, R., Tejeda, A., Human bioclimatic atlas for México. *Solar Energy Journal,* 76: 781–792, 2004.
Morillón, G. D., Rosas, F. D., Castañeda, N. G., Resendis, P. O. *Materiales y Sistemas Constructivos Usados en Techos y Muros de la Vivienda en México.* Memorias del XXV Congreso Nacional de Energía Solar, ANES, Chihuahua, 2011.
NOM-020-ENER-2011, *Eficiencia energética en edificaciones. Envolvente de edificios para uso habitacional,* Secretaria de Energía, México, DF, 2011.
Rosas, F. J., Rosas, F. D., Morillón, G. D., Saturation, energy consumption, CO_2 emission and energy efficiency from urban and rural households appliances in Mexico. *Energy and Buildings,* 43: 10–18, 2011.
Sartori, I., Napolitano, A., Voss, K., Net zero energy buildings: a consistent definition framework. *Energy and Buildings,* 42: 220–232, 2012.

Index

9 780367 572679